Radically Different

Radically Different—A Themed Issue in Honor of Professor Bernd Giese on the Occasion of His 80th Birthday

Special Issue Editor

Katharina M. Fromm

MDPI • Basel • Beijing • Wuhan • Barcelona • Belgrade • Manchester • Tokyo • Cluj • Tianjin

Special Issue Editor
Katharina M. Fromm
University of Fribourg
Switzerland

Editorial Office
MDPI
St. Alban-Anlage 66
4052 Basel, Switzerland

This is a reprint of articles from the Special Issue published online in the open access journal *Chemistry* (ISSN) (available at: https://www.mdpi.com/journal/chemistry/special_issues/themed_issue_bernd).

For citation purposes, cite each article independently as indicated on the article page online and as indicated below:

LastName, A.A.; LastName, B.B.; LastName, C.C. Article Title. *Journal Name* **Year**, *Article Number*, Page Range.

ISBN 978-3-03936-308-7 (Hbk)
ISBN 978-3-03936-309-4 (PDF)

© 2020 by the authors. Articles in this book are Open Access and distributed under the Creative Commons Attribution (CC BY) license, which allows users to download, copy and build upon published articles, as long as the author and publisher are properly credited, which ensures maximum dissemination and a wider impact of our publications.

The book as a whole is distributed by MDPI under the terms and conditions of the Creative Commons license CC BY-NC-ND.

Contents

About the Special Issue Editor vii

Preface to "Radically Different—A Themed Issue in Honor of Professor Bernd Giese on the Occasion of His 80th Birthday" ix

Edwin C. Constable and Catherine E. Housecroft
Before Radicals Were Free – the *Radical Particulier* of de Morveau
Reprinted from: *Chemistry* 2020, 2, 293–304, doi:10.3390/chemistry2020019 1

Torben Duden and Ulrich Lüning
Towards a Real Knotaxane
Reprinted from: *Chemistry* 2020, 2, 305–321, doi:10.3390/chemistry2020020 13

Torsten Linker
Addition of Heteroatom Radicals to *endo*-Glycals
Reprinted from: *Chemistry* 2020, 2, 80–92, doi:10.3390/chemistry2010008 29

Silvia Hristova, Fadhil S. Kamounah, Aurelien Crochet, Nikolay Vassilev, Katharina M. Fromm and Liudmil Antonov
OH Group Effect in the Stator of β-Diketones Arylhydrazone Rotary Switches
Reprinted from: *Chemistry* 2020, 2, 374–389, doi:10.3390/chemistry2020024 43

Ned A. Porter, Libin Xu and Derek A. Pratt
Reactive Sterol Electrophiles: Mechanisms of Formation and Reactions with Proteins and Amino Acid Nucleophiles
Reprinted from: *Chemistry* 2020, 2, 390–417, doi:10.3390/chemistry2020025 59

Tomas Haddad, Joses G. Nathanael, Jonathan M. White and Uta Wille
Oxidative Repair of Pyrimidine Cyclobutane Dimers by Nitrate Radicals ($NO_3^{\bullet}$): A Kinetic and Computational Study
Reprinted from: *Chemistry* 2020, 2, 453–469, doi:10.3390/chemistry2020027 87

David Bossert, Christoph Geers, Maria Inés Placencia Peña, Thomas Volkmer, Barbara Rothen-Rutishauser and Alke Petri-Fink
Size and Surface Charge Dependent Impregnation of Nanoparticles in Soft- and Hardwood
Reprinted from: *Chemistry* 2020, 2, 361–373, doi:10.3390/chemistry2020023 105

Sara Nasiri Sovari and Fabio Zobi
Recent Studies on the Antimicrobial Activity of Transition Metal Complexes of Groups 6–12
Reprinted from: *Chemistry* 2020, 2, 418–452, doi:10.3390/chemistry2020026 119

Helmut Quast, Georg Gescheidt and Martin Spichty
Topological Dynamics of a Radical Ion Pair: Experimental and Computational Assessment at the Relevant Nanosecond Timescale
Reprinted from: *Chemistry* 2020, 2, 219–230, doi:10.3390/chemistry2020014 155

Paul R. Rablen
A Procedure for Computing Hydrocarbon Strain Energies Using Computational Group Equivalents, with Application to 66 Molecules
Reprinted from: *Chemistry* 2020, 2, 347–360, doi:10.3390/chemistry2020022 167

Tina P. Andrejević, Biljana Đ. Glišić and Miloš I. Djuran
Amino Acids and Peptides as Versatile Ligands in the Synthesis of Antiproliferative Gold Complexes
Reprinted from: *Chemistry* **2020**, 2, 203–218, doi:10.3390/chemistry2020013 179

Wenyu Gu and Ross D. Milton
Natural and Engineered Electron Transfer of Nitrogenase
Reprinted from: *Chemistry* **2020**, 2, 322–346, doi:10.3390/chemistry2020021 195

Lucille Babel, Soledad Bonnet-Gómez and Katharina M. Fromm
Appropriate Buffers for Studying the Bioinorganic Chemistry of Silver(I)
Reprinted from: *Chemistry* **2020**, 2, 193–202, doi:10.3390/chemistry2010012 219

Elena C. dos Santos, Alessandro Angelini, Dimitri Hürlimann, Wolfgang Meier and Cornelia G. Palivan
Giant Polymer Compartments for Confined Reactions
Reprinted from: *Chemistry* **2020**, 2, 470–489, doi:doi:10.3390/chemistry2020028 229

Moritz Welter and Andreas Marx
Combining the Sensitivity of LAMP and Simplicity of Primer Extension via a DNA-Modified Nucleotide
Reprinted from: *Chemistry* **2020**, 2, 490–498, doi:10.3390/chemistry2020029 249

Adriana Edenharter, Lucie Ryckewaert, Daniela Cintulová, Juan Estévez-Gallego, José Fernando Díaz and Karl-Heinz Altmann
On the Importance of the Thiazole Nitrogen in Epothilones: Semisynthesis and Microtubule-Binding Affinity of Deaza-Epothilone C
Reprinted from: *Chemistry* **2020**, 2, 499–509, doi:10.3390/chemistry2020030 259

About the Special Issue Editor

Katharina M. Fromm, Prof. Dr., was raised and educated in Germany, France, and the United States. After receiving a PhD in metal–organic chemistry from the University of Karlsruhe in Germany in 1994, she joined the group of Prof. Joachim Strähle in Tübingen (solid-state chemistry) and Nobel-Prize winner Prof. Jean-Marie Lehn (supramolecular chemistry) for her postdoctoral studies. In 1998, she moved to the University of Geneva for her habilitation, which she received in 2002. After a short intermission at the University of Karlsruhe with an Emmy Noether Stipend II, she was awarded a Swiss National Science Foundation Professorship at the University of Basel, allowing her to expand her research at the University of Fribourg, taking over the chair from Prof. Alexander von Zelewsky. From 2010 to end of 2019, she served as a Research Councilor for the Division "Programs" of the Swiss National Science Foundation, of which she became president in mid-2015 before being elected Vice-President of the Research Council in 2016. After her time with SNSF, she was nominated Vice-Rector for Research and Innovation in January 2020. Her main research interests deal with s-block elements and silver bioinorganic chemistry, although her activities go beyond that and include mechano-responsive polymers, nanocapsules and batteries. In 2013, she was named Fellow of the American Chemical Society (first in Europe), and became a member of the European Academy of Sciences in 2018, the same year she was also announced winner of the Prix Jaubert of the University of Geneva. In 2019, she was elected member of the Swiss Academy of Technical Sciences.

Preface to "Radically Different—A Themed Issue in Honor of Professor Bernd Giese on the Occasion of His 80th Birthday"

Dear Bernd,

Upon the editor's prompt, your colleagues and friends, took steps and would like to dedicate this issue of Chemistry to you for your 80th birthday. We all felt that your birthday was a unique opportunity to convey our deep appreciation to an exceptional scientist, a great teacher and a man of culture. Many more would have liked to join the Special Issue but a nasty combination of ARN, proteins and lipids called COVID-19 prevented them from meeting the deadline.

Three words come to our mind when we look at each step of your scientific career: eclecticism, curiosity, and rigor.

Eclecticism: Your research covers a wide range of subjects, from bridged cations, selectivity–reactivity correlations of reactive intermediates, and the polar and steric effects of radical addition reactions, stereoselectivity of radical carbon–carbon bond formations and conformation determinations of chiral radicals by ESR to important problems of chemical biology such as radical-induced DNA strand cleavage, electron transfer through DNA, peptides, and, more recently, the electron transfer mechanism used by bacteria to adapt to the presence of metal ions in their environment. We should not forget important interludes such as the total synthesis of macrolides or the development of photocleavable protective groups.

Curiosity: Your desire to learn by exploring the unknown has obviously been the driving force of your research. It was and is still served by your unique ability to select important and fundamental questions.

Rigor: The way you approach scientific projects obviously originates from your education in Munich, where you were nurtured in the principles of physical organic chemistry by your former mentor Rolf Huisgen. You certainly belong to a small group of creative physical chemists who use the deep understanding of molecular properties to devise new reactions or new molecules with important properties. The acclaimed "Giese Reaction" is a textbook example of this interplay between "understanding" and "making".

You once said to me, "Whatever we have done in research is probably less important than our contribution as teacher." Many young chemists owe you thanks for the stimulating and sound mentorship you provided. There is undoubtedly a Giese school in the community of chemists. Each of us has enjoyed listening to your stimulating lectures and chatting with you about chemical problems, not only because of your knowledge of the field but also because you are a man with an original and profound vision of modern society.

For this, your colleagues and friends who took part in this Special Issue would like to give thanks.

Congratulations and happy birthday!

Léon
Prof. Emeritus Dr. L. Ghosez
Visiting scientist at the Institut de Chimie et Biologie, Université de Bordeaux

Professor Giese is a pioneer in selective radical chemistry, electron transfer through biomolecules, and, recently, electron transfer through bacterial membranes

Professor Bernd Giese is Professor Emeritus of Organic Chemistry at the University of Basel, Switzerland, and is now "postprof" in the group of Katharina M. Fromm at the University of Fribourg, Switzerland.

He was born in Hamburg, Germany, in 1940, studied in Heidelberg, Hamburg, and Munich, and received his Ph.D. in 1969 while working in the group of the late Rolf Huisgen. After two years in a pharmaceutical research group at the BASF, Ludwigshafen, he started his independent research at the University of Münster and received his Habilitation at the University of Freiburg in 1976. One year later, he became a Full Professor at TU Darmstadt, Germany, and accepted the position of Chair at the University of Basel in 1989. He served as dean at TU Darmstadt and as head of the department at the University of Basel. He is a member of the Editorial Advisory Board of several journals and institutes and has acted as a regional editor of SYNLETT from its beginning. Professor Giese has published more than 300 papers and has authored or co-authored three books on radical chemistry. He is a member of the Deutsche Akademie der Naturforscher Leopoldina and the American Academy of Arts and Sciences. His awards are numerous and include the Gottfried Wilhelm Leibniz Prize in 1987, the Tetrahedron Prize in 2005, the Emil Fischer Medal in 2006, and the Paracelsus Prize of the Swiss Chemical Society in 2012. In 2019, on the occasion of his 50th Ph.D. anniversary, his Ph.D. diploma from the Ludwig Maximilian University was renewed in the presence of his Ph.D. supervisor Rolf Huisgen, then aged 99.

From left to right: Katharina M. Fromm, Rolf Huisgen, and Bernd Giese on the occasion of Bernd Giese's 50th Ph.D. jubilee and renewal at LMU on June 18th, 2019.

Professor Giese's research encompasses studies on bridged cations, selectivity–reactivity correlations of reactive intermediates, polar and steric effects of radical addition reactions, the stereoselectivity of radical C–C bond formations, conformation determinations of chiral radicals by ESR, the total synthesis of macrolides, radical-induced DNA strand cleavage, photocleavable protecting groups, as well as electron transfer through DNA, peptides, and proteins.

He developed a new synthetic method that involves alkyl halides, metal hydrides, and alkenes. This three-component radical chain reaction was one of the starting points of modern synthesis with carbon-centered radicals. He applied this method—referred to today as the "Giese Reaction" in textbooks and much of the recent literature—to the synthesis of several target molecules. Bernd has thus developed important concepts for the understanding of kinetics and the selectivity of complex reactions. He has pioneered the introduction of radical reactions as powerful synthesis methods and contributed substantially to the area of physical–organic chemistry.

Today, modern physical–organic chemistry plays a major role in biochemistry. In his bioorganic studies, Bernd Giese's experiments were crucial in elucidating the controversial problem of long-distance electron transfer through DNA. He showed that electrons migrate through DNA in a multistep hopping reaction, where each single hopping step depends strongly on the distance, using appropriate DNA bases as stepping stones. He also proposed new mechanisms for DNA strand breaks via intermediate radicals. Another topical case is the study of electron transfer through proteins that connect distant molecule parts and enable redox reactions, for example, ribonucleotide reductase—the only enzyme that makes deoxyribonucleotide (DNA) available from ribonucleotide (RNA). The production site of the reactive intermediate is 3.5 nm from the reduction site, and the intervening protein is the medium for long-distance electron transfer. Bernd Giese was able to show here that amino acid side groups are used as stepping stones in this process by generating radical cations in the ground state at one end of a peptide model and studying the kinetic of electron transfer as a function the amino acid sequences and further charges at the end groups.

From model systems, Bernd Giese recently moved to living microorganisms, in particular, Geobacter sulfurreducens, which is able to reduce metal ions outside of the cell and can produce metal nanoparticles in aqueous solutions. Here, electrons can migrate from the inside to the outside of the cell using either filaments (pili) of aggregated proteins or c-type cytochromes, which transport electrons through the periplasm and the inner and outer membrane. These studies are important for understanding basic processes in life. They can also lead to enzyme inhibitors and nanoelectronic devices or help to clean polluted water. The interplay between the understanding of molecular behavior and the creation of new materials or devices is crucial, and Bernd Giese will further contribute to these exciting research activities.

Bernd Giese's seminal contributions have thus not only shaped organic synthesis but also had a profound impact on chemical biology research. Happy Birthday, and many more important results to come!

Katharina M. Fromm
Guest Editor

Review

Before Radicals Were Free – the *Radical Particulier* of de Morveau

Edwin C. Constable * and Catherine E. Housecroft

Department of Chemistry, University of Basel, BPR 1096, Mattenstrasse 24a, CH-4058 Basel, Switzerland; catherine.housecroft@unibas.ch
* Correspondence: edwin.constable@unibas.ch; Tel.: +41-61-207-1001

Received: 31 March 2020; Accepted: 17 April 2020; Published: 20 April 2020

Abstract: Today, we universally understand radicals to be chemical species with an unpaired electron. It was not always so, and this article traces the evolution of the term radical and in this journey, monitors the development of some of the great theories of organic chemistry.

Keywords: radicals; history of chemistry; theory of types; valence; free radicals

1. Introduction

The understanding of chemistry is characterized by a precision in language such that a single word or phrase can evoke an entire back-story of understanding and comprehension. When we use the term "transition element", the listener is drawn into an entire world of memes [1] ranging from the periodic table, colour, synthesis, spectroscopy and magnetism to theory and computational chemistry. Key to this subliminal linking of the word or phrase to the broader context is a defined precision of terminology and a commonality of meaning. This is particularly important in science and chemistry, where the precision of meaning is usually prescribed (or, maybe, proscribed) by international bodies such as the International Union of Pure and Applied Chemistry [2]. Nevertheless, words and concepts can change with time and to understand the language of our discipline is to learn more about the discipline itself. The etymology of chemistry is a complex and rewarding subject which is discussed eloquently and in detail elsewhere [3–5]. One word which has had its meaning refined and modified to an extent that its original intent has been almost lost is *radical*, the topic of this special issue.

This article has two origins: firstly and most importantly, on the occasion of his 80th birthday, it is an opportunity to express our gratitude and thanks for the friendship and assistance of Bernd Giese in our years together in Basel, and secondly to acknowledge a shared interest with Bernd in the history of our chosen discipline.

2. Modern Understanding

It seems relevant to present the IUPAC definition of a radical in full at this point in the text as it both provides a precision for modern usage and also contains hints of the historical meaning:

"A molecular entity such as ·CH$_3$, ·SnH$_3$, Cl· possessing an unpaired electron. (In these formulae the dot, symbolizing the unpaired electron, should be placed so as to indicate the atom of highest spin density, if this is possible.) Paramagnetic metal ions are not normally regarded as radicals. However, in the 'isolobal analogy', the similarity between certain paramagnetic metal ions and radicals becomes apparent. At least in the context of physical organic chemistry, it seems desirable to cease using the adjective 'free' in the general name of this type of chemical species and molecular entity, so that the term 'free radical' may in future be restricted to those radicals which do not form parts of radical pairs. Depending upon the core atom that possesses the unpaired electron, the radicals can be described as carbon-, oxygen-, nitrogen-, metal-centered radicals. If the unpaired electron occupies an orbital having considerable s or more or less pure p character, the respective radicals are termed σ- or π-radicals. In the past, the term 'radical' was used to designate a substituent group bound to a molecular entity, as opposed to 'free radical', which nowadays is simply called radical. The bound entities may be called groups or substituents, but should no longer be called radicals" [6].

To summarize, in accepted modern usage, a radical possesses an unpaired electron.

3. A Radical Birth

3.1. de Morveau's Introduction

The word radical was introduced by the French politician and chemist, Louis-Bernard Guyton, Baron de Morveau (1737–1816, prudently identified after the French revolution without the aristocratic rank as Louis-Bernard Guyton-Morveau, Figure 1) [7]. In 1782, de Morveau published an article entitled *Sur les Dénominations Chymiques, La nécessité d'en perfectioner le systême, et les règles pour y parvenir* in which he identified the need for a new systematic nomenclature in chemistry [8]. In this paper, he not only formulated his five principles of nomenclature which later became embodied in the *Méthode de Nomenclature Chimique* [9,10], but also introduced the word *radical* to describe a multiatomic entity; in his own words "Having found the adjectives arsenical and acetic consecrated by usage, it was necessary to preserve them and form only such close nouns to the radicals of these terms that they could be understood without explanation. Arseniates and acetates seemed to me to fulfil this condition." He makes no further comment on the term in this paper, which also includes a table which lists acids, the generic names of salts derived from these acids, bases or substances that bind to acids. This table also confirms that he was still a phlogistonist [11,12] in 1782, as phlogiston is listed amongst the bases or substances that bind to acids. The word *radical* itself seems to derive from the Latin word radix (root).

Figure 1. Louis-Bernard Guyton, Baron de Morveau (1737–1816, subsequently Louis-Bernard Guyton-Morveau) was a French chemist and politician who introduced the word radical in 1782. (Public domain image. Source https://en.wikipedia.org/wiki/Louis-Bernard_Guyton_de_Morveau#/media/File:Louis-Bernard_Guyton_de_Morveau.jpg).

By the time of the publication of the *Méthode*, the concept of radicals was embedded in the core of the model in five classes of substances which had not been decomposed into simpler materials (the second class includes all the acidifiable bases or radical principles of the acids) [9,10]. In this work, the "radical of the acid" was precisely defined as "the expression of acidifiable base". The explanations given in the text are difficult for the modern reader to follow as the conversion of the radical (such as nitrate or acetate) to the parent acid did not involve the addition of protons but rather oxygen.

Although the credit for the discovery of oxygen should be shared between William Scheele, Joseph Priestley and Antoine Laurent de Lavoisier [13,14], Lavoisier's contribution included the name *oxygène*, from the Greek ὀξύς (acid, sharp) and -γενής (producer, begetter), on the basis of his belief that oxygen was a constituent of all acids. On this basis, the *Méthode* continues to clarify the nomenclature of radicals defining known acids as arising from the addition of oxygen to "pure charcoal, carbon or carbonic radical ... Sulphur or sulphuric ... radical and phosphorus or phosphoric radical". The identification of oxygen as the essential component of an acid was not without its difficulties and for elements such as sulfur, with variable oxidation states, it was necessary to state that "it is evident that the sulphur is at the same time sulphuric radical, and sulphureous radical". Additional problems arose with nitrogen derivatives, with de Morveau using both *Azote* and *Radical Nitrique* for the parent radical. It took Jean Antoine Chaptal [15] to introduce the name *nitrogène* in his 1790 work *Eléments de chimie* [16,17].

The text of the *Méthode* uses the term radical extensively to describe acids and their salts and the construction of the names is illustrated in the extensive tables correlating the old names with the ones which are newly proposed. One of the most important features of the *Méthode* was the folding table of substances in which the core radicals are identified.

One aspect of the establishment of the concept of radicals is reminiscent of the later work of Mendeleev, who proposed missing elements from the periodic table and identified their likely properties. In the same way, the *Méthode* recognizes that muriatic acid (modern name hydrochloric acid) contained an unknown radical, described as muriatic radical or muriatic radical principle. The extention of the radical concept to organic chemistry was also pre-empted by de Morveau when he noted that the reaction of sucrose with nitric acid to give ethanedioic acid (*acide saccharin*), which is a combination of oxygen and *radical saccharin*.

3.2. Lavoisier's Adoption

The use of the term radical in the original sense of de Morveau was broadly adopted by Antoine-Laurent de Lavoisier and his wife Marie-Anne Pierrette Paulze Lavoisier [18–21] in a number of subsequent and influential texts (Figure 2). The *Méthode* was republished and expanded [22], but the most influential was the *Traité Élémentaire de Chimie, Présenté dans un Ordre Nouveau, et d'Après des Découvertes Modernes* [23–25]. This also served to further bring the changes in nomenclature and philosophy to the attention of the anglophone world, which received the first translation of the *Méthode* in 1788 and was able to delight in the English translation of the *Traité* from 1791 onwards [22,26–30]. The radical concept is intrinsic to the book and is also clearly defined "The word acid, being used as a generic term, each acid falls to be distinguished in language, as in nature, by the name of its base or radical. Thus, we give the generic names of acids to the products of the combustion or oxygenation of phosphorus, of sulphur, and of charcoal; and these products are respectively named, phosphoric acid, sulphuric acid, and carbonic acid". In his list of elements in the Traité, Lavoisier lists *Radical muriatique*, *Radical fluorique* and *Radical boracique* (the elements chlorine, fluorine and boron respectively) as unknown (*Inconnu*). In the context of organic chemistry, Lavoisier recognized that organic compounds contained compound radicals which could combine with oxygen to form more complex substances, such as ethanol or ethanoic acid. We are fortunate that not only was Marie-Anne Pierrette Paulze Lavoisier an enthusiastic and gifted co-worker (and according to the *mores* of the times, not listed as a co-author), but that she also actively contributed to the *Traité* and preserved many of Antoine Lavoisier's writings, including his notebooks, for the benefit of future generations.

Figure 2. Antoine-Laurent de Lavoisier (1743–1794, subsequently Antoine Lavoisier) popularized the use of the term radical (Public domain image. Source https://commons.wikimedia.org/wiki/File: Antoine_Laurent_de_Lavoisier.png).

4. From *Radical Particulier* to the Radical Theory and the Theory of Types

4.1. Gay-Lussac and the CN Radical

The next player in our drama of radicals should be Joseph Louis Gay-Lussac [31] (Figure 3a) and, in particular, his work on cyanides. Although HCN (hydrocyanic acid, prussic acid) was a known compound, Gay-Lussac established its formula and showed that it contained no oxygen, another of the nails in the coffin of Lavoisier's theory that all acids contained oxygen. By 1815, he had prepared metal cyanide salts as well as ClCN and cyanogen and correctly identified that the CN unit was retained throughout chemical transformations. His publication *Recherches sur l'acide prussique*, repeatedly refers to the *radical de l'acide prussique* [32–35]. This, in turn, necessitates a subsequent and consequent linguistic distinction between "simple radicals" (iron, sulphur, nitrogen, phosphorus and carbon) and "compound radicals"; containing multiple elements bonded together but which behave as distinct (and inseparable) units. As Gay-Lussac wrote "Here, then, is a very great analogy between prussic acid and muriatic and hydriodic acids. Like them, it contains half its volume of hydrogen; and, like them, it contains a radical which combines with the potassium, and forms a compound quite analogous to the chloride and iodide of potassium. The only difference is, that this radical is compound, while those of the chloride and iodide are simple" [36]. In isolating cyanogen, Gay-Lussac claimed to have isolated the first compound radical (actually the dimer, $(CN)_2$).

The identification of compound radicals was further expanded by Jöns Jacob Berzelius in 1817. Berzelius (Figure 3b) was the leading exponent of the electrochemical dualism theory which considered that all compounds are salts derived from basic and acidic oxides [37,38]. As one of the most respected chemists of the time, Berzelius' support for this model resulted in its widespread acceptance. For example, Berzelius would regard the compound potassium sulfate, K_2SO_4, as arising from the combination of the positively charged metal oxide K_2O and negatively charged SO_3. The radical theory as applied to inorganic compounds meshed well with his views, but he had difficulties in extending these to organic species. Nevertheless, he considered that the new concept of simple and compound radicals would clarify the differences between the inorganic acids with simple radicals and the organic acids with compound radicals "In inorganic nature all oxidized bodies contain a simple radical, while all organic substances are oxides of compound radicals. The radicals of vegetable substances consist generally of carbon and hydrogen, and those of animal substances of carbon, hydrogen and nitrogen" [39]. In reality, Berzelius refused to accept the possibility that a radical could contain oxygen and this, ultimately, led to the discrediting of the theory. In the intermediate period, however, the compound radical model was the origin of a new radical theory for organic chemistry and ultimately the modern functional group model.

(a) (b)

Figure 3. (**a**) Joseph Louis Gay-Lussac (1778 – 1850) showed that CN was a compound radical and opened the doors to the Radical Theory of organic chemistry. (Public domain image. Source https://en.wikipedia.org/wiki/Joseph_Louis_Gay-Lussac#/media/File:Gaylussac.jpg) (**b**) Jöns Jacob Berzelius (1779 – 1848) was one of the leading chemists of his age and in 1817 he laid the basis for the Radical Theory in organic chemistry. (Public domain image. Source https://en.wikipedia.org/wiki/Jöns_Jacob_Berzelius#/media/File:Jöns_Jacob_Berzelius.jpg).

4.2. The General Radical Theory

The stage is now set for the generalization of the radical theory. The major players in this were Friedrich Wöhler (Figure 4a) [40], Justus Freiherr von Liebig (Figure 4b) [41,42] and (at least for a period) Jean Baptiste André Dumas (Figure 4c) [43]. The three had a vision of radicals as collections of atoms that behaved like elements and persisted through chemical reactions, although Dumas subsequently shifted his allegiance to the theory of types (Section 4.3).

(a) (b) (c)

Figure 4. (**a**) Friedrich Wöhler (1800–1882) showed that CN was a compound radical and opened the doors to the Radical Theory of organic chemistry. (Public domain image. Source https://en.wikipedia.org/wiki/Friedrich_Wöhler#/media/File:Friedrich_Wöhler_Litho.jpg) (**b**) Justus Freiherr von Liebig (1803–1873) was one of the leading chemists of his age and in 1817 he laid the basis for the Radical Theory in organic chemistry. (Public domain image. Source https://en.wikipedia.org/wiki/Justus_von_Liebig#/media/File:Justus_von_Liebig_NIH.jpg) (**c**) Jean Baptiste André Dumas (1800–1884).

One of the critical publications was *Untersuchungen über das Radikal der Benzoesäure* by Liebig and Wöhler in 1832 [44], which introduces synthetic chemistry in a manner that we rarely see today "If it is possible to find a bright point in the dark area of organic nature, which seems to us to be one of the entrances through which we can perhaps reach true paths of exploration and recognition. From this point of view, one may consider the following attempts, which, as far as their extent and their connection with other phenomena is concerned, leave a wide, fertile field to cultivate". In a way, this publication was somewhat heretical, at least in the eyes of Berzelius, as Wöhler and Liebig maintained that a radical could be more than just the base of an acid. Specifically, Wöhler and Liebig showed that the benzoyl radical (C_6H_5CO in modern formulation) persisted in the compounds C_6H_5CO-H, C_6H_5CO-OH, C_6H_5CO-Cl, C_6H_5CO-I, C_6H_5CO-NH_2, C_6H_5CO-Br, and $(C_6H_5CO$-$)_2S$. The conclusion was that the benzoyl radical behaved in a similar manner to an inorganic radical and persisted unchanged through multiple reactions.

The impact of this publication on the organic chemistry community cannot be underestimated and resulted in an explosive reporting of new radicals over the next few years, including acetyl, methyl, ethyl, cacodyl (Me_2As), cinnamoyl (C_6H_5CH=CH), and n-$C_{16}H_{33}$. Originally, Dumas was opposed to the radical theory but eventually became convinced by Liebig's arguments. Dumas was responsible for the recognition of the methyl, cinnamoyl and n-$C_{16}H_{33}$ radicals. Although the radical theory has not survived, the nomenclature introduced is still in use today. Berzelius himself was responsible for the identification of the ethyl radical [37,45]. The state-of-the-art in radical theory in the Berzelius spirit is found in another publication of Liebig which interprets a large number of experimental results on ethers in terms of the Berzelius radical model [46].

By 1837, although Dumas and Liebig still disagreed in detail on which groups of atoms were to be considered radicals, they were sufficiently confident in the universality of their radical model, that they published their "Note on the present state of organic chemistry", which is a comprehensive overview of the radical theory at that time [47]. It appears that Liebig was given to flights of purple prose "and that, we are convinced, is the whole secret of organic chemistry. Thus, organic chemistry possesses its own elements which at one time play the role belonging to chlorine or to oxygen in mineral chemistry and at another time, on the contrary, play the role of metals. Cyanogen, amide, benzoyl, the radicals of ammonia, the fatty substances, the alcohols and analogous compounds—these are the true elements on which organic chemistry is founded and not at all the final elements, carbon, hydrogen, oxygen, and nitrogen elements which appear only when all trace of organic origin has disappeared. For us, mineral chemistry embraces all substances which result from the direct combination of the elements as such. Organic chemistry, on the contrary, should comprise all substances formed by compound bodies functioning as elements would function. In mineral chemistry, the radicals are simple; in organic chemistry, the radicals are compound; that is all the difference One year later, in 1838, Liebig clearly defined what he understood by the term radical, in the context of the CN radical: "So we call cyanogen a radical, because 1) it is the non-changing constituent in a series of compounds, because 2) it can be replaced in them by other simple bodies, because 3) it can be found in its connections with a simple body of the latter, and represented by equivalents of other simple bodies. Of these three main conditions for the characteristic of a composite radical, at least two must always be fulfilled if we are to regard it in fact as a radical" [48].

The proposals of Liebig were not universally accepted. Robert Hare in the United States of America published a number of articles dismissing the commonality of the oxoacids and "simple" acids such as the hydrogen halides, well summarized in his monograph "An attempt to refute the reasoning of Liebig in favor of the salt radical theory" [49]. Berzelius, in particular, came to have difficulties with the radical theory of Wöhler and Liebig because it directly challenged his electrochemical dualism theory [50]. For example, the relationship between benzaldehyde C_6H_5CO-H and benzoyl chloride C_6H_5CO-Cl could not possibly be correct because the hydrogen which has a positive charge cannot be replaced by a negative chlorine.

Not only were ever more radicals being identified, but they were also being isolated as chemical species. A few highlights serve to exemplify this. Robert Wilhelm Bunsen (1811–1899) reinvestigated some arsenic compounds first reported by Cadet and obtained a foul-smelling and highly toxic liquid which he called *Alkarsin*, although Berzelius suggested that cacodyl (or kakodyl) was more appropriate. The compound, formulated $(CH_3)_2As$ [51] was obtained from the reaction of $(CH_3)_2AsCl$ with zinc and was widely thought to be the free cacodyl radical. This compound was subsequently shown to be the dimer, $(CH_3)_2AsAs(CH_3)_2$. Similarly, Kolbe isolated the free methyl radical [52] and Frankland the free ethyl radical [53], although both were actually the dimers (ethane and butane, respectively).

4.3. The Theory of Types

The theory of types is rather a difficult concept for the modern chemist to appreciate. Put simply, it retains the fundamentals of the radical theory, but allows the replacement of elements and groups within a radical. With hindsight, it is possible to see the origins of the functional group model of organic chemistry within this approach. The development leading to the theory of types came from Dumas, who in 1838 described the chlorination of acetic acid to give trichloroacetic acid [54–57]. The substitution of hydrogen by chlorine generated a new radical (trichloroacetyl or trichloromethyl rather than acetyl or methyl) but did not change the molecular *type*. The chemical properties of acetic acid and trichloroacetic acid were very similar, indicating the same molecular type. Dumas published two papers which enunciated his theory of types [55,56] The level of vitriol and animosity in the debate is well exemplified by the spoof publication by S. C. H. Windler (actually written by Wöhler) in *Annalen* in which he rather wickedly parodies the substitution theories of Dumas and collagues [58]. He describes sequentially replacing atoms in manganese(II) acetate (his formulation, $MnO + C_4H_6O_3$) with chlorine, initially producing manganese(II) trichloroacetate and eventually, $Cl_2Cl_2 + Cl_8Cl_6Cl_6$ (i.e., Cl_{24}). This compound was a yellow solid resembling the original manganese(II) acetate, because "hydrogen, manganese, and oxygen may be replaced by chlorine, there is nothing surprising in this substitution". In a footnote, he adds "I have just learned that there is already in the London shops a cloth of chlorine thread, which is very much sought after and preferred above all others for night caps, underwear, etc."

By 1853, primarily due to the work of Charles Adolphe Wurtz, Hoffman, Williamson and Gerhardt, four different types had been identified; the water type, the hydrogen type, the hydrogen chloride type and the ammonia type. The water type included water, alcohols, ethers and carboxylic acids, the hydrogen type, dihydrogen, and alkanes, the hydrogen chloride type included organohalogen compounds such as C_2H_5Cl and finally, the ammonia type which included all primary, secondary and tertiary amines [59].

4.4. Laurent and the Theory of Types

Auguste Laurent (1807–1853) also studied substitution reactions and from 1834 onwards described numerous examples in which hydrogen atoms within radicals were replaced by halogens or oxygen [60–62]. Probably, the credit for the theory of types should be shared by Laurent with Dumas, because the former clearly recognized that the fundamental properties of the compound were not significantly changed by the substitution [63–65]. His theories are clearly stated in his book *Méthode de Chimie* from 1854 [66] but the ideas are clearly formulated (and seen to be almost identical to those of Dumas) as early as 1836 "All organic compounds are derived from a hydrocarbon, a fundamental radical, which often does not exist in its compounds but which may be represented by a derived radical containing the same number of equivalents" [67]. It appears that Dumas deliberately underplayed the importance of Laurent and over-emphasized the relevance of his protegé Henri Victor Regnault. On occasion, Laurent expressed his feelings in plain rather than scientific language " … others, pretend that I have taken some ideas of M. Dumas. M. Dumas. … has done much for the science; his part is sufficiently great that one should not snatch from me the fruit of my labors and present the offering to him" [68]. And concerning radicals, he wrote "I claim with a conviction most profound that to me belongs, and to me alone, the most part of the ideas developed by M. Dumas" [69]. The arguments continued!

In 1837, Laurent developed a theory of fundamental and derived radicals, subsequently known as his nucleus theory, which was based upon an obscure geometrical argument and attempted to rationalize the carbon core of radicals undergoing substitution. Like much of his work, this was an interesting and novel attempt to bring order to organic chemistry [70]. Nevertheless, the theory of Laurent was anathema to Liebig, who in his usual offensive manner discussed it "not because he found something in it worthy of mention, not in order to admit its having an influence on the development of chemistry but in order to demonstrate that it is unscientific, good for nothing".

4.5. Dualities, Inconsistencies and Ambiguities within the Radical Theory

Even at the time of its greatest success, there were many inconsistencies and dualities within the radical theory. Today, we would understand the term acetyl radical to refer to the species CH_3CO. Unfortunately, this was not the case in the 19th Century CE. In 1835, Henri Victor Regnault (1810–1878) [71,72] reported a new radical C_2H_3 (formulated C_4H_6 at the time) which he termed *aldehydène* [73]. This radical was present in the compounds $H_2C=CHCl$, $H_2C=CHBr$, $BrCH_2CH_2Br$ and many others that he isolated. He also linked the radical aldehydène to ethanal and ethanoic acid, which Regnault formulated as $\{C_4H_6O + H_2O\}$ and $\{C_4H_6O_3 + H_2O\}$, respectively. In 1839, Liebig suggested that the radical C_2H_3 should be called acetyl, in accord with his own system of nomenclature [74]. This 1839 paper of Liebig served to link together in a more-or-less coherent manner the various radicals and radical theories which had been proposed for C_2 compounds (although with the atomic weight confusion at the time many of these were formulated C_4 species). The Aetherin (or etherin) theory was proposed by Dumas and Boullay in 1828 and considered that C_2H_4 (formulated C_4H_8 at the time) was the common radical in C_2 compounds: thus, C_2H_5OH, $C_2H_5OC_2H_5$ and C_2H_5Cl were the aetherin radical with water, ethanol and HCl, respectively [75]. In contrast, Berzelius formulated these compounds in terms of the C_2H_5 (ethyl) radical [37,45].

5. Valency Displaces Radicals

The real death of the old radical theory and the theory of types came in 1852 when Edward Frankland formulated what was to become the concept of valency, "When the formulae of inorganic chemical compounds are considered, even a superficial observer is struck with the general symmetry of their construction ... it is sufficiently evident ... no matter what the character of the uniting atoms may be, the combining power of the attracting element, if I may be allowed the term, is always satisfied by the same number of these atoms" [76]. Frankland's combining power was the first formulation of the basic idea of valence and the entry to the electronic view that has dominated chemistry ever since.

A few years later, in 1858, Kekulé proposed a fixed valence for elements; although he did not equate the combining power with valence [77]. Kekulé successfully rationalized the structures of organic compounds by assuming a fixed valence of four for carbon, and extended this fixed valence idea to the elements nitrogen and oxygen which had fixed valences of three and two, respectively. The fixed valence of four for carbon necessitated multiple bonds (or free valences) in appropriate compounds. And so modern organic chemistry was born—or rather, as we have seen on a number of occasions in this article, we can testify to another of its births!

It is one of the pleasures associated with the study of the development of chemistry in the 19th Century CE, to read not only the contemporary primary literature, but also the textbooks and monographs of the period. These often provide a unique view of the way in which views changed and also give an understanding of the tensions and controversies in the science of the time. One of the lesser known works of this period is "A Short History of the Progress of Scientific Chemistry in Our Own Times" by William Tilden, which gives a detailed account of the evolution of chemistry to the last year of the 19th Century CE. The sections on the development of the Theory of Types and the subsequent Valency Model are excellent and also document a number of the poorly documented highways and by-ways associated with the scientific journey to the Valency Model [78]. An excellent contemporary (1867) overview of the Theory of Types and the relationship to the atomicity of the radicals is given by Adolphe Wurtz [79].

An interesting historical overview of the development of the subject written after the triumph of valency theory is to be found in the books by von Meyer [80] and Venable [81].

6. The Freeing of the Radical—the First Modern Radicals

Although transition metal compounds with unpaired electrons were well-known, and "simple" inorganic substances, such as Frémy's salt ($K_4[ON(SO_3)_2]_2$) [82], NO or NO_2, which fulfill our modern definition of a radical had been long established, the dominance and success of the valence theory in organic chemistry, based upon the invariable and inviolable tetravalency of carbon led to the widely accepted opinion that organic radicals (modern sense) could not exist. The confidence in the tetravalency of carbon and the complacency of the organic community was shattered in 1900, when Moses Gomberg at the University of Michigan reported the preparation of triphenylmethyl radical, Ph_3C, as the product from the attempted preparation of hexaphenylethane from the reaction of chlorotriphenylmethane with zinc [83]. The title of the paper, "An instance of trivalent carbon: triphenylmethyl" hints at the supremacy of the "tetravalent carbon" dogma [84].

The rest, dear reader, is history.

7. Final Words

In this short article, we have presented a story which describes the evolution of organic chemistry and which laid the basis for our modern understanding based on the electronic, molecular orbital and functional group approaches. Perhaps surprising for the modern reader is the passion with which the debate was conducted and the manner in which the personalities of the individual involved come though and, indeed, the personalization of the rhetoric. The well-known *Schwindler* article has already been referred to. The correspondence between Berzelius, Liebig, Dumas and Wöhler is a wonderful introduction to the art and science of denigrating your rivals in language that is rarely found in the scientific literature [85]. The discourse was not limited to scientific matters, but also to the character and nationality of the players, for example, Liebig described Dumas on various occasions as a swindler, charlatan, tightrope dancer, Jesuit, highwayman, and a thief, like "nearly all Frenchmen" [86]. As he became older, Berzelius became increasingly cantankerous, and writes of Liebig "I will say nothing of Liebig's ruthless, thoughtless and unjustified criticism, ... it just disappoints and saddens me ... with the manner of a dictator, who wishes to abolish an old constitution and create a new one ... I hold it unlikely that he will take the slightest notice of my advice" [85]. Berzelius again, talking of Liebig "Either Liebig is mad, which I already began to painfully fear a year ago, in which case he deserves the pity of everyone and needs to be treated accordingly, or he is an unwise, inflated fool" [85]. Wentrup has recently published an assessment of some aspects of the debate in the context of Zeise's discovery of his eponymous salt, $K[Pt(C_2H_4)Cl_3]$ which also documents the acrimonious exchanges between the players [86].

Author Contributions: This article was conceived and written jointly by E.C.C. and C.E.H. All authors have read and agreed to the published version of the manuscript.

Funding: This work received no external funding.

Acknowledgments: As always, we give our thanks to the various library and abstracting services which have aided us in identifying and sourcing material and, in particular, to the online sources of historical materials which have made the life of scientists interested in the origins of their subject so much easier in the 21st Century C.E. Finally, we acknowledge the friendship and scholarship of Bernd Giese on the occasion of his 80th birthday.

Conflicts of Interest: The authors declare no conflict of interest.

References

1. Dawkins, R. *The Selfish Gene*; Oxford University Press: Oxford, UK, 1976.
2. International Union of Pure and Applied Chemistry. Available online: https://iupac.org (accessed on 20 March 2020).
3. Crosland, M.P. *Historical Studies in the Language of Chemistry*; Dover Publications: Mineola, NY, USA, 2004.
4. Senning, A. *Elsevier's Dictionary of Chemoetymology: The Whies and Whences of Chemical Nomenclature and Terminology*; Elsevier: Amsterdam, The Netherlands, 2007.
5. Senning, A. *The Etymology of Chemical Names: Tradition and Convenience Vs. Rationality in Chemical Nomenclature*; De Gruyter: Berlin, Germany, 2019.
6. Odinga, E.S.; Waigi, M.G.; Gudda, F.O.; Wang, J.; Yang, B.; Hu, X.; Li, S.; Gao, Y. Occurrence, formation, environmental fate and risks of environmentally persistent free radicals in biochars. *Environ. Int.* **2020**, *134*, 105172. [CrossRef]
7. Bouchard, G. *Guyton-Morveau Chimiste Et Conventionnel (1737–1816)*; Libr. académique Perrin: Paris, France, 1938.
8. De Morveau, L.-B.G. Sur Les Dénominations Chymiques, La Nécessité D'en Perfectioner Le Systême, Et Les Règles Pour Y Parvenir. *Obs. Phys. Chim. Hist. Nat. Arts* **1782**, *19*, 370–383.
9. De Morveau, L.-B.G.; Lavoisier, A.L.; Bertholet, P.E.M.; de Fourcroy, A.F. *Méthode De Nomenclature Chimique. Proposé Par Mm. De Morveau, Lavoisier, Bertholet, & De Fourcroy. On Y a Joint Un Nouveau Système De Caractères Chimiques, Adaptés À Cette Nomenclature, Par Mm. Hassenfratz & Adet*; Cuchet: Paris, France, 1787.
10. De Morveau, L.-B.G.; Lavoisier, A.L.; Bertholet, P.E.M.; de Fourcroy, A.F. *Method of Chymical Nomenclature, Proposed by Messrs. De Morveau, Lavoisier, Bertholet, and De Fourcroy. To Which is Added, a New System of Chymical Characters, Adapted to the Nomenclature by Mess. Hassenfratz and Adet. Translated from the French, and the New Chymical Names Adapted to the Genius of the English Language by James St. John*; Kearsley: Flint, MI, USA, 1788.
11. White, J.H. *The History of the Phlogiston Theory*; Edward Arnold: London, UK, 1932.
12. Conant, J.B. *Overthrow of the Phlogiston Theory: Chemical Revolution of 1775–89*; Harvard University Press: Cambridge, MA, USA, 1950.
13. Rodwell, G.F. Lavoisier, Priestley, and the Discovery of Oxygen. *Nature* **1882**, *27*, 8–11. [CrossRef]
14. Brown, O.R. Discovery and History of Oxygen. In *Oxygen, the Breath of Life: Boon and Bane in Human Health, Disease, and Therapy*; Brown, O.R., Ed.; Bentham Science Publishers: Sharjah, UAE, 2017; pp. 54–77.
15. Bolado, E.; Argemí, L. Jean Antoine Chaptal: From Chemistry to Political Economy. *Eur. J. Hist. Econ. Thought* **2005**, *12*, 215–239. [CrossRef]
16. Chaptal, J.A. *Élémens De Chimie, 3 Volumes*; J.-F. Picot: Montpellier, France, 1790.
17. Chaptal, J.A. *Elements of Chemistry. Third American Edition. Three Volumes in One, Translated from the French*; J. T. Buckingham: Boston, MA, USA, 1806.
18. Guerlac, H. *Antoine-Laurent Lavoisier Chemist and Revolutionary*; Scribner: New York, NY, USA, 1975.
19. McKie, D. *Antoine Lavoisier Scientist, Economist, Social Reformer*; Constable: London, UK, 1952.
20. Szabadváry, F. *Antoine Laurent Lavoisier*; B. G. Teubner: Leipzig, Germany, 1987.
21. Donovan, A. *Antoine Lavoisier*; Cambridge University Press: Cambridge, UK, 1996.
22. Lavoisier, A.L.; de Fourcroy, A.F.; de Morveau, L.; de Cadet, L.C.C.; Baumé, A.; d'Arcet, J.; Sage, B.-G. *Nomenclature Chimique Ou Synonymie Ancienne Et Moderne, Pour Server À L'intelligence Des Auteurs. Nouvelle Édition. A Laquelle on a Joint Différens Mémoires & Rapports De Mm. Lavoisier, Fourcroy, Morveau, Cadet, Baumé, D'Arcet & Sage, Sur La Nécessité De Réformer Et De Perféctionner La Nomenclature*; Cuchet: Paris, France, 1789.
23. Lavoisier, A.L. *Traité Élémentaire De Chimie, Présenté Dans Un Ordre Nouveau, Et D'après Des Découvertes Modernes*; Cuchet: Paris, France, 1789.
24. Lavoisier, A.L. *Traité Élémentaire De Chimie, Présenté Dans Un Ordre Nouveau, Et D'après Des Découvertes Modernes*, 2nd ed.; Cuchet: Paris, France, 1793.
25. Lavoisier, A.L. *Traité Élémentaire De Chimie, Présenté Dans Un Ordre Nouveau, Et D'après Des Découvertes Modernes*, 3rd ed.; Deterville Libraire: Paris, France, 1801.
26. Lavoisier, A.L. *Elements of Chemistry in a New Systematic Order Containing All the Modern Discoveries. Translated from the French by Robert Kerr*; William Creech: Edinburgh, UK, 1790.

27. Lavoisier, A.L. *Elements of Chemistry in a New Systematic Order Containing All the Modern Discoveries. Translated from the French by Robert Kerr*, 2nd ed.; With Notes, Tables, and Considerable Additions; William Creech: Edinburgh, UK, 1793.
28. Lavoisier, A.L. *Elements of Chemistry in a New Systematic Order Containing All the Modern Discoveries. Translated from the French by Robert Kerr*, 3rd ed.; With Notes, Tables, and Considerable Additions; William Creech: Edinburgh, UK, 1796.
29. Lavoisier, A.L. *Elements of Chemistry in a New Systematic Order Containing All the Modern Discoveries. Translated from the French by Robert Kerr*, 4th ed.; With Notes, Tables, and Considerable Additions; William Creech: Edinburgh, UK, 1799.
30. Lavoisier, A.L. *Elements of Chemistry in a New Systematic Order Containing All the Modern Discoveries. Translated from the French by Robert Kerr*, 5th ed.; With Notes, Tables, and Considerable Additions; William Creech: Edinburgh, UK, 1802; Volume 2.
31. Goldwhite, H. Gay-Lussac After 200 Years. *J. Chem. Educ.* **1978**, *55*, 366. [CrossRef]
32. Gay-Lussac, J.L. Recherches Sur L'acide Prussique. *Ann. Chim.* **1815**, *95*, 136–231.
33. Gay-Lussac, J.L. Experiments on Prussic Acid. *Ann. Philos.* **1815**, *7*, 350–364.
34. Gay-Lussac, J.L. Experiments on Prussic Acid. *Ann. Philos.* **1816**, *8*, 108–115.
35. Gay-Lussac, J.L. Experiments on Prussic Acid. *Ann. Philos.* **1816**, *8*, 37–52.
36. Gay-Lussac, J.L. Experiments on Prussic Acid. *Ann. Philos.* **1816**, *8*, 350–364.
37. Berzelius, J.J. *Lehrbuch Der Chemie, Vol. 1 Translated by F. Wöhler*; Arnold: Dresden, Germany, 1825.
38. Hudson, J. *Electrochemistry and the Dualistic Theory the History of Chemistry*; Hudson, J., Ed.; Macmillan Education UK: London, UK, 1992; pp. 92–103.
39. Berzelius, J.J. *Elemente Der Chemie Der Unorganischen Natur. Aufs Neue Durchgesehen Vom Verfasser. Aus De. Schwedischen Übersetzt, Und Mit Eingen Anmerkungen Begleitet Von Dr. Johann Georg Ludolph Blumhof Part 1*; Johann Ambrosius Barth: Leipzig, Germany, 1816.
40. Keen, R. *The Life and Works of Friedrich Wöhler (1800–1882)*; Wiley-VCH Verlag GmbH & Co., KGaA: Weinheim, Germany, 2005.
41. Brock, W.H. *Justus Von Liebig*; Cambridge University Press: Cambridge, UK, 2002.
42. Shenstone, W.A. *Justus Von Liebig, His Life and Work: 1803–1873*; Macmillan: New York, NY, USA, 1895.
43. Cooke, J. Jean-Baptiste-andré Dumas. *Science* **1884**, *3*, 750–752.
44. Wöhler, F.; Liebig, J. Untersuchungen Über Das Radikal Der Benzoesäure. *Ann. Pharm.* **1832**, *3*, 249–282. [CrossRef]
45. Berzelius, J.J. Ueber Die Constitution Organischer Zusammensetzungen. *Ann. Pharm.* **1833**, *6*, 173–176. [CrossRef]
46. Liebig, J. Ueber Die Constitution Des Aethers Und Seiner Verbindungen. *Ann. Pharm.* **1834**, *9*, 1–39. [CrossRef]
47. Dumas, J.-B.-A.; von Liebig, J. Note on the Present State of Organic Chemistry. *C. R. Hebd. Seances Acad. Sci.* **1837**, *5*, 567–572.
48. Liebig, J. Ueber Laurent's Theorie Der Organischen Verbindungen. *Ann. Pharm.* **1838**, *25*, 1–31. [CrossRef]
49. Hare, R. *An Attempt to Refute the Reasoning of Liebig in Favor of the Salt Radical Theory*; B.L. Hamlen: New Haven, CT, USA, 1846.
50. Berzélius, J.J. Letter from Berzélius to Pelouze. *C. R. Hebd. Seances Acad. Sci.* **1838**, *6*, 629–649.
51. Bunsen, R.; von Baeyer, A. *Untersuchungen Über Die Kakodylreihie*; W. Engelmann: Leipzig, Germany, 1891.
52. Kolbe, H. Untersuchungen Über Die Elektrolyse Organischer Verbindungen. *Ann. Chem. Pharm.* **1849**, *69*, 257–294. [CrossRef]
53. Frankland, E. Xxvii—On the Isolation of the Organic Radicals. *Q. J. Chem. Soc.* **1850**, *2*, 263–296. [CrossRef]
54. Dumas, J.-B.A. Acide Produit Par L'action Du Chlore Sur L'acide Acétique. *C. R. Hebd. Seances Acad. Sci.* **1838**, *7*, 474.
55. Dumas, J.-B. Mé_moire Sur La Constitution De Quelques Corps Organiques Et. Sur La Thé_orie Des Substitutions. *C. R. Hebd. Seances Acad. Sci.* **1839**, *8*, 609–633.
56. Dumas, J.-B. Mémoire Sur La Loi Des Substitutions Et La Théorie Des Types. *C. R. Hebd. Seances Acad. Sci.* **1840**, *10*, 149–178.
57. Dumas, J. Ueber Das Gesetz Der Substitutionen Und Die Theorie Der Typen. *Justus Liebigs Ann. Der Chem.* **1840**, *33*, 259–300. [CrossRef]

58. Windler, S.C.H. Ueber Das Substitutionsgesetz Und Die Theorie Der Typen. *Ann. Chem. Pharm.* **1840**, *33*, 308.
59. Gerhardt, C. Recherches Sur Les Acides Organiques Anhydres. *Ann. Chim. Phys. Ser. 3* **1853**, *37*, 285–342.
60. Wisniak, J. Auguste Laurent. Radical and Radicals. *Educ. Quim.* **2009**, *20*, 166–175. [CrossRef]
61. DeMilt, C. Auguste Laurent, Founder of Modern Organic Chemistry. *Chymia* **1953**, *4*, 85–114. [CrossRef]
62. DeMilt, C. Auguste Laurent—Guide and Inspiration of Gerhardt. *J. Chem. Educ.* **1951**, *28*, 198–204. [CrossRef]
63. Laurent, A. Sur De Nouveaux Chlorures Et Brômures D'Hydrogène Carboné. *L'Institut* **1834**, *2*, 30.
64. Laurent, A. Sur De Nouveaux Chlorures Et Brômures D'Hydrogène Carboné. *Ann. Chim. Phys. Ser. 2* **1835**, *59*, 196–220.
65. Laurent, A. Sur Le Benzoïle Et La Préparation De La Benzimide; Analyse De L'essence D'amandes Amères. *Ann. Chim. Phys. Ser. 2* **1835**, *60*, 220–223.
66. Laurent, A. *Méthode De Chimie*; Mallet-Bachelier, Gendre et Successeur de Bachelier: Paris, France, 1854.
67. Laurent, A. Théorie Des Combinaisons Chimiques. *Ann. Chim. Phys. Ser.* **1836**, *61*, 125–146.
68. Laurent, A. Suite De Recherches Diverses De Chimie Organique. *Ann. Chim. Phys. Ser.* **1837**, *66*, 326–337.
69. Laurent, A. Réclamation De Priorité Relativement À La Théorie Des Substitutions, Et a Celle Des Types Ou Radicaux Dérivés. *C. R. Hebd. Seances Acad. Sci.* **1840**, *10*, 409–417.
70. Kapoor, S.C. The Origins of Laurent's Organic Classification. *Isis* **1969**, *60*, 477–527. [CrossRef]
71. Reif-Acherman, S. Henri Victor Regnault: Experimentalist of the Science of Heat. *Phys. Perspect.* **2010**, *12*, 396–442. [CrossRef]
72. Reif-Acherman, S. The Contributions of Henri Victor Regnault in the Context of Organic Chemistry of the First Half of the Nineteenth Century. *Quím. Nova* **2012**, *35*, 438–443. [CrossRef]
73. Regnault, H.V. Recherches De Chimie Organique. *Ann. Chim. Phys. Ser.* **1835**, *59*, 358–375.
74. Liebig, J. Vermischte Notizen. *Ann. Der Pharm.* **1839**, *30*, 129–150. [CrossRef]
75. Dumas, J.; Boullay, F.-P. Mémoire Sur Les Ethers Composés. *Ann. Chim. Phys. Ser. 2* **1828**, *37*, 15–52.
76. Frankland, E. On a New Series of Organic Bodies Containing Metals. *Phil. Trans.* **1852**, *142*, 417–444.
77. Kekulé, A. Ueber Die Constitution Und Die Metamorphosen Der Chemischen Verbindungen Und Über Die Chemische Natur Des Kohlenstoffs. *Ann. Chem. Pharm.* **1858**, *106*, 129–159. [CrossRef]
78. Tilden, W.A. *A Short History of the Progress of Scientific Chemistry in Our Own Times*; Longmans, Green, and Co.: London, UK, 1899.
79. Wurtz, A.C. *An Introduction to Chemical Philosophy According to the Modern Theories. Translated from the French, By Permission of the Author, by William Crookes, F.r.s.*; J. H. Dutton: London, UK, 1867.
80. Von Meyer, E. *A History of Chemistry from Earliest Times to the Present Day Being Also an Introduction to the Study of the Science Translated with the Author's Sanction by George Mcgowan, Ph.d*, 3rd ed.; Translated from the Third German Edition, With Various Additions and Alterations; Macmillan and Co.: New York, NY, USA, 1906.
81. Venable, F.P. *A Short History of Chemistry*, 3rd ed.; D.C. Heath and Co.: Boston, MA, USA, 1907.
82. Frémy, E. Sur Une Nouvelle Série D'acides Formés D'oxygène, De Soufre, D'hydrogène Et D'azote. *Ann. Chim. Phys. Ser. 3* **1845**, *15*, 408–488.
83. Gomberg, M. Radicals in Chemistry, Past and Present. *Ind. Eng. Chem.* **1928**, *20*, 159–164. [CrossRef]
84. Gomberg, M. An Instance of Trivalent Carbon: Triphenylmethyl. *J. Am. Chem. Soc.* **1900**, *22*, 757–771. [CrossRef]
85. Berzelius, J.J.; Liebig, C.; Wöhler, J.; Berzelius Und Liebig, F. *Ihre Briefe Von 1831–1845. Mit Erläuterenden Einschaltungen Aus Gleichzeitigen Briefen Von Liebig Und Wöhler Sowie Wissenschaftlichen Nachweisen. Hrsg. Mit Unterstützung Der Kgl. Bayer. Akademie Der Wissenschaften Von Justus Carrière*; Munich and Leipzig; J.F. Lehman: New York, NY, USA, 1893.
86. Wentrup, C. Zeise, Liebig, Jensen, Hückel, Dewar, and the Olefin Π-Complex Bonds. *Angew. Chem. Int. Ed. Engl.* **2019**. [CrossRef]

© 2020 by the authors. Licensee MDPI, Basel, Switzerland. This article is an open access article distributed under the terms and conditions of the Creative Commons Attribution (CC BY) license (http://creativecommons.org/licenses/by/4.0/).

Article

Towards a Real Knotaxane

Torben Duden and Ulrich Lüning *

Otto-Diels-Institut für Organische Chemie, Christian-Albrechts-Universität zu Kiel, Olshausenstr. 40, D-24098 Kiel, Germany; tduden@oc.uni-kiel.de
* Correspondence: luening@oc.uni-kiel.de

Received: 12 March 2020; Accepted: 10 April 2020; Published: 26 April 2020

Abstract: Two classes of mechanically interlocked molecules, [3]rotaxanes and knotted [1]rotaxanes, were the subject of this investigation. The necessary building blocks, alkyne-terminated axles containing two ammonium ions and azide-terminated stoppers, and azide-containing substituted macrocycles, have been synthesized and characterized. Different [3]rotaxanes were synthesized by copper-catalyzed "click" reactions between the azide stoppers and [3]pseudorotaxanes formed from the dialkyne axles and crown ethers (DB24C8). Methylation of the triazoles formed by the "click" reaction introduced a second binding site, and switching via deprotonation/protonation was investigated. In preliminary tests for the synthesis of a knotted [1]rotaxane, pseudorotaxanes were formed from azide-containing substituted macrocycles and dialkyne substituted diammonium axles, and copper-catalyzed "click" reactions were carried out. Mass spectral analyses showed successful double "click" reactions between two modified macrocycles and one axle. Whether a knotted [1]rotaxane was formed could not be determined.

Keywords: mechanically interlocked molecules; knot; rotaxane; macrocycle; click reaction; switching; shuttle

---

1. Introduction

Chemical elements bind each other through metallic, ionic and covalent bonds. Furthermore, molecules may bind each other supramolecularly, and in addition, for half a century, the mechanical bond has been known of [1]. The latter is responsible for the formation of rotaxanes [2], catenanes [3,4], molecular knots [5] and other mechanically interlocked molecules (MIM). Although only known since the second half of last century, a plethora of unusual MIMs have been synthesized and investigated [1], and some MIM types have even been combined. In 2003, Vögtle and co-workers described a MIM they called knotaxane because it is a rotaxane with molecular knots as stoppers [6].

In this work, we would like to discuss a "real" knotaxane. In contrast to Vögtle's MIM, it is not a combination of a rotaxane and a knot but a molecule in which the rotaxane property arises from a knotting. The fundamental feature of a "normal" rotaxane is the fact that the stoppers prevent the ring from slipping off the axle. However, from a mathematical-topological point of view, a rotaxane is topologically not special because by deformation (shrinking of the stoppers or enlargement of the ring), the ring may slip off—as it does in pseudorotaxanes [7–9]. Figure 1 (right) shows a sketch of a "real" knotaxane. The rings sit on a central axle as in a [3]rotaxane (Figure 1, left), but connections between the rings and the ends of the axle rather than stoppers prevent their slipping-off. This special MIM is a knotted molecule but also contains rings on an axle, as in a rotaxane. However, in contrast to standard rotaxanes, *all* atoms in this MIM are covalently connected. It may therefore be called a "knotted [1]rotaxane".

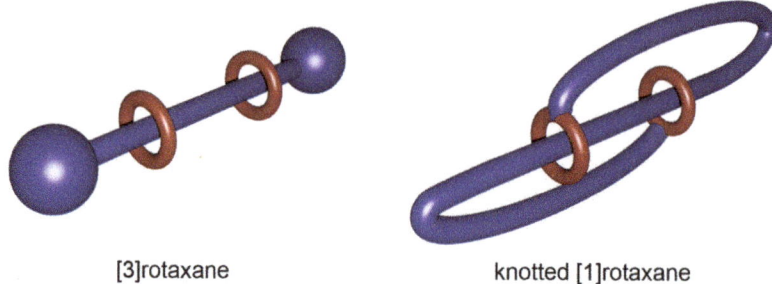

Figure 1. Comparison of the geometry of a [3]rotaxane with that of a knotted [1]rotaxane.

By introducing different binding sites into the axle of a rotaxane, the position of rings on the central axle may be switched, as demonstrated first by Stoddart [10]. Introduced into the knotaxane, this switching will result in a breathing of the molecule, or if suspended at its ends, it may act as a type of spring. We chose crown ethers as macrocyclic rings and ammonium and triazolium ions as binding sites (see below; for instance, **15** in Figure 6 and **17** in Figure 7). Such a pair of binding sites allows the controlled shuttling of a ring [11].

In acidic media, the ammonium nitrogen atoms are protonated. In comparison to ammonium ions, triazolium ions are poorer binding sites, since the charge is not localized and there are no hydrogen bonds. Therefore, a crown ether binds to an ammonium ion preferentially due to stronger Coulomb interactions and hydrogen bonds. Upon deprotonation, the interactions of the positive charge and one hydrogen bond vanish. In basic media, the triazolium ion is the only positively charged site and is, therefore, the better binding site, and the ring binds there after deprotonation.

Retrosynthetically, triazolium ions call for a "click" reaction between an azide and an alkyne (copper-catalyzed alkyne-azide cycloaddition, CuAAC) followed by alkylation. We chose to place the azides on the stopper side. Therefore, bis-alkyne terminated, diammonium-containing central parts of the axles were needed. In order to study the CuAAC and to allow the investigation of the switching, we first synthesized [3]rotaxanes by using simple azide stoppers. For the synthesis of the knotaxane, the azide function had to be connected to the crown ether. Therefore, the following tasks had to be accomplished: synthesis of the central axles, the alkyne terminated diammonium ions and crown ethers, each with an azide-terminated side chain; syntheses of [3]rotaxanes from the central axles, crown ethers and azide stoppers; alkylation of the triazoles; subsequent study of switching of the [3]rotaxanes; and finally, the study of the connection between the central axles with the azide-terminated crown ethers to give the knotted [1]rotaxane (Figure 2). As depicted in Figure 2, after formation of the [3]pseudorotaxane, there are two possibilities for "click" reactions. Either the azide connected to ring **a** reacts with alkyne end **b** or it reacts with **b′**. In the latter case, the desired knotaxane is formed (Figure 2, top right); in the other case a different topology is produced in which the rings may even slip off each end to form a handcuff topology (Figure 2, bottom right). The exact dimensions of the axle and the lengths of the tethers at the macrocycles needed to obtain the knotaxane cannot be forecasted. Therefore, different axles and tethers had to be synthesized.

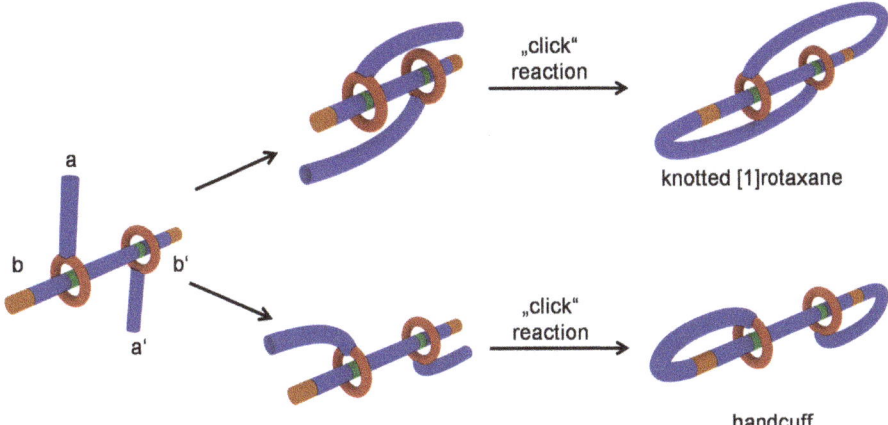

Figure 2. A double click reaction between macrocyclic rings carrying an azide function **a** and **a'**, and the alkyne ends **b** and **b'** of the axle in a [3]pseudorotaxane (**left**) may generate a knotted [1]rotaxane (**top, right**) by reaction of **a** with **b'** and **a'** with **b**, respectively, or a handcuff molecule (**bottom, right**) by reaction of **a** with **b** and **a'** with **b'**, respectively.

2. Materials and Methods

2.1. Methods

^{1}H and ^{13}C NMR spectra were recorded with Bruker DRX 500 MHz or Bruker Avance 600 MHz spectrometers (Bruker, Billerica, MA, USA) Mass spectrometric analysis was performed with AccuTOFGCv4G (HR-MS, electron ionization) from Jeol (Tokyo, Japan); a Q Exactive Plus mass spectrometer (HR-MS, electrospray ionization, positive mode) from Thermo Scientific (Waltham, MA, USA); and Autoflex speed (MALDI) from Bruker (Billerica, MA, USA). HPLC-MS experiments were carried out with VWR-Hitachi HPLC system Elite LaChrom coupled to an expression CMS mass spectrometer (electrospray ionization, positive mode) from Advion (Ithaca, NY, USA). IR spectra were recorded with Perkin–Elmer Spectrum100 FT-IR spectrometer (Perkin-Elmer, Waltham, MA, USA). The spectra were recorded using a MKII Golden GateTM Single Reflection ATR A531-G system from Specac (Orpington, UK). The elemental analyses were performed with the CHNS-O elemental analyzer EURO EA 3000 Series from Euro Vector (Pavia, Italy) and vario MICRO CUBE from Elementar at the Institut für Anorganische Chemie of the Christian-Albrechts-Universität zu Kiel (Germany). For this purpose, the samples were burned in zinc containers in a stream of oxygen.

2.2. General Synthetic Procedure

The detailed synthetic procedures and all analytical data for molecules not known in the literature are described in the Supplementary Materials. General procedures:

2.2.1. Reductive Amination of Aldehydes

Under a nitrogen atmosphere, a diamine (1,8-diaminooctane, 1,10-diaminodecane) and two equivalents of aldehyde **1** were dissolved in dry methanol. Sodium borohydride was added in portions while cooling with an ice bath followed by stirring at room temperature. Water was added and the solvent was removed in vacuo. The aqueous layer was extracted three times with dichloromethane. The combined organic layer was dried with magnesium sulfate and the solvent was removed in vacuo. Pale yellow solid diamines **2a/b** were obtained in good yields. Diamine **6** was synthesized analogously by starting with one equivalent of **4** and two equivalents of amine **5**.

2.2.2. Protonation and Ion Exchange

The secondary amines **2a/b** or **6** were dissolved in ethanol and then mixed with concentrated aqueous hydrochloric acid. The suspensions were stirred at room temperature and then filtered. The hydrochlorides were suspended in acetone and mixed with a saturated aqueous ammonium hexafluorophosphate solution. The solvent was removed in vacuo and the residue was stirred in water. The colorless solids **3a/b** and **7** were filtered and dried in vacuo.

2.2.3. Click Reaction to Form [3]rotaxanes

Under nitrogen atmosphere, an axle **3a/b** or **7** and dibenzo-24-crown-8 (DB24C8, **13**) were suspended in dry dichloromethane. The mixture was stirred at room temperature until a clear solution was obtained. Then, stopper **11**, tetrakis(acetonitrile)copper(I) hexafluorophosphate and 2,6-dimethylpyridine were added. The solution was stirred for 1 d at room temperature and then water was added. The aqueous layer was extracted with dichloromethane and the solvent was removed in vacuo. The residue was filtered through silica gel (dichloromethane:methanol (80:20)). The solvent of the filtrate was removed in vacuo and the residue was purified by chromatography on silica gel (dichloromethane:methanol (1:0) → (24:1)). The resulting yellow solid was dissolved in ethyl acetate and excess DB24C8 (**13**) precipitated. The solid was filtered off and the process was repeated. Finally, after concentration, the residue was again purified by chromatography on silica gel (dichloromethane:methanol (1:0) → (24:1)). Pale yellow solid rotaxanes **14a/b** or **16** were obtained.

2.2.4. Methylation of the Triazole

A rotaxane **14a/b** or **16** was dissolved in methyl iodide and stirred for 4 days at room temperature. The solvent was removed in vacuo and the residue was dissolved in dichloromethane. An identical volume of saturated aqueous ammonium hexafluorophosphate solution was added and the mixture was stirred vigorously. The aqueous layer was extracted with dichloromethane. The organic layer was dried with sodium sulfate and filtered, and the solvent was removed in vacuo. Pale yellow solids **15a/b** or **17** were obtained in good yields.

2.2.5. Switching of the Rotaxanes (Deprotonation and Protonation)

Deprotonation: The rotaxane **15a/b** or **17** was dissolved in chloroform (5 mL), mixed with 1 N sodium hydroxide solution (5 mL) and shaken vigorously. The layers were separated, the organic layer was dried with sodium sulfate and the solvent was removed in vacuo. Pale yellow solids were obtained.

Protonation: The residue of the deprotonation experiment was dissolved in chloroform (5 mL) and treated with a 0.2 N solution of trifluoroacetic acid in chloroform (3 mL). The solution was shaken vigorously and was then mixed with a saturated, aqueous ammonium hexafluorophosphate solution (3 mL). The mixture was again shaken vigorously followed by separation of the layers. The organic layer was dried with sodium sulfate, and the solvent was removed in vacuo. Pale yellow solids were obtained.

2.2.6. Etherification of 4-Benzyloxyphenol (25)

Under nitrogen, 4-benzyloxyphenol (**25**) and potassium carbonate were suspended in acetonitrile and a dibromoalkane was added. The suspension was stirred under reflux for 20 h and then filtered. The solvent was removed in vacuo and the residue was dissolved in dichloromethane. The organic layer was washed with an aqueous sodium hydroxide solution (10%) several times, dried with magnesium sulfate and filtered. The solvent was removed in vacuo and the residue was recrystallized from *n*-hexane. Colorless solids **26a–d** were obtained.

2.2.7. Reductive Deprotection of the Benzyl Protecting Group

Under a hydrogen atmosphere, the benzyl-protected hydroquinone derivative (**26a–d**) was dissolved in chloroform, and palladium on activated carbon (10% Pd content, 0.1 equivalent) was added. The suspension was stirred for 16 h at room temperature, then filtered through Celite, and the solvent was removed in vacuo. Pale grey solids **27a–d** were obtained.

2.2.8. Substitution of Bromide by Azide

A bromide (**27a–d**) was dissolved in dimethyl sulfoxide, sodium azide was added and the mixture was stirred at room temperature for 18 h. Water was added, and the aqueous layer was extracted with diethyl ether. The organic layer was dried with magnesium sulfate and filtered, and the solvent was removed in vacuo. The crude product was purified by chromatography on silica gel [dichloromethane → dichloromethane:methanol (93:7)]. Colorless oils **28a–d** were obtained.

3. Results and Discussion

3.1. Syntheses of the Axles

All dialkyne substituted central axles contain two secondary amine functions. These were synthesized by condensation of aldehydes with primary amines followed by reduction.

Alkyne substituted aldehyde **1** was synthesized following a literature-known synthesis [12]. In a ratio of 2:1, it was reacted with commercially available diamines (1,8-diaminooctane and 1,10-diaminodecane) (Figure 3). Quickly, the diimines precipitated as colorless solids. The imines were not isolated, but were directly reduced using an excess of sodium borohydride to obtain the secondary amines **2a** or **2b**. **2a/b** were obtained in very good yields after aqueous work-up and without further purification. Next, the secondary amines **2a/b** were dissolved in ethanol and mixed with concentrated hydrochloric acid. The hydrochlorides precipitated directly. In a mixture of acetone/water, the chloride ions were then exchanged by hexafluorophosphate ions in good yields.

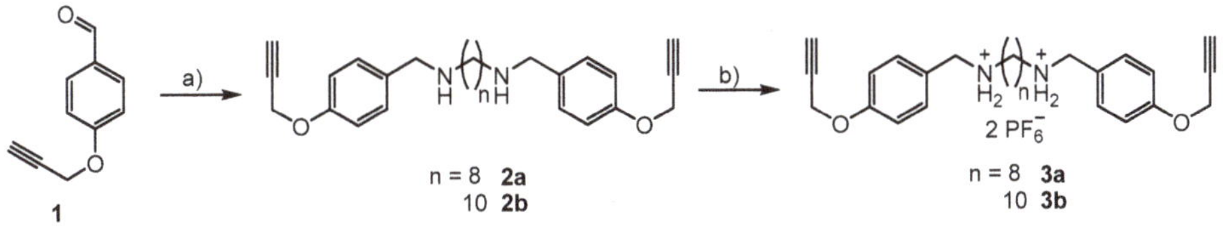

Figure 3. Synthesis of the axles **3a/b**. (a) 1. 1,8-diaminooctane or 1,10-diaminodecane, MeOH, 2 h, r.t.; 2. NaBH$_4$, 16 h, 95% (**2a**), 90% (**2b**). (b) 1. HCl, EtOH, 2 h, r.t.; 2. NH$_4$PF$_6$, acetone/water, 2 h, r.t., 74% (**3a**), 86% (**3b**).

The starting materials for the synthesis of axle **7**, dialdehyde **4** and amine **5**, were prepared following synthetic procedures from the literature [13–15]. As in the reactions to give **2a/b**, the reductive amination of **4** was performed with sodium borohydride in methanol (Figure 4). Due to the poor solubility of dialdehyde **4**, the reaction was carried out under reflux. The secondary amine **6** was obtained after aqueous work-up in a yield of 89%. Chromatographic purification was not possible. Therefore, the crude diamine **6** was directly protonated with hydrochloric acid to give the ammonium salt. The dihydrochloride was obtained after precipitation from ethanol in a yield of 87%. The subsequent ion exchange was again carried out in a solvent mixture of acetone and water with ammonium hexafluorophosphate. The bishexafluorophosphate salt **7** was obtained with a yield of 99%.

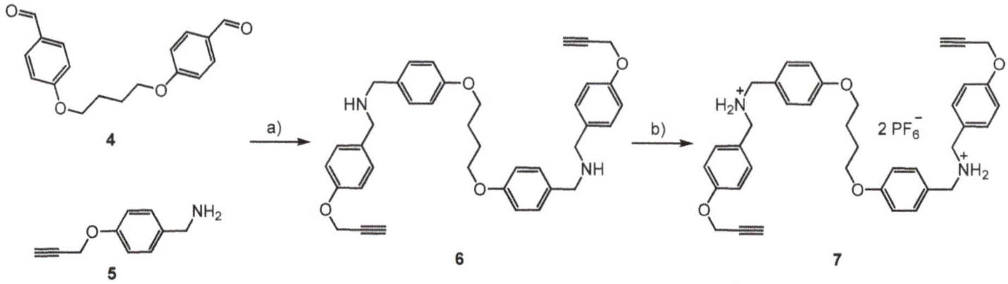

Figure 4. Synthesis of axle **7**. (a) MeOH, 3 h, reflux; 2. NaBH$_4$, 16 h, 89%. (b) 1. HCl, EtOH, 2 h, r.t.; 2. NH$_4$PF$_6$, acetone/water, 2 h, r.t., 99%.

3.2. Syntheses of the [3]rotaxanes

In order to synthesize a rotaxane by the capping method, an axle, a macrocycle and stoppers are needed. In this work, the capping was performed by a "click" reaction between the dialkyne terminated central axles and stoppers carrying an azide function. Azide **11**, whose synthesis is literature-known [16,17], was chosen as the stopper (Figure 5). It was synthesized from the respective trityl alcohol **8**, first connecting it with phenol, and then forming the aryl alkyl ether **10**. The synthesis of the phenol is possible following two different routes: direct reaction of the trityl alcohol **8** with phenol under acidic conditions or generation of the trityl chloride **12** first. It turned out that the route via the chloro intermediate **12** gave better yields [18]. Finally, bromide **10** was converted to azide **11** in 90% yield by reaction with sodium azide.

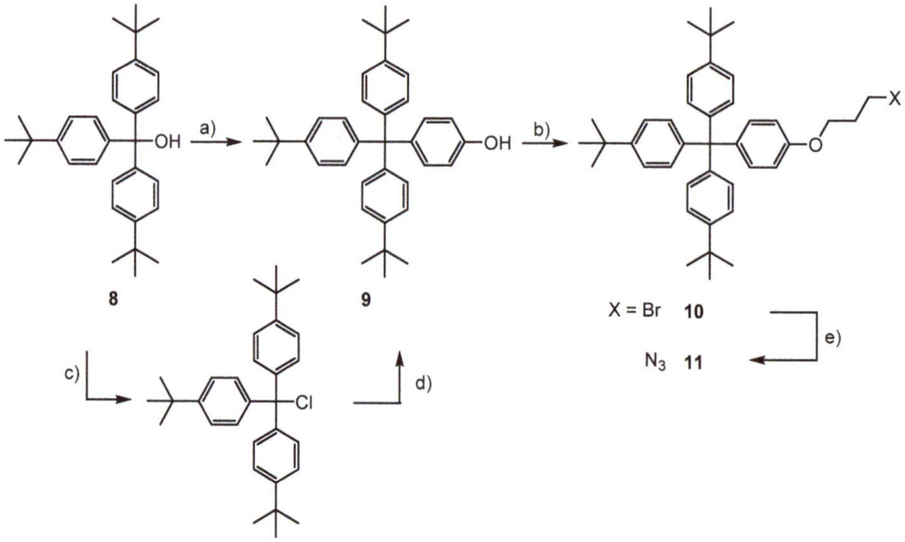

Figure 5. Synthesis of stopper **11**. (a) Phenol, HCl, 24 h, 160 °C, 63%; (b) 3-bromopropanol, PPh$_3$, DIAD, THF, 17 h, r.t., 73%; (c) acetyl chloride, toluene, 1 h, reflux, 92%; (d) phenol, HCl, 4 d, 120 °C, 98%; (e) NaN$_3$, DMF, 20 h, 80 °C, 90%.

For a successful synthesis of [3]rotaxanes, it is beneficial to generate a pseudorotaxane first before stoppering takes place. For this purpose, the axles **3a/b** or **7** and two equivalents of macrocycle **13** were mixed in dichloromethane. Due to the insolubility of the diammonium ions in non-polar solvents, first, a suspension was formed. Upon stirring, the mixture turned into a clear solution, indicating that the pseudorotaxanes formed. Then, two equivalents of stopper **11**, a copper(I) salt as a catalyst for the CuAAC "click" reaction and a catalytic amount of 2,6-dimethylpyridine were added. All reactions were carried out in the non-polar solvent dichloromethane to allow strong ion-dipole interactions and hydrogen bonds between the ammonium ions of the axles and the macrocycles. After hydrolytic work-up and extraction, the copper salt was removed by filtration and the product was purified by chromatography on silica gel. Remaining stopper **11** could be removed, indicating that the conversion had not been complete. By addition of ethyl acetate, excess crown ether precipitated. This process was repeated several times until no more solid precipitated. Finally, the products were purified again by chromatography on silica gel.

[3]Rotaxane **14a** was isolated as a pale yellow solid in a yield of 34% (Figure 6). [3]Rotaxane **14b** was synthesized analogously. NMR analysis of **14b** showed more than two crown ethers per rotaxane. Two sets of NMR signals for the macrocycle were found in an approximate ratio of 2:1, both sets of signals differing from those of the free macrocycle DB24C8 (**13**). Due to overlapping signals, the amount of the additional macrocycle cannot be determined precisely. As discussed below, additional macrocycle was also found in the next product, the dimethylated [3]rotaxane **15b**. After deprotonation, the amount of excess **13** was determined to be one equivalent per rotaxane. A possible explanation for additional macrocycle **13** in the rotaxanes **14b** and **15b** could be that this macrocycle adheres to the outside of the [3]rotaxanes **14b** and **15b**. The interaction to rotaxane **14b** must have been strong enough to survive the chromatography. [3]Rotaxane **14b** was obtained in a yield of 39%.

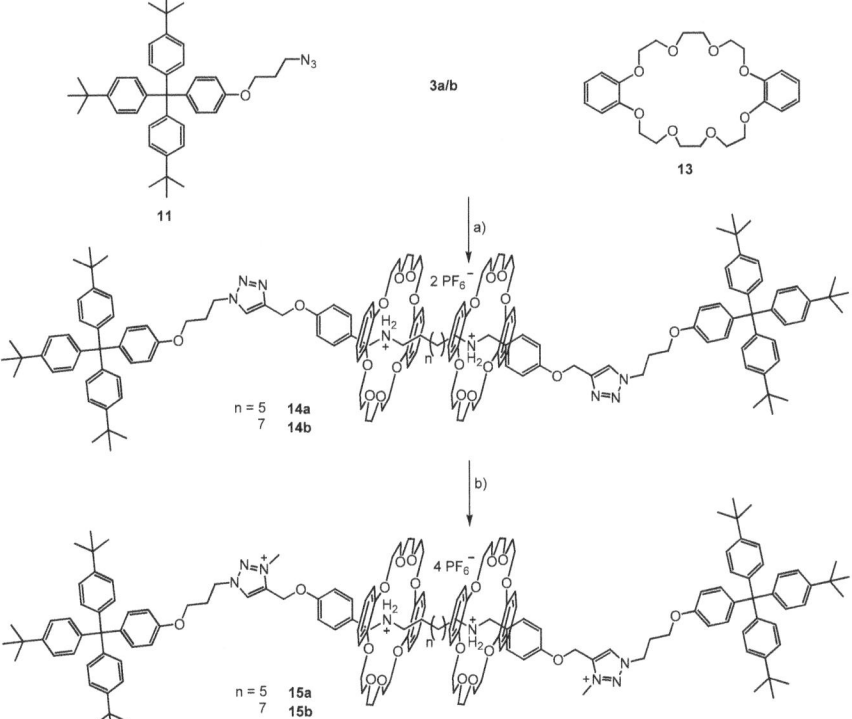

Figure 6. Syntheses of the rotaxanes **15a/b**. (a) CH_2Cl_2, $Cu(MeCN)_4PF_6$, 2,6-dimethylpyridine, 1 d, r.t., 34% (**14a**), 39% (**14b**); (b) MeI, 4 d, r.t,. quantitative (**15a**), 98% (**15b**).

Due to the two positive charges of the two ammonium functions, the [3]rotaxanes **14a/b** possess one binding site for each of the two macrocycles. Since, as discussed in the introduction, the rotaxanes should be switched by addition of a base or an acid, additional binding sites had to be introduced; i.e., by methylation of the triazole rings. The resulting triazolium ions are permanently positively charged and will serve as secondary binding sites, and the only ones after deprotonation. The bis(triazolium ion) **15a** was formed by reacting [3]rotaxane **14a** with excess methyl iodide. After aqueous work-up, methylated [3]rotaxane **15a** was obtained in quantitative yield. The same conditions were used for the analogous [3]rotaxane **14b**. The methylated [3]rotaxane **15b** was obtained in a yield of 98%. Unfortunately, the excess crown ether in the **14b** sample could also not be separated from the dimethylated product **15b**.

Axle **7** differs from the axles **3a/b** in the substitution pattern of the ammonium ions: dibenzyl versus benzyl-alkyl. Using the same synthetic procedure as for the rotaxanes **14a/b**, [3]rotaxane **16** with two benzyl units in the proximity of each of the ammonium ions was obtained in a yield of 6% after purification (Figure 7). NMR spectroscopy showed again that the ring starting material, crown ether **13**, also could not be completely separated from **16**. The subsequent methylation of **16** was carried out identically to the methylation of **14a/b**. After aqueous work-up, the methylated [3]rotaxane **17** was obtained with a quantitative yield. Again, the excess crown ether was detected and could not be removed.

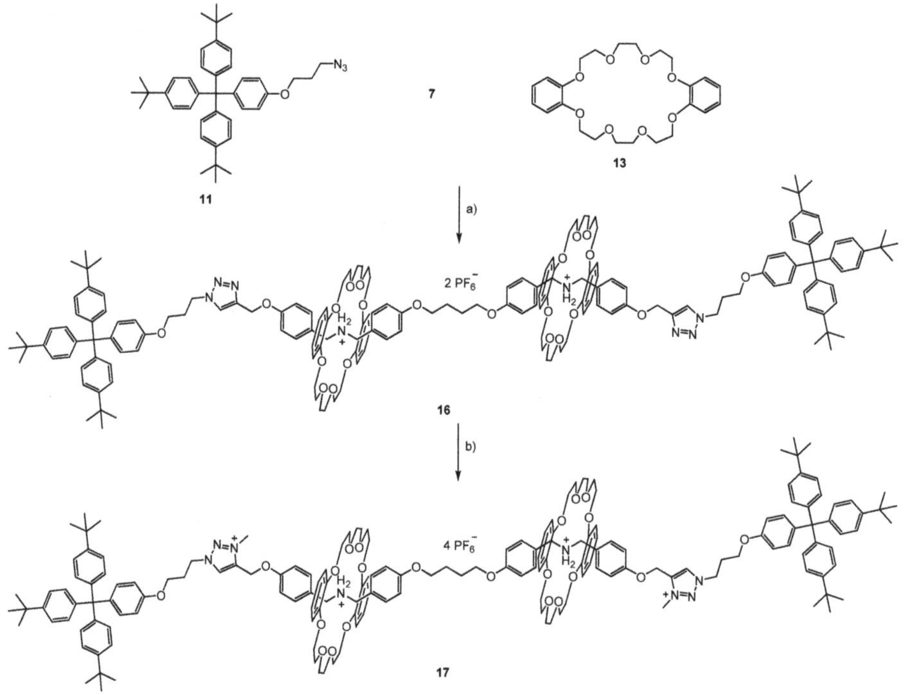

Figure 7. Synthesis of rotaxane **17**. (a) CH_2Cl_2, $Cu(MeCN)_4PF_6$, 2,6-dimethylpyridine, 1 d, r.t., 6%; (b) MeI, 4 d, r.t. quantitative.

3.3. Switching of the [3]rotaxanes

Having introduced the additional binding sites by methylation of the triazoles, the base/acid switching of the methylated [3]rotaxanes **15a/b** and **17** was investigated. By addition of sodium hydroxide, the ammonium ions were deprotonated, leaving only the triazolium ions as positively charged binding sites. Protonation by trifluoroacetic acid regenerated the ammonium functions. The switching process was followed by NMR spectroscopy.

3.3.1. Switching of [3]rotaxane 15a

In the aromatic region of the ^{1}H NMR spectrum of rotaxane **15a**, deprotonation caused a low field shift of almost 0.6 ppm for triazolium proton at 8.51 ppm (Figure 8, green). Binding of the crown ether to the triazolium ion leads to a deshielding. All other aromatic signals show much smaller shift changes, but no signal has an identical chemical shift as in the previous spectrum (Figure 8, blue). The multiplet at 6.8 ppm has split to become three individual signals. In the aliphatic range, more distinct differences can be observed. The two methylene groups (5.22 ppm and 4.76 ppm) near the triazolium unit exchange their positions. Additionally, the methylene groups at 3.0 ppm and 2.5 ppm switch positions. In this region, most signals experience a high field shift after deprotonation, except for the signal at 3.4 ppm, which has low field shifted. In the deprotonated [3]rotaxane (Figure 8, green), the signals for the central methylene groups appear between 1.2 and 1.5 ppm, as expected for oligomethylene H atoms. In the protonated cases (Figure 8, blue and red), however, these signals are high field shifted (1.3, 0.8 and 0.75 ppm), indicating the vicinity of the aromatic benzene rings of the macrocycles to these methylene groups.

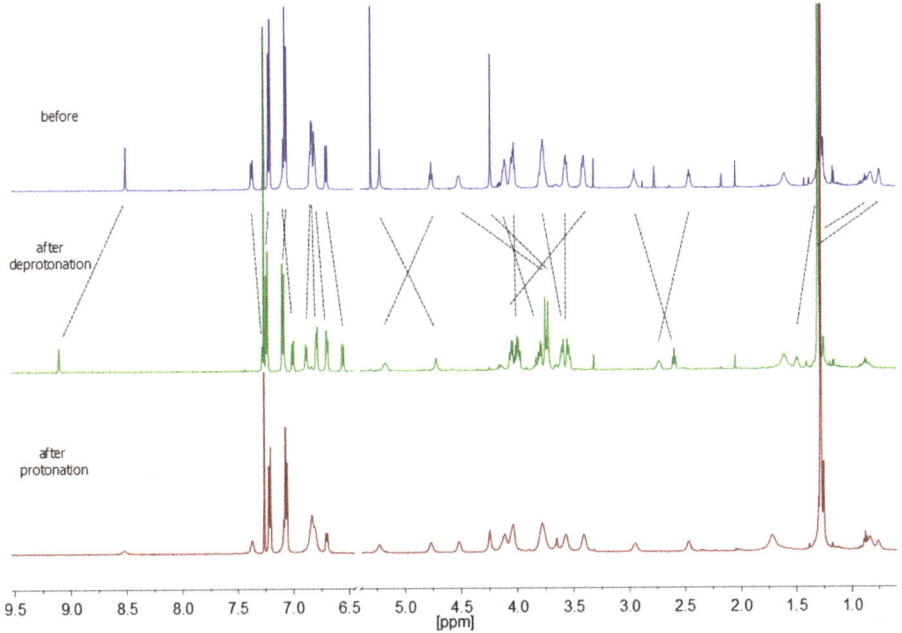

Figure 8. ^{1}H NMR spectra (500 MHz, CDCl$_3$) of the base/acid switching of rotaxane **15a**. Blue: before; green: after deprotonation; red: after protonation.

3.3.2. Switching of [3]rotaxane 15b

Analogous to the previous experiment, a pH-dependent NMR investigation was also carried out with rotaxane **15b** (Figure 9). In the aromatic range, the triazolium proton (8.51 ppm) shows a low field shift of almost 0.4 ppm after deprotonation. Compared to the switching of the analogous [3]rotaxane **15a**, this shift is smaller by 0.2 ppm. Reasons for this difference are unclear. But overall, the same trend can be observed in the aromatic sector as for [3]rotaxane **15a**. In the aliphatic sector, changes similar to those observed with rotaxane **15a** can be observed as well. Additionally, the two signals of the methylene groups near the triazolium group exchange positions. It is noticeable, however, that signals which can be assigned to the free macrocycle **13** are observed after deprotonation (two centered multiplets at 4.14 and 3.92, and a singlet at 3.84 ppm). This supports the postulated formation of a complex that was formed during the synthesis of rotaxane **14b**. Deprotonation released the ring from the complex, resulting in the observation of signals for the free ring **13**. After reprotonation, the signals of the free macrocycle **9** cannot be detected anymore. This supports the assumption that the free macrocycle **13** is attached to [3]rotaxane **15b** in the protonated form. After protonation, all other signals are also found at their original positions. Repeated deprotonation/protonation cycles resulted in the identical changes of the chemical shifts.

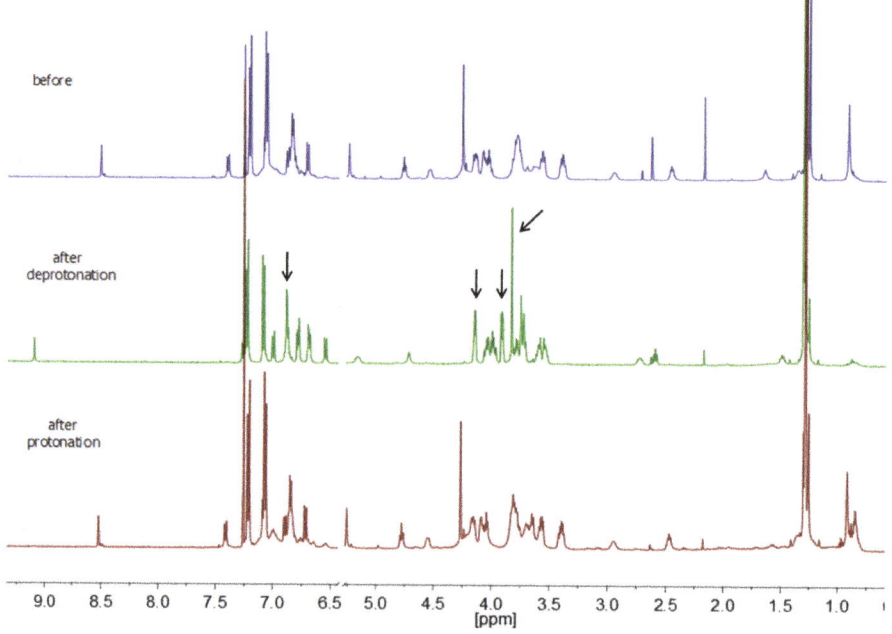

Figure 9. ^{1}H NMR spectra (500 MHz, CDCl$_3$) of the base/acid switching of rotaxane **15b**. Blue: before; green: after deprotonation (the arrows highlight the signals of additional crown ether **13**); red: after protonation.

3.3.3. Switching of [3]rotaxane 17

Additionally, dimethylated [3]rotaxane **17** was deprotonated and reprotonated, and the switching was investigated by ^{1}H NMR spectroscopy (Figure 10). Additionally, in this case, the spectra are in accordance with a base/acid shuttling of the rings from the ammonium ions to the triazolium ions and back. For the proton of the triazolium ion, a low field shift of the signal from 8.49 to 9.10 ppm was detected. The changes in the aromatic and the aliphatic regions of the spectra are comparable to what was observed when rotaxanes **15a** and **15b** were switched. For example, again, the methylene groups near the triazolium ions reverse positions, from 5.21 or 4.75 ppm to 4.73 or 5.16 ppm, respectively. The signals of the methylene groups near the ammonium ions exhibit a high field shift from 4.48 to 3.72 ppm. Additionally, in this case, signals for free macrocycle (**13**) can be observed after deprotonation (4.16, 3.90 and 3.83 ppm).

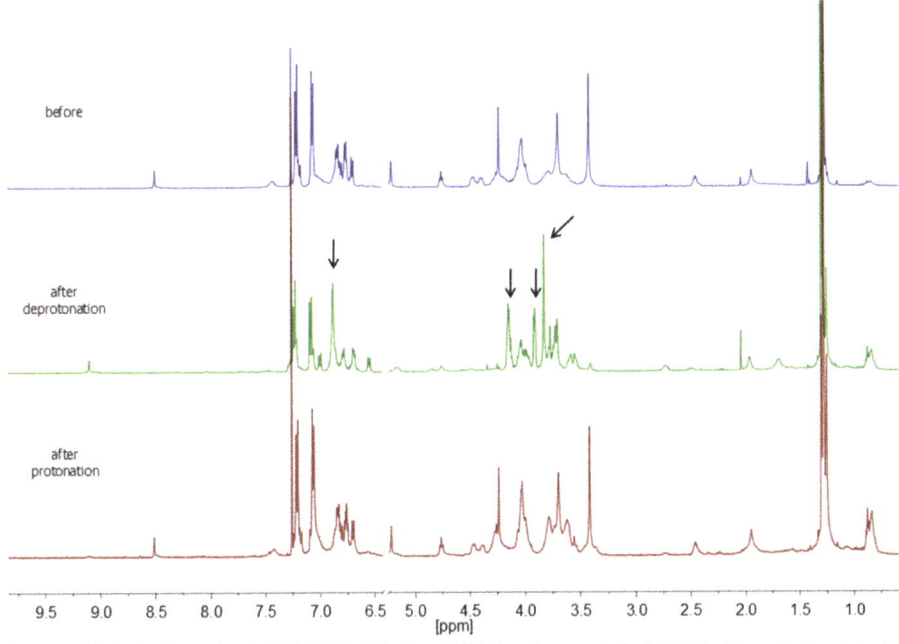

Figure 10. ^{1}H NMR spectra (500 MHz, CDCl$_3$) of the base/acid switching of rotaxane **17**. Blue: before; green: after deprotonation (the arrows highlight the signals of additional crown ether **13**); red: after protonation.

After reprotonation and counter ion exchange, all signals are back at their original positions, and no more free macrocycle **13** can be detected, arguing for a strong binding of an additional macrocycle to the "outside" of the [3]rotaxane **17**, its release in basic media and re-binding in acidic media.

3.4. Synthesis of Crown Ether DB24C8-CH$_2$Br (**24**)

In the envisaged knotted [1]rotaxane, no separate stoppers are needed, as the macrocycles themselves prevent the slipping-off. For this purpose, the macrocyclic rings must be connected to azide functions which are then reacted with the alkynes by the CuAAC "click" reaction. An obvious position to connect an azide tether to the macrocycle is position 4 of one of the benzene rings of DB24C8 (**13**). The synthesis of a respective hydroxymethylated DB24C8 **23** is described in the literature [19–21]. Tosylation of triethylene glycol (**18**), connection with catechol, tosylation of the remaining hydroxyl groups of **20**, macrocyclization with methyl 3,4-dihydroxybenzoate and reduction of the ester group of **22** were reproduced in good yields (total yield: 68% over five steps) (Figure 11). Finally, the hydroxysubtituted macrocycle **23** was converted to bromide **24** in quantitative yield using phosphorus tribromide.

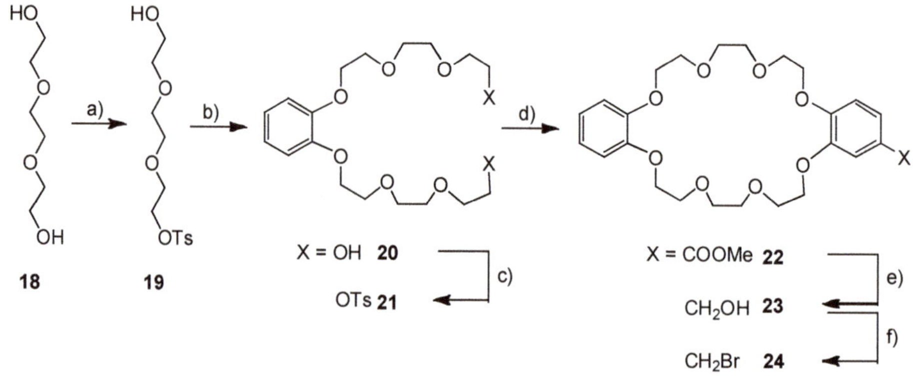

Figure 11. Synthesis of the bromomethyl macrocycle **24**. (a) pTsCl, NaOH, THF/H$_2$O, 18 h, r.t., 91%; (b) 1,2-dihydroxybenzene, K$_2$CO$_3$, MeCN, 60 h, reflux, 98%; (c) pTsCl, NaOH, THF/H$_2$O, 18 h, r.t., 90%; (d) methyl 3,4-dihydroxybenzoate, K$_2$CO$_3$, MeCN, 60 h, reflux, 86%; (e) LiAlH$_4$, THF, 2 h, reflux, 99%; (f) PBr$_3$, CH$_2$Cl$_2$, 2 h, r.t., quant.

3.5. Syntheses of Azide Substituted Phenols 28a–d

Next, the azide tethers had to be synthesized. The three step syntheses of azides **28a–d** begin with an ether formation to give mono-protected hydroquinone **25** (Figure 12). For this purpose, conditions from the literature were chosen [22]. The resulting 4-(benzyloxy)phenol (**25**) was dissolved in acetonitrile and was reacted under alkaline conditions with an excess of dibromoalkanes. After alkaline work-up, bromoalkyl ethers **26a–d** were obtained with yields of 14% to 75%. Deprotection of the benzyl ethers **26a–d** adapting a literature protocol [23] yielded the phenols **27a–d** in yields of 65% to quant. The work-up was easy. Filtration of the suspensions through Celite and evaporation of the solvent yielded pure products **27a–d**; no further purification was needed. In the last step, the phenols **27a–d** were dissolved in dimethyl sulfoxide and sodium azide was added. After stirring at room temperature for 18 h and chromatographic purification, azides **28a–d** were obtained in yields of 65%–93% which is comparable to the literature yield of the already known azide **28a** (67%) [24].

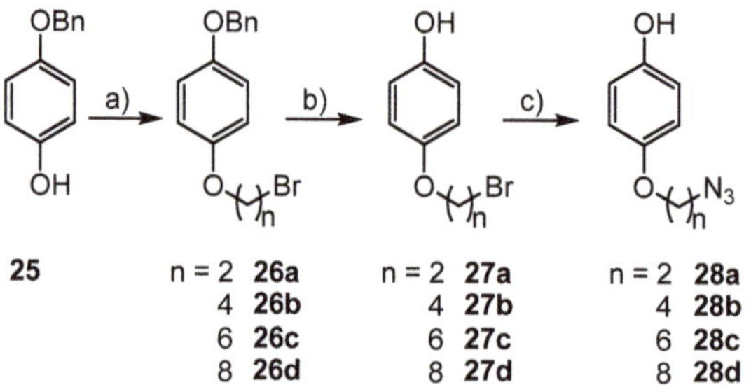

Figure 12. Syntheses of the azide compounds **28a–d**. (a) Dibromoalkane, MeCN, K$_2$CO$_3$, 20 h, reflux, 14% (**26a**), 46% (**26b**), 36% (**26c**), 75% (**26d**); (b) Pd/C, H$_2$, CHCl$_3$, 18 h, r.t, 94% (**27a**), 98% (**27b**), 65% (**27c**), quant. (**27d**); (c) NaN$_3$, DMSO, 18 h, r.t., 93% (**28a**), 69% (**28b**), 65% (**28c**), 90% (**28d**).

All azides **28a–d** possess oligomethylene chains. To allow a different rotational flexibility in the chain, an additional azide **32** with a triethylene glycol chain could be made. The length of the tether in **32** is identical to that of the octamethylene derivative **28d**, but two methylene groups are replaced by oxygen atoms. The first three reactions were described in the literature [25–27]. In the last step, a substitution reaction with sodium azide gave azide **32** in a yield of 84% (Figure 13).

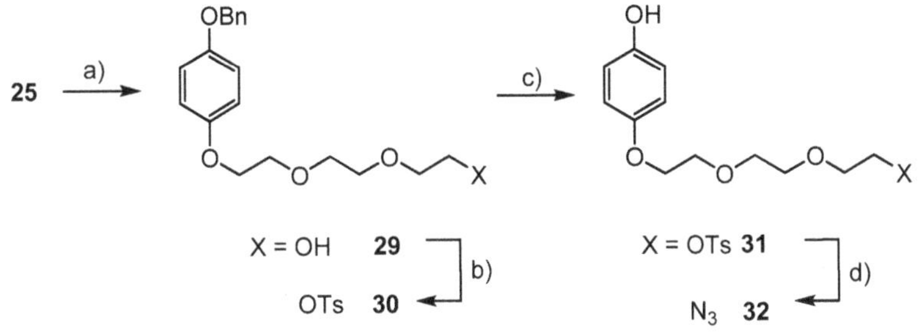

Figure 13. Synthesis of the azide **32**. (a) **19**, DMF, 90 °C, 24 h, 61%; (b) pTsCl, NEt₃, DMAP, CH₂Cl₂, 24 h, r.t., 86%; (c) Pd/C, H₂, CHCl₃, 16 h, r.t., 97%; (d) NaN₃, DMF, 2 d, 70 °C, 84%.

3.6. Syntheses of the Azide Containing Crown Ethers

The final reaction to give the azide substituted macrocycles **33a–e** was the connection of the macrocyclic bromide precursor **24** with the azide-terminated phenols **28a–d** and **32**. Cesium carbonate in acetone was used as deprotonating reagent. The yields varied strongly. For the aliphatic azides **33a–d**, the yields were between 26% and 75% (Figure 14). It shall be noted that in all four reactions a double purification had to be performed to obtain the crown ethers in crystalline form. Chromatography was performed first followed by crystallization. Macrocycle **34** containing a triethyleneglycol chain was obtained after recrystallization in a yield of 37% (Figure 15).

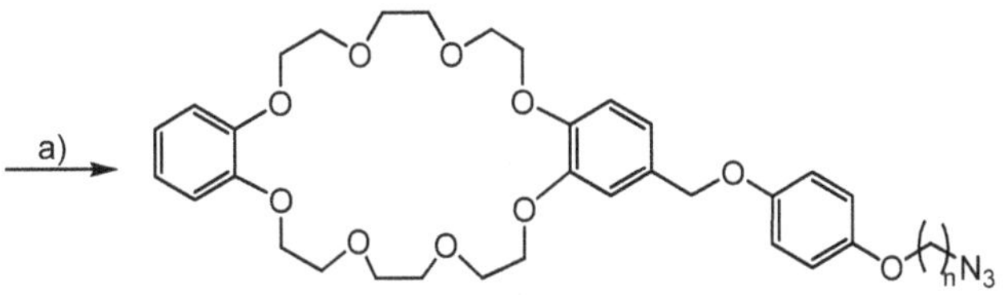

n = 2 **33a**
4 **33b**
6 **33c**
8 **33d**

Figure 14. Synthesis of azide-terminated macrocycles **33a–d**. (a) DB24C8-CH₂Br (**24**), Cs₂CO₃, KI, acetone, 16 h, reflux, 57% (**33a**), 34% (**33b**), 75% (**33c**), 26% (**33d**).

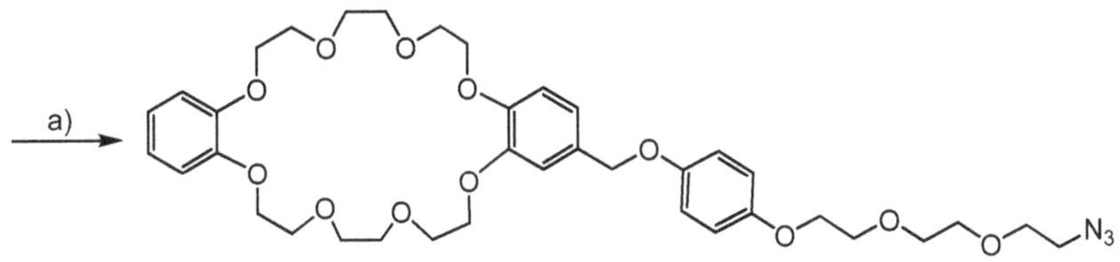

Figure 15. Synthesis of the azide-terminated macrocycle **34**. (a) DB24C8-CH₂Br (**24**), Cs₂CO₃, KI, acetone, 16 h, reflux, 37%.

3.7. Knotting Attempts

For first orientational experiments, two of the azides (**33c/d**) were selected, primarily due to their better availabilities. First, an axle and the macrocycles were dissolved in dichloromethane and pseudorotaxane formation was allowed. Then, the resulting solution was added very slowly to a highly diluted solution of copper(I) hexafluorophosphate and 2,6-dimethylpyridine. The resulting suspension was filtered through silica gel. After evaporation of the solvent, the residue was analyzed by MALDI mass spectrometry. The results for azide **33c** are shown in Table 1. No mass signal matching a double click reaction could be found when axles **3a** or **3b** were used. But with axle **7**, signals in the mass spectrum could be observed which were in accordance to a successful double click reaction. But please note that mass spectra cannot differentiate between the desired knotted [1]rotaxane and the alternative product in which the azides and alkynes have reacted "wrongly" with one another (handcuff, see Figure 2).

Table 1. MALDI mass spectra signals for the "click" reactions with azide-terminated macrocycle **33c**.

Axle	m/z	Product
3a	-	-
3b	-	-
7	2127.2 and 1981.2	7•2 33c − PF_6^- 7•2 33c − 2 PF_6^- − H^+

Azide-terminated macrocycle **33d** was also investigated. The knotting attempts were performed using the optimized conditions of the previous experiments. After purification on silica gel, mass spectra were also recorded (Table 2). In all experiments, signals for doubly charged ions without counter ions for 2:1 adducts could be observed. This indicates a successful reaction, and thus the formation of a [1]rotaxane. Again though, the question of whether the [1]rotaxane is knotted or not cannot be answered from these data.

Table 2. ESI mass spectral signals for the experiments with the macrocycle **33d**.

Axle	m/z	Molecule
3a	940.5175	3a•2 33d − 2 PF_6^-
3b	955.0352	3b•2 33d − 2 PF_6^-
7	1019.0300	7•2 33d − 2 PF_6^-

Next, the triethyleneglycol analog of **33d**, **34**, was tested by applying the same reaction conditions. Additionally, with **34**, signals matching a [1]rotaxane were found in the MS (3a•2 34 − 2 PF_6^- − H^+, m/z = 1889.2).

In summary, with azide-terminated macrocycle **33c**, a mass signal corresponding to a [1]rotaxane was only found in one case, but the longer tethers (**33d** and **34**) showed respective signals in all tested combinations.

The reaction mixtures were then investigated by reverse-phase HPLC coupled with ESI-MS. The HPLC runs revealed numerous reaction products probably also arising from intermolecular "click" reactions. The fact that no mass spectral signals were found for larger parts of the HPLC traces argues for oligomers. However, also in the HPLC runs, mass spectral signals corresponding to the products of a double "click" reaction could be found.

4. Conclusions

Dialkyne-substituted axles which contain two ammonium ions as primary binding sites for macrocyclic crown ethers can be used to form [3]pseudorotaxanes with these rings followed by locking of the rotaxane structure by two-fold copper-catalyzed "click" reaction between the alkynes and azides. In the [3]rotaxanes, the resulting triazoles can be methylated, which introduces a second type of binding site. Upon deprotonation and reprotonation, the macrocycles can be shuttled between the two binding sites, the ammonium ions and the triazolium ions. The click reaction is also possible when the azide function is connected to the DB24C8 macrocycle. But the isolation of the coupling products from by-products is still a challenge. Preparative HPLC might be the solution. Then, whether the applied dimensions of the tethers and the axles were allowing the formation of the desired knotaxane or whether axles and tethers have to be modified (other lengths and rigidity) can be studied.

Supplementary Materials: Supplementary material with synthetic procedures and analyses is available online at http://www.mdpi.com/2624-8549/2/2/305\T1\textendash321/s1.

Author Contributions: Conceptualization, U.L. and T.D.; methodology, U.L. and T.D.; validation, U.L. and T.D.; formal analysis, T.D.; investigation, T.D.; resources, U.L.; writing—original draft preparation, U.L. and T.D.; writing—review and editing, U.L. and T.D.; visualization, T.D.; supervision, U.L.; project administration, U.L.; funding acquisition, U.L. All authors have read and agreed to the published version of the manuscript.

Funding: This research received no external funding.

Acknowledgments: We would like to thank Tobias Paschelke, and Sven Schultzke, for their synthetic support in the production of the azide containing phenols. Furthermore, we thank Dennis Stöter, for the support in the syntheses of the axles. We would also like to thank Vanessa Nowatschin, for her support in the optimization of the reproduction of the stopper.

Conflicts of Interest: The authors declare no conflict of interest.

References and Note

1. Bruns, C.J.; Stoddart, J.F. *The Nature of the Mechanical Bond: From Molecules to Machines*; John Wiley & Sons: Hoboken, NJ, USA, 2017.
2. Schill, G.; Zollenkopf, H. Rotaxan-Verbindungen. *Liebigs Ann. Chem.* **1969**, *721*, 53–74. [CrossRef]
3. Wasserman, E. The preparation of interlocking rings: A catenane. *J. Am. Chem. Soc.* **1960**, *82*, 4433–4434. [CrossRef]
4. Schill, G.; Lüttringhaus, A. The preparation of catena compounds by directed synthesis. *Angew. Chem.* **1964**, *76*, 567–568; Internation version in *Angew. Chem. Int. Ed. Engl.* **1964**, *3*, 546–547. [CrossRef]
5. Dietrich-Buchecker, C.O.; Sauvage, J.-P. A synthetic molecular trefoil knot. *Angew. Chem.* **1989**, *101*, 192–194; Internation version in *Angew. Chem. Int. Ed. Engl.* **1989**, *28*, 189–192. [CrossRef]
6. Lukin, O.; Kubota, T.; Okamoto, Y.; Schelhase, F.; Yoneva, A.; Walter, M.M.; Müller, U.; Vögtle, F. Knotaxanes—Rotaxanes with knots as stoppers. *Angew. Chem.* **2003**, *115*, 4681–4684; Internation version in *Angew. Chem. Int. Ed.* **2003**, *42*, 4542–4545. [CrossRef]
7. Whether a rotaxane is stable or not depends on the relative sizes of ring and stopper but also on the temperature: [8,9] and reference cited.
8. Harrison, I.T. The effect of ring size on threading reactions of macrocycles. *J. Chem. Soc. Chem. Commun.* **1972**, 231–232. [CrossRef]
9. Saito, S.; Takahashi, E.; Wakatsuki, K.; Inoue, K.; Orikasa, T.; Sakai, K.; Yamasaki, R.; Mutoh, Y.; Kasama, T. Synthesis of large [2]rotaxanes. The relationship between the size of the blocking group and the stability of the rotaxane. *J. Org. Chem.* **2013**, *78*, 3553–3560. [CrossRef]
10. Bissell, R.A.; Cordova, E.; Kaifer, A.E.; Stoddart, J.F. A chemically and electrochemically switchable molecular shuttle. *Nature* **1994**, *369*, 131–137. [CrossRef]
11. Coutrot, F.; Busseron, E. A new glycorotaxane molecular machine based on an anilinium and a triazolium station. *Chem. Eur. J.* **2008**, *14*, 4784–4787. [CrossRef]
12. Chavez-Acevedo, L.; Miranda, L.D. Synthesis of novel tryptamine-based macrocycles using an Ugi 4-CR/microwave assisted click-cycloaddition reaction protocol. *Org. Biomol. Chem.* **2015**, *13*, 4408–4412. [CrossRef]
13. Talotta, C.; Gaeta, C.; Pierro, T.; Neri, P. Sequence stereoisomerism in calixarene-based pseudo[3]rotaxanes. *Org. Lett.* **2011**, *13*, 2098–2101. [CrossRef] [PubMed]
14. Zhang, Y.; Lai, L.; Cai, P.; Cheng, G.-Z.; Xuc, X.-M.; Liua, Y. Synthesis, characterization and anticancer activity of dinuclear ruthenium(ii) complexes linked by an alkyl chain. *New J. Chem.* **2015**, *39*, 5805–5812. [CrossRef]
15. Zhang, Z.-J.; Zhang, H.-Y.; Wang, H.; Liu, Y. A twin-axial hetero[7]rotaxane. *Angew. Chem.* **2011**, *123*, 11026–11030; Internation version in *Angew. Chem. Int. Ed.* **2011**, *50*, 10834–10838. [CrossRef]
16. Aucagne, V.; Hänni, K.D.; Leigh, D.A.; Lusby, P.J.; Walke, D.B. Catalytic click rotaxanes: A substoichiometric metal-template pathway to mechanically interlocked architectures. *J. Am. Chem. Soc.* **2006**, *128*, 2186–2187. [CrossRef]

17. Zheng, H.; Zhou, W.; Lv, J.; Yin, X.; Li, Y.; Liu, H.; Li, Y. A dual-response [2]rotaxane based on a 1,2,3-triazole ring as a novel recognition station. *Chem. Eur. J.* **2009**, *15*, 13253–13262. [CrossRef]
18. Ashton, P.R.; Ballardini, R.; Balzani, V.; Bělohradský, M.; Gandolfi, M.T.; Philp, D.; Prodi, L.; Raymo, F.M.; Reddington, M.V.; Spencer, N.; et al. Self-assembly, spectroscopic, and electrochemical properties of [n]rotaxanes. *J. Am. Chem. Soc.* **1996**, *118*, 4931–4951. [CrossRef]
19. Wang, X.; Ervithayasuporn, V.; Zhang, Y.; Kawakami, Y. Reversible self-assembly of dendrimer based on polyhedral oligomeric silsesquioxanes (POSS). *Chem. Commun.* **2011**, 1282–1284. [CrossRef]
20. Wu, L.; He, Y.-M.; Fan, Q.-H. Controlled reversible anchoring of η6-arene/TsDPEN- ruthenium(II) complex onto magnetic nanoparticles: A new strategy for catalyst separation and recycling. *Adv. Synth. Catal.* **2011**, *353*, 2915–2919. [CrossRef]
21. Yamaguchi, N.; Gibson, H.W. Formation of supramolecular polymers from homoditopic molecules containing secondary ammonium ions and crown ether moieties. *Angew. Chem.* **1999**, *111*, 195–199; Internation version in *Angew. Chem. Int. Ed.* **1999**, *38*, 143–147. [CrossRef]
22. Davidson, A.B.; Guthrie, R.W.; Kierstead, R.W.; Ziering, A. Hoffmann-La Roche Inc. U.S. Patent 4471116, 11 September 1984.
23. Grice, C.A.; Tays, K.L.; Savall, B.M.; Wei, J.; Butler, C.R.; Axe, F.U.; Bembenek, S.D.; Fouriem, A.M.; Dunford, P.J.; Lundeen, K.; et al. Identification of a potent, selective, and orally active leukotriene A4 hydrolase inhibitor with anti-inflammatory activity. *J. Med. Chem.* **2008**, *51*, 4150–4169. [CrossRef]
24. Lim, J.Y.C.; Marques, I.; Thompson, A.L.; Christensen, K.E.; Félix, V.; Beer, P.D. Chalcogen bonding macrocycles and [2]rotaxanes for anion recognition. *J. Am. Chem. Soc.* **2017**, *139*, 3122–3133. [CrossRef] [PubMed]
25. Hodyl, J.A.Z.; Lincoln, S.F.; Wainwright, K.P. Silica-attached molecular receptor complexes for benzoates and naphthoates. *J. Incl. Phenom. Macrocycl. Chem.* **2010**, *68*, 261–270. [CrossRef]
26. Amabilino, D.B.; Ashton, P.R.; Boyd, S.E.; Gomez-Lopez, M.; Hayes, W.; Stoddart, J.F. Translational isomerism in some two- and three-station [2]rotaxanes. *J. Org. Chem.* **1997**, *62*, 3062–3075. [CrossRef] [PubMed]
27. Yang, Y.; Fu, H.; Cui, M.; Peng, C.; Liang, Z.; Dai, J.; Zhang, Z.; Lin, C.; Liu, B. Preliminary evaluation of fluoro-pegylated benzyloxybenzenes for quantification of β-amyloid plaques by positron emission tomography. *Eur. J. Med. Chem.* **2015**, *104*, 86–96. [CrossRef] [PubMed]

© 2020 by the authors. Licensee MDPI, Basel, Switzerland. This article is an open access article distributed under the terms and conditions of the Creative Commons Attribution (CC BY) license (http://creativecommons.org/licenses/by/4.0/).

Review
Addition of Heteroatom Radicals to *endo*-Glycals †

Torsten Linker

Department of Chemistry, University of Potsdam, Karl-Liebknecht-Str. 24-25, 14476 Golm, Germany; linker@uni-potsdam.de; Tel.: +49 331 9775212

† Dedicated to Bernd Giese on the occasion of his 80th birthday and his pioneering work on radicals in carbohydrate chemistry.

Received: 6 February 2020; Accepted: 18 February 2020; Published: 20 February 2020

Abstract: Radical reactions have found many applications in carbohydrate chemistry, especially in the construction of carbon–carbon bonds. The formation of carbon–heteroatom bonds has been less intensively studied. This mini-review will summarize the efforts to add heteroatom radicals to unsaturated carbohydrates like *endo*-glycals. Starting from early examples, developed more than 50 years ago, the importance of such reactions for carbohydrate chemistry and recent applications will be discussed. After a short introduction, the mini-review is divided in sub-chapters according to the heteroatoms halogen, nitrogen, phosphorus, and sulfur. The mechanisms of radical generation by chemical or photochemical processes and the subsequent reactions of the radicals at the 1-position will be discussed. This mini-review cannot cover all aspects of heteroatom-centered radicals in carbohydrate chemistry, but should provide an overview of the various strategies and future perspectives.

Keywords: radicals; carbohydrates; heteroatoms; synthesis

1. Introduction

Radical reactions of carbohydrates are important for chemistry, biology, and medicine. For example, free radicals are involved in several biosynthetic pathways or are used for cancer-treatment [1–4]. On the other hand, radiation can cause DNA strand break by H atom abstraction and radical generation at the sugar backbone [5–7]. Bernd Giese's group proved that radical cations are involved in the mechanism [8] and investigated how the positive charge is transferred through the DNA [9]. For synthetic applications, Bernd Giese's group also made carbohydrate radicals available for the formation of carbon–carbon bonds under mild conditions [10]. Due to steric interactions, carbohydrates provide high stereoselectivities [11,12], and the importance of such reactions has been reviewed many times [13–18]. Furthermore, under appropriate conditions, 2-deoxy sugars can be synthesized from bromo sugars in only one step [19].

To develop efficient radical reactions, it is important to understand the reactivities of the corresponding radicals [12,20,21]. Thus, such reactive species can have a nucleophilic or an electrophilic character [22], which controls their addition to alkenes. Applied to the anomeric center of carbohydrates, the radicals **1** exhibit a nucleophilic character due to the adjacent oxygen atom, and add preferentially to electron poor double bonds with electron withdrawing (EWG) and nondonating (EDG) groups (Scheme 1a). If the carbohydrate is used as radical acceptor, unsaturated carbohydrates like *endo*-glycals **2** become attractive substrates, which can be easily synthesized on a large scale [23]. However, once the double bond becomes electron rich the reaction proceeds only with electrophilic radicals (Scheme 1b).

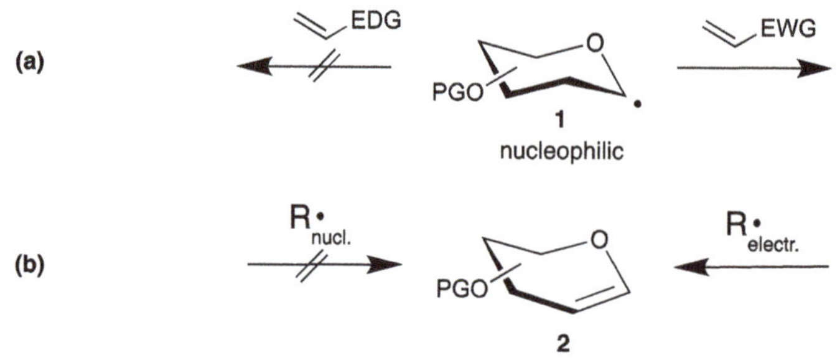

Scheme 1. Examples for radical reactions in carbohydrate chemistry: (a) starting from a carbohydrate radical **1** and (b) radical additions to glycals **2**.

During the last 25 years, we developed C-C bond formations by the addition of electrophilic radicals, mainly derived from malonates, to glycals **2** with various further synthetic transformations [24–27]. However, the addition of heteroatom radicals to glycals is very attractive as well, since such radicals exhibit the required electrophilic character [22,28]. On the other hand, heteroatom radicals are prone to undergo H atom abstraction, which is problematic in carbohydrate chemistry with various functional groups. Thus, although alkoxyl radicals [29] can be generated from carbohydrates and undergo fast fragmentations [30,31], of the way in which such radicals can be added to glycals is unknown. Since halogen atoms, nitrogen-, phosphorus-, and sulfur-centered radicals are less prone to undergo H atom abstraction, this mini-review will focus on the addition of such radicals to *endo*-glycals.

2. Addition of Halogen Atoms

The halogenation of *endo*-glycals **2** is one of the oldest transformations of such unsaturated carbohydrates, already described by Lemieux in 1965 [32]. Thus, tri-*O*-acetyl-D-glucal (**2a**) or the corresponding isomer tri-*O*-acetyl-D-galactal (**2b**) reacted with chlorine or bromine in high yields to the main products **3a** and **3b** (Scheme 2).

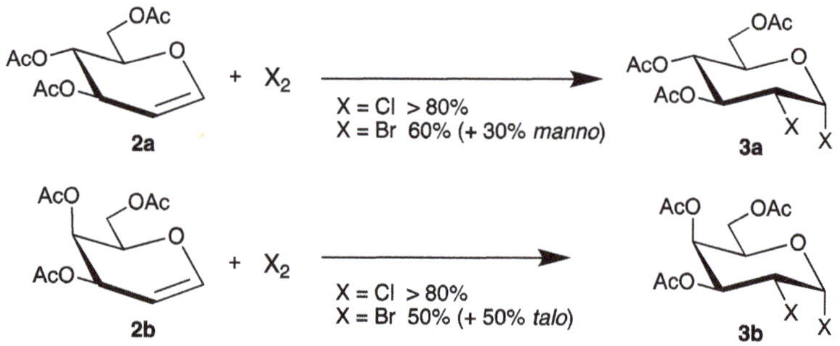

Scheme 2. Halogenation of glycals **2a** and **2b**.

Although the authors proposed an ionic pathway via halonium ions, the 1,2-*cis*-configurations might be explained by homolysis of the labile halogen bonds and addition of the resulting electrophilic radicals. To distinguish between a radical or an ionic pathway, halogen azides are attractive precursors because they easily undergo homolysis and are used in regio- and stereoselective syntheses [33]. Thus, reaction of tri-*O*-acetyl-D-glucal (**2a**) with chlorine azide afforded regioisomers **4** and **5**, depending on the reaction conditions (Scheme 3) [34].

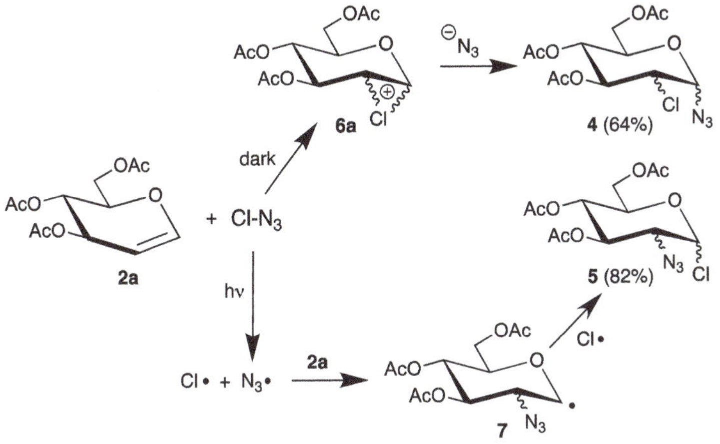

Scheme 3. Reaction of tri-O-acetyl-D-glucal (2a) with chlorine azide.

In the dark, 2-chloro-2-deoxy sugars **4** were isolated as main products, whereas irradiation gave the 2-azido-2-deoxy isomer **5**. Such different regioselectivities were explained by an ionic pathway via chloronium ions **6a** in the dark and a radical mechanism during irradiation via radical **7**. However, it was not possible to add the generated chlorine atom to the glucal, because the azide radical is more reactive (see Section 3).

More recently, Vankar developed a reagent system based on oxalyl chloride and silver nitrate to activate the carbon-chlorine bond [35]. An intermediate **8** was proposed, which cleaves into nitrate and carbon monoxide and transfers chlorine to the double bond of tri-O-acetyl-D-galactal (**2b**). The chloronium ion **6b** is subsequently trapped by the solvent acetonitrile/water to afford the 2-chloro-2-deoxy sugar **9** in high yield (Scheme 4).

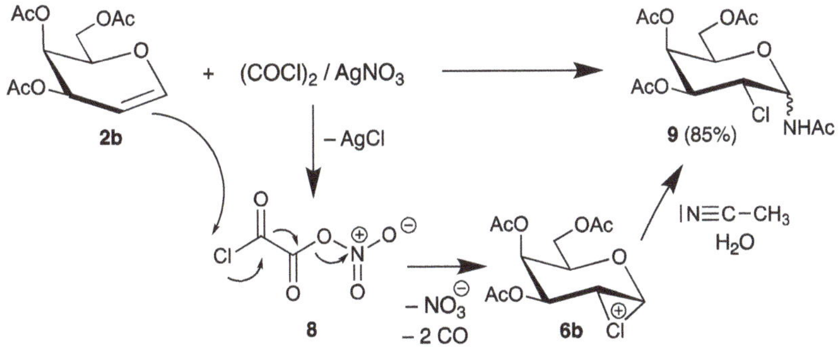

Scheme 4. Reaction of tri-O-acetyl-D-galactal (2b) with oxalyl choride/silver nitrate.

Compared to 2-chloro derivatives, the corresponding iodides are even more attractive because the carbon-iodine bond can be easily reduced to 2-deoxy sugars, important building blocks for carbohydrate chemistry. Thus, various strategies have been developed by oxidation of iodides by hypervalent iodine(III) [36,37] or sodium periodate [38] in the presence of *endo*-glycals **2**, affording 2-iodo-2-deoxy sugars **10** in very good yields (Scheme 5). The mechanism proceeds by oxidation of iodide to iodine in the first step, formation of an iodonium ion similar to intermediate **6a** (Scheme 3), and trapping of the 1-position with the carboxylate with high 1,2-*trans* selectivity. In summary, halogen atoms can be easily introduced at the 2-position of glycals. However, the reactions proceed mainly by ionic pathways, and irradiation of halogen azides results in the formation of C-N bonds in the 2-position.

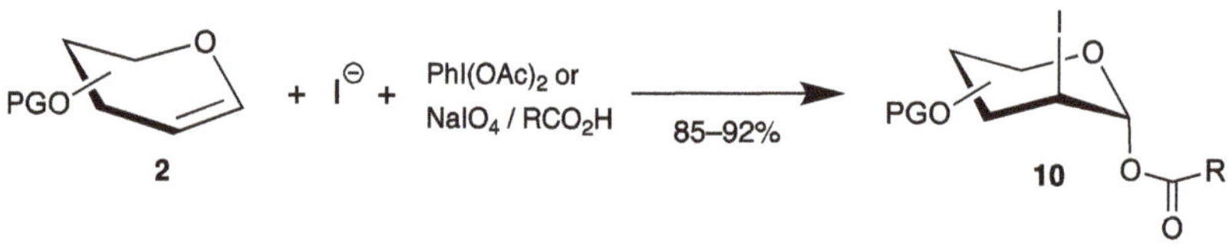

Scheme 5. Synthesis of 2-iodo-2-deoxy sugars **10** from *endo*-glycals **2**.

3. Addition of Nitrogen-Centered Radicals

In contrast to halogenations (chapter 2), the oxidative addition of azides to *endo*-glycals clearly proceeds by a radical pathway. The azidohalogenation was one example (Scheme 3); however, it is more attractive to generate the radicals by electron transfer. Cerium(IV) ammonium nitrate (CAN) is a very versatile single-electron oxidant, which can oxidize anions efficiently and has found many applications in organic synthesis [39]. We used this reagent for the generation of malonyl radicals and investigated the mechanism of their reactions with glycals **2** [25,26,40]. The pioneer Lemieux described the first application of azide oxidation in carbohydrate chemistry by addition of sodium azide to tri-*O*-acetyl-D-galactal (**2b**) in the presence of CAN (Scheme 6) [41].

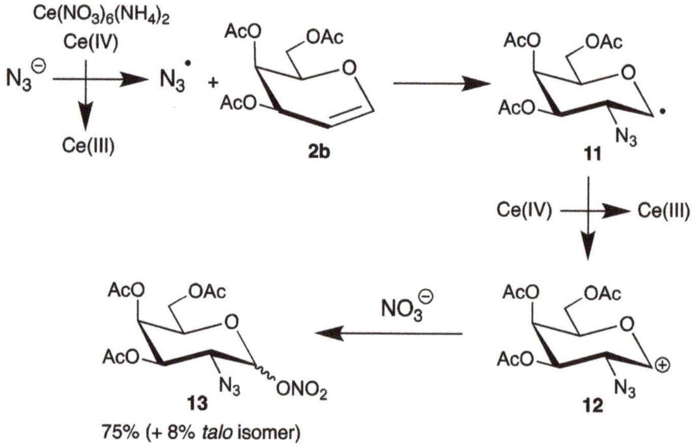

Scheme 6. Addition of sodium azide to tri-*O*-acetyl-D-galactal (**2b**) in the presence of cerium(IV) ammonium nitrate (CAN).

In the first step, CAN oxidizes the azide anion to the corresponding radical, which has electrophilic character and adds to the double bond exclusively at the 2-position. The preferential formation of the equatorial product (only 8% of the *talo* isomer is formed as well) can be explained by the steric demands of the substituents in the 3- and 4-position. The resulting C-1 radical **11** is further oxidized by CAN to the carbenium ion **12**, which is finally trapped by the nitrate ligand from both faces to afford the 2-azido-2-deoxy sugar **13** in 75% yield. Thus, this addition is not a typical radical chain-reaction [13] because more than two equivalents of CAN are required.

The azidonitration of glycals was later extended to tri-*O*-acetyl-D-glucal (**2a**), but with lower stereoselectivity because not all substituents shield the same face. Furthermore, Paulsen [42] and Schmidt [43] found with this glucal **2a** different selectivities depending on the reaction conditions and temperature. However, up to now the azidonitration of glycals has been the best method to synthesize 2-amino sugars by simple reduction of the azide group, and has found many applications in carbohydrate chemistry, like in a very recent synthesis of a bisphosphorylated trisaccharide [44].

However, a disadvantage of the azidonitration of glycals is the lability of the nitrate group at the anomeric center, which can be easily hydrolyzed. Although it is possible to use glycosyl nitrates directly for glycosidations [45], they usually have to be transformed into suitable glycosyl donors. To overcome this problem, an interesting azidophenylselenylation has been developed [46–48]. Now, sodium azide is oxidized by (diacetoxyiodo)benzene to the corresponding radical, which adds regioselectively to glycals like tri-O-acetyl-D-galactal (2b) (Scheme 7). In the presence of diphenyldiselenide, the C-1 radical is trapped to afford directly selenoglycoside 14 in high yield and steroeselectivity [47].

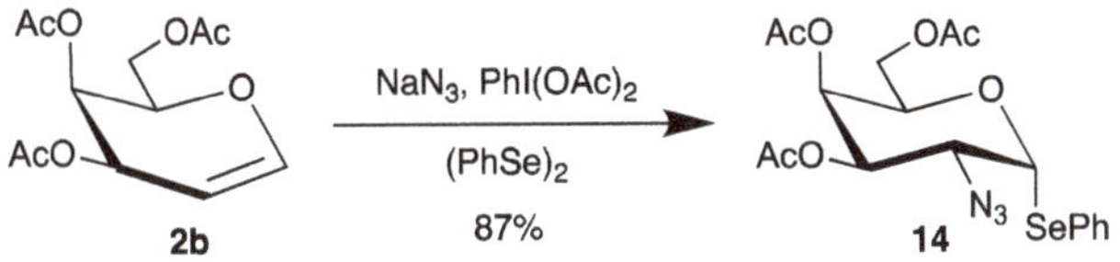

Scheme 7. Azidophenylselenylation of tri-O-acetyl-D-galactal (2b).

An interesting intramolecular version of a radical C-N bond formation was developed by Rojas (Scheme 8) [49] In the first step, azidoformate 2c reacts with FeCl2 under extrusion of nitrogen to intermediate 15, which can be discussed as a Fe-complexed nitrogen-centered radical. Addition to the double bond affords C-1 radical 16, which is trapped by chlorine to the labile complex 17; after work-up, tricycle 18 is formed in moderate yield.

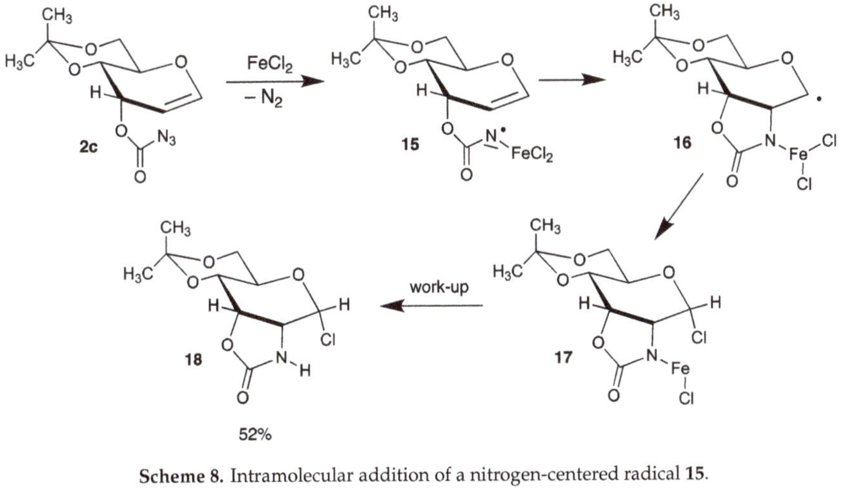

Scheme 8. Intramolecular addition of a nitrogen-centered radical 15.

A similar intermolecular addition of hydroxylamines as radical precursors to glycals was described recently as well [50] In summary, the formation of C-N bonds in the 2-position of carbohydrates can be easily accomplished by the addition of nitrogen-centered radicals to glycals. The best method is azidonitration in the presence of cerium(IV) ammonium nitrate, or azidophenylselenylation, which has found many applications in carbohydrate chemistry.

4. Addition of Phosphorus-Centered Radicals

The reaction of phosphorus-centered radicals is well-established and has many synthetic applications, summarized in several reviews [28,51,52]. Because phosphorus can exist in different oxidation states, it is possible to generate phosphinyl, phosphinoyl, or phosphonyl radicals.

Furthermore, the lability of the phosphorus-hydrogen bond allows for efficient chain-reactions with only catalytic amounts of radical initiator or under photochemical conditions. However, in contrast to nitrogen, only a few examples of the addition of phosphorus-centered radicals to glycals have been described in literature. Already in 1969, Inokawa demonstrated that diethyl thiophosphite reacts with unprotected glucal **2d** under UV irradiation with a high-pressure mercury lamp to the 2-deoxy-2-phosphorus analogue **21** in high yield and stereoselectivity (Scheme 9) [53].

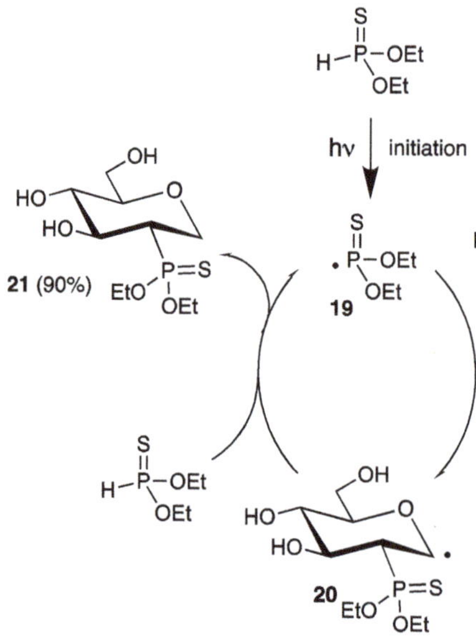

Scheme 9. Addition of diethyl thiophosphite to D-glucal (**2d**) under irradiation.

After the radical initiation step, the thiophosphonyl radical **19** adds regioselectively to the 2-position of the carbohydrate, due to its electrophilic character. The resulting C-1 radical **20** abstracts a hydrogen atom from diethyl thiophosphite, regenerating the phosphorus-centered radical **19**, closing the chain.

A very similar approach with protected tri-O-acetyl-D-glucal (**2a**) was published more recently (Scheme 10) [54]. This time, the radical chain was initiated by triethylborane/air, which generates ethyl radicals, and the additions of diethyl thiophosphite and diethyl phosphite were realized. However, the reactions afforded products **22a** and **22b** in somewhat lower yields compared to the photochemical process.

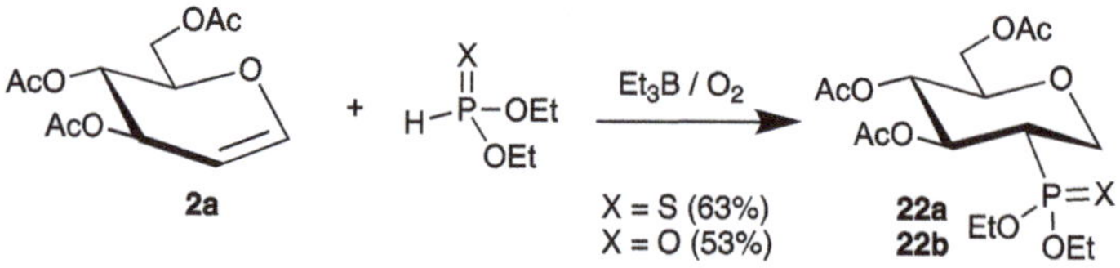

Scheme 10. Addition of diethyl phosphites to tri-O-acetyl-D-glucal (**2a**) initiated by BEt$_3$/air.

Recently, phosphinoyl radicals were added to tri-O-acetyl-D-glucal (**2a**) by a similar mechanism. The radicals were generated from diphenylphosphine oxide and manganese(II) acetate and air, affording the 2-deoxy-2-phosphorus analogue **23** in high yield and stereoselectivity (Scheme 11). The authors could extend this reaction to various other *endo*-glycals **2** as well [55].

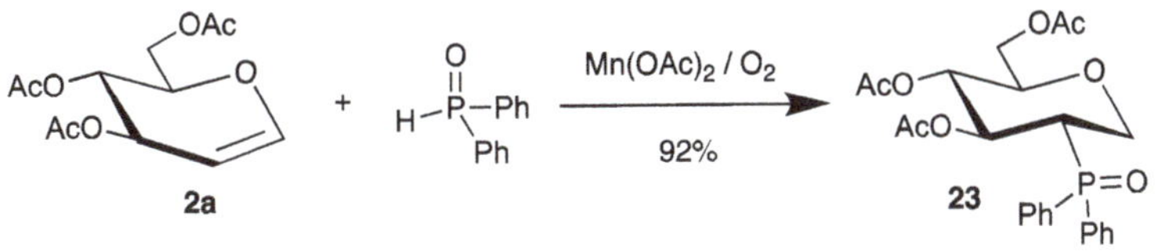

Scheme 11. Addition of diphenylphosphine oxide to tri-O-acetyl-D-glucal (**2a**) in the presence of manganese(II) acetate and air.

However, all methods have the disadvantage that the 1-position is reduced under the reaction conditions. Therefore, we investigated the addition of dimethyl phosphite to various benzyl-protected glycals **2e** in the presence of cerium(IV) ammonium nitrate (CAN) (Scheme 11) [56]. Now, the C-1 radical is further oxidized to a carbenium ion (see Scheme 6), which is trapped by the solvent methanol, generating the anomeric center of carbohydrates. The yields of the 2-deoxy-2-phosphorus analogues **24** are good, but stereoisomers had to be separated. Subsequent Horner–Emmons reaction with benzaldehyde afforded unsaturated carbohydrates **25** as E/Z isomers in only one step (Scheme 12) [56].

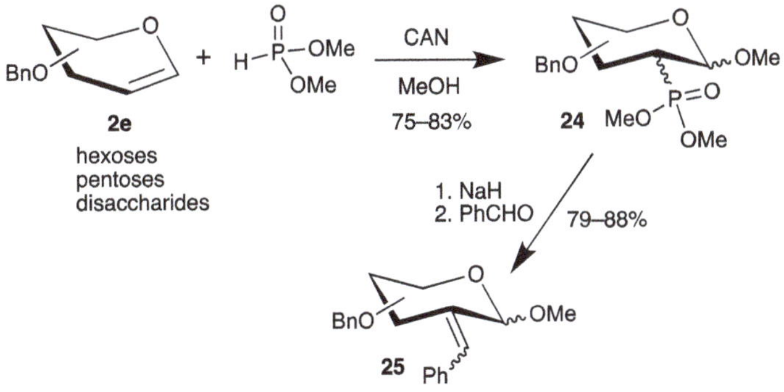

Scheme 12. Addition of dimethyl phosphite to benzyl-protected glycals **2e** in the presence of CAN and subsequent Horner–Emmons reaction.

5. Addition of Sulfur-Centered Radicals

Sulfur-centered radicals can be easily generated from thiols by chemical or photochemical processes, because the S-H bond is much weaker than the corresponding O-H bond [28]. Subsequent addition to alkenes can initiate efficient chain reactions by hydrogen atom abstraction (thiol-ene reaction) or polymerizations. Indeed, the application of thiyl radicals in organic synthesis [52,57] or polymer chemistry [58] has been reviewed extensively. Even thio sugars are suitable radical precursors, and have been used for cyclizations and additions to other unsaturated carbohydrates at various positions [59,60] Therefore, this mini-review will focus only on the additions of sulfur-centered radicals to endo-glycals.

The first example of a C-S bond formation by radical addition to glycals was published in 1970 [61]. Thus, the chain-reaction was initiated by cumene hydroperoxide (CHP) with thioacetic acid as radical precursor. The 2-thiocarbohydrates **26** were isolated in high yields with the manno isomer **26a** as main product (Scheme 13).

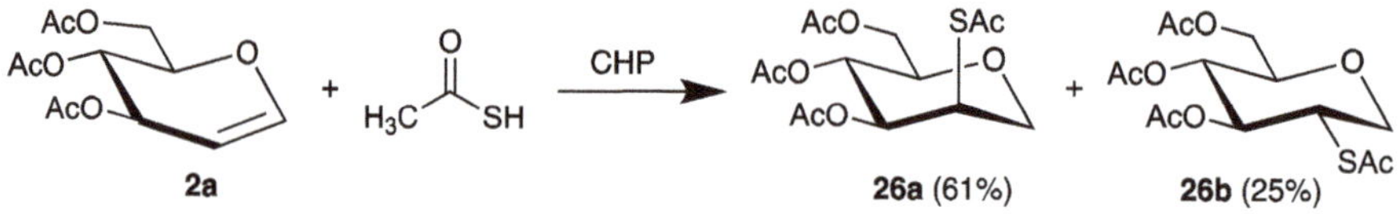

Scheme 13. Addition of thioacetic acid to tri-O-acetyl-D-glucal (2a), initiated by CHP.

The addition of alkyl thiols to tri-O-acetyl-D-glucal (2a) was realized by photochemical initiation with acetone as sensitizer [62]. The 2-S analogues 27a and 27b were isolated in even higher yields but with lower stereoselectivities (Scheme 14).

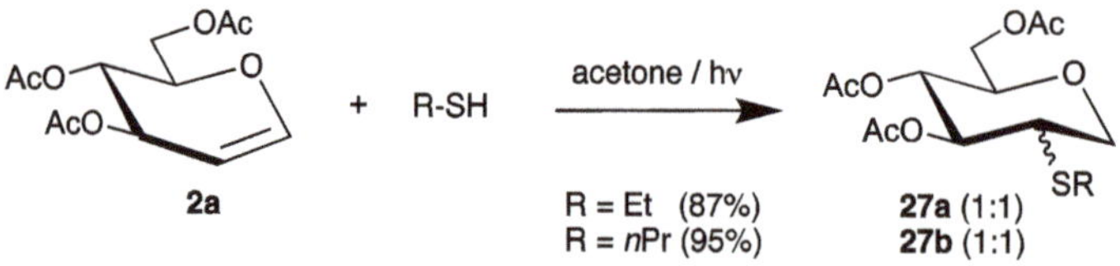

Scheme 14. Photochemical addition of alkyl thiols to tri-O-acetyl-D-glucal (2a).

More recently, 2,2-dimethoxy-2-phenylacetophenone (DPAP 28) became more attractive as radical initiator, which was developed for polymerizations and fragments under UV irradiation by an interesting mechanism (Scheme 15) [58]. Thus, in the first step a carbon–carbon bond is cleaved to generate a benzoyl radical 29, which can abstract hydrogen atoms from thiols to initiate the chain reaction. The second dimethoxybenzyl radical 30 can fragment into benzoate 31 and methyl radicals 32, which act as initiators as well.

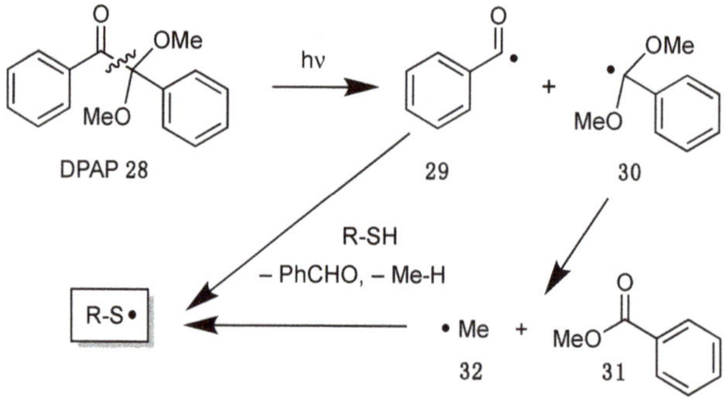

Scheme 15. Mechanism of the decomposition of 2,2-dimethoxy-2-phenylacetophenone (DPAP 28).

Dondoni applied this initiator for the synthesis of S-disaccharides 33 [63]. Starting from thiosugar 34 and tri-O-acetyl-D-glucal (2a), the products 33 were isolated in high yield as a 1:1 mixture of epimers (Scheme 16).

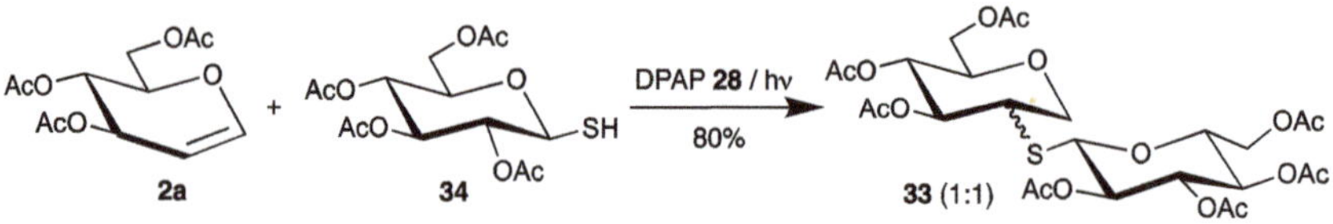

Scheme 16. Addition of thiosugar 34 to tri-O-acetyl-D-glucal (2a), initiated by DPAP 28.

In all reactions described above (Schemes 13–16), the C-S bond is formed selectively at the 2-position of the carbohydrates, due to the enol structure of the glycals. To obtain this bond at the 1-position of sugars, another strategy was developed by Borbas [64,65]. Thus, 2-acetoxy-3,4,6-tri-O-acetyl-D-glucal (**2f**) was used as radical acceptor, which reacted with various thiols in the presence of DPAP. Because of the additional oxygen substituent in the 2-position, orbital interactions allow the attack of electrophilic radicals from the 1- and 2-position. However, steric interactions result in the sole formation of 1-thiosugars **35** (Scheme 17, only one example with thiosugar **34** is shown).

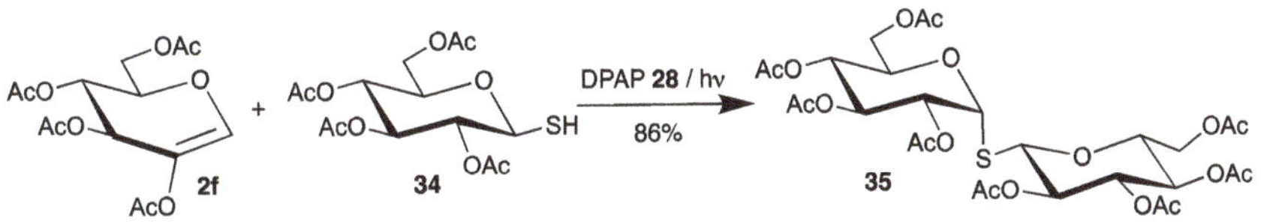

Scheme 17. Addition of thiosugar **34** to 2-acetoxy-3,4,6- tri-O-acetyl-D-glucal (**2f**).

The addition of thiols by radical chain reactions to the 2-position of glycals has only one disadvantage: that the 1-position is reduced under the reaction conditions. Therefore, we investigated the oxidation of ammonium thiocyanate by cerium(IV) ammonium nitrate (CAN) and addition of the generated sulfur-centered radicals to various benzyl-protected glycals **2e** (Scheme 18) [66]. Similarly to the reaction of dimethyl phosphite (Scheme 12), the C-1 radical is further oxidized to a carbenium ion, which is trapped by the solvent methanol, generating the anomeric center of carbohydrates. The yields of the 2-deoxy-2-sulfur analogues **36** are moderate to good, but stereoisomers have to be separated. The thiocyanate groups can be cleaved to the corresponding thiols, which can bind to concanavalin A [66] or gold nanoparticles [67].

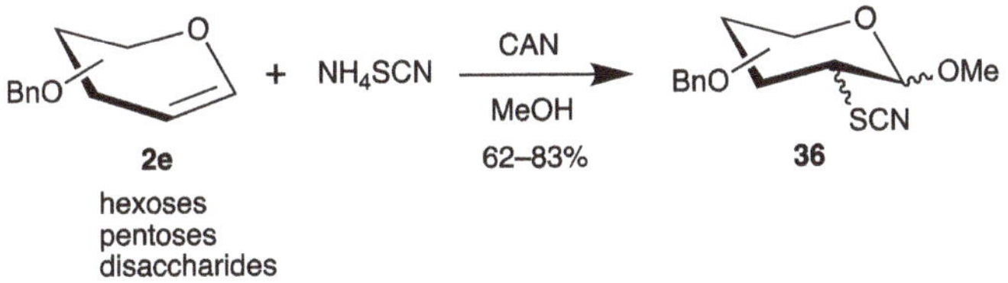

Scheme 18. Addition of thiocyanate to benzyl-protected glycals **2e** in the presence of CAN.

6. Summary and Perspectives

The addition of heteroatom radicals to glycals has been known for more than 50 years and has found various applications. The aim of this mini-review was to highlight early examples and discuss recent developments. Heteroatom radicals can be easily generated by initiators, photochemical processes, or by electron transfer. They exhibit electrophilic character and add regioselectively to the 2-position of the electron-rich double bond of *endo*-glycals. On the other hand, they are prone to H atom abstraction, which limits, especially for alkoxyl radicals, their applications in carbohydrate chemistry. The simple reaction of unsaturated sugars with halogens is possible, but proceeds mainly by ionic pathways. Nitrogen-centered radicals can be generated by oxidation of azides with cerium(IV) ammonium nitrate and add readily to glycals, which is still the best method to synthesize glycosamines. Phosphorus-centered radicals have been less intensively studied in carbohydrate chemistry, but addition products can be used for further transformations. On the other hand, the addition of sulfur-centered radicals to glycals has become very attractive for the synthesis of thio-disaccharides [68–72]. Photochemical initiators based on ketones have been developed for thiol-ene-reactions with unsaturated sugars, affording products in high yields. Finally, simple 2-thio sugars were synthesized by oxidation of thiocyanate and addition to glycals.

In conclusion, many methods for the introduction of heteroatoms in the 2-position of carbohydrates by radical processes exist in the literature. However, the addition to *endo*-glycals has been limited to nitrogen-, phosphorus-, or sulfur-centered radicals until now. Therefore, there is still space for new developments for other heteroatom additions, like boryl radicals, which can be easily generated [73–75], or future applications of such radical reactions in carbohydrate chemistry.

Funding: We acknowledge the support of the Open Access Publishing Fund of the University of Potsdam.

Conflicts of Interest: The author declares no conflict of interest.

References and Notes

1. Fehér, J.; Csomós, G.; Vereckei, A. *Free Radical Reactions in Medicine*; Springer Science & Business Media: Heidelberg, Germany, 1987.
2. Nicolaou, K.C.; Dai, W.-M. Chemistry and Biology of the Enediyne Anticancer Antibiotics. *Angew. Chem. Int. Ed.* **1991**, *30*, 1387–1416. [CrossRef]
3. Joshi, M.C.; Rawat, D.S. Recent Developments in Enediyne Chemistry. *Chem. Biodivers.* **2012**, *9*, 459–498. [CrossRef] [PubMed]
4. Romeo, R.; Glofre, S.V.; Chiacchio, M.A. Synthesis and Biological Activity of Unnatural Enediynes. *Curr. Med. Chem.* **2017**, *24*, 3433–3484. [CrossRef] [PubMed]
5. Early book: von Sonntag, C. *Radiation Chemistry of Carbohydrates and of the Sugar Moiety in DNA*; Elsevier Scientific: Amsterdam, The Netherlands, 1979.
6. Von Sonntag, C. Weiss Lecture: Carbohydrate Radicals: From Ethylene Glycol to DNA Strand Breakage. *Int. J. Radiat. Biol.* **1984**, *46*, 507–519. [CrossRef]
7. Very recent book: Chadwick, K.H. *Understanding Radiation Biology*; CRC Press: Boca Raton, FL, USA, 2019. [CrossRef]
8. Giese, B.B.; Be, X.; Burger, J.; Kesselheim, C.; Senn, M.; Schayer, T. The Mechanism of Anaerobic, Radical-Induced DNA Strand Scission. *Angew. Chem. Int. Ed.* **1993**, *2*, 1742–1743. [CrossRef]
9. Giese, B. *Hole Injection and Hole Transfer through DNA: The Hopping Mechanism BT - Long-Range Charge Transfer in DNA I*; Schuster, G.B., Ed.; Springer Berlin Heidelberg: Berlin/Heidelberg, Germany, 2004; pp. 27–44. [CrossRef]
10. First example: Giese, B.; Dupuis, J. Diastereoselective Syntheses of C-Glycopyranosides. *Angew. Chem. Int. Ed.* **1983**, *22*, 622–623. [CrossRef]
11. Giese, B. The Stereoselectivity of Intermolecular Free Radical Reactions. *Angew. Chem. Int. Ed.* **1989**, *8*, 969–1146. [CrossRef]
12. Giese, B.; Dupuis, J.; Gröninger, K.; Haßkerl, T.; Nix, M.; Witzel, T. Orbital Effects in Carbohydrate Radicals. In *Substituent Effects in Radical Chemistry*; Viehe, H.G., Janousek, Z., Merényi, R., Eds.; Springer Netherlands: Dordrecht, The Netherlands, 1986; pp. 283–296. [CrossRef]
13. Giese, B. Syntheses with Radicals—C-C Bond Formation via Organotin and Organomercury Compounds. *Angew. Chem. Int. Ed.* **1985**, *24*, 553–565. [CrossRef]
14. Giese, B. *Radicals in Organic Synthesis: Formation of Carbon–carbon Bonds*; Organic Chemistry Series; Pergamon Press: Oxford, UK, 1986.
15. Giese, B. Stereoselective Syntheses with Carbohydrate Radicals. *Pure Appl. Chem.* **1988**, *60*, 1655–1658. [CrossRef]
16. Hansen, S.G.; Skrydstrup, T. Modification of Amino Acids, Peptides, and Carbohydrates through Radical Chemistry. In *Radicals in Synthesis II*; Gansäuer, A., Ed.; Springer Berlin Heidelberg: Berlin/Heidelberg, Germnay, 2006; pp. 135–162. [CrossRef]
17. Pérez-Martín, I.; Suárez, E. Radicals and Carbohydrates. In *Encyclopedia of Radicals in Chemistry, Biology and Materials*; American Cancer Society: Chichester, UK, 2012. [CrossRef]
18. Very recent book: Binkley, R.W.; Binkley, E.R. *Radical Reactions of Carbohydrates*. 2019. Available online: http://www.carborad.com (accessed on 20 February 2020).
19. Giese, B.; Gröninger, K.S.; Witzel, T.; Korth, H.-G.; Sustmann, R. Synthesis of 2-Deoxy Sugars. *Angew. Chem. Int. Ed.* **1987**, *26*, 233–234. [CrossRef]

20. Giese, B. Formation of CC Bonds by Addition of Free Radicals to Alkenes. *Angew. Chemie Int. Ed.* **1983**, *22*, 753–764. [CrossRef]
21. Praly, J.-P. Structure of Anomeric Glycosyl Radicals and Their Transformations under Reductive Conditions. In *Advances in Carbohydrate Chemistry and Biochemistry*; Academic Press: London, UK, 2000; Volume 56, pp. 65–151. [CrossRef]
22. Very recent review: De Vleeschouwer, F.; Van Speybroeck, V.; Waroquier, M.; Geerlings, P.; De Proft, F. Electrophilicity and Nucleophilicity Index for Radicals. *Org. Lett.* **2007**, *9*, 2720–2724. [CrossRef]
23. Kinfe, H.H. Versatility of Glycals in Synthetic Organic Chemistry: Coupling Reactions, Diversity Oriented Synthesis and Natural Product Synthesis. *Org. Biomol. Chem.* **2019**, *17*, 4153–4182. [CrossRef] [PubMed]
24. First example: Linker, T.; Hartmann, K.; Sommermann, T.; Scheutzow, D.; Ruckdeschel, E. Transition-Metal-Mediated Radical Reactions as an Easy Route to 2-C-Analogues of Carbohydrates. *Angew. Chem. Int. Ed.* **1996**, *35*, 1730–1732. [CrossRef]
25. Linker, T.; Sommermann, T.; Kahlenberg, F. The Addition of Malonates to Glycals: A General and Convenient Method for the Synthesis of 2-C-Branched Carbohydrates. *J. Am. Chem. Soc.* **1997**, *119*, 9377–9384. [CrossRef]
26. Review: Elamparuthi, E.; Kim, B.G.; Yin, J.; Maurer, M.; Linker, T. Cerium(IV)-Mediated C–C Bond Formations in Carbohydrate Chemistry. *Tetrahedron* **2008**, *64*, 11925–11937. [CrossRef]
27. Vankar, Y.D.; Linker, T. Recent Developments in the Synthesis of 2-C-Branched and 1,2-Annulated Carbohydrates. *Eur. J. Org. Chem.* **2015**, *2015*, 7633–7642. [CrossRef]
28. Taniguchi, T. Recent Advances in Reactions of Heteroatom-Centered Radicals. *Synthesis* **2017**, *49*, 3511–3534. [CrossRef]
29. Very recent example: Wu, X.; Zhu, C. Recent Advances in Alkoxy Radical-Promoted C-C and C-H Bond Functionalization Starting from Free Alcohols. *Chem. Commun.* **2019**, *55*, 9747–9756. [CrossRef]
30. De Armas, P.; Francisco, C.G.; Suárez, E. Reagents with Hypervalent Iodine: Formation of Convenient Chiral Synthetic Intermediates by Fragmentation of Carbohydrate Anomeric Alkoxy Radicals. *Angew. Chem. Int. Ed.* **1992**, *31*, 772–774. [CrossRef]
31. Hernández-Guerra, D.; Rodríguez, M.S.; Suárez, E. Fragmentation of Carbohydrate Anomeric Alkoxyl Radicals: Synthesis of Chiral Polyhydroxylated β-Iodo- and Alkenylorganophosphorus(V) Compounds. *Eur. J. Org. Chem.* **2014**, *2014*, 5033–5055. [CrossRef]
32. Lemieux, R.U.; Fraser-Reid, B. The Mechanisms of the Halogenations and Halogenomethoxylations of D-Glucal Triacetate, D-Galactal Triacetate, and 3,4-Dihydropyran. *Can. J. Chem.* **1965**, *43*, 1460–1475. [CrossRef]
33. Hassner, A. Regiospecific and Stereospecific Introduction of Azide Functions into Organic Molecules. *Acc. Chem. Res.* **1971**, *4*, 9–16. [CrossRef]
34. Bovin, N.V.; Zurabyan, S.É.; Khorlin, A.Y. Addition of Halogenoazides to Glycals. *Carbohydr. Res.* **1981**, *98*, 25–35. [CrossRef]
35. Rawal, G.K.; Kumar, A.; Tawar, U.; Vankar, Y.D. New Method for Chloroamidation of Olefins. Application in the Synthesis of N-Glycopeptides and Anticancer Agents. *Org. Lett.* **2007**, *9*, 5171–5174. [CrossRef]
36. Kirschning, A.; Jesberger, M.; Schönberger, A. The First Polymer-Assisted Solution-Phase Synthesis of Deoxyglycosides. *Org. Lett.* **2001**, *3*, 3623–3626. [CrossRef]
37. Islam, M.; Tirukoti, N.D.; Nandi, S.; Hotha, S. Hypervalent Iodine Mediated Synthesis of C-2 Deoxy Glycosides and Amino Acid Glycoconjugates. *J. Org. Chem.* **2014**, *79*, 4470–4476. [CrossRef]
38. Kundoor, G.; Rao, D.S.; Kashyap, S. Regioselective Direct Difunctionalization of Glycals: Convenient Access to 2-Deoxyglycoconjugates Mediated by Tetra-n-Butylammonium Iodide/Sodium Periodate. *Asian J. Org. Chem.* **2016**, *5*, 264–270. [CrossRef]
39. Nair, V.; Deepthi, A. Cerium(IV) Ammonium Nitrate—A Versatile Single-Electron Oxidant. *Chem. Rev.* **2007**, *107*, 1862–1891. [CrossRef]
40. Linker, T.; Schanzenbach, D.; Elamparuthi, E.; Sommermann, T.; Fudickar, W.; Gyóllai, V.; Somsák, L.; Demuth, W.; Schmittel, M. Remarkable Oxidation Stability of Glycals: Excellent Substrates for Cerium(IV)-Mediated Radical Reactions. *J. Am. Chem. Soc.* **2008**, *130*, 16003–16010. [CrossRef]
41. Ratcliffe, R.M.; Lemieux, R.U. The Azidonitration of Tri-O-Acetyl-D-Galactal. *Can. J. Chem.* **1979**, *57*, 1244–1251.

42. Paulsen, H.; Lorentzen, J.P.; Kutschker, W. Erprobte Synthese von 2-Azido-2-Desoxy-D-Mannose Und 2-Azido-2-Desoxy-D-Mannuronsäure Als Baustein Zum Aufbau von Bakterien-Polysaccharid-Sequenzen. *Carbohydr. Res.* **1985**, *136*, 153–176. [CrossRef]
43. Kinzy, W.; Schmidt, R.R. Glycosylimidate, 16. Synthese Des Trisaccharids Aus Der "Repeating Unit" Des Kapselpolysaccharids von Neisseria Meningitidis (Serogruppe L). *Liebigs Ann. Chem.* **1985**, *1985*, 1537–1545. [CrossRef]
44. Keith, D.J.; Townsend, S.D. Total Synthesis of the Congested, Bisphosphorylated Morganella Morganii Zwitterionic Trisaccharide Repeating Unit. *J. Am. Chem. Soc.* **2019**, *141*, 12939–12945. [CrossRef]
45. Singh, Y.; Wang, T.; Demchenko, A.V. Direct Glycosidation of 2-Azido-2-Deoxyglycosyl Nitrates. *Eur. J. Org. Chem.* **2019**, *2019*, 6413–6416. [CrossRef]
46. Czernecki, S.; Randriamandimby, D. Azido-Phenylselenylation of Protected Glycals. *Tetrahedron Lett.* **1993**, *34*, 7915–7916. [CrossRef]
47. Santoyo-González, F.; Calvo-Flores, F.G.; García-Mendoza, P.; Hernández-Mateo, F.; Isac-García, J.; Robles-Díaz, R. Synthesis of Phenyl 2-Azido-2-Deoxy-1-Selenoglycosides from Glycals. *J. Org. Chem.* **1993**, *58*, 6122–6125. [CrossRef]
48. Jiaang, W.T.; Chang, M.Y.; Tseng, P.H.; Chen, S.T. A Concise Synthesis of the O-Glycosylated Amino Acid Building Block; Using Phenyl Selenoglycoside as a Glycosyl Donor. *Tetrahedron Lett.* **2000**, *41*, 3127–3130. [CrossRef]
49. Churchill, D.G.; Rojas, C.M. Iron(II)-Promoted Amidoglycosylation and Amidochlorination of an Allal C3-Azidoformate. *Tetrahedron Lett.* **2002**, *43*, 7225–7228. [CrossRef]
50. Lu, D.F.; Zhu, C.L.; Jia, Z.X.; Xu, H. Iron(II)-Catalyzed Intermolecular Amino-Oxygenation of Olefins through the N-O Bond Cleavage of Functionalized Hydroxylamines. *J. Am. Chem. Soc.* **2014**, *136*, 13186–13189. [CrossRef]
51. Leca, D.; Fensterbank, L.; Lacôte, E.; Malacria, M. Recent Advances in the Use of Phosphorus-Centered Radicals in Organic Chemistry. *Chem. Soc. Rev.* **2005**, *34*, 858–865. [CrossRef] [PubMed]
52. Pan, X.-Q.; Zou, J.-P.; Yi, W.-B.; Zhang, W. Recent Advances in Sulfur- and Phosphorous-Centered Radical Reactions for the Formation of S–C and P–C Bonds. *Tetrahedron* **2015**, *71*, 7481–7529. [CrossRef]
53. Kazuo, K.; Yoshida, H.; Ogata, T.; Inokawa, S. Sugars Containing a Carbon-Phosphorus Bond. I. Photochemical Addition of Diethyl Thiophosphonate to Unsaturated Sugars. *Bull. Chem. Soc. Jpn.* **1969**, *42*, 3245–3248.
54. Jessop, C.M.; Parsons, A.F.; Routledge, A.; Irvine, D.J. Phosphonyl Radical Addition to Enol Ethers. The Stereoselective Synthesis of Cyclic Ethers. *Tetrahedron Lett.* **2004**, *45*, 5095–5098. [CrossRef]
55. Zhang, F.; Wang, L.; Zhang, C.; Zhao, Y. Novel Regio- and Stereoselective Phosphonyl Radical Addition to Glycals Promoted by Mn(II)-Air: Syntheses of 1,2-Dideoxy 2-C- Diphenylphosphinylglycopyranosides. *Chem. Commun.* **2014**, *50*, 2046–2048. [CrossRef] [PubMed]
56. Elamparuthi, E.; Linker, T. Carbohydrate-2-Deoxy-2-Phosphonates: Simple Synthesis and Horner-Emmons Reaction. *Angew. Chem. Int. Ed.* **2009**, *48*, 1853–1855. [CrossRef]
57. Dénès, F.; Pichowicz, M.; Povie, G.; Renaud, P. Thiyl Radicals in Organic Synthesis. *Chem. Rev.* **2014**, *114*, 2587–2693. [CrossRef]
58. Hoyle, C.E.; Lee, T.Y.; Roper, T. Thiol-Enes: Chemistry of the Past with Promise for the Future. *J. Polym. Sci. Part A Polym. Chem.* **2004**, *42*, 5301–5338. [CrossRef]
59. Fiore, M.; Marra, A.; Dondoni, A. Photoinduced Thiol-Ene Coupling as a Click Ligation Tool for Thiodisaccharide Synthesis. *J. Org. Chem.* **2009**, *74*, 4422–4425. [CrossRef]
60. McSweeney, L.; Dénès, F.; Scanlan, E.M. Thiyl-Radical Reactions in Carbohydrate Chemistry: From Thiosugars to Glycoconjugate Synthesis. *Eur. J. Org. Chem.* **2016**, *2016*, 2080–2095. [CrossRef]
61. Igarashi, K.; Honma, T. Addition Reactions of Glycals. IV. Free-Radical Addition of Thiolacetic Acid to D-Glucal Triacetate. *J. Org. Chem.* **1970**, *35*, 606–610. [CrossRef]
62. Araki, Y.; Matsuura, K.; Ishido, Y.; Kushida, K. Synthetic Studies of Carbohydrate Derivatives with Photochemical Reaction. VII. Photochemical Addition of Ethanethiol and 1-Propanethiol to Enoses. *Chem. Lett.* **1973**, *2*, 383–386. [CrossRef]
63. Staderini, S.; Chambery, A.; Marra, A.; Dondoni, A. Free-Radical Hydrothiolation of Glycals: A Thiol-Ene-Based Synthesis of S-Disaccharides. *Tetrahedron Lett.* **2012**, *53*, 702–704. [CrossRef]
64. Lázár, L.; Csávás, M.; Herczeg, M.; Herczegh, P.; Borbás, A. Synthesis of S-Linked Glycoconjugates and S-Disaccharides by Thiol–Ene Coupling Reaction of Enoses. *Org. Lett.* **2012**, *14*, 4650–4653. [CrossRef]

65. Lázár, L.; Csávás, M.; Hadházi, Á.; Herczeg, M.; Tóth, M.; Somsák, L.; Barna, T.; Herczegh, P.; Borbás, A. Systematic Study on Free Radical Hydrothiolation of Unsaturated Monosaccharide Derivatives with Exo- and Endocyclic Double Bonds. *Org. Biomol. Chem.* **2013**, *11*, 5339–5350. [CrossRef] [PubMed]
66. Pavashe, P.; Elamparuthi, E.; Hettrich, C.; Möller, H.M.; Linker, T. Synthesis of 2-Thiocarbohydrates and Their Binding to Concanavalin A. *J. Org. Chem.* **2016**, *81*, 8595–8603. [CrossRef]
67. Fudickar, W.; Pavashe, P.; Linker, T. Thiocarbohydrates on Gold Nanoparticles: Strong Influence of Stereocenters on Binding Affinity and Interparticle Forces. *Chem. Eur. J.* **2017**, *23*, 8685–8693. [CrossRef]
68. Lázár, L.; Borbás, A.; Somsák, L. Synthesis of Thiomaltooligosaccharides by a Thio-Click Approach. *Carbohydr. Res.* **2018**, *470*, 8–12. [CrossRef]
69. Eszenyi, D.; Kelemen, V.; Balogh, F.; Bege, M.; Csávás, M.; Herczegh, P.; Borbás, A. Promotion of a Reaction by Cooling: Stereoselective 1,2-Cis-α-Thioglycoconjugation by Thiol-Ene Coupling at −80 °C. *Chem. Eur. J.* **2018**, *24*, 4532–4536. [CrossRef]
70. Review: Dondoni, A.; Marra, A. Recent Applications of Thiol-Ene Coupling as a Click Process for Glycoconjugation. *Chem. Soc. Rev.* **2012**, *41*, 573–586. [CrossRef]
71. Lázár, L.; Juhász, L.; Batta, G.; Borbás, A.; Somsák, L. Unprecedented β-Manno Type Thiodisaccharides with a C-Glycosylic Function by Photoinitiated Hydrothiolation of 1-C-Substituted Glycals. *New J. Chem.* **2017**, *41*, 1284–1292. [CrossRef]
72. Very recent example: Kelemen, V.; Bege, M.; Eszenyi, D.; Debreczeni, N.; Bényei, A.; Stürzer, T.; Herczegh, P.; Borbás, A. Stereoselective Thioconjugation by Photoinduced Thiol-Ene Coupling Reactions of Hexo- and Pentopyranosyl D- and L-Glycals at Low-Temperature—Reactivity and Stereoselectivity Study. *Chem. Eur. J.* **2019**, *25*, 14555–14571. [CrossRef] [PubMed]
73. Ollivier, C.; Renaud, P. Organoboranes as a Source of Radicals. *Chem. Rev.* **2001**, *101*, 3415–3434. [CrossRef] [PubMed]
74. Renaud, P. Boron in Radical Chemistry. In *Encyclopedia of Radicals in Chemistry, Biology and Materials*; Chatgilialoglu, C., Studer, A., Eds.; Wiley: Chichester, UK, 2012. [CrossRef]
75. Taniguchi, T. Boryl Radical Addition to Multiple Bonds in Organic Synthesis. *Eur. J. Org. Chem.* **2019**, *2019*, 6308–6319. [CrossRef]

© 2020 by the author. Licensee MDPI, Basel, Switzerland. This article is an open access article distributed under the terms and conditions of the Creative Commons Attribution (CC BY) license (http://creativecommons.org/licenses/by/4.0/).

Article

OH Group Effect in the Stator of β-Diketones Arylhydrazone Rotary Switches [†]

Silvia Hristova [1], Fadhil S. Kamounah [2], Aurelien Crochet [3], Nikolay Vassilev [1], Katharina M. Fromm [3] and Liudmil Antonov [1,*]

[1] Institute of Organic Chemistry with Centre of Phytochemistry, Bulgarian Academy of Sciences, Acad. G. Bonchev str., bldg. 9, 1113 Sofia, Bulgaria; hristowa.silvia@gmail.com (S.H.); niki@orgchm.bas.bg (N.V.)
[2] Department of Chemistry, University of Copenhagen, Universitetsparken 5, DK-2100 Copenhagen Ø, Denmark; fadil@chem.ku.dk
[3] Department of Chemistry, University of Fribourg, Chemin du Musée 9, CH-1700 Fribourg, Switzerland; aurelien.crochet@unifr.ch (A.C.); katharina.fromm@unifr.ch (K.M.F.)
* Correspondence: Lantonov@orgchm.bas.bg
† Dedicated to Prof. Bernd Giese on Behalf of His 80th Birthday.

Received: 14 April 2020; Accepted: 30 April 2020; Published: 4 May 2020

Abstract: The properties of several hydrazon-diketone rotary switches with OH groups in the stators (2-(2-(2-hydroxy-4-nitrophenyl)hydrazono)-1-phenylbutane-1,3-dione, 2-(2-(2-hydroxyphenyl)hydrazono)-1-phenylbutane-1,3-dione and 2-(2-(4-hydroxyphenyl)hydrazono)-1-phenylbutane-1,3-dione) were investigated by molecular spectroscopy (UV-Vis and NMR), DFT calculations (M06-2X/TZVP) and X-ray analysis. The results show that, when the OH group is in *ortho* position, the E' and Z' isomers are preferred in DMSO as a result of a stabilizing intermolecular hydrogen bonding with the solvent. The availability, in addition, of a nitro group in *para* position increases the possibility of deprotonation of the OH group in the absence of water. All studied compounds showed a tendency towards formation of associates. The structure of the aggregates was revealed by theoretical calculation and confirmed by X-ray analysis.

Keywords: rotary switch; UV-Vis spectroscopy; NMR; DFT; X-ray; aggregation

1. Introduction

The hydrazone functional group has found extensive use in medicine [1–6], supramolecular chemistry (molecular switches and chelate ligands) [7–10] and in combinatory chemistry [11–13]. One important facet of hydrazine-group-containing compounds is the fact that upon appropriate substitution they can exist in solution as a mixture of isomers. 1,2,3-tricarbonyl-2-arylhydrazones are a typical example—they are presented in solution as an equilibrated mixture of intramolecularly H-bonded E and Z isomers [14–16]. The position of the isomerization equilibrium can be altered by catalytic amounts of acid or base. Upon external stimulation, a controlled switching between the isomers is possible through C-N bond rotation, giving the name "rotary switches". The position of the equilibrium and the switching can be strongly affected by structural modifications, as has already been shown [17–21].

Recently, the spectral properties of **1** (Schemes 1 and 2) were studied in respect of the possible tautomerism and E/Z isomerization in solution [17,22]. The results show that the availability of the OH group in the stator does not lead to azo-hydrazone tautomerism as could be expected at a first glimpse. The compound exists as a mixture of isomers of the single ketohydrazone tautomer, as shown in Scheme 1. In DMSO, due to the specific stabilizing effect of the solvent, only the E' and Z' forms are presented [22,23]. Moreover, the availability of the OH group leads to some side effects, according to the spectral and crystallographic data [22,24]: compound **1** deprotonates at low concentrations in DMSO and aggregates at high concentrations (10^{-4} M and higher), forming linear (E'-E') aggregates.

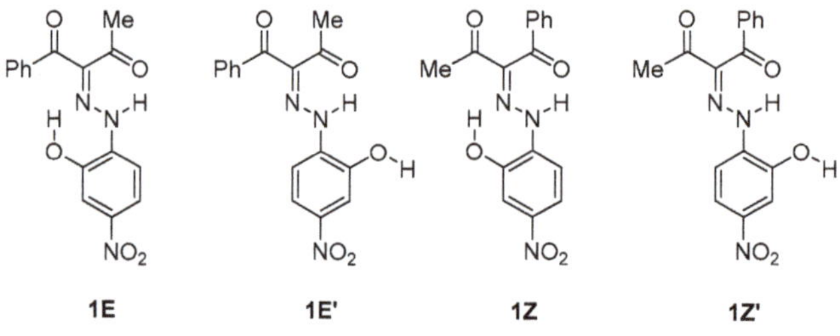

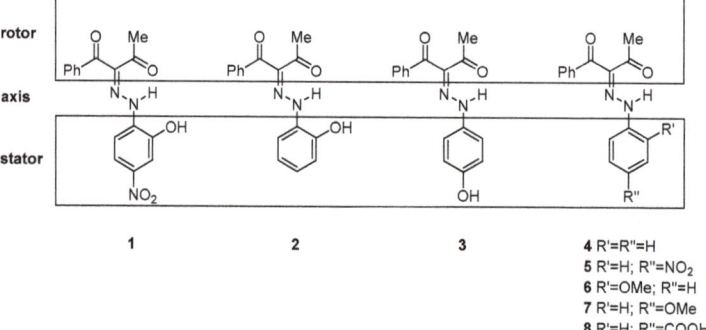

Scheme 1. Conformational isomers of **1**.

Scheme 2. Sketch of the investigated compounds.

It is an interesting question whether the lack of tautomerism, the existence of the " ' " isomers (indicating E' and Z' forms) and the side effects could be attributed to the strong electron acceptor ability of the nitro group. To answer it, we have studied compounds **2** and **3**, in which two effects could be clarified: the role of the existence of a nitro group (**2** vs. **1**) and of the position of the OH group by itself (**2** vs. **3**). A combined approach (theoretical calculation, UV-Vis and NMR investigations in solution, and X-ray analysis in solid state) was applied, and the results obtained are presented in the current communication. To best of our knowledge, no such comparative study of **1**–**3** has been performed before.

2. Results and Discussion

The 1,2,3-tricarbonyl-2-arylhydrazones are potentially tautomeric compounds even without an OH group in the stator. The possible tautomers include ketohydrazone, azoketone and azoenol forms, depending on the substitution [25]. The availability, in addition, of an *ortho* or *para* OH group in the stator makes the tautomeric situation even more complex. The possible tautomers of **2** are sketched in Scheme 3 as an example. Theoretical prediction of the stabilities of the individual forms is complicated by the large number of possible conformers. For instance, the tautomer **I** of **2** can be presented as 24 possible isomers (Scheme SI).

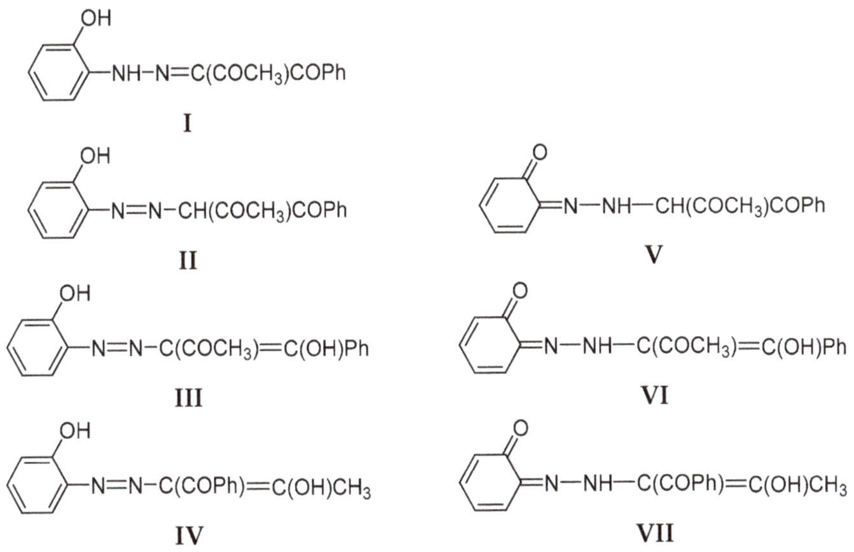

Scheme 3

I: 2-HO-C₆H₄-NH-N=C(COCH₃)COPh

II: 2-HO-C₆H₄-N=N-CH(COCH₃)COPh

III: 2-HO-C₆H₄-N=N-C(COCH₃)=C(OH)Ph

IV: 2-HO-C₆H₄-N=N-C(COPh)=C(OH)CH₃

V: =N-NH-CH(COCH₃)COPh (quinoid)

VI: =N-NH-C(COCH₃)=C(OH)Ph (quinoid)

VII: =N-NH-C(COPh)=C(OH)CH₃ (quinoid)

Scheme 3. Sketch of the possible tautomeric forms of **2** (The same is valid for **1** and **3**).

We have shown theoretically that, in the case of **1**, only tautomer **I** could be present [22]. As shown in Table S1, the same conclusion can be drawn for **2** and **3**. The theoretical calculations are in agreement with the NMR data in solution [22] and the recent crystallographic data [21]. In analogy, compounds **2** and **3** also should exist as the same single tautomeric form, stabilized as a mixture of E and Z conformers. The most stable isomers of **2** and **3**, as predicted in DMSO as an environment, are shown in Figures 1 and 2, respectively.

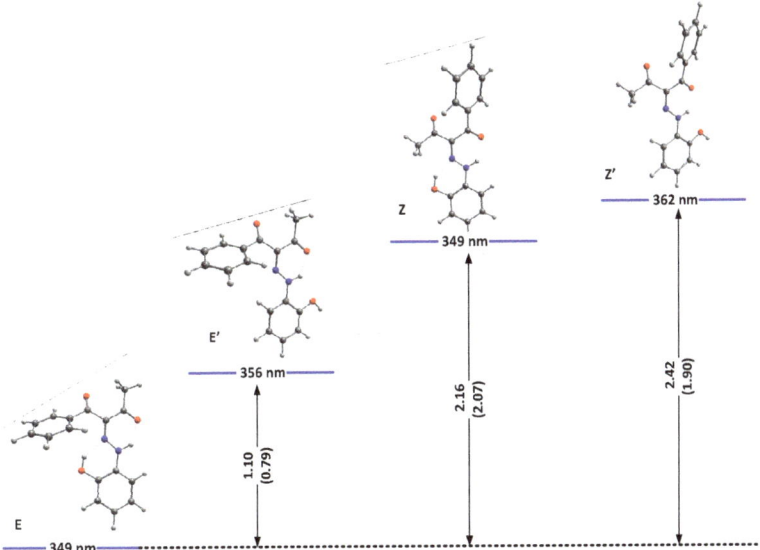

Figure 1. Relative energies (in kcal/mole units) and predicted positions of the long-wavelength bands of the most stable isomers of **2(I)** in DMSO. The corresponding relative energies for **1**, taken from [24], are given in brackets.

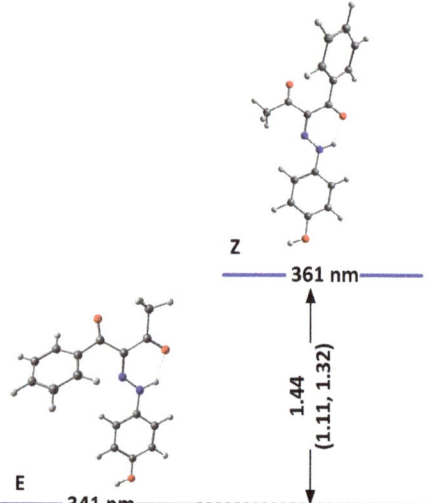

Figure 2. Relative energies (in kcal/mole units) and predicted positions of the long-wavelength bands of the most stable isomers of **3(I)** in DMSO. The corresponding relative energies for **5**, taken from [24] and **4**, are given in brackets.

As seen from the figures, the stabilization is a result of the strength of the formed intramolecular hydrogen bonds. While better proton-attracting ability of MeCO through the NH..OMe determines a better stabilization of the E isomer, additional stabilization through OH..N bonding makes the E/Z pairs more stable compared to E'/Z' in **1** and **2**. The effect of the nitro group in **1** leads to an overall weak stabilization in the E/Z forms and a more pronounced stability of the " ' " isomers. In the case of **3**, the effect of the OH group is limited to a non-hydrogen bonding substituent and leads to stabilization of the Z isomer. The predicted stabilization effect in the series **5** [22], **4** and **3** follows the experimentally observed trend for a destabilization of the Z isomer (molar fractions of 15%, 10% and 5%, correspondingly) going from electron acceptor to electron donor substituents in *para* position in the stator [19]. Most probably, the absence of the OH..N hydrogen bonding in E'/Z' of **2** and in **3** reduces the steric hindrance between the rotor and the stator, leading to an overall stabilization of the corresponding isomers.

The solvation model used so far describes the solvent as a dielectric medium and does not take into account the possible specific solute–solvent interactions. As known previously in the case of **1**, the proton of the OH group interacts with proton-acceptor solvents (such as DMSO), when it is not involved in the intramolecular hydrogen bond with the rotor part, which leads to strong additional stabilization of the " ' " isomers. The model of this specific solvent effect is illustrated in Figures 3 and 4, showing the most stable complexes with DMSO. As can be seen, the interaction between the solvent molecule and the free OH proton in E' and Z' leads to their stabilization. Moreover, in **2E** and **2Z**, there are no conditions for the formation of any OH..O = SMe$_2$ hydrogen bond, and the formed NH ... O = SMe$_2$ is weak due to the low acidity of the NH proton and steric effect from adjacent functional groups (Figure 3). The changes in the case of **3** are caused by reducing the electron donor ability of the OH group and hence to a rise in the polarization of the N-H bond, leading finally to the stabilization of the Z isomer.

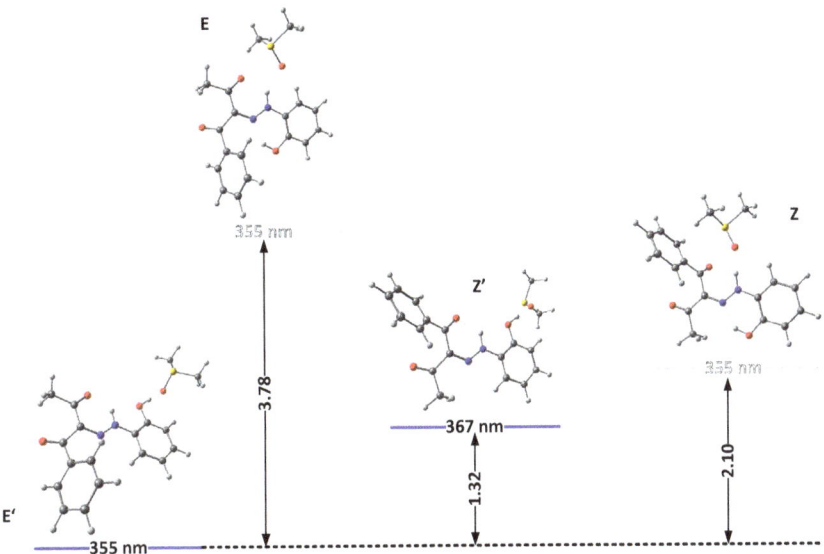

Figure 3. Relative energies (in kcal/mole units) and predicted positions of the long-wavelength bands of the most stable isomers of **2(I)** in DMSO, accounting for the specific solute–solvent interactions.

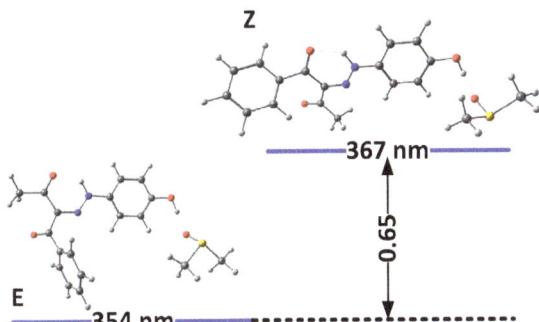

Figure 4. Relative energies (in kcal/mole units) and predicted positions of the long-wavelength bands of the most stable isomers of **3(I)** in DMSO, accounting for the specific solute–solvent interactions.

In addition to the relative stability of the isomers, the predicted positions of the long-wavelength bands in the absorption spectra are shown in Figures 1–4 as well. The absolute values should be considered with care due to the systematic blue shift of the used M06-2X functional. The relative changes indicate, as expected, that it is practically impossible to distinguish between the most stable isomers by means of UV-Vis spectroscopy. The absorption spectra of **2** and **3**, shown in Figure 5, indicate that there are no substantial changes in the spectral shape upon changing the solvent. This figure strongly supports the hypothesis, in analogy to **1**; there is no tautomeric equilibrium, because the tautomers of **1**–**3** have different conjugated systems, and substantially different spectra could be expected upon changing the solvent [26]. It is seen that the observed long-wavelength absorption band consists of two sub-bands, whose intensity slightly varies with the solvent. They can be associated with the most stable isomers according to the theoretical predictions. However, the strong overlapping between them does not allow either precise estimation of the positions of the bands by derivative spectroscopy (Figure S1) nor a quantitative estimation of the isomers' molar fractions [27].

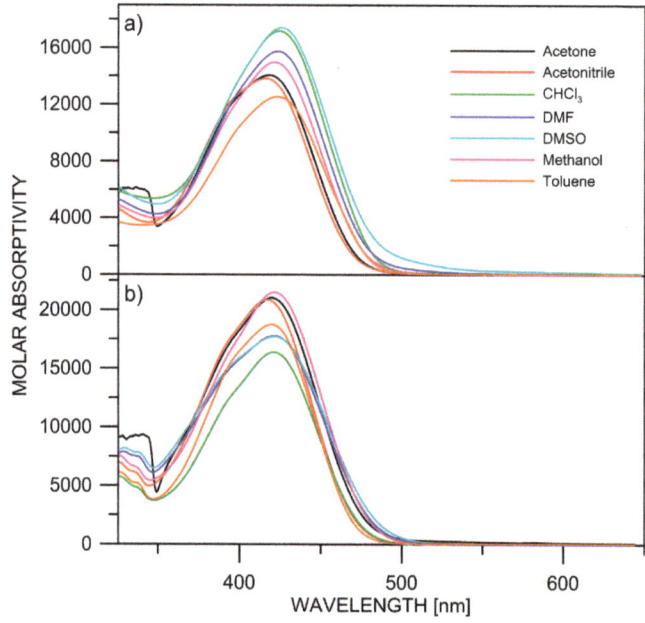

Figure 5. Experimental absorption spectra of **2** (**a**) and **3** (**b**) in various solvents.

The conformational isomers existing in solution can be identified and quantified using NMR. Due to their low solubility in acetonitrile, the investigations were performed in DMSO-d_6. The corresponding ^{1}H NMR spectra of **2** and **3** in DMSO-d_6 are shown in Figures S2 and S3. The data from the NMR measurements are summarized in Table 1. The obtained data for chemical shifts can be compared with those for **1** [22]. From the ^{1}H NMR spectra of **2** and **3**, it can be seen that, in both cases, two isomers are present in DMSO-d_6. The chemical shifts in DMSO-d_6 for NH for the major and minor form of **2** are at 14.55 ppm and 12.62 ppm, and for **3** at 14.48 ppm and 11.42 ppm. Based on the NH signals, the ratio between the isomers is 80%/20% and 45%/55%, respectively for **2** and **3** (65%/35% for **1**), which corresponds to ΔG values (RT) of 0.36, 0.82 and −0.11 kcal/mol for **1**–**3**.

Table 1. ^{1}H and ^{13}C chemical shifts[f] of the major form (minor form) of **1**, **2** and **3** in DMSO-d_6.

Compound	1	2	3
Conformers (%)	65/35	80/20	45/55
Protons		^{1}H chemical shift	
1	2.51 (2.56)	2.52 ([a])	2.50 (2.44)
2′	-	-	7.16 (7.27)
3′	7.72 (7.70)	6.93 ([ab])	6.74 (6.76)
4′	-	7.00 ([ab])	-
5′	7.75 (7.85)	6.79 ([ab])	6.74 (6.76)
6′	7.21 (7.78)	7.09 ([ab])	7.16 (7.27)
2″	7.91 (7.72)	7.82 ([ab])	7.78 (7.73)
3″	7.55 (7.51)	7.51 ([ab])	7.50 (7.52)
4″	7.67 (7.66)	7.61([ab])	7.59(7.65)
NH	14.14 (11.56) [d]	14.55 (12.62)	14.46 (11.42)
OH	11.56 (11.56) [d]	10.56 (10.56) [c]	9.58 (9.29)

Table 1. Cont.

Compound	1	2	3
Carbons		^{13}C chemical shift	
1	30.67 (25.92)	30.62 (b)	30.48 (25.31)
2	198.08 (b)	197.49 (b)	197.03 (195.86) [e]
3	135.66 (b)	133.27 (b)	134.21 (132.15) [e]
4	191.95 (b)	192.25 (b)	194.96 (192.13) [e]
1'	136.25 (b)	139.06 (b)	136.67 (135.54) [e]
2'	145.99 (b)	146.76 (b)	118.31 (116.94)
3'	110.60 (b)	116.19 (b)	116.50 (116.27) [e]
4'	143.78 (b)	126.38 (b)	156.15 (154.43) [e]
5'	116.78 (b)	120.53 (b)	116.50 (116.27) [e]
6'	114.08 (b)	114.98 (b)	118.31 (116.94)
1"	137.96 (b)	133.27 (b)	139.34 (137.06) [e]
2"	130.75 (b)	130.49 (b)	130.32 (128.99)
3"	128.50 (b)	128.20 (b)	128.16 (129.14)
4"	133.10 (b)	132.30 (b)	132.05 (134.03)

[a] Could not be assigned due to overlap; [b] Could not be assigned due to low intensity; [c] The hydroxyl signals of the two forms are overlapped; [d] The hydroxyl signals of two forms and the NH signal of minor form are overlapped; [e] May be interchanged; [f] Numbering according to the Scheme 4.

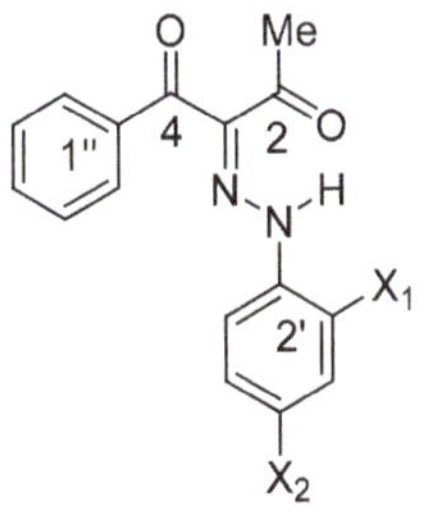

Scheme 4. Numbering of the carbon atoms in **1** (X_1=OH, X_2=NO$_2$), **2** (X_1=OH, X_2=H) and **3** (X_1=H, X_2=OH).

In analogy to **1** [22,24] and following the theoretically predicted relative stabilization, it can be concluded that in DMSO there is an equilibrium between E' (major) and Z' (minor) forms in **2** and between E and Z of **3**. The theoretically predicted values for the chemical shifts of the NH proton of the major and minor form at **2** and **3**, respectively, are 14.61 ppm (2E') and 13.67 ppm (2Z') and 14.51 ppm (3E) and 13.44 ppm (3Z), i.e., the theoretical results are consistent with the experimental ones. Although the NMR determined Gibb's free energies are lower comparing the predicted relative energies (ΔE), the latter correctly predict the general trend of stabilization of the Z' isomers (**3** > **1** > **2**) with a good linearity ($\Delta G = 1.28 \times \Delta E - 0.98$, $R^2 = 0.92$).

Three additional factors influence the conformational equilibrium in **2** and **3** in solution, namely the temperature, the concentration and the water content of the used solvent. As previously shown in the case of **1**, a spontaneous deprotonation (loss of OH proton) occurs in diluted solutions of dry proton acceptor solvents [22], leading to a new red-shifted band. Actually, **1** is almost fully deprotonated in dry DMSO (Figure S4), while, as seen in Figure 5, deprotonation in **2** is weak and negligible in **3**. Obviously, the effect of the nitro group is decisive in this case. Upon addition of water, the equilibrium is fully shifted towards the neutral isomers (Figure S5). According to the theoretical calculations (Table S2), deprotonation does not substantially change the isomers' ratios.

The concentration is an essential factor determining the deprotonation of the investigated compounds. In **2**, as in **1**, the increase in the concentration leads to a decrease in the content of the deprotonated form, as shown in Figure 6. No such effect is observed in **3**.

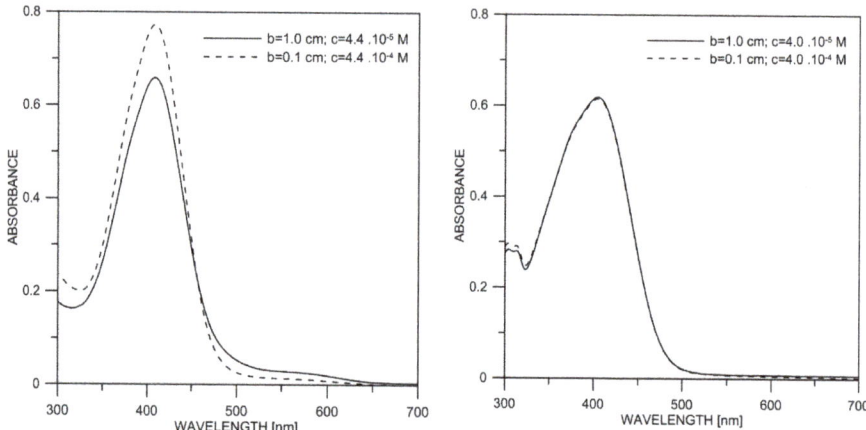

Figure 6. Experimental absorption spectra of **2** (**left**) and **3** (**right**) in DMSO as a function of the concentration, keeping the cell thickness (b) × concentration (c) constant.

The results above indicate that the association is a possible reason for the observed changes. The theoretical calculations and X-ray data (Figure 7 and Scheme 5) suggest that cyclic aggregates are formed in the case of **2**. In the solid state, compound **2** exists as an E' conformer, stabilized via a cyclic dimer. The major difference with **1** is in the shape of the aggregate—again, the E' form is stabilized in **1**, but in form of a linear aggregate [24]. The stability of the E'-E' cyclic dimer in **2** is probably due to the stronger proton acceptor properties of the CH_3CO group compared to the PhCO moiety of **1**. The formation of aggregates, as in **1** [22,24], limits the deprotonation, which explains the observed concentration effects.

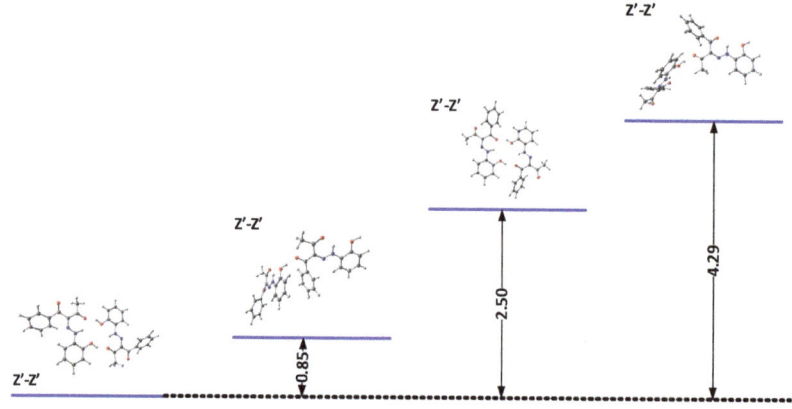

Figure 7. Relative energies (in kcal/mol) of the most stable dimers of **2** in DMSO.

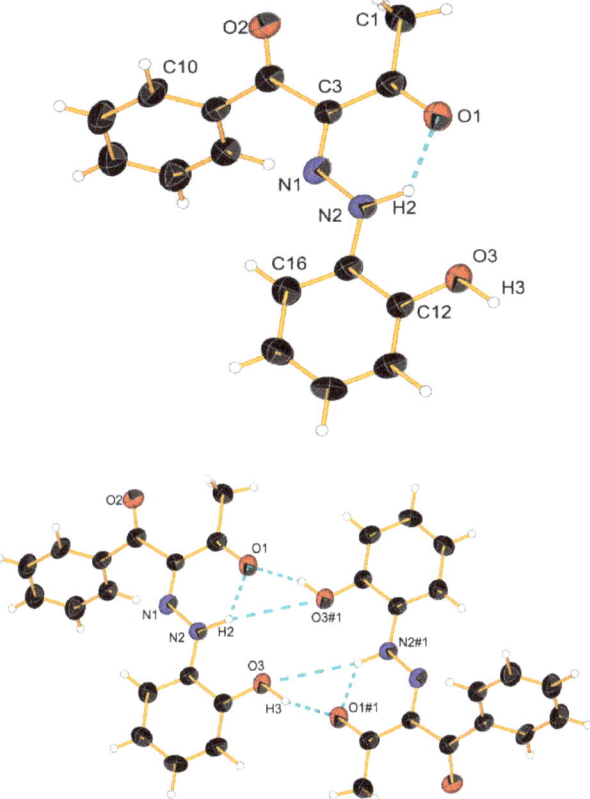

Scheme 5. Crystal structure of **2** and a cyclic dimer model via an intermolecular hydrogen bond. Ellipsoids are drawn with a probability of 50% and H-bonds are represented as dashed blue lines, #1: 2 − x, 1 − y, 2 − z.

The increase in the concentration did not lead to significant spectral changes in **3** (Figure 6). This is also expected since the OH group is not polar enough in this particular case, but some linear aggregation cannot be excluded. According to the theoretical simulations (Figure S6) and the crystal structure (Scheme 6), linear *E-E* aggregates are expected in solution.

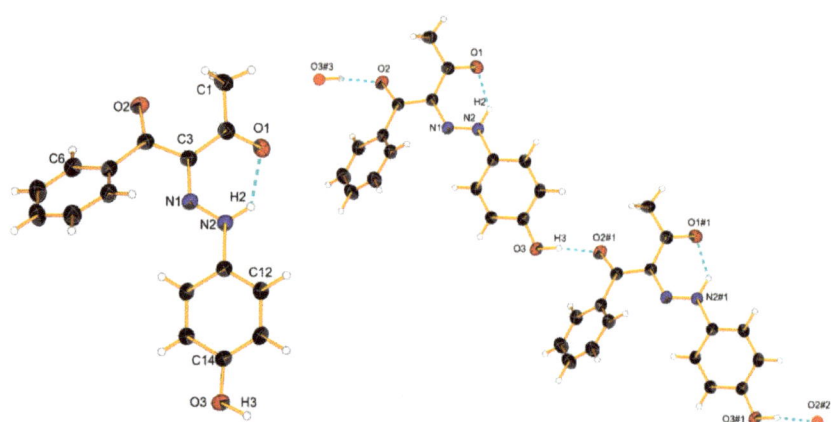

Scheme 6. Crystal structure of **3** and a cyclic dimer model via an intermolecular hydrogen bond. Ellipsoids are drawn with a probability of 30% and H-bonds are represented as dashed blue lines, #1: x − 1, 1 + y, z; #2: x − 2, y + 2, z; #3: x + 1, y − 1, z.

The availability of crystallographic data for the series **1–8** allows qualitative estimation of the strength of the existing intra- and intermolecular hydrogen bonding according to Steiner [28] and Jeffrey [29]. The corresponding bond length and angles are collected in Table 2. According to the classification given in [28], the existing N-H … O hydrogen bonds are classified as moderate ones using the H … A and D … .A distances and the D-H … A angle. It seems that the strength (at least in solid state) of this bond is almost independent on the substitution in the stator. In the cases where this bond is bifurcated, namely **1**, **2** and **6**, the contribution from N-H … O(=C) is the dominant one. This explains why, according to the theoretical calculations using the solvent only as media, the *E* and *Z* isomers are always more stable compared to the " ' " ones. The data in Table 2 show clearly that the formation of associates through intramolecular hydrogen bonding with strong directionality (D-H … A angle > 160°) has a noticeable stabilizing effect.

Table 2. Parameters of the hydrogen bonds of the studied compounds, taken from their crystallographic data.

Comp.	Type D-H … A Bond		Distances [Å]			D-H … A Angle, [°]	Ref.
			D-H	H … A	D … A		
1	intramolecular	N-H … O(=C)	0.896(9)	1.890(7)	2.565(1)	130.6(7)	[24]
	intermolecular	N-H … O(-H)		2.266(9)	2.619(1)	103.1(6)	
	(linear dimer)	O-H … O(=C)	0.85(2)	1.89(2)	2.738(1)	176(1)	
2	intramolecular	N-H … O(=C)	1.03(3)	1.79(2)	2.567(4)	128(1)	current work
	intermolecular	N-H … O(-H)		2.242(8)	2.609(4)	99(1)	
	(cyclic dimer)	O-H … O(=C)	0.92(6)	1.79(7)	2.677(4)	161(6)	
2	intramolecular	N-H … O(=C)	0.873	1.898	2.572(1)	132.5	[30]
	intermolecular	N-H … O(-H)		2.257	2.617(1)	104.5	
	(complex with water)	O-H … OH$_2$	0.908	1.736	2.678	170.2	
3	intramolecular intermolecular (linear dimer)	N-H … O(=C)	0.95(2)	1.83(2)	2.565(2)	132(2)	current work
		O-H … O(=C)	0.93(3)	1.75(3)	2.667(2)	168(2)	
4	intramolecular	N-H … O(=C)	0.82(2)	1.93(2)	2.581(2)	135(2)	
6	intramolecular	N-H … O(=C)	0.91(2)	1.85(2)	2.561(2)	133(2)	[31]
		N-H … O(-Me)		2.26(2)	2.609(2)	102(1)	
7	intramolecular	N-H … O(=C)	0.86(3)	1.89(3)	2.559(3)	133(2)	[32]
8	intramolecular	N-H … O(=C)	0.935	1.866	2.594(2)	132(8)	

The absorption spectra of **2** and **3** in DMSO in the temperature range of 20–70 °C are shown in Figure 8 and Figure S7. In both compounds, it can be seen that with increasing temperature, the absorption maximum of the neutral form (~400 nm) decreases slightly, while the amount of the deprotonated form absorbing at ~500 nm increases.

The result can be interpreted in analogy to **1** [22]. Increasing the temperature leads to the destruction of existing aggregates, which subsequently facilitates the deprotonation of the monomers. The temperature effect was strongest in **1** (Figure S8), followed by **2** (Figure 8) and lowest for **3** (Figure S7). This indicates that **1** has the highest tendency to aggregate again related to the effect of the nitro group. The percentage contents (in %) of the two neutral isomers were determined by measuring NMR spectra as a function of the temperature (Table S3). It could be thereby shown that the ratio does not change significantly (Figures S9–S11), which probably indicates that they exist in the form of associates in the concentration range used in NMR.

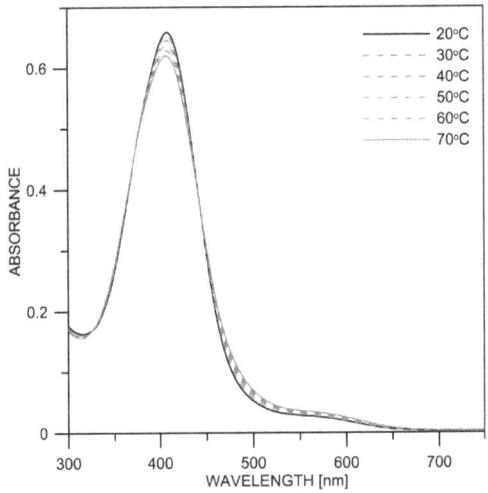

Figure 8. Experimental absorption spectra of **2** in dry DMSO at different temperatures.

3. Materials and Methods

3.1. Synthesis

Reagents and solvents were analytical grade, purchased from Sigma-Aldrich Chemical Co. and used as received unless otherwise stated. Fluka silica gel/TLC-cards 60778 with fluorescence indicator 254 nm were used for TLC chromatography. Merck silica gel 60 (0.040–0.063 mm, Merck, Darmstadt, Germany) was used for flash chromatography purification of the products. LC/MS was carried out on a Bruker MicrOTOF-QIII-system with an ESI source with nebulizer 1.2 bar, dry gas 10.0 L/min, dry temperature 220 °C, capillary 4500 V, and end plate offset −500 V, Bruker, Hamburg, Germany. The title 2-(2(2-Hydroxy-4-nitrophenyl)hydrazono)-1-phenylbutane-1,2,3-trione **1**, 2-[2-(2-Hydroxyphenyl)diazenyl]-1-phenyl-1,3-Butanedione **2** and 2-[2-(4-Hydroxyphenyl)diazenyl]-1-phenyl-1,3-Butanedione **3** were synthesized as follows.

3.1.1. 2-(2(2-Hydroxy-4-nitrophenyl)hydrazono)-1-phenylbutane-1,2,3-trione **1**

The synthesis of the compound **1** has been described previously [22].

3.1.2. 2-[2-(2-Hydroxyphenyl)diazenyl]-1-phenyl-1,3-butanedione **2**

A solution of 2-aminophenol (0.2728 g, 2.5 mmol) in concentrated hydrochloric acid (1.5 mL) was cooled in an ice bath for 30 min with stirring. Sodium nitrite (0.1725 g, 2.5 mmol) was added gradually in small portions over 30 min. The diazonium salt was left stirring in the cold for 45 min and added slowly over 30 min into a well-cooled stirred mixture of sodium acetate (1.5 g, 18.3 mmol) and 1-phenylbutane-1,3-dione (0.4061 g, 2.5 mmol) in absolute ethanol (15 mL). The mixture was stirred in an ice-bath for 3 h, and then left to warm to room temperature over 12 h. The solid was filtered off and washed with water (3 × 25 mL) to give a light brown solid. The ethanol portion was evaporated under reduced pressure to give dark brown solid. The combined solids were dissolved in dichloromethane (30 mL) and washed with water (3 × 50 mL) and the organic portion dried over anhydrous sodium sulfate, filtered and the solvent evaporated under reduced pressure to obtain a crude brown solid. This was purified by flash column chromatography using a silica flash column and ethyl acetate-dichloromethane (1:20) as eluent to give the pure compound as a light brown solid (0.48 g, 68%). HRMS-ESI (m/z): (M + H) calculated for $C_{16}H_{14}N_2O_3$, 282.1004; found 282.1084.

3.1.3. 2-[2-(4-Hydroxyphenyl)diazenyl]-1-phenyl-1,3-butanedione 3

A solution of 4-aminophenol (0.2728 g, 2.5 mmol) in concentrated hydrochloric acid (1.5 mL) was cooled in an ice bath for 30 min with stirring. Sodium nitrite (0.1725 g, 2.5 mmol) was added gradually in small portions over 30 min. The diazonium salt was left stirring in the cold for 45 min and added slowly over 30 min into a well-cooled stirred mixture of sodium acetate (1.5 g, 18.3 mmol) and 1-phenylbutane-1,3-dion (0.4061 g, 2.5 mmol) in absolute ethanol (15 mL). The mixture was stirred in an ice bath for 3 h, and then warmed to room temperature over 12 h. The solid was filtered off and washed with water (3 × 25 mL) to give a yellow solid. The ethanol portion was evaporated under reduced pressure to give a dark yellow solid. The combined solids were dissolved in dichloromethane (30 mL) and washed with water (3 × 50 mL) and the organic portion was dried over anhydrous sodium sulfate, filtered, and the solvent evaporated under reduced pressure to obtain a crude brown solid. This was purified by flash column chromatography using a silica flash column and ethyl acetate-dichloromethane (1:20) as eluent to give the pure compound as a light-yellow solid (0.52 g, 74%). HRMS-ESI (m/z): (M + H) calculated for $C_{16}H_{14}N_2O_3$, 282.1004; found 282.1066.

3.2. Spectral Measurements

Spectral measurements were performed on a Jasco V-570 UV-Vis-NIR spectrophotometer, equipped with a thermostatic cell holder (using Huber MPC-K6 thermostat with precision 1 °C), in spectral grade solvents.

The ^{1}H-NMR and ^{13}C-NMR spectra were recorded at 600 MHz and 150 MHz or 400 MHz and 100 MHz on a Bruker Avance II+ 600 or Bruker Avance III 400 spectrometer using $CDCl_3$ or DMSO-d_6 as a solvent and TMS as internal standard.

3.3. X-ray Measurements

Single crystals of $C_{16}H_{14}N_2O_3$ (2) were obtained from a mixture of ethanol:water (5:1) by slow evaporation. A suitable single crystal was selected and mounted on a loop with oil and measured on a STOE IPDS 2T diffractometer. The crystal was kept at 200 K during data collection. Using Olex2 [33], the structure was solved with the ShelXT [34] structure solution program using intrinsic phasing and refined with the ShelXL [35] refinement package using least squares minimization. All the crystal structures have been deposited at the CCDC 1858058 (1), 1993960 (2) and 1993961 (3).

Crystal Data for $C_{16}H_{14}N_2O_3$ (M = 282.29 g/mol): triclinic, space group P-1 (no. 2), a = 5.8853(12) Å, b = 8.0633(16) Å, c = 15.944(3) Å, α = 77.760(15)°, β = 88.887(16)°, γ = 68.978(15)°, V = 688.9(2) Å^3, Z = 2, T = 200 K, μ(Cu Kα) = 0.785 mm^{-1}, Dcalc = 1.361 g/cm^3, 5325 reflections measured (14.428° ≤ 2θ ≤ 138.228°), 5365 unique (R_{int} = 0.0571, R_{sigma} = 0.0392) which were used in all calculations. The final R_1 was 0.0712 (I > 4 u(I)) and wR_2 was 0.2505 (all data).

Single crystals of $C_{16}H_{14}N_2O_3$ (3) were obtained from a mixture of ethanol:chloroform:water (10:1:2) by slow evaporation. A suitable single crystal was selected and mounted on a loop with oil and measured on a STOE IPDS 2T diffractometer. The crystal was kept at 298(2) K during data collection. Using Olex2 [33], the structure was solved with the ShelXT [34] structure solution program using intrinsic phasing and refined with the ShelXL [35] refinement package using least squares minimization.

Crystal Data for $C_{16}H_{14}N_2O_3$ (M = 282.29 g/mol): triclinic, space group P-1 (no. 2), a = 6.3843(4) Å, b = 8.0018(5) Å, c = 15.0129(10) Å, α = 96.235(5)°, β = 101.298(5)°, γ = 107.252(5)°, V = 706.78(8) Å^3, Z = 2, T = 298(2) K, μ(MoKα) = 0.093 mm^{-1}, $Dcalc$ = 1.326 g/cm^3, 8242 reflections measured (2.812° ≤ 2θ ≤ 52.618°), 2808 unique (R_{int} = 0.0164, R_{sigma} = 0.0146) which were used in all calculations. The final R_1 was 0.0390 (I > 2σ(I)) and wR_2 was 0.1142 (all data).

3.4. Quantum-Chemical Calculations

Quantum-chemical calculations were performed using the Gaussian 09 program suite [36]. M06-2X functional [37,38] was used with the TZVP basis set [39]. This fitted hybrid meta-GGA functional with 54% HF exchange has especially been developed to describe main-group thermochemistry and non-covalent interactions, showing very good results in the prediction of the position of the tautomeric equilibrium in azonaphthols possessing intramolecular hydrogen bonds [40] and in the description of the proton transfer reactions in naphthols [41,42]. All structures were optimized without restrictions, using tight optimization criteria and an ultrafine grid in the computation of two-electron integrals and their derivatives, and the true minima were verified by performing frequency calculations in the corresponding environment. Solvent effects are described by using the Polarizable Continuum Model (the integral equation formalism variant, IEFPCM, as implemented in Gaussian 09) [43]. The absorption spectra of the compounds were predicted using the TD-DFT formalism. TD-DFT calculations were carried out at the same functional and basis set, which is in accordance with conclusions about the effect of the basis set size and the reliability of the spectral predictions [44–46].

4. Conclusions

The effect of the position of the OH group and the availability of a nitro group substitution in the stator was investigated in solution by means of DFT calculations, NMR and UV-Vis spectroscopy. The results indicate that, when the OH group is in *ortho* position, the E' and Z' isomers are present in DMSO, stabilized by intermolecular hydrogen bonding with the solvent. The availability, in addition, of a nitro group in *para* position increases the possibility for deprotonation of the OH group in the absence of water. In all studied compounds, a clear tendency towards formation of associates is evident. The obtained X-ray data explain the types of the possible homo-aggregates and correspond very well to the theoretical predictions. The obtained results, revealing the effect of the structural modifications in the stator and the influence of the environment, could be useful in the design of new rotary switches.

Supplementary Materials: The following are available online at http://www.mdpi.com/2624-8549/2/2/374\T1\textendash389/s1. Figure S1: Second derivative spectra of a) **2** and b) **3** in various solvents, Figure S2: ^{1}H NMR spectrum of **2** in DMSO-d_6, Figure S3: ^{1}H NMR spectrum of **3** in DMSO-d_6, Figure S4: Absorption spectra of **1**, **2** and **3** in dry DMSO and upon base (TEA) addition, Figure S5: Experimental absorption spectra of **1** and **2** upon water addition, Figure S6: Relative energies (in kcal/mol) of the most stable dimers of **3** in DMSO, Figure S7: Absorption spectra of **3** in DMSO in different temperatures, Figure S8: Absorption spectra of **1** in DMSO in different temperatures, Figure S9: ^{1}H NMR spectrum of **1** in DMSO at a temperature range of 20 °C to 70 °C, Figure S10: ^{1}H NMR spectrum of **2** in DMSO at a temperature range of 20 °C to 70 °C, Figure S11: ^{1}H NMR spectrum of **3** in DMSO at a temperature range of 20 °C to 70 °C, Table S1: The most stable isomers of I-VII of **2** in gas phase, Table S2: Relative energies (M06-2X/TZVP) of the most stable neutral and deprotonated forms of **1, 2, 3** in kcal/mol units, Table S3: Ratio between E(E')/Z(Z') forms in **1, 2** and **3** in the temperature range of 293 K–343 K, Scheme SI: Possible conformers of **I** and their relative energies (in kcal/mol, according to quantum-chemical calculations) in acetonitrile.

Author Contributions: Conceptualization, L.A.; methodology, L.A. and N.V.; validation, S.H. and A.C.; formal analysis, S.H. and L.A.; investigation, S.H., F.S.K., N.V. and A.C.; resources, L.A., F.S.K. and K.M.F; data curation, L.A. and K.M.F.; writing—original draft preparation, S.H.; writing—review and editing, L.A., N.V. and K.M.F; supervision, L.A. and K.M.F.; project administration, L.A.; funding acquisition, K.M.F. and L.A. All authors have read and agreed to the published version of the manuscript.

Funding: This research was funded by The Swiss National Science Foundation, Institutional Partnership project IZ74Z0_160515, Bulgarian National Science Fund, project DN09/10 MolRobot, and Bulgarian Ministry of Educations, project DFNP-17-66/26.07.2017.

Acknowledgments: The financial support by Bulgarian Ministry of Educations (project DFNP-17-66/26.07.2017 for support of young scientists), Bulgarian National Science Fund (project DN09/10 MolRobot) and The Swiss National Science Foundation (SCOPES Institutional Partnership project IZ74Z0_160515 SupraMedChem@Balkans.Net) is gratefully acknowledged.

Conflicts of Interest: No conflict of interest.

References

1. Vicini, P.; Zani, F.; Cozzini, P.; Doytchinova, I. Hydrazones of 1,2-benzisothiazole hydrazides: Synthesis, antimicrobial activity and QSAR investigations. *Eur. J. Med. Chem.* **2002**, *37*, 553–564. [CrossRef]
2. Loncle, C.; Brunel, M.; Vidal, N.; Dherbomez, M.; Letourneux, Y. Synthesis and antifungal activity of cholesterol-hydrazone derivatives. *Eur. J. Med. Chem.* **2004**, *39*, 1067–1071. [CrossRef] [PubMed]
3. Savini, L.; Chiasserini, L.; Travagli, V.; Pellerano, C.; Novellino, E.; Cosentino, S.; Pisano, B. New α-(N)-heterocyclichydrazones: Evaluation of anticancer, anti-HIV and antimicrobial activity. *Eur. J. Med. Chem.* **2004**, *39*, 113–122. [CrossRef] [PubMed]
4. Cocco, T.; Congiu, C.; Lilliu, V.; Onnis, V. Synthesis and in vitro antitumoral activity of new hydrazinopyrimidine-5-carbonitrile derivatives. *Bioorganic Med. Chem.* **2006**, *14*, 366–372. [CrossRef]
5. Masunari, A.; Tavares, C. A new class of nifuroxazide analogues: Synthesis of 5-nitrothiophene derivatives with antimicrobial activity against multidrug-resistant Staphylococcus aureus. *Bioorganic Med. Chem.* **2007**, *15*, 4229–4236. [CrossRef]
6. Vicini, P.; Incerti, M.; La Colla, P.; Loddo, R. Anti-HIV evaluation of benzo[d]isothiazole hydrazones. *Eur. J. Med. Chem.* **2009**, *44*, 1801–1807. [CrossRef]
7. Lehn, J. From supramolecular chemistry towards constitutional dynamic chemistry and adaptive chemistry. *Chem. Soc. Rev.* **2007**, *36*, 151–160. [CrossRef]
8. Lehn, J. *Constitutional Dynamic Chemistry: Bridge from Supramolecular Chemistry to Adaptive Chemistry*; Barboiu, M., Ed.; Springer: Berlin/Heidelberg, Germany, 2012; Volume 322, pp. 1–32.
9. Lehn, J. Perspectives in Chemistry-Steps towards Complex Matter. *Angew. Chem. Int. Ed.* **2013**, *52*, 2836–2850. [CrossRef]
10. Lygaitis, R.; Getautis, V.; Grazulevicius, J.V. Hole-transporting hydrazones. *Chem. Soc. Rev.* **2008**, *37*, 770–788. [CrossRef]
11. Rowan, J.; Cantrill, J.; Cousins, L.; Sanders, M.; Stoddart, F. Dynamic Covalent Chemistry. *Angew. Chem. Int. Ed.* **2002**, *41*, 898–952. [CrossRef]
12. Corbett, T.; Leclaire, J.; Vial, L.; West, R.; Wietor, L.; Sanders, M.; Otto, S. Dynamic Combinatorial Chemistry. *Chem. Rev.* **2006**, *106*, 3652–3711. [CrossRef] [PubMed]
13. Wilson, J. Dynamic Combinatorial Chemistry. In Drug Discovery, Bioorganic Chemistry, and Materials Science. Edited by Benjamin, L. Miller. *Angew. Chem. Int. Ed.* **2010**, *49*, 4011. [CrossRef]
14. Courtot, P.; Pichon, R.; Le Saint, J. Determination du site de chelation chez les arylhydrazones de tricetones et D' α-dicetones substituees. *Tetrahedron Lett.* **1976**, *17*, 1177–1180. [CrossRef]
15. Courtot, P.; Pichon, R.; Le Saint, J. Photochromisme par isomerisation syn-anti de phenylhydrazones-2- de tricetones-1,2,3 et de dicetones-1,2 substituees. *Tetrahedron Lett.* **1976**, *17*, 1181–1184. [CrossRef]
16. Pichon, R.; Le Saint, J.; Courtot, P. Photoisomerisation d'arylhydrazones-2 de dicetones-1,2 substituees en 2.: Mecanisme d'isomerisation thermique de la double liaison C=N. *Tetrahedron* **1981**, *37*, 1517–1524. [CrossRef]
17. Mahmudov, K.T.; Rahimov, R.A.; Babanly, M.B.; Hasanov, P.Q.; Pashaev, F.G.; Gasanov, A.G.; Kopylovich, M.N.; Pombeiro, A.J.L. Tautomery and acid–base properties of some azoderivatives of benzoylacetone. *J. Mol. Liq.* **2011**, *162*, 84–88. [CrossRef]
18. Kuznik, W.; Kopylovich, M.N.; Amanullayeva, G.I.; Pombeiro, A.J.L.; Reshak, A.; Mahmudov, K.T.; Kityk, I. Role of tautomerism and solvatochromism in UV–VIS spectra of arylhydrazones of β-diketones. *J. Mol. Liq.* **2012**, *171*, 11–15. [CrossRef]
19. Mitchell, A.; Nonhebel, D.C. Spectroscopic studies of tautomeric systems—III. *Tetrahedron* **1979**, *35*, 2013–2019.
20. Su, X.; Aprahamian, I. Hydrazone-based switches, metallo-assemblies and sensors. *Chem. Soc. Rev.* **2014**, *43*, 1963–1981. [CrossRef]

21. Gurbanov, A.V.; Kuznetsov, M.L.; Demukhamedova, S.D.; Aliyeva, I.N.; Godjaev, N.M.; Zubkov, F.I.; Mahmudov, K.T.; Pombeiro, A.J.L. Role of substituents on resonance assisted hydrogen bonding vs. intermolecular hydrogen bonding. *CrystEngComm* **2020**, *22*, 628–633. [CrossRef]
22. Hristova, S.; Kamounah, F.S.; Molla, N.; Hansen, P.E.; Nedeltcheva, D.; Antonov, L. The possible tautomerism of the potential rotary switch 2-(2-(2-Hydroxy-4-nitrophenyl)hydrazono)-1-phenylbutane-1,3-dione. *Dye. Pigment.* **2017**, *144*, 249–261. [CrossRef]
23. Lycka, A. 15N NMR study of (E)- and (Z)-2-(2-(2-hydroxy-4-nitrophenyl)hydrazono)-1-phenylbutane-1,3-diones. A suitable method for analysis of hydrazone isomers. *Dye. Pigment.* **2018**, *150*, 181–184. [CrossRef]
24. Hristova, S.; Kamounah, F.S.; Crochet, A.; Hansen, P.E.; Fromm, K.M.; Nedeltcheva, D.; Antonov, L. Isomerization and aggregation of 2-(2-(2-hydroxy-4-nitrophenyl)hydrazono)-1-phenylbutane-1,3-dione: Recent evidences from theory and experiment. *J. Mol. Liq.* **2019**, *283*, 242–248. [CrossRef]
25. Gawinecki, R.; Kolehmainen, E.; Janota, H.; Kauppinen, R.; Nissinen, M.; Osmialowski, B. Predominance of 2-arylhydrazones of 1,3-diphenylpropane-1,2,3-trione over its proton-transfer products. *J. Phys. Org. Chem.* **2001**, *14*, 797–803. [CrossRef]
26. Antonov, L. Absorption UV-Vis Spectroscopy and Chemometrics: From Qualitative Conclusions to Quantitative Analysis. In *Tautomerism*; Antonov, L., Ed.; Wiley-VCH Verlag GmbH & Co. KGaA: Weinheim, Germany, 2013; pp. 25–47.
27. Nedeltcheva, D.; Antonov, L.; Lycka, A.; Damyanova, B.; Popov, S. Chemometric Models For Quantitative Analysis of Tautomeric Schiff Bases and Azo Dyes. *Curr. Org. Chem.* **2009**, *13*, 217–240. [CrossRef]
28. Steiner, T. The Hydrogen Bond in the Solid State. *Angew. Chem. Int. Ed.* **2002**, *41*, 48–76. [CrossRef]
29. Jeffrey, G. *An Introduction to Hydrogen Bonding*; Oxford University Press: Oxford, UK, 1997.
30. Kopylovich, M.N.; Mahmudov, K.T.; Haukka, M.; Luzyanin, K.V.; Pombeiro, A.J.L. (E)-2-(2-(2-hydroxyphenyl)hydrazono)-1-phenylbutane-1,3-dione: Tautomery and coordination to copper(II). *Inorg. Chim. Acta* **2011**, *374*, 175–180. [CrossRef]
31. Bertolasi, V.; Nanni, L.; Gilli, P.; Ferretti, V.; Gilli, G.; Issa, Y.; Sherif, O. Intramolecular N-H. O=C hydrogen-bonding assisted by resonance—Intercorrelation between structural and spectroscopic data for 6 beta-diketo-arylhydrazones derived from benzoylacetone or acetylacetone. *New J. Chem.* **1994**, *18*, 251–261.
32. Eliseeva, S.V.; Minacheva, L.K.; Kuz'Mina, N.P.; Sergienko, V.S. Crystal structure of p-carboxyphenylhydrazone benzoylacetone. *Crystallogr. Rep.* **2005**, *50*, 85–88. [CrossRef]
33. Dolomanov, O.; Bourhis, L.J.; Gildea, R.; Howard, J.A.; Puschmann, H. OLEX2: A complete structure solution, refinement and analysis program. *J. Appl. Crystallogr.* **2009**, *42*, 339–341. [CrossRef]
34. Sheldrick, G.M. SHELXT - integrated space-group and crystal-structure determination. *Acta Crystallogr. Sect. A Found. Adv.* **2015**, *A71*, 3–8. [CrossRef] [PubMed]
35. Sheldrick, G.M. Crystal structure refinement with SHELXL. *Acta Crystallogr. Sect. C Struct. Chem.* **2015**, *C71*, 3–8. [CrossRef] [PubMed]
36. Frisch, M.; Trucks, G.; Schlegel, H.; Scuseria, G.; Robb, M.; Cheeseman, J.; Scalmani, G.; Barone, V.; Mennucci, B.; Petersson, G.; et al. *Gaussian 09 Revision D.01*; Gaussian, Inc.: Wallingford, CT, USA, 2013.
37. Zhao, Y.; Truhlar, D. Density Functionals with Broad Applicability in Chemistry. *Acc. Chem. Res.* **2008**, *41*, 157–167. [CrossRef] [PubMed]
38. Zhao, Y.; Truhlar, D. The M06 suite of density functionals for main group thermochemistry, thermochemical kinetics, noncovalent interactions, excited states, and transition elements: Two new functionals and systematic testing of four M06 functionals and 12 other functionals. *Theor. Chem. Acc.* **2008**, *120*, 215–241. [CrossRef]
39. Weigend, F.; Ahlrichs, R. Balanced basis sets of split valence, triple zeta valence and quadruple zeta valence quality for H to Rn: Design and assessment of accuracy. *Phys. Chem. Chem. Phys.* **2005**, *7*, 3297–3305. [CrossRef]
40. Kawauchi, S.; Antonov, L. Description of the Tautomerism in Some Azonaphthols. *J. Phys. Org. Chem.* **2013**, *26*, 643–652. [CrossRef]
41. Manolova, Y.; Kurteva, V.B.; Antonov, L.; Marciniak, H.; Lochbrunner, S.; Crochet, A.; Fromm, K.M.; Kamounah, F.S.; Hansen, P.E. 4-Hydroxy-1-naphthaldehydes: Proton transfer or deprotonation. *Phys. Chem. Chem. Phys.* **2015**, *17*, 10238–10249. [CrossRef]
42. Manolova, Y.; Marciniak, H.; Tschierlei, S.; Fennel, F.; Kamounah, F.S.; Lochbrunner, S.; Antonov, L. Solvent control of intramolecular proton transfer: Is 4-hydroxy-3-(piperidin-1-ylmethyl)-1-naphthaldehyde a proton crane? *Phys. Chem. Chem. Phys.* **2017**, *19*, 7316–7325. [CrossRef]

43. Tomasi, J.; Mennucci, B.; Cammi, R. Quantum Mechanical Continuum Solvation Models. *Chem. Rev.* **2005**, *105*, 2999–3094. [CrossRef]
44. Improta, R. UV-Visible Absorption and Emission Energies in Condensed Phase by PCM/TD-DFT Methods. In *Computational Strategies for Spectroscopy*; Barone, V., Ed.; John Wiley & Sons, Inc.: Hoboken, NJ, USA, 2011; pp. 37–75.
45. Antonov, L.; Kawauchi, S.; Okuno, Y. Prediction of the Color of Dyes by Using Time-Dependent Density Functional Theory. *Bulg. Chem. Commun.* **2014**, *46*, 228–237.
46. Jacquemin, D.; Mennucci, B.; Adamo, C. Excited-state calculations with TD-DFT: From benchmarks to simulations in complex environments. *Phys. Chem. Chem. Phys.* **2011**, *13*, 16987–16998. [CrossRef] [PubMed]

© 2020 by the authors. Licensee MDPI, Basel, Switzerland. This article is an open access article distributed under the terms and conditions of the Creative Commons Attribution (CC BY) license (http://creativecommons.org/licenses/by/4.0/).

Review

Reactive Sterol Electrophiles: Mechanisms of Formation and Reactions with Proteins and Amino Acid Nucleophiles †

Ned A. Porter [1],*, Libin Xu [2] and Derek A. Pratt [3]

1. Department of Chemistry and Vanderbilt Institute of Chemical Biology, Vanderbilt University, Nashville, TN 37235, USA
2. Department of Medicinal Chemistry, University of Washington, Seattle, WA 98195, USA; libinxu@uw.edu
3. Department of Chemistry and Biomolecular Sciences, University of Ottawa, 10 Marie Curie Pvt., Ottawa, ON K1N 6N5, Ontario, Canada; dpratt@uottawa.ca
* Correspondence: n.porter@vanderbilt.edu
† Dedicated to Prof. Bernd Giese in Recognition of His Contributions to the Understanding of Selectivity and Reactivity in Free Radical Chemistry, on the Occasion of His 80th Birthday.

Received: 6 April 2020; Accepted: 30 April 2020; Published: 6 May 2020

Abstract: Radical-mediated lipid oxidation and the formation of lipid hydroperoxides has been a focal point in the investigation of a number of human pathologies. Lipid peroxidation has long been linked to the inflammatory response and more recently, has been identified as the central tenet of the oxidative cell death mechanism known as ferroptosis. The formation of lipid electrophile-protein adducts has been associated with many of the disorders that involve perturbations of the cellular redox status, but the identities of adducted proteins and the effects of adduction on protein function are mostly unknown. Both cholesterol and 7-dehydrocholesterol (7-DHC), which is the immediate biosynthetic precursor to cholesterol, are oxidizable by species such as ozone and oxygen-centered free radicals. Product mixtures from radical chain processes are particularly complex, with recent studies having expanded the sets of electrophilic compounds formed. Here, we describe recent developments related to the formation of sterol-derived electrophiles and the adduction of these electrophiles to proteins. A framework for understanding sterol peroxidation mechanisms, which has significantly advanced in recent years, as well as the methods for the study of sterol electrophile-protein adduction, are presented in this review.

Keywords: peroxidation; free radical; sterol; cholesterol; lipid electrophiles

1. Introduction

Unsaturated lipids are prone to undergoing reactions with oxidizing species, such as oxygen-centered free radicals, singlet molecular oxygen, and ozone. The high abundance of these vulnerable lipids in humans is associated with pathologies that result from exposure to such reactive oxidants. Ozone, for example, is the most widespread air pollutant found in the U.S. and contributes to a growing variety of health problems, all of which potentially increase the risk of premature death [1,2]. Reactive oxygen species (ROS), such as alkoxyl and peroxyl free radicals, are generally linked to oxidative stress, which has been associated with many human disorders. Indeed, ROS and free radical lipid peroxidation has been invoked as a cause or consequence in diseases such as asthma [3], cardiovascular disease [4–6], diabetes [5,7,8], Alzheimer's [9–11], Parkinson's [12,13], cancer [14,15], and macular degeneration [16]. More recently, a form of regulated necrosis associated with the accumulation of lipid hydroperoxides has been characterized, which may link specific oxidative events and tissue dysfunction in these pathological contexts. This process, coined ferroptosis, which may also serve as a vulnerability that may be exploited for cancer treatment, implicates a labile iron pool and lipid hydroperoxides as causal agents and a glutathione-dependent enzyme (GPX4), radical trapping antioxidants (RTAs), and iron chelating compounds as protective agents [17–19].

Reactive electrophiles have been suggested to be products of lipid peroxidation since the 1940s, when thiobarbituric acid (TBA) was shown to give a characteristic red-orange color with animal tissue that had been exposed to air [20]. The colored lipid-derived species was subsequently identified as a 2:1 complex of TBA with malondialdehyde (MDA) [21,22]—an electrophilic byproduct of the free radical peroxidation of polyunsaturated fatty acids (PUFAs) and esters with three or more double bonds. While a number of methods have been developed over the decades to assay the overall levels of peroxidation and identify specific products of lipid oxidation, the TBARS assay (TBA 'reactive species') is still used to give a semi-quantitative measure of MDA levels.

At the time that reactive lipid-derived electrophiles like MDA were being associated with lipid oxidation reactions in animal tissues, the chemistry of free radical chain reactions responsible for their formation was also drawing significant attention. While early work on the mechanism of autoxidation was principally centered on the degradation of commercially important hydrocarbons, lipid peroxidation drew increased interest during the latter half of the 20th century, with efforts to describe mechanisms for the autoxidation of polyunsaturated fatty acids (PUFAs) [23–30] and sterols [27,31–42] attracting interest that has continued to this day.

The chemistry and biology of PUFA-derived electrophiles has drawn attention since MDA was shown to be present in cells, fluids, and tissues under conditions of oxidative stress. Mechanisms for its formation were proposed [43,44], and routes were suggested for the formation of other fatty acid-derived electrophiles, such as the cytotoxins 4-hydroxy-2-nonenal (4-HNE) and 5-oxo-2-nonenal (4-ONE) [45–49]. Another set of reactive electrophiles with a core 4-oxo-pentanal structure, the isolevuglandins (IsoLGs), were also found in cells, tissues, and fluids undergoing oxidative stress. Mechanisms for the formation of IsoLGs centered on the decomposition of endoperoxide intermediates formed in the peroxidation of arachidonate esters [50–53]. In fact, the same unstable endoperoxide intermediate serves as a precursor to both MDA and the IsoLGs.

In the decades following the identification of MDA as a byproduct of lipid peroxidation, the formation and repair of DNA adducts caused by MDA was the subject of extensive investigation [14]. Adducts of 4-HNE and 4-ONE with DNA have also drawn interest, leading to suggestions that small-molecule electrophile adduction to nucleic acids may be a major cause of cancers linked to lifestyle and dietary factors [15].

Reactions between PUFA-derived electrophiles and protein nucleophiles have also been studied in detail. MDA-promoted crosslinking of proteins has been linked to the formation of the fluorescent age-related pigment lipofuscin [54], and the protein adduction of 4-HNE, 4-ONE, and IsoLGs has been the subject of extensive investigation and several excellent reviews [55–57]. MDA and IsoLG adduction to proteins generally involves initial reversible imine formation, which is generally followed by molecular rearrangement to give stable end-products. The α,β-unsaturated aldehydes (4-HNE and 4-ONE) usually undergo Michael addition with cysteines, histidines, and lysines.

Free radical-, singlet oxygen-, and ozone-promoted cholesterol oxidation also leads to several reactive compounds [31–38], but the reactions of these sterol-derived electrophiles with biological nucleophiles has drawn relatively little attention compared to the many studies on PUFA-derived compounds such as MDA and 4-HNE. Cholesterol levels are particularly high in the brain and central nervous system and neurons are among the most vulnerable cells to reactive oxygen species (ROS), due to their elevated metabolic activity, highly unsaturated neuronal lipid composition (such as docosahexaenoic acid, arachidonic acid, and cholesterol), high level of transition metals, and modest antioxidant defense systems [58,59]. This background, along with the suggestion that reactive oxysterols may promote protein aggregation [10,11,60], has led to an increase in interest in the chemistry and biology of cholesterol-derived electrophiles.

Recent discoveries have stimulated interest in post-lanosterol sterols other than cholesterol. One of the immediate biosynthetic precursors to cholesterol, 7-dehydrocholesterol (7-DHC, see Figure 1), is highly vulnerable to free radical chain oxidation [61] and reactive electrophiles are major products of its peroxidation [62,63]. Elevated levels of 7-DHC are found in tissues and fluids of patients with the genetic disorder Smith-Lemli-Opitz syndrome (SLOS) [64], which is caused by mutations in the gene encoding 7-dehydrocholesterol reductase (DHCR7)—the enzyme that converts 7-DHC to cholesterol [65]. The high levels of 7-DHC found in SLOS tissues, the proclivity of this sterol to participate in radical chain oxidation reactions, and the formation of highly reactive sterol electrophiles in the process have led to recent suggestions that SLOS is a disorder driven by lipid peroxidation [63,66–69].

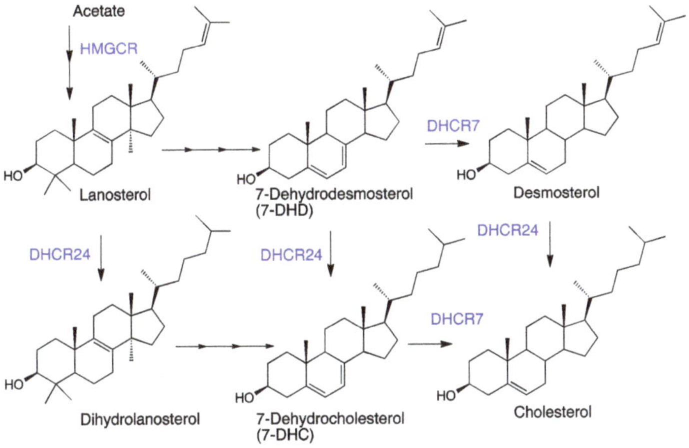

Figure 1. Selected cholesterol biosynthetic precursors and enzymes (in blue) that promote the transformations shown.

The increased interest in sterol-derived electrophiles, their reactions with biologically important nucleophiles, and the link between these reactive species and human disorders, including neurodegenerative disease, were the principal stimuli for undertaking this review. In this work, we describe recent advances in the chemistry leading to sterol-derived electrophiles and the reaction of these species with proteins, peptides, and amino acids. An introduction to the chemical tools developed to date to study electrophile-protein adduction and a discussion of what insights these have provided are a part of this contribution, as are the potential biological consequences of the formation and transformations of these species.

2. Primary Reactions

The oxidation of cholesterol by singlet molecular oxygen, ozone, and peroxyl free radicals has been particularly well-studied. While hydroxyl radicals will also undoubtedly react with cholesterol, these species will react with virtually any C-H bond encountered, eliminating these reactions as likely sources of oxysterols in a complex cellular environment. Isolated olefins like the $\Delta^{5,6}$ double bond in cholesterol will undergo "ene"-type reactions with singlet oxygen to primarily give chol 5α-OOH, and to a lesser extent, the chol 6-OOHs, and conjugated dienes like those present in 7-DHC give cyclic peroxides by Diels-Alder type transformations (see Figure 2A). Ozone's primary reaction with isolated double bonds is a [3+2] cycloaddition, producing an unstable adduct (ozonide). Subsequent decomposition of this (primary) ozonide eventually results in carbonyls as the major products, although, under some conditions, α-substituted hydroperoxides are formed, as are epoxides (Figure 2B). The link between primary ozone exposure and deleterious health consequences has been a topic of continued interest.

When inhaled, ozone reacts with cholesterol in airway epithelial cells by mechanisms such as those presented in Figure 2B [70–74].

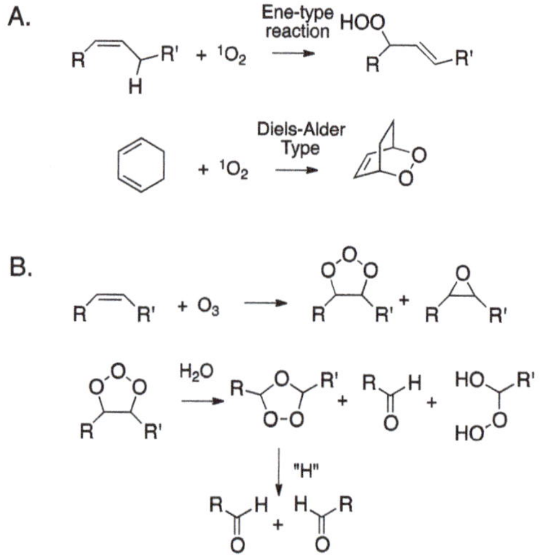

Figure 2. Common reactions of olefins with (**A**) singlet oxygen and (**B**) ozone.

Free radical peroxidation (autoxidation) is a chain reaction mediated by peroxyl free radicals. The two primary propagation steps in the process are rate-limiting hydrogen atom transfer (HAT) from an organic substrate to a peroxyl radical and the near diffusion controlled addition of oxygen to an intermediate carbon radical (see Figure 3) [75]. The events that initiate the chain sequence have been of interest for as long as the process has been studied with a general understanding that the peroxide products of the reaction can themselves serve as initiators of the process. As a consequence, a single radical-generating event can lead to the cascade formation of peroxide initiators and a dramatic increase in the rate of "auto" oxidation.

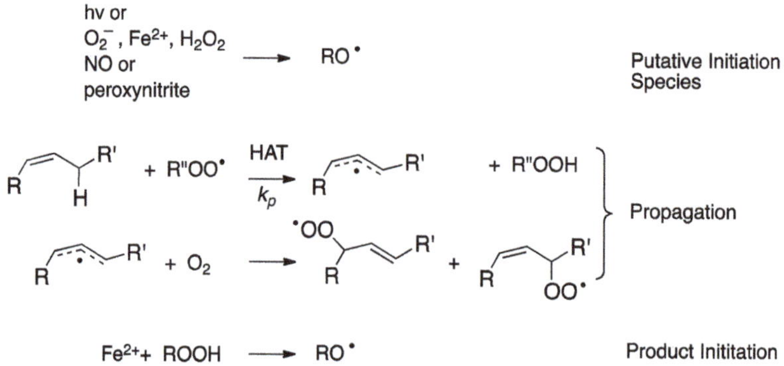

Figure 3. Mechanisms of initiation and propagation for free radical chain oxidation (peroxidation or autoxidation). Propagation steps are illustrated with an isolated alkene reactive substructure.

The nature of important radical-generating events in biology has been the topic of some debate, with McCord's suggestion of a metal-mediated Haber–Weiss reaction [76] and the iron-promoted decomposition of lipid hydroperoxides [77–81] having been suggested some four decades ago. The latter has gained prominence with the recent characterization of ferroptosis as the oxidative cell death modality associated with the accumulation of lipid peroxidation products [17–19].

3. Cholesterol Autoxidation

As a monounsaturated lipid, cholesterol may be expected to autoxidize by a mechanism analogous to that of oleic acid—the prototypical monounsaturated fatty acid. Oleic acid autoxidizes via an analogous mechanism to more highly unsaturated fatty acids (vide infra, Figure 4). Specifically, H-atom abstraction from one of the two allylic positions (C8 or C11) leads to two isomeric allylic radicals to which O_2 can add, and the resultant peroxyl radicals propagate the chain reaction to yield a mixture of regio- and stereoisomeric hydroperoxides [82]. The autoxidation chemistry of cholesterol differs from that of oleic acid because the reactivity of the two allylic positions is not identical. The proximity of the electronegative 3-OH substituent and the integration of the allylic positions in two distinct carbocyclic rings have a substantial impact on both the rates of H-atom abstraction and the stability of the peroxyl radicals that are formed upon the addition of O_2 to the resultant allylic radicals. In oleic acid, the electronegative carboxylic acid moiety is seven carbon atoms away and both allylic positions are part of acyclic aliphatic carbon chains, such that they are essentially identical in reactivity.

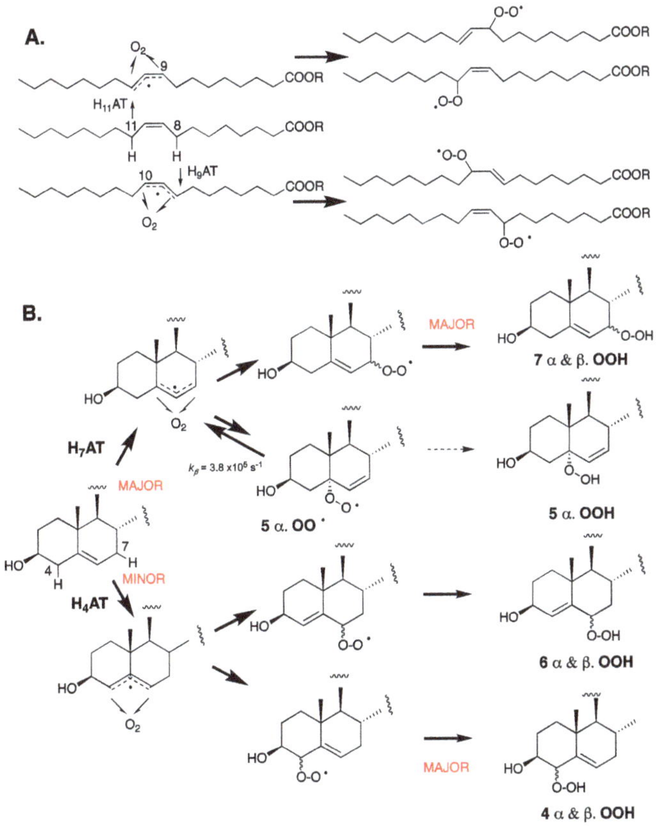

Figure 4. Propagation mechanisms for the autoxidation of (**A**) oleate, where H-atoms at C8 and C11 have similar reactivities toward hydrogen atom transfer (HAT), and (**B**) cholesterol, where H-atoms at C7 are significantly more reactive than C4 hydrogens.

The C-H bond dissociation enthalpies (BDEs) of the two allylic positions at C4 and C7 have been computed to be 89.0 and 83.2 kcal/mol, respectively, whereas the allylic positions in oleate have been computed to have a C-H BDE value of 83.4 kcal/mol [41]. The stronger bond at C4 results from (1) the electron-withdrawing effect of the neighboring C3-OH, (2) the lower substitution of the terminal ends of the allylic radical where the unpaired electron spin density is localized, and (3) the associated planarization of both the A and B rings of the steroid required to maximize radical delocalization. As a result, H-atom abstraction from C7 is favored by a significant margin. The resultant allylic radical can, in principle, be oxygenated at C5 and C7 to yield peroxyl radicals that can propagate the chain reaction, However, since the C5-peroxyl radical is relatively unstable, it generally undergoes rapid β-fragmentation instead of propagation, leaving the C7-peroxyl-derived product, C7-OOH, as the major autoxidation product. C4-OOH and C6-OOH resulting from initial H-atom abstraction from C4 are also formed in measurable amounts because H-atom abstraction from C4 is faster than predicted by its relatively strong bond due to favorable interactions in the H-atom transfer transition state [41].

Radical trapping antioxidants (RTAs) can have a profound effect on the profile of peroxide products formed in the course of inhibited free radical lipid oxidations [83–86]. The autoxidation of cholesterol in solution and in the presence of high concentrations of a very good H-atom donor, such as pentamethylchromanol (PMC), which is a truncated version of α-tocopherol, yields a significant amount of C5α-OOH [39,41]. PMC traps peroxyl radicals in solution with a rate constant in excess of 10^6 M^{-1} s^{-1}, making hydroperoxide formation from C5α-OO$^\bullet$ competitive with β-fragmentation—a unimolecular process that occurs with a rate constant of $k_\beta = 3.8 \times 10^5$ s^{-1}. No C5β-OOH is observed, presumably due to a much larger k_β of the precursor peroxyl radical arising from repulsion from the C9β-methyl substituent.

The autoxidation of cholesterol incorporated into phosphatidylcholine liposomes gives the same set of products that are found in solution, but the dynamics of reactions in membrane-like vesicles have a significant effect on the product profiles observed when compared to the product distributions found in isotropic solutions. One notable feature of cholesterol autoxidation in phosphatidylcholine liposome is that no evidence of C5α-OOH formation was found when α-tocopherol or PMC was incorporated into the vesicles [40]. This can be understood when consideration is given to the fact that α-tocopherol is a much poorer H-atom donor in phospholipid bilayers ($k_{inh}= 4.7 \times 10^3$ M^{-1} s^{-1}) compared to organic solutions ($k_{inh}= 3.6 \times 10^6$ M^{-1} s^{-1}) due to phenolic hydrogen bonding with the polar headgroup of the phospholipid (see Figure 5A) [87]. Phenolic RTAs have thus been proven to be much less effective at trapping short-lived peroxyl radicals in membrane-like vesicles than in isotropic media, which is a conclusion that may have consequences for the ultimate success of phenolic antioxidant therapies. In contrast to the results obtained with α-tocopherol, the use of aromatic amine RTAs, such as phenoxazine [87], in the oxidation of cholesterol in liposomes leads to the formation of C5α-OOH (see Figure 5B). Phenoxazine has an apparent rate constant for reactions with peroxyl radicals in liposomes ($k_{inh}= 2.4 \times 10^5$ M^{-1} s^{-1}) that is some 50 times greater than α-tocopherol, making H-atom transfer from the amine competitive with β-fragmentation of the C5α peroxyl radical [40].

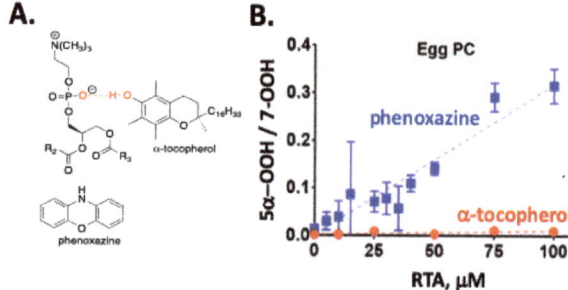

Figure 5. (**A**) Hydrogen bonding in membrane bilayers reduces the effect of phenolic antioxidants like α-tocopherol. (**B**) Phenoxazine scavenges 5α-OOH in bilayers, while α-tocopherol does not.

In competition with H-abstraction from the allylic positions in cholesterol, peroxyl radicals add to the C5-C6 double bond. Addition yields a short-lived alkyl radical that can undergo an intramolecular homolytic substitution ($S_H{}^i$) reaction on the peroxide moiety to yield an epoxide or undergo O_2 addition to produce a different chain-propagating peroxyl radical. Epoxides make up a significant fraction of the product mixture formed in autoxidation, with the α:β epoxide ratio of ~3:1 having been found. Epoxide formation can be eliminated by antioxidants, since the process involves an intermolecular addition of the peroxyl radical to cholesterol. Antioxidants trap peroxyl radicals, completely suppressing the radical addition pathway, while the formation of product hydroperoxides is reduced, but not eliminated. Antioxidant-mediated peroxidation has been suggested to account for the differential effect of antioxidants on epoxide and hydroperoxide products [41]. The tocopheryl radical, for example, can mediate peroxidation by H-atom abstraction, but it cannot facilitate reactions leading to epoxides.

4. Cholesterol-Derived Electrophiles

Although epoxides are nominally electrophilic, the cholesterol 5,6-epoxide is too highly substituted for efficient nucleophilic substitution. C5 is fully substituted and thus unreactive to bimolecular nucleophilic substitution (S_N2) and, although C6 is a tertiary center, the adjacent fully substituted C5 and the C19β-methyl substituent hinder the addition of nucleophiles to the α-epoxide, while the adjacent axial C-H bond at C9 hinders the addition to the β-epoxide [88]. The epoxides are, however, substrates for hydrolase-catalyzed ring-opening hydration [89–93], as well as non-enzymatic reactions with thiol and amine nucleophiles [89,94–101]. Dendrogenin A, the product of the enzymically-promoted ring-opening reaction of the α-epoxide with histamine shown in Figure 6, was recently discovered in mammalian tissues [94–101]. This sterol alkaloid has a specific potency with regards to inducing cell differentiation at low doses, suggesting its possible existence as a cholesterol metabolite.

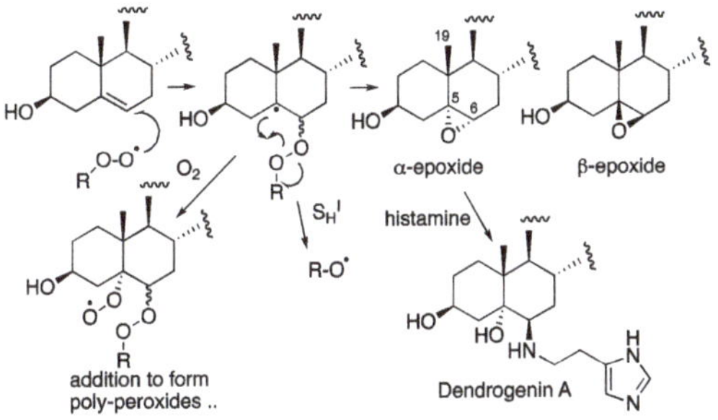

Figure 6. Epoxide formation and ring opening reactions. Intermolecular addition of a peroxyl radical followed by intramolecular homolytic substitution ($S_H{}^i$) attack of the intermediate carbon radical on the peroxide gives an alkoxyl and the α- and β-epoxides. Oxygen addition can compete with $S_H{}^i$. Dendrogenin A is a product of histamine and the α-epoxide.

Dehydration or further oxidation of the cholesterol α- and β-stereoisomers of C4-OOH, C6-OOH, and C7-OOH yields the corresponding ketones. Although these are α,β-unsaturated carbonyls—the quintessential motif of Michael acceptors—studies with the most abundant of these, 7-ketocholesterol, have indicated that it is not particularly electrophilic [102]. It remains unclear whether sufficient levels of 4-ketocholesterol or 6-ketocholesterol are formed in vivo to be relevant. However, should they be, they are expected to be more reactive than 7-ketocholesterol, because the β-carbons are monosubstituted and less sterically hindered than the disubstituted and sterically-hindered β-carbon in 7-ketocholesterol.

The fragmentation of cholesterol hydroperoxides leads to highly electrophilic aldehydic species. Both C5α-OOH and C6β-OOH readily undergo acid-catalyzed (Hock) fragmentation [103–105] to give the B-ring cleavage product secosterol A, and its aldolized product, secosterol B (see Figure 6) [39–41,106]. These derivatives are precisely the same compounds as those formed from the reaction of cholesterol with ozone [70,71,74,107,108]; their character as reactive electrophiles will be subsequently discussed in this review. The C5α-OOH and C6β-OOH compounds are minor products in the free radical oxidation of cholesterol compared to products formed with hydroperoxide functionality at C7, so the propensity for Hock fragmentation and the products derived from the C7 hydroperoxides are of interest.

Recent studies showed that C7α-OOH undergoes Hock fragmentation readily, while C7β-OOH is unreactive [40]. This rearrangement does not follow the typical Hock mechanism shown in Figure 7A; instead, an intermediate epoxy carbocation is formed, followed by water entrapment or fragmentation, to give an allylic epoxide that should be highly electrophilic. Indeed, under the conditions of Hock treatment and workup in the presence of ethanol, epoxide hydrolysis and/or ethanolysis products are observed, as shown in the figure. Presumably, amine and thiol nucleophiles would add to either the C4 or C6 centers, opening the possibility of protein adduction from this reactive species. The C4-OOH compound is a minor product of free radical oxidation and the identities of the products formed from this precursor have yet to be reported. Potential pathways for the reaction of C4-OOH under Hock-like conditions are suggested in Figure 6. The Grob-like pathway outlined in the figure would provide an electrophilic di-carbonyl compound similar to secosterol A that would adduct to proteins in either aldol form.

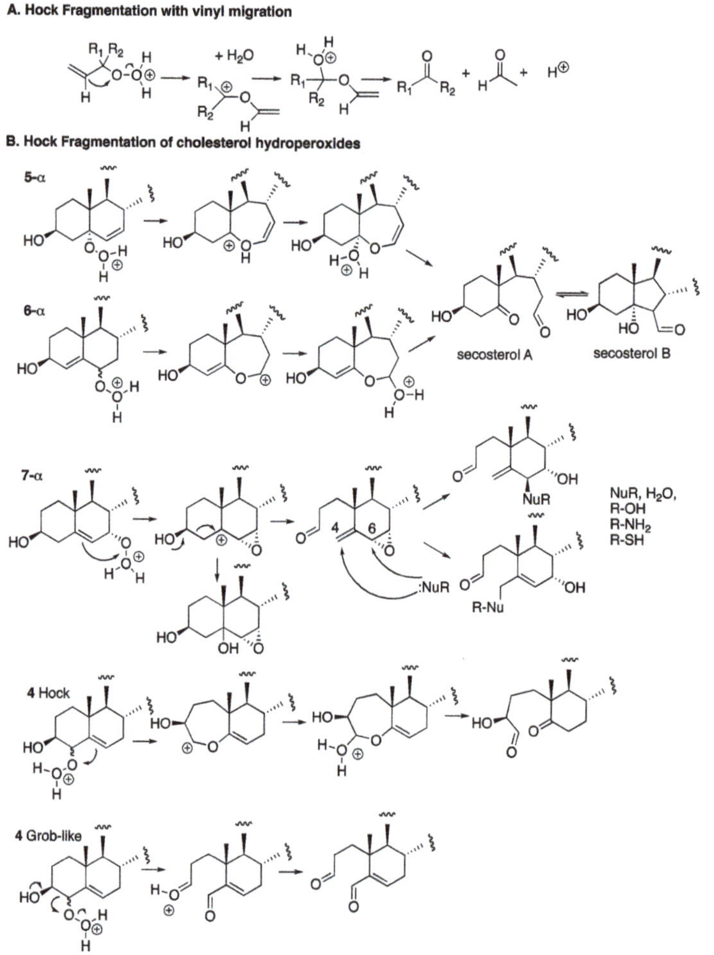

Figure 7. Hock fragmentation mechanisms. (**A**) General mechanism for the acid-promoted fragmentation of allylic hydroperoxides. (**B**) Acid-catalyzed fragmentation of cholesterol hydroperoxides.

It is noteworthy that glutathione peroxidase-4 (GPX4), which is primarily responsible for the detoxification of lipid hydroperoxides [109], reacts with the different regio- and stereoisomers of cholesterol hydroperoxides at different rates (i.e., 5α-OOH < 6αR-OOH $\approx$ $7\alpha/\beta 7$-OOH < 6β-OOH) [110]. As such, although 5α-OOH may form at a slower rate than the other regio/stereoisomers, it may accumulate due to its slower reduction by GPX4, and fragment to afford secosterol A and B. To the best of our knowledge, the relative reactivity of C4-OOHs as substrates for GPX4 has not yet been determined.

5. Autoxidation of 7- and 8-Dehydrocholesterols

7-DHC, the immediate biosynthetic precursor to cholesterol, is usually found in human tissues and fluids at very low levels compared to cholesterol. The isomer 8-DHC, which has a homo-conjugated 5,8-diene [111], is found at comparably low concentrations in tissues and fluids, with an isomerase (EBP) able to interconvert them. 7-DHC occupies a branchpoint in isoprenoid biosynthesis between cholesterol and vitamin D_3 [112]—the B-ring diene undergoing a photochemically-promoted ring opening on the pathway to the vitamin (see Figure 8) [113,114].

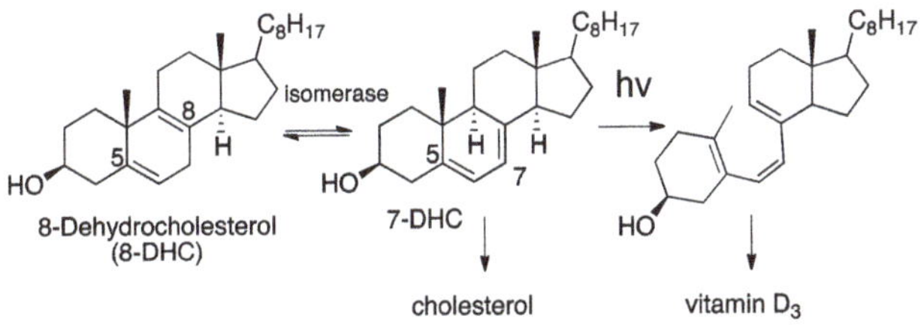

Figure 8. 7-dehydrocholesterol (7-DHC) is a biosynthetic branchpoint between cholesterol and vitamin D_3; 7- and 8-DHC are equilibrated by an isomerase enzyme.

The autoxidation of 7- and 8-DHC has drawn increased attention in recent years [61,62], but the susceptibility to the oxidation of ergosterol, which is a 5,7-diene analog of 7-DHC found in fungi and protozoa, was noted over a century ago [115–117]. The 1933 publication of Meyer on ergosterol is particularly noteworthy [117]. Three mechanisms for oxidation were suggested, with two involving photolysis in the presence or absence of dyes that likely involve the intermediacy of singlet molecular oxygen. A third mechanism that has all of the characteristics of a free radical chain reaction consuming over 2 moles of oxygen per ergosterol, is promoted by heme in the dark and is diminished if ergosterol is carefully purified, thus removing peroxides that could initiate autoxidation.

Nearly 50 years after Meyer, the enhanced susceptibility of 7-DHC to oxidation was noted in studies of its free radical co-oxidation in liposomes made up of unsaturated linoleate phospholipids [118]. Linoleate gives *trans-cis* and *trans-trans* conjugated dienes under conditions of free radical chain oxidation and the ratio of these products was found to reflect the H-donor character of the medium undergoing oxidation (see Figure 9) [119]. The oxidation of liposomes of phospholipids bearing a linoleate and a palmitate on the glyceryl headgroup, for example, gave linoleate conjugated diene products at a *trans-cis/trans-trans* ratio of 0.69, but if the liposomes were made up as a mixture of 0.7 moles of the linoleate phospholipid to 0.3 moles of 7-DHC, the *trans-cis/trans-trans* ratio determined was nearly 3.0. In the presence of good H-atom donors, more *trans-cis* products were formed, and in the absence of H-atom donors, more *trans-trans* products were found.

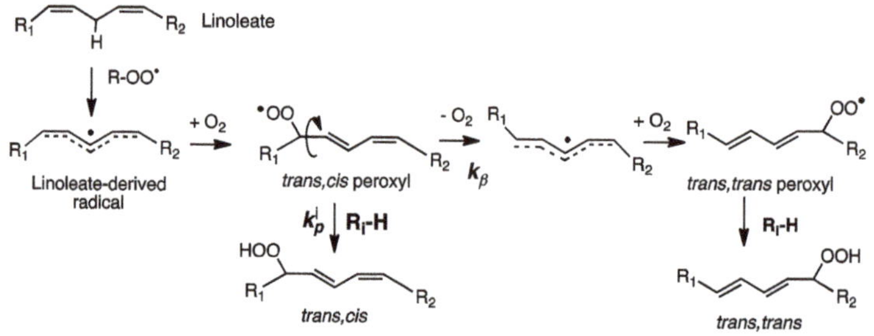

Figure 9. Linoleate oxidation mechanism and basis of a free radical clock for determining the propagation rate constant k_p for the autoxidation of any R_i-H.

The linoleate mechanism describes a unimolecular process—the loss of oxygen from the *trans-cis* peroxyl (k_β) in competition with bimolecular H-atom transfer from R_i-H to a peroxyl radical that occurs with rate constant k_p. This kinetic competition is the basis for a radical clock approach to determine propagation rate constants for the autoxidation of a number of small molecules, fatty esters, and sterols [61,83,120]. The rate constant originally determined in benzene at 37 °C for cholesterol by the clock method was 11 M^{-1} s^{-1}, and a more recent value of 8.4 M^{-1} s^{-1} confirms that cholesterol is roughly 10-times more reactive than oleic acid [40]. These rate constants represent the sum of the values for H-atom transfer from C-7 and C-4 of the molecule, although, since C7-derived products dominate HAT, from C7, it is presumably much faster than from C4, as predicted by computations. For comparison, the k_p of 7-DHC was found to be ca. 2260 M^{-1} s^{-1}, some 200 times the cholesterol value. Indeed, 7-DHC has the largest rate constant for free radical propagation found to date for a lipid molecule. The values of k_p determined for other lipids of interest were 197 M^{-1} s^{-1} for arachidonate and 960 M^{-1} s^{-1} for 8-DHC [121]. The kinetics of the autoxidation of linoleate phospholipid liposomes has also been examined and the relative reactivity of lipids studied in isotropic media is mirrored in liposomal bilayers, with 7-DHC having the greatest effect on the linoleate hydroperoxide *trans-cis/trans-trans* product ratio of any oxidizable co-substrate [61].

Studies on the products formed from the autoxidation of 7-DHC suggest that it is reactive as both an H-atom donor and a peroxyl radical addition acceptor [29,61,67]. Based on product and mechanistic studies, the high reactivity of 7-DHC was rationalized by the planarity of the conjugated system, the perfectly aligned allylic C-H bonds at C9 and C14 for hydrogen abstraction, the highly substituted pentadienyl radical after H-atom removal, and the formation of a stabilized allylic radical after peroxyl radical addition. Therefore, both H-atom transfer (loss of the H-atom at C9 or C14) and peroxyl radical addition (to the conjugated diene) contributes to the free radical oxidation of 7-DHC (Figure 10). The loss of H-9 from 7-DHC (or H-7 from 8-DHC) leads to a pentadienyl radical in ring-B, which then undergoes oxygen addition [122] and a series of intramolecular radical rearrangements to give a number of oxysterol products, including compounds **1** (5α,6α-epoxycholest-7-en-3β,9α- diol or 9-OH-7DHCep), **2a** (5,9-endoperoxy-cholest-7-en-3β,6α(β)-diol or EPCD-a), **2b** (EPCD-b), and **3** (5α,9α-Epidioxy-8α,14α-epoxycholesta-6-en-3β-ol or EnPep). The loss of H-14 leads to a pentadienyl radical across rings B and C and eventually to compounds **3** and **4**, as major products. On the other hand, peroxyl radical addition to 7-DHC results in the formation of 7-DHC 5α,6α-epoxide (7DHCep).

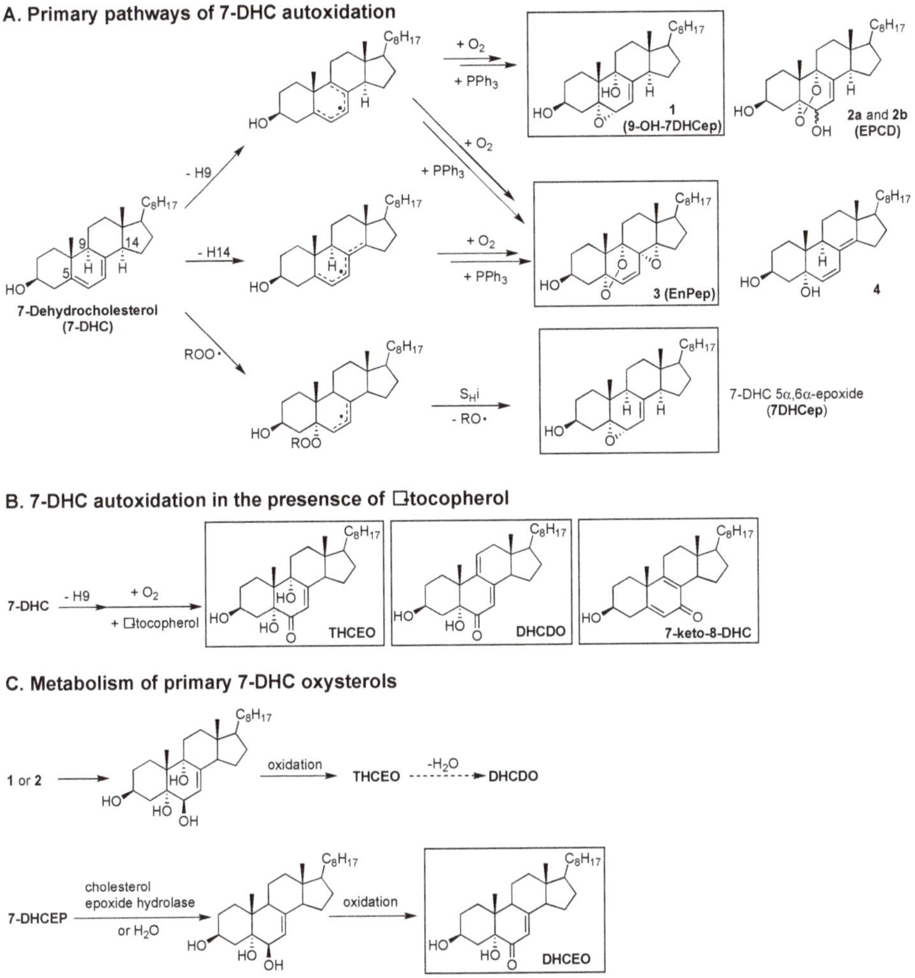

Figure 10. Mechanisms of the formation of 7-DHC-derived oxysterols in solution (**A,B**) and in biological systems (**C**). (**A**) Primary pathways of 7-DHC autoxidation in organic solution at 37 °C; (**B**) 7-DHC autoxidation in the presence of α-tocopherol (only loss of the H-9 pathway is shown); (**C**) metabolism of primary 7-DHC oxysterols in cells. Compounds highlighted in the rectangle are potential electrophiles.

As was the case for cholesterol, when the oxidation of 7-DHC was carried out in the presence of tocopherol, the peroxyl radical addition pathway was completely suppressed while other oxidation products were still observed, most likely by tocopheryl-mediated H-atom transfer. The abstraction of H-9 leads to the formation of a number of oxysterols containing the enone moiety, including 3β,5α,9α-trihydroxycholest-7-en-6-one (THCEO), 3β,5α-dihydroxycholesta-7,9(11)-dien-6-one (DHCDO), and 7-keto-8-dehydrocholesterol (7-keto-8-DHC), and the H-14 pathway gives peroxyl radicals that lead to simple dienol products (not shown) [121]. It should be noted that compounds 1–4 have not been observed in cell and animal models of SLOS, but metabolites of **1**, **2a**, and **2b**, including THCEO and DHCDO, have been observed in vivo along with 7-keto-8-DHC. Furthermore, a metabolite of 7DHCep, 3β,5α-dihydroxycholest-7-en-6-one(DHCEO), has been observed at high levels in cell and animal models of SLOS, particularly in fibroblasts and mice brains [68,123].

The observation of DHCEO is interesting because it suggests that the level and distribution of α-tocopherol in vivo is not sufficient to completely suppress the peroxyl radical addition pathway.

6. 7- and 8-Dehydrocholesterol-Derived Electrophiles

The oxysterols **1** and 7DHCep derived from 7- and 8-DHC oxidation are excellent electrophiles based on the presence of their substructure allylic epoxide moiety, while DHCEO, THCEO, and DHCDO may also be electrophiles due to their α,β-unsaturated enone. Indeed, evidence will be presented in a subsequent section of this review that 7DHCep or 7-DHC readily adduct proteins. Compound **1** appears to be as reactive as 7DHCep toward nucleophilic adduction, while the enone-containing oxysterols, DHCEO and THCEO, are hindered toward Michael addition due to the γ-alkyl substituents and the axial C-H or C-OH bonds at C9 and C14, while such addition for DHCDO is less hindered and a likely Michael acceptor. Although all of these enone moieties can potentially form imine adducts with protein lysine residues, this reactivity typically requires acid catalysis and is reversible.

7-DHC-derived oxysterols exert varied cytotoxicity in vitro that is dependent on their specific structures. Specifically, oxysterols EPCD-a and -b are toxic in Neuro2a cells and retina-derived cell lines, with EnPep and DHCEO showing a lower toxicity in these cells [124,125]. Among all of the cells tested, primary neurons appear to be the most susceptible to oxysterol exposure, with an IC50 of approximately 0.75 µM for DHCEO—the most abundant autoxidation-derived oxysterol observed in the brain of SLOS rodent models. Interestingly, at a physiological concentration (5 µM) [63], DHCEO accelerates the formation of neuronal processes from primary neurons, such as dendrites [68]. On the other hand, the electrophilic 9-OH-7DHCep (compound **1**) did not display any toxicity to Neuro2a cells, indicating that electrophilicity may not be the most important determinant of cytotoxicity [124]. Cytotoxicity of the precursor of DHCEO, 7DHCep, has not been examined, but a large number of protein adducts with this oxysterol have been reported in *Dhcr7*-knock down Neuro2a cells [69].

It is interesting to note that the concentrations of DHCEO in the brain of SLOS rodent models are much higher than those in the matching liver and retina. For example, in AY9944-treated rats, even after normalization by the levels of 7-DHC (DHCEO/7-DHC), the amount of DHCEO observed in the brain is 2.7-fold of that in the liver and 11-fold of that in the retina. We speculate that several factors could account for such tissue-specific variation. The metabolism of 7DHCep may be tissue-specific, as the ring-opening of the epoxide is likely an enzymatic process based on the precedent of the soluble cholesterol 5,6-epoxide hydrolase (ChEH) [126] More recently, ChEH was identified as a hetero-oligomer of the cholesterol biosynthesis enzymes DHCR7 and 3β-hydroxysterol-Δ^8-Δ^7- isomerase (EBP), and the same hetero-oligomer also serves as the antiestrogen binding site (AEBS) [89]. No study has been carried out to determine if 7-DHCep can be a substrate of ChEH, but an earlier report concluded that 7-DHC 5β,6β-epoxide can serve as a mechanism-based inhibitor of rat microsomal ChEH via covalent modification of the active site by the reaction intermediate [91]. 7DHCep (epoxide on the α-face of the sterol ring) was not tested in that study, but the stereochemistry of the C5-hydroxyl group of the cationic intermediate resulting from α- or β-epoxide could potentially lead to different fates of this intermediate, i.e., covalent modification of the enzyme active site vs. nucleophilic attack by water to give the diol on C5 and C6.

It should also be noted that the amount of DHCEO is likely dependent on the extent of reactions between 7DHCep and nucleophiles in a particular tissue. Therefore, nucleophiles could be protein residues or glutathione (GSH), which is an abundant nucleophile in some tissues. The more adduction of 7DHCep that occurs in any organ, the less DHCEO that is found in that tissue. GSH conjugation with electrophiles is normally catalyzed by glutathione S-transferases (GST) [127]. Indeed, a rat microsomal GST (isoform B, which is equivalent to GST A1 in current nomenclature) has been found to catalyze the conjugation between cholesterol 5α,6α-epoxide and GSH [93,128].

It is known that the human liver contains much higher (2.6-fold) levels of various isoforms of GSTs (with A1 as the predominant isoform), than those in the human brain (with P1 being the major isoform, followed by M3 and M2) [127], which indicates that 7DHCep could be more readily detoxified in the liver than in the brain. The levels of GSTs in the human retina have not been reported, but the GST isoform M1 is expressed at the highest level in photoreceptor cells in the rat retina, where most of the degeneration occurs in the AY9944-rat model, followed by isoforms A4 and P1. Therefore, the variation in the level of DHCEO could arise from the different expression patterns and levels of GST isoforms in each tissue.

Although the level of DHCEO in the retina is low and other 7-DHC autoxidation-derived oxysterols were not observed, retinal degeneration and increased lipid peroxidation are hallmarks of the AY9944-rat model of SLOS, suggesting that the protein adduction of electrophilic oxysterols could be a significant factor contributing to the retinal pathophysiology. Indeed, in a recent pre-clinical therapeutic study using the AY9944-rat model, a combination of cholesterol and antioxidant (vitamin E, vitamin C, and selenite) completely prevented retinal degeneration in this model, while cholesterol supplementation alone only partially prevented this phenotype [129]. An analysis of protein-oxysterol adducts with and without antioxidant treatment has not been accomplished due to current limitations in antibody availability and in vivo pull-down methodology, but such a study would presumably reveal whether protein adduction is indeed underlying the retinal degeneration pathobiology. On the other hand, protein adducts with lipid electrophiles, particularly 4-HNE, have been found to be significantly (9-fold) higher in the retinas of AY9944-treated rats than in matching controls [130], supporting the general elevation of lipid peroxidation and protein-lipid electrophile adducts in this SLOS model.

7. Protein Adduction of Lipid-Derived Electrophiles

7.1. PUFA-Derived Protein Adducts

The formation of protein adducts with fatty acid-derived electrophiles has been the subject of extensive investigations, with 4-HNE (Figure 11) [48,131]—the electrophile generated from the peroxidation of ω-6 fatty acids [132–135]—being a principal focus of interest [49,132–138]. Michael addition of protein cysteines, lysines, or histidines is the most common means of protein covalent attachment to 4-HNE. Lysine also undergoes reversible imine formation that may lead to cyclodehydration with the irreversible formation of a pyrrole protein adduct. Strategies have been developed to isolate and identify protein adducts from PUFA-derived electrophiles [55–57,139–145] and excellent reviews of these topics have been published [57,146,147].

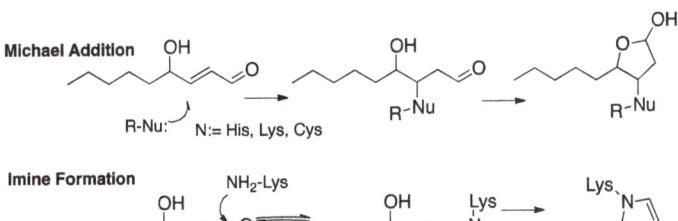

Figure 11. Mechanisms of 4-hydroxy-2-nonenal (4-HNE) protein adduction. Michael-lactol and imine-pyrrole formation.

7.2. Cholesterol-Derived Protein Adducts

In contrast to the extensive effort to characterize protein modification by 4-HNE and other PUFA-derived species, protein adduction by sterol-derived electrophiles has received much less attention. The reports of Wentworth, Kelly, and collaborators [11,60,148–151] suggested that secosterol electrophiles are present in human atherosclerotic and neurodegenerative tissue, stimulating interest in the field, and subsequent studies showed that protein misfolding is a consequence of protein adduction.

The structural elucidation of protein adducts formed from cholesterol-derived electrophiles has been a topic of extensive research in recent years. While the secosterols have been of particular interest, it should be noted that other cholesterol-derived electrophiles have the same mass as the secosterols (see Figure 7), making the unambiguous structural assignment of adducts found in vivo difficult. For secosterols A and B, the assignment of the structure is further complicated by the fact that these two electrophiles are present in an aldol-retroaldol equilibrium.

Both secosterol A and B have a free aldehyde that can react with protein lysines (see Figure 12), causing some ambiguity about the nature of the adduct or adduct mixture formed. Reversible imine formation is the presumed initial step of adduction and in most studies of secosterol-protein reactions, imine reduction with borohydride or cyanoborohydride has been used to stabilize the adduct. This reduction strategy is required to "fix" the secosterol-protein covalent bond, since imine bond formation is reversible. The basic conditions of reduction likely minimize this problem; nevertheless, it should not be overlooked. It is also worth emphasizing that the mixture formed after imine reduction is only a close approximation of the authentic adducts formed in a biological setting, with the difference being a labile imine bond for adducts in cells or tissues and a stable sterol protein amine bond after reduction and isolation.

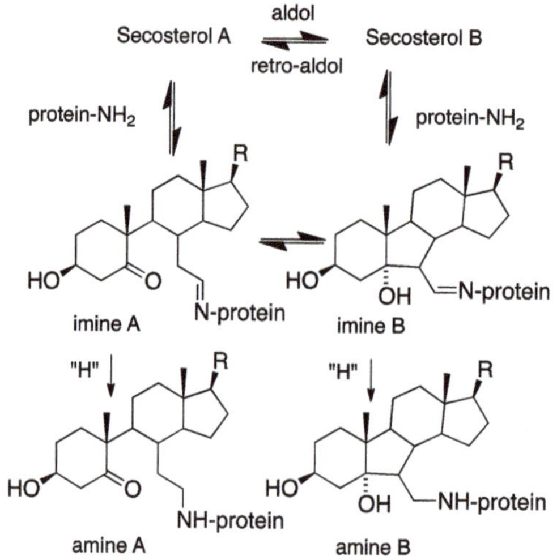

Figure 12. Mechanisms of secosterol-protein reactions resulting in imine reduction to stabilize adducts.

As an example of an important early study of oxysterol conjugates of Alzheimer's amyloid Aβ-peptides, Usui et. al. used solid phase synthesis to prepare secosterol adducts of specific Aβ peptide lysines, as well as the conjugate of the peptide terminal Asp amine [10]. Reduction of the imine A/B mixture with cyanoborohydride was used to fix the secosterol conjugates at Lys-16 and Lys-28, as well as the N-terminal Asp-1 amine of the Aβ peptide. It is of interest that the secosterol conjugates at Lys-16 and Lys-28 significantly increased the kinetics of Aβ peptide aggregation, while the terminal amine Asp-1 adduct had no measurable effect on the process. The aggregates formed from the Lys-16 secosterol A/B adducts were also found to be highly toxic to cultured cortical neurons.

In a detailed study of adduction, Windsor et. al. reported on the reactions of a mixture of secosterol A/B with amino acids, peptides, and isolated proteins [102]. The reaction of secosterol A with lysine under mild conditions (pH 7.4 buffer) gave multiple products, including m/z = lysine + secosterol A/B, lysine + secosterol A/B-H$_2$O, and lysine + secosterol A/B-2H$_2$O. Dehydration apparently competes with the aldol cyclization of secosterol A, affording multiple electrophilic species that can react with lysine from the single secosterol A precursor, as shown in Figure 13. Two of the dehydration products have structures that make them likely Michael acceptors and thus capable of adduction with cysteines and histidines and, indeed, histidine adducts are formed when either secosterol A or B is reacted with the model peptide Ac-Ala-Val-Ala-Gly-**His**-Ala-Gly-Ala-Arg.

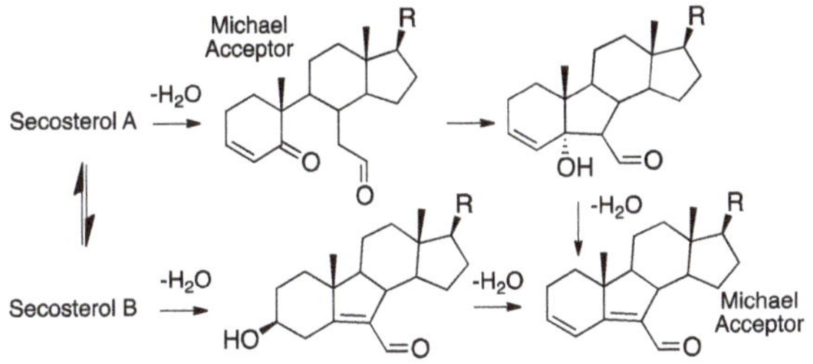

Figure 13. Dehydration of secosterols gives a complex mixture of electrophiles, including two Michael acceptors.

The exposure of cytochrome c to secosterol A gave evidence of significant adduction, as measured by MALDI-TOF MS analysis of the product mixture, which showed the addition of up to five secosterols to the protein [152]. Following borohydride reduction to fix lysine-derived imines, tryptic digestion and a proteomics assay of the major peptides indicated the presence of both lysine imine and histidine Michael adducts.

Tryptic peptides of lysine-secosterol adducts undergo characteristic mass fragmentations that have proved useful in determining the specific site of secosterol adduction. This was demonstrated in reactions of secosterol A or B with the model peptide Ac-Ala-Val-Ala-Gly-**Lys**-Ala-Gly-Ala-Arg and is shown in Figure 14, where the neutral loss of a sterol fragment gives the peptide + 12 Da at the site of lysine modification. Subsequent fragmentation of the +12 Da ion gives b and y ions that indicate the site of the modified lysine on the tryptic peptide, making identification of the protein adduct straightforward.

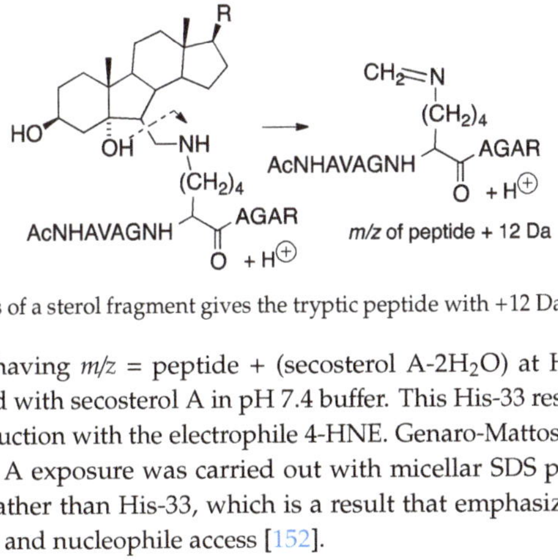

Figure 14. Neutral loss of a sterol fragment gives the tryptic peptide with +12 Da at the modified lysine.

Histidine adducts having m/z = peptide + (secosterol A-2H$_2$O) at His-33 were found when cytochrome c was reacted with secosterol A in pH 7.4 buffer. This His-33 residue has also been found to be a major site for adduction with the electrophile 4-HNE. Genaro-Mattos et al. reported that when cytochrome c-secosterol A exposure was carried out with micellar SDS present, the major adduct formed was at Lys-22, rather than His-33, which is a result that emphasizes the importance of the protein tertiary structure and nucleophile access [152].

8. Alkynyl-Sterol Probes

The development of bio-orthogonal reagents and their use in probing mechanistic pathways and metabolism have seen widespread application in recent years. Sharpless–Huisgen or "click" cycloaddition [153,154] (see Figure 15A), has been applied in a variety of settings to monitor cellular processes and the application of this strategy in studies of electrophile-protein adduction has been particularly useful. Alkynyl versions of PUFAs and their derived electrophiles, including 4-HNE, have been prepared and their use as surrogates for endogenous species has been employed for over a decade [55–57,139–145]. The synthesis and study of alkynyl surrogates of sterols (see Figure 15B), and their derived electrophiles, have occurred more recently, but the sterol compounds have a utility comparable to those in PUFA series.

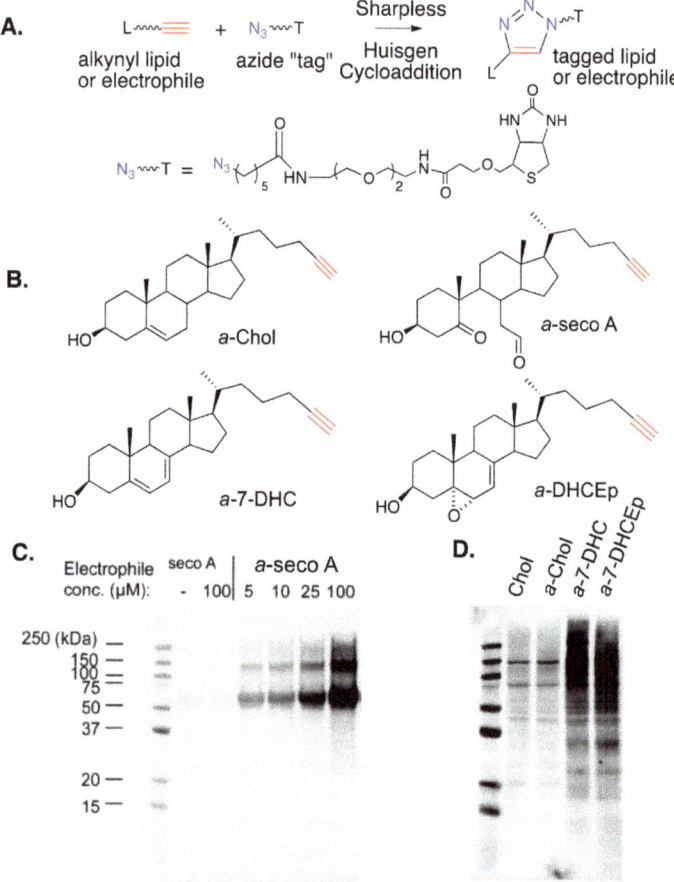

Figure 15. (**A**) Click cycloaddition of an alkynyl lipid and an azide "Tag". (**B**) Illustrative alkynyl lipids or lipid electrophiles. (**C**) Products of adduction of alkynyl-secosterol A (a-seco A) with human serum albumin (HSA) in pH 7.4 buffer, followed by "click" cyclo-addition with an azido-biotin. Gel visualized with a streptavidin fluorophore. (**D**) Protein-lipid adducts of Neuro2a cells treated with 20 µM of cholesterol, a-Chol, a-7-DHC, and a-DHCEp for 24 h.

Figure 15C shows an SDS gel for the lipid-adducted protein products that are formed from the reaction of human serum albumin (HSA) with alkynyl-secosterol A (a-seco A) at concentrations from 5 to 100 µM. After borohydride reduction to fix any adducts, Sharpless–Huisgen cycloaddition (click reaction) was carried out on the protein extracts, with the azide (shown in the figure) having an ethylene glycol linked to biotin. Gel electrophoresis of the protein product (western blotting using the streptavidin-AlexaFluor 680 conjugate) showed that a-seco A forms adducts with HSA, with protein aggregates being one consequence of adduction. The nature of the protein aggregates has not been established, but it is worth noting that the secosterol A and B dehydration product (secosterol – 2 H$_2$O) could serve as a protein crosslinking agent, since it is both a Michael acceptor and an imine precursor.

Figure 15D shows a western blot comparison of the proteome modification by alkynyl sterols obtained from protein extracts of Neuro2a cells that were treated with 20 µM of either cholesterol (as a control), *a*-Chol, *a*-7-DHC, or *a*-DHCEp for 24 h. Proteins adducted with alkynyl lipids were ligated with biotin via a click reaction and adduction was determined as in Figure 15C. As shown in Figure 15D, *a*-Chol gives only background levels of adduction, with a blot intensity comparable to the cholesterol control. In contrast, both *a*-7-DHC and *a*-DHCEp show substantial levels of protein adduction covering a range of protein molecular weights. Since *a*-DHCEp is a reactive electrophile, the observation of significant protein adduction with this oxysterol is not surprising, but *a*-7-DHC is not itself an electrophile. The results obtained with this sterol imply that a significant conversion of *a*-7-DHC to alkynyl electrophilic species occurs over the course of the exposure, generating protein adducts in situ. This conclusion is consistent with the fact that 7-DHC is extremely vulnerable to free radical peroxidation, with electrophiles like 7-DHCEp being formed in the process.

Miyamoto and collaborators recently reported that Cu, Zn superoxide dismutase (SOD1) formed high molecular weight aggregates when the apo-enzyme was exposed to either seco A or seco B [155]. This observation is of interest since the accumulation of SOD1 aggregates has been associated with the development of familial amyotrophic lateral sclerosis ALS [12,156]. MALDI-TOF MS analysis showed that seco A and seco B react at multiple lysine sites on SOD1 with as many as five secosterols attached to the protein. The application of click methods similar to those described above for cytochrome c revealed that SOD1-secosterol adducts were primarily associated with high molecular weight aggregates. Therefore, click ligation of the secosterol-SOD1 product mixtures to a fluorophore and an SDS gel showed that the highest level of protein adduction was in the high molecular weight region of the gel. The protein adduction of highly hydrophobic sterol electrophiles will affect the protein structure, and it was suggested that protein aggregation is initiated by hydrophobic-hydrophobic interactions of sterol adducts. When SOD1 was exposed to *a*-4-HNE, dimers, trimers, and multimers were formed, but there was no evidence of very high molecular weight aggregates from this less hydrophobic electrophile.

Speen et al. recently reported on the use of alkynyl sterols and secosterols to study protein adduction in human epithelial cells [157]. Cultured cells were exposed to alkynyl seco A or B and, after reduction with sodium borohydride, cellular proteins were treated with a biotin azide (photo-azide), as shown in Figure 16A. The specially designed photo-azide had a photo-cleavable linker insert between the azide and biotin functional groups so that the "catch and photo-release" sequence shown in Figure 16B could be applied. In the experiment, the mixture of un-modified proteins and biotinylated adducts was treated with a slurry of streptavidin beads, binding the adducted proteins to the beads and, after the unmodified proteins were removed from the beads by filtration, the protein adducts were released by photolysis and eluted from the beads. SDS gels of the protein input to the streptavidin beads and the photo-released (eluted) protein adducts are shown for the treated (exposure) and control cells in Figure 16C. The blue gels on the left of the figure show that photolysis of the beads released adducts with a range of molecular weights in the eluted/exposure lane. Adducts of specific proteins can be identified if selective antibodies for a protein are available, as shown in Figure 16C for the chaperone protein HSP90, which is an important therapeutic target for the treatment of a variety of cancers dependent on the chaperone-mediated stabilization of oncogenic proteins [158,159]. Levels of this protein input to the streptavidin beads were comparable for the control and treated cells, but no HSP90 was found in the photo-release fraction from the control cells, while this protein was evident in the release fraction of the treated cells, confirming that HSP90 is adducted by seco A/B in epithelial cells. In the same way, the liver X receptors LXRα and LXRβ were identified as targets for seco A/B adduction, as was the peroxisome proliferator-activated receptor PPAR.

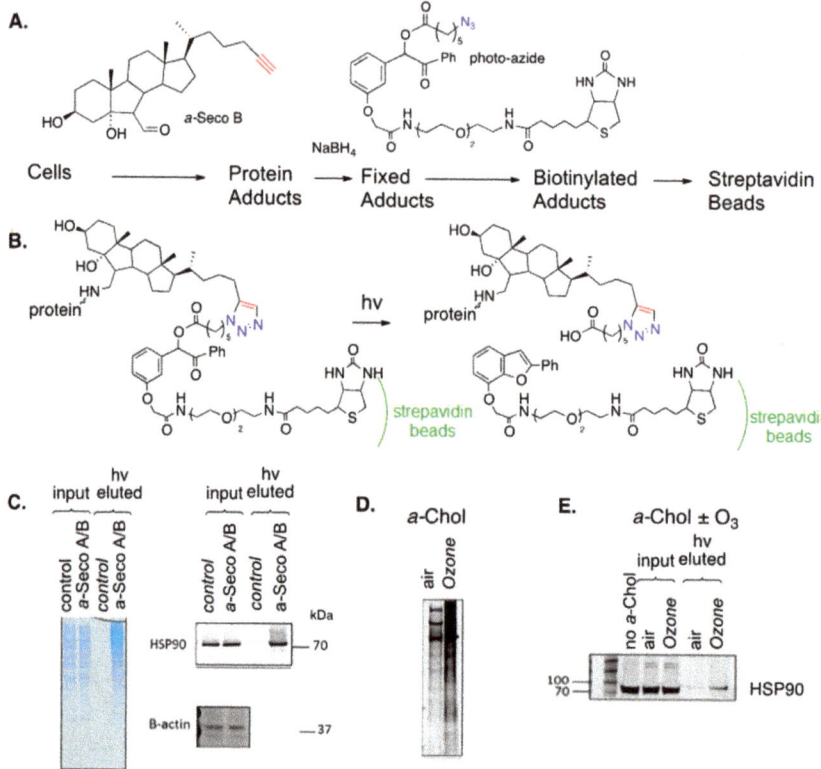

Figure 16. (**A**) Work-flow for the treatment of epithelial cells with *a*-seco A. (**B**) *a*-seco A adducted proteins were pulled-down on streptavidin beads. Beads were washed and adducted proteins photo-released (hv eluted). (**C**) SDS gels of control and *a*-seco A-treated cells were input to streptavidin beads and photo-released from the beads. Total adducted proteins are shown in blue, and HSP90 adducted proteins are presented in black and white. (**D**) Epithelial cells treated with *a*-Chol under air or ozone. Total protein adduct detected with anti-biotin fluorophore. (**E**) Pull-down on streptavidin beads and photo-release shows HSP90 adduction with *a*-Chol under air or ozone.

Figure 16D,E show the results of an experiment in which *a*-Chol was incorporated into epithelial cells, followed by exposure of those cells to ozone. This experiment parallels the seco A/B study described above, but in this case, the electrophiles that formed adducts were generated in situ. Overall protein adduction from cellular treatment with *a*-Chol and ozone is shown in Figure 16D and the adduction of HSP90 by this same combination treatment is demonstrated in Figure 16E.

The photo-azide strategy for adduct pull-down and photo-release was also used to define the adductome for 7-DHC-derived electrophiles in Neuro2a cells (see Figure 17) [160]. In this study, Neuro2a and *dhcr7*-deficient Neuro2a cells were incubated with alkynyl lanosterol (*a*-Lan) for 24 h and the alkynyl sterols present in the cells were assayed by HPLC-MS. In Neuro2a cells, most of the *a*-Lan was converted into *a*-Chol, demonstrating that the biosynthetic apparatus tolerated the alkynyl modification in the tail of the sterol. The same experiment carried out in *dhcr7*-deficient Neuro2a cells gave *a*-7-DHC as the major product, since the critical enzyme that carries out the last step in cholesterol biosynthesis is missing in these cells.

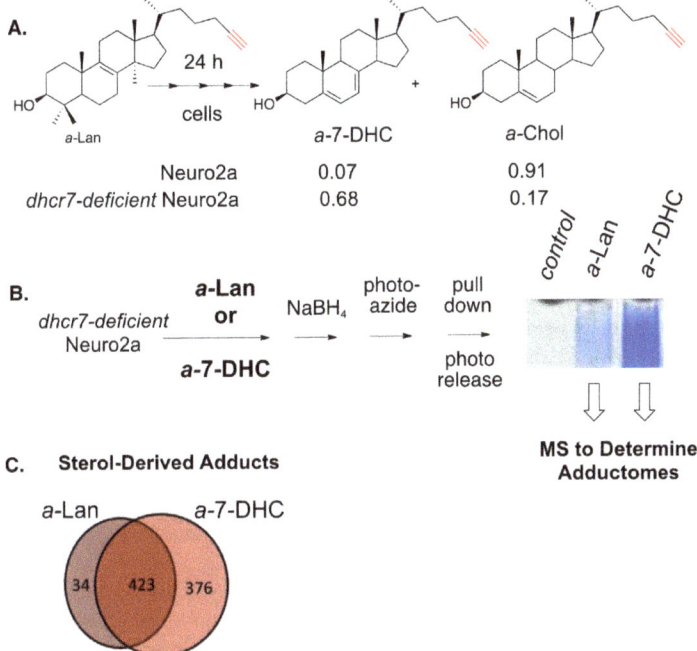

Figure 17. (**A**) Alkynyl sterols are viable surrogates for endogenous sterols in cell culture. Alkynyl lanosterol (*a*-Lan) undergoes multiple biosynthetic steps to give *a*-Chol in Neuro2a. In *dhcr7*-deficient Neuro2a, the biosynthesis is terminated at *a*-7-DHC. (**B**) *dhcr7*-deficient Neuro2a cells were treated with either *a*-Lan or *a*-7-DHC. Adducted proteins were pulled-down on streptavidin beads. Beads were washed and adducted proteins photo-released (hv eluted). SDS gels of control, *a*-Lan-treated, and *a*-7-DHC-treated cells were input to streptavidin beads and photo-released from the beads. Total adducted proteins, shown in blue, were subjected to proteomics assays. (**C**) *a*-7-DHC-derived electrophiles adducted nearly 800 proteins, whilst *a*-Lan-derived electrophiles adducted only 457 proteins, and 423 proteins were common to adduction by both *a*-Lan- and *a*-7-DHC-derived electrophiles.

9. Questions and Prospects

HPLC-MS has been particularly helpful in defining mechanistic pathways and providing product profiles for the oxidation of cholesterol, 7-DHC, and other sterols. Product mixtures from radical chain processes are particularly complex, with recent reports expanding the sets of known electrophilic compounds. Cellular protein adduction by specific sterol-derived electrophiles has also been established, as have methods to identify the adductomes of various sterol-derived electrophiles. Indeed, oxysterol protein adduction appears to be a common outcome of many cellular oxidative exposures. Therefore, pieces of the puzzle linking oxidative stress exposure with electrophile formation and protein adduction are in place, but the picture remains blurry. In the simplest example, it was established that cholesterol reacts with ozone to yield electrophilic secosterols and that cellular exposure to secosterols gives protein adducts. The evidence that cellular ozone exposures leads to secosterol protein adducts has not yet been confirmed by a proteomics analysis from in vivo exposures. Radical chain oxidation provides an even more circumstantial picture, with the nature of the active electrophilic species in doubt for the radical chain-promoted oxidation of both cholesterol and 7-DHC.

In spite of the lack of detail outlined above, it seems highly likely that sterol-protein adduct formation occurs, raising general questions about the consequences of and control mechanisms for the process. The inhibition of oxidation serves as a control mechanism against adduct formation, with natural antioxidants forming a primary line of defense. If no electrophiles are generated, no adducts will be formed.

The disposal of protein adducts once formed would also appear to be a plausible defense mechanism. A 300 m/z hydrophobic sterol mass decorating any protein would be a significant structural perturbation and it seems likely that mechanisms exist to repair adducts and recover the native protein. In this regard, it seems worth mentioning that protein adduction should be reversible for many electrophiles. Secosterol adduction occurs by initial imine formation, which is a process that is chemically reversible. The reversible nature of adduct formation opens the possibility of an equilibrium distribution of a given secosterol among a set of available proteins in the locus of electrophile generation. Given this dynamic, it seems reasonable to speculate that a mechanism exists for the equilibration of a hydrophobic adduct from protein to protein, until a sink is found for disposal. We note that this idea is speculative and while it is conceptually pleasing, no evidence to support this suggestion has, to our knowledge, been presented.

Reversible electrophile-protein adduction, which can be considered a type of post-translational modification, can also serve as a signaling mechanism, because many of the protein targets of lipid electrophiles are involved in stress and inflammatory responses, such as Keap1/Nrf2, HSF1, PPARγ, and NF-κB [161]. Therefore, a small amount of electrophilic adduction likely serves as a protective mechanism in response to elevated oxidative stress. However, it remains to be elucidated whether sterol-derived electrophiles can also play the same roles in inducing protective cellular responses.

Given the recent recognition that the accumulation of phospholipid hydroperoxides drives the oxidative cell death modality now known as ferroptosis, it is compelling to suggest that sterol oxidation may contribute to either the initiation or execution of this process. As far as we are aware, all attention to date has been focused on (phospho)lipids. However, given the abundance of cholesterol and the integral structural role it plays in the lipid bilayers that are compromised during ferroptotic cell death, sterol oxidation and the products derived therefrom may be (the) key players. Along these lines, Birsoy and co-workers recently reported that cells devoid of squalene monooxygenase activity and that accumulate squalene at the expense of cholesterol are resistant to ferroptosis [162].

Many questions remain, but research on sterol peroxidation and sterol-derived electrophiles has advanced rapidly in recent years, with many tools now available to allow progress in the field. It seems likely that the links between oxidative stress, oxidizable sterols, oxysterol electrophiles, and the lipid-protein adductome will provide a fertile ground for exploration for years to come.

Funding: Support for this research from NIH R01HD092659 (LX), (NICHD R01 HD064727(NAP), NIEHS R01 ES024133 (NAP), R21 ES024666(NAP), and NSF CHE-1664851 (LX) is gratefully acknowledged.

Acknowledgments: The authors acknowledge the dedicated work of many co-workers who have been involved with the various projects. Professor Zeljka Korade and Dr. Thiago Genaro-Mattos, first at Vanderbilt and then at the University of Nebraska Medical School Omaha, have provided an important link to cell culture and animal models of the human disorders described in these studies. More recent collaborations with Professor Ilona Jaspers at the University of North Carolina and her in depth knowledge of ozone environmental exposure and cholesterol oxidation products in the lung has been stimulating and productive. At Vanderbilt, Keri Tallman, Hye-Young Kim, Wei Liu, Hubert Muchalski, Katherine Windsor, Connor Lamberson, and Phillip Wages have made important contributions to this effort.

Conflicts of Interest: The authors declare no conflict of interest.

References

1. EPA. *National Ambient Air Quality Standards (NAAQS)—Ozone (O3)*; EPA: Washington, DC, USA, 2014.
2. Hollingsworth, J.W.; Kleeberger, S.R.; Foster, W.M. Ozone and Pulmonary Innate Immunity. *Proc. Am. Thorac. Soc.* **2007**, *4*, 240–246. [CrossRef]
3. Antczak, A.; Nowak, D.; Shariati, B.; Król, M.; Piasecka, G.; Kurmanowska, Z. Increased hydrogen peroxide and thiobarbituric acid-reactive products in expired breath condensate of asthmatic patients. *Eur. Respir. J.* **1997**, *10*, 1235–1241. [CrossRef]

4. Halliwell, B. Lipid peroxidation, antioxidants and cardiovascular disease: How should we move forward? *Cardiovasc. Res.* **2000**, *47*, 410–418. [CrossRef]
5. Bouhajja, H.; Kacem, F.H.; Abdelhedi, R.; Ncir, M.; Dimitrov, J.D.; Marrakchi, R.; Jamoussi, K.; Rebaï, A.; El Feki, A.; Abid, M.; et al. Potential Predictive Role of Lipid Peroxidation Markers for Type 2 Diabetes in the Adult Tunisian Population. *Can. J. Diabetes* **2018**, *42*, 263–271. [CrossRef]
6. Polidori, M.C.; Praticó, D.; Savino, K.; Rokach, J.; Stahl, W.; Mecocci, P. Increased F2 isoprostane plasma levels in patients with congestive heart failure are correlated with antioxidant status and disease severity. *J. Card. Fail.* **2004**, *10*, 334–338. [CrossRef]
7. Bastos, A.D.S.; Graves, D.T.; Loureiro, A.P.D.M.; Júnior, C.R.; Corbi, S.; Frizzera, F.; Scarel-Caminaga, R.; Câmara, N.O.S.; Andriankaja, O.M.; Hiyane, M.I.; et al. Diabetes and increased lipid peroxidation are associated with systemic inflammation even in well-controlled patients. *J. Diabetes Complicat.* **2016**, *30*, 1593–1599. [CrossRef]
8. Mishra, S.; Mishra, B.B. Study of Lipid Peroxidation, Nitric Oxide End Product, and Trace Element Status in Type 2 Diabetes Mellitus with and without Complications. *Int. J. Appl. Basic Med Res.* **2017**, *7*, 88–93. [CrossRef]
9. Di Domenico, F.; Tramutola, A.; Butterfield, D.A. Role of 4-hydroxy-2-nonenal (HNE) in the pathogenesis of alzheimer disease and other selected age-related neurodegenerative disorders. *Free. Radic. Boil. Med.* **2017**, *111*, 253–261. [CrossRef]
10. Usui, K.; Hulleman, J.D.; Paulsson, J.F.; Siegel, S.J.; Powers, E.; Kelly, J.W. Site-specific modification of Alzheimer's peptides by cholesterol oxidation products enhances aggregation energetics and neurotoxicity. *Proc. Natl. Acad. Sci. USA* **2009**, *106*, 18563–18568. [CrossRef]
11. Zhang, Q.; Powers, E.; Nieva, J.; Huff, M.E.; Dendle, M.A.; Bieschke, J.; Glabe, C.G.; Eschenmoser, A.; Wentworth, P.; Lerner, R.A.; et al. Metabolite-initiated protein misfolding may trigger Alzheimer's disease. *Proc. Natl. Acad. Sci. USA* **2004**, *101*, 4752–4757. [CrossRef]
12. Simonian, N.A.; Coyle, J.T. Oxidative stress in neurodegenerative diseases. *Annu. Rev. Pharmacol. Toxicol.* **1996**, *36*, 83–106. [CrossRef]
13. Björkhem, I.; Cedazo-Mínguez, A.; Leoni, V.; Meaney, S. Oxysterols and neurodegenerative diseases. *Mol. Asp. Med.* **2009**, *30*, 171–179. [CrossRef]
14. Marnett, L.J. Lipid peroxidation—DNA damage by malondialdehyde. *Mutat. Res. Mol. Mech. Mutagen.* **1999**, *424*, 83–95. [CrossRef]
15. Marnett, L.J. Oxyradicals and DNA damage. *Carcinogenesis* **2000**, *21*, 361–370. [CrossRef]
16. Rodriguez, I.; Larrayoz, I. Cholesterol oxidation in the retina: Implications of 7KCh formation in chronic inflammation and age-related macular degeneration. *J. Lipid Res.* **2010**, *51*, 2847–2862. [CrossRef]
17. Dixon, S.J.; Lemberg, K.M.; Lamprecht, M.R.; Skouta, R.; Zaitsev, E.M.; Gleason, C.E.; Patel, D.N.; Bauer, A.J.; Cantley, A.M.; Yang, W.S.; et al. Ferroptosis: An Iron-Dependent Form of Nonapoptotic Cell Death. *Cell* **2012**, *149*, 1060–1072. [CrossRef]
18. Friedmann Angeli, J.P.; Schneider, M.; Proneth, B.; Tyurina, Y.Y.; Ryurin, V.A.; Hammond, V.J.; Herback, N.; Aichler, M.; Walch, A.; Eggenhofer, E.; et al. Inactivation of the ferrotopsis regulator Gpx4 triggers acute renal failure in mice. *Nat. Cell Biol.* **2014**, *16*, 1180–1191. [CrossRef]
19. Conrad, M.; Pratt, D.A. The chemical basis of ferroptosis. *Nat. Methods* **2019**, *15*, 1137–1147. [CrossRef]
20. Kohn, H.I.; Liversedge, N. On a New Aerobic Metabolite Whose Production by Brain is Inhibited by Apomorphine, Emetine, Ergotamine, Epinephrine and Menadione. *J. Pharmacol. Exper. Therap.* **1944**, *82*, 292–300.
21. Nair, V.; Turner, G.A. The thiobarbituric acid test for lipid peroxidation: Structure of the adduct with malondialdehyde. *Lipids* **1984**, *19*, 804–805. [CrossRef]
22. Bernheim, F.; Bernheim, M.L.C.; Wilbur, K.M. The reaction between thiobarbituric acid and the oxidation products of certain lipides. *J. Boil. Chem.* **1948**, *174*, 257–264.
23. Ingold, K.U. Peroxy radicals. *Accounts Chem. Res.* **1969**, *2*, 1–9. [CrossRef]
24. Ingold, K.U. *60 Years of Research on Free Radical Physical Organic Chemistry*; American Chemical Society: Washington, DC, USA, 2015; pp. 223–250.
25. Porter, N.A. Mechanisms for the autoxidation of polyunsaturated lipids. *Accounts Chem. Res.* **1986**, *19*, 262–268. [CrossRef]

26. Niki, E. Biomarkers of lipid peroxidation in clinical material. *Biochim. Biophys. Acta* **2014**, *1840*, 809–817. [CrossRef] [PubMed]
27. Zielinski, Z.; Pratt, D.A. Lipid Peroxidation: Kinetics, Mechanisms, and Products. *J. Org. Chem.* **2017**, *82*, 2817–2825. [CrossRef] [PubMed]
28. Yin, H.; Xu, L.; Porter, N.A.; Yin, H. Free Radical Lipid Peroxidation: Mechanisms and Analysis. *Chem. Rev.* **2011**, *111*, 5944–5972. [CrossRef]
29. Xu, L.; Porter, N.A. Free radical oxidation of cholesterol and its precursors: Implications in cholesterol biosynthesis disorders. *Free. Radic. Res.* **2015**, *49*, 835–849. [CrossRef] [PubMed]
30. Porter, N.A. A Perspective on Free Radical Autoxidation: The Physical Organic Chemistry of Polyunsaturated Fatty Acid and Sterol Peroxidation. *J. Org. Chem.* **2013**, *78*, 3511–3524. [CrossRef] [PubMed]
31. Smith, L.L. *Cholesterol Autoxidation*; Plenum Press: New York, NY, USA, 1981.
32. Smith, L.L. Cholesterol autoxidation 1981–1986. *Chem. Phys. Lipids* **1987**, *44*, 87–125. [CrossRef]
33. Smith, L.L. Oxygen, oxysterols, ouabain, and ozone: A cautionary tale. *Free Radic. Biol. Med.* **2004**, *37*, 318–324. [CrossRef]
34. Girotti, A.W. Mechanisms of lipid peroxidation. *J. Free. Radicals Boil. Med.* **1985**, *1*, 87–95. [CrossRef]
35. Girotti, A.W. Lipid hydroperoxide generation, turnover, and effector action in biological systems. *J. Lipid Res.* **1998**, *39*, 1529–1542.
36. Girotti, A.W.; Korytowski, W. Cholesterol as a natural probe for free radical-mediated lipid peroxidation in biological membranes and lipoproteins. *J. Chromatogr. B* **2016**, *1019*, 202–209. [CrossRef]
37. Girotti, A.W.; Korytowski, W. Cholesterol Hydroperoxide Generation, Translocation, and Reductive Turnover in Biological Systems. *Cell Biophys.* **2017**, *75*, 413–419. [CrossRef]
38. Girotti, A.W.; Korytowski, W. Cholesterol Peroxidation as a Special Type of Lipid Oxidation in Photodynamic Systems. *Photochem. Photobiol.* **2018**, *95*, 73–82. [CrossRef]
39. Zielinski, Z.; Pratt, D.A. Cholesterol Autoxidation Revisited: Debunking the Dogma Associated with the Most Vilified of Lipids. *J. Am. Chem. Soc.* **2016**, *138*, 6932–6935. [CrossRef]
40. Schaefer, E.L.; Zopyrus, N.; Zielinski, Z.A.M.; Facey, G.A.; Pratt, D.A. On the Products of Cholesterol Autoxidation in Phospholipid Bilayers and the Formation of Secosterols Derived Therefrom. *Angew. Chem. Int. Ed.* **2020**, *59*, 2089–2094. [CrossRef]
41. Zielinski, Z.A.M.; Pratt, D.A. H-Atom Abstraction vs Addition: Accounting for the Diverse Product Distribution in the Autoxidation of Cholesterol and Its Esters. *J. Am. Chem. Soc.* **2019**, *141*, 3037–3051. [CrossRef] [PubMed]
42. Iuliano, L. Pathways of cholesterol oxidation via non-enzymatic mechanisms. *Chem. Phys. Lipids* **2011**, *164*, 457–468. [CrossRef] [PubMed]
43. Pryor, W.A.; Stanley, J.P. Letter: A suggested mechanism for the production of malonaldehyde during the autoxidation of polyunsaturated fatty acids. Nonenzymatic production of prostaglandin endoperoxides during autoxidation. *J. Org. Chem.* **1975**, *40*, 3615–3617. [CrossRef] [PubMed]
44. Porter, N.A.; Funk, M.O. Peroxy radical cyclization as a model for prostaglandin biosynthesis. *J. Org. Chem.* **1975**, *40*, 3614–3615. [CrossRef] [PubMed]
45. Pryor, W.A.; Porter, N.A. Suggested mechanisms for the production of 4-hydroxy-2-nonenal from the autoxidation of polyunsaturated fatty acids. *Free. Radic. Boil. Med.* **1990**, *8*, 541–543. [CrossRef]
46. Schneider, C.; Porter, N.A.; Brash, A.R. Routes to 4-hydroxynonenal: Fundamental issues in the mechanisms of lipid peroxidation. *J. Boil. Chem.* **2008**, *283*, 15539–15543. [CrossRef] [PubMed]
47. Benedetti, A.; Comporti, M.; Esterbauer, H. Identification of 4-hydroxynonenal as a cytotoxic product originating from the peroxidation of liver microsomal lipids. *Biochim. Biophys. Acta* **1980**, *620*, 281–296. [CrossRef]
48. Esterbauer, H.; Zollern, H. Methods for determination of aldehydic lipid peroxidation products. *Free. Radic. Boil. Med.* **1989**, *7*, 197–203. [CrossRef]
49. Zhong, H.; Yin, H. Role of lipid peroxidation derived 4-hydroxynonenal (4-HNE) in cancer: Focusing on mitochondria. *Redox Boil.* **2014**, *4*, 193–199. [CrossRef]
50. Salomon, R.G.; Bi, W. Isolevuglandin Adducts in Disease. *Antioxid. Redox Signal.* **2015**, *22*, 1703–1718. [CrossRef]

51. Brame, C.J.; Salomon, R.G.; Morrow, J.D.; Roberts, L.J. Identification of extremely reactive gamma-ketoaldehydes (isolevuglandins) as products of the isoprostane pathway and characterization of their lysyl protein adducts. *J. Boil. Chem.* **1999**, *274*, 13139–13146. [CrossRef]
52. Davies, S.S.; Zhang, L. Isolevuglandins and cardiovascular disease. *Prostaglandins Lipid Mediat.* **2018**, *139*, 29–35. [CrossRef]
53. Zhang, L.; Yermalitsky, V.; Huang, J.; Pleasent, T.; Borja, M.S.; Oda, M.; Jerome, W.G.; Yancey, P.G.; Linton, E.F.; Davies, S.S. Modification by isolevuglandins, highly reactive γ-ketoaldehydes, deleteriously alters high-density lipoprotein structure and function. *J. Boil. Chem.* **2018**, *293*, 9176–9187. [CrossRef]
54. Uchida, K. Lipofuscin-like fluorophores originated from malondialdehyde. *Free. Radic. Res.* **2006**, *40*, 1335–1338. [CrossRef] [PubMed]
55. Liebler, D.; Zimmerman, L.J. Targeted Quantitation of Proteins by Mass Spectrometry. *Biochemistry* **2013**, *52*, 3797–3806. [CrossRef] [PubMed]
56. Connor, R.E.; Codreanu, S.G.; Marnett, L.J.; Liebler, D. Targeted protein capture for analysis of electrophile-protein adducts. *Breast Cancer* **2013**, *987*, 163–176.
57. Codreanu, S.G.; Liebler, D. Novel approaches to identify protein adducts produced by lipid peroxidation. *Free. Radic. Res.* **2015**, *49*, 881–887. [CrossRef] [PubMed]
58. Cobley, J.; Fiorello, M.L.; Bailey, D.M. 13 reasons why the brain is susceptible to oxidative stress. *Redox Boil.* **2018**, *15*, 490–503. [CrossRef] [PubMed]
59. Carrié, I.; Clément, M.; De Javel, D.; Francès, H.; Bourre, J.M. Specific phospholipid fatty acid composition of brain regions in mice. Effects of n-3 polyunsaturated fatty acid deficiency and phospholipid supplementation. *J. Lipid Res.* **2000**, *41*, 465–472. [PubMed]
60. Bosco, D.A.; Fowler, D.M.; Zhang, Q.; Nieva, J.; Powers, E.; Wentworth, A.D.; Lerner, R.A.; Kelly, J.W. Elevated levels of oxidized cholesterol metabolites in Lewy body disease brains accelerate α-synuclein fibrilization. *Nat. Methods* **2006**, *2*, 249–253. [CrossRef]
61. Xu, L.; Davis, T.A.; Porter, N.A. Rate Constants for Peroxidation of Polyunsaturated Fatty Acids and Sterols in Solution and in Liposomes. *J. Am. Chem. Soc.* **2009**, *131*, 13037–13044. [CrossRef]
62. Xu, L.; Korade, Z.; Porter, N.A. Oxysterols from Free Radical Chain Oxidation of 7-Dehydrocholesterol: Product and Mechanistic Studies. *J. Am. Chem. Soc.* **2010**, *132*, 2222–2232. [CrossRef]
63. Xu, L.; Korade, Z.; Rosado, D.A.; Liu, W.; Lamberson, C.R.; Porter, N.A. An oxysterol biomarker for 7-dehydrocholesterol oxidation in cell/mouse models for Smith-Lemli-Opitz syndrome[S]. *J. Lipid Res.* **2011**, *52*, 1222–1233. [CrossRef]
64. Smith, D.W.; Lemli, L.; Opitz, J.M. A newly recognized syndrome of multiple congenital anomalies. *J. Pediatr.* **1964**, *64*, 210–217. [CrossRef]
65. Porter, F.D. Smith–Lemli–Opitz syndrome: Pathogenesis, diagnosis and management. *Eur. J. Hum. Genet.* **2008**, *16*, 535–541. [CrossRef] [PubMed]
66. Xu, L.; Korade, Z.; Rosado, D.A.; Mirnics, K.; Porter, N.A. Metabolism of oxysterols derived from nonenzymatic oxidation of 7-dehydrocholesterol in cells. *J. Lipid Res.* **2013**, *54*, 1135–1143. [CrossRef]
67. Xu, L.; Liu, W.; Sheflin, L.G.; Fliesler, S.J.; Porter, N.A. Novel oxysterols observed in tissues and fluids of AY9944-treated rats: A model for Smith-Lemli-Opitz syndrome. *J. Lipid Res.* **2011**, *52*, 1810–1820. [CrossRef] [PubMed]
68. Xu, L.; Mirnics, K.; Bowman, A.B.; Liu, W.; Da, J.; Porter, N.A.; Korade, Z. DHCEO accumulation is a critical mediator of pathophysiology in a Smith–Lemli–Opitz syndrome model. *Neurobiol. Dis.* **2012**, *45*, 923–929. [CrossRef] [PubMed]
69. Windsor, K.; Genaro-Mattos, T.; Kim, H.-Y.H.; Liu, W.; Tallman, K.A.; Miyamoto, S.; Korade, Z.; Porter, N.A. Probing lipid-protein adduction with alkynyl surrogates: Application to Smith-Lemli-Opitz syndrome. *J. Lipid Res.* **2013**, *54*, 2842–2850. [CrossRef] [PubMed]
70. Pryor, W.A.; Squadrito, G.L.; Friedman, M. The cascade mechanism to explain ozone toxicity: The role of lipid ozonation products. *Free. Radic. Boil. Med.* **1995**, *19*, 935–941. [CrossRef]
71. Uppu, R.; Cueto, R.; Squadrito, G.; Pryor, W. What Does Ozone React with at the Air Lung Interface? Model Studies Using Human Red Blood Cell Membranes. *Arch. Biochem. Biophys.* **1995**, *319*, 257–266. [CrossRef] [PubMed]

72. Pulfer, M.K.; Harrison, K.; Murphy, R.C. Direct electrospray tandem mass spectrometry of the unstable hydroperoxy bishemiacetal product derived from cholesterol ozonolysis. *J. Am. Soc. Mass Spectrom.* **2004**, *15*, 194–202. [CrossRef]
73. Pulfer, M.K.; Murphy, R.C.; Ishida, J.; Nishiwaki, S.; Iguchi, T.; Matsuzaki, H.; Shiota, N.; Okunishi, H.; Sugiyama, F.; Kasuya, Y.; et al. Formation of Biologically Active Oxysterols during Ozonolysis of Cholesterol Present in Lung Surfactant. *J. Boil. Chem.* **2004**, *279*, 26331–26338. [CrossRef]
74. Pulfer, M.K.; Taube, C.; Gelfand, E.; Murphy, R.C. Ozone Exposure in Vivo and Formation of Biologically Active Oxysterols in the Lung. *J. Pharmacol. Exp. Ther.* **2005**, *312*, 256–264. [CrossRef] [PubMed]
75. Maillard, B.; Ingold, K.U.; Scaiano, J.T. Rate constants for the reactions of free radicals with oxygen in solution. *J. Am. Chem. Soc.* **1983**, *105*, 5095–5099. [CrossRef]
76. Mccord, J.M.; Day, E.D. Superoxide-dependent production of hydroxyl radical catalyzed by iron-EDTA complex. *FEBS Lett.* **1978**, *86*, 139–142. [CrossRef]
77. Gardner, H.W.; Kleiman, R.; Weisleder, D. Homolytic decomposition of linoleic acid hydroperoxide: Identification of fatty acid products. *Lipids* **1974**, *9*, 696–706. [CrossRef]
78. Gardner, H.W.; Weisleder, D.; Kleiman, R. Addition of N-acetylcysteine to linoleic acid hydroperoxide. *Lipids* **1976**, *11*, 127–134. [CrossRef]
79. Gardner, H.W.; Kleiman, R. Degradation of linoleic acid hydroperoxides by a cysteine. FeCl3 catalyst as a model for similar biochemical reactions. II. Specificity in formation of fatty acid epoxides. *Biochim. Biophys. Acta* **1981**, *665*, 113–125. [CrossRef]
80. Dix, T.A.; Marnett, L.J. Conversion of linoleic acid hydroperoxide to hydroxy, keto, epoxyhydroxy, and trihydroxy fatty acids by hematin. *J. Boil. Chem.* **1985**, *260*, 5351–5357.
81. Schaich, K.M. Metals and lipid oxidation. Contemporary issues. *Lipids* **1992**, *27*, 209–218. [CrossRef]
82. Porter, N.A.; Mills, K.A.; Carter, R.L. A Mechanistic Study of Oleate Autoxidation: Competing Peroxyl H-Atom Abstraction and Rearrangement. *J. Am. Chem. Soc.* **1994**, *116*, 6690–6696. [CrossRef]
83. Pratt, D.A.; Tallman, K.A.; Porter, N.A. Free Radical Oxidation of Polyunsaturated Lipids: New Mechanistic Insights and the Development of Peroxyl Radical Clocks. *Accounts Chem. Res.* **2011**, *44*, 458–467. [CrossRef]
84. Tallman, K.A.; Pratt, D.A.; Porter, N.A. Kinetic products of linoleate peroxidation: Rapid beta-fragmentation of nonconjugated peroxyls. *J. Am. Chem. Soc.* **2001**, *123*, 11827–11828. [CrossRef] [PubMed]
85. Tallman, K.A.; Rector, C.L.; Porter, N.A. Substituent Effects on Regioselectivity in the Autoxidation of Nonconjugated Dienes. *J. Am. Chem. Soc.* **2009**, *131*, 5635–5641. [CrossRef]
86. Tallman, K.A.; Roschek, B.; Porter, N.A. Factors Influencing the Autoxidation of Fatty Acids: Effect of Olefin Geometry of the Nonconjugated Diene. *J. Am. Chem. Soc.* **2004**, *126*, 9240–9247. [CrossRef] [PubMed]
87. Shah, R.; Farmer, L.A.; Zilka, O.; Van Kessel, A.T.; Pratt, D.A. Beyond DPPH: Use of Fluorescence-Enabled Inhibited Autoxidation to Predict Oxidative Cell Death Rescue. *Cell Chem. Boil.* **2019**, *26*, 1594–1607. [CrossRef] [PubMed]
88. Paillasse, M.R.; Saffon, N.; Gornitzka, H.; Silvente-Poirot, S.; Poirot, M.; De Medina, P. Surprising unreactivity of cholesterol-5,6-epoxides towards nucleophiles. *J. Lipid Res.* **2012**, *53*, 718–725. [CrossRef] [PubMed]
89. De Medina, P.; Paillasse, M.R.; Segala, G.; Poirot, M.; Silvente-Poirot, S. Identification and pharmacological characterization of cholesterol-5,6-epoxide hydrolase as a target for tamoxifen and AEBS ligands. *Proc. Natl. Acad. Sci. USA* **2010**, *107*, 13520–13525. [CrossRef] [PubMed]
90. Meijer, J.; DePierre, J.W.; Jörnvall, H. Cytosolic epoxide hydrolase from liver of control and clofibrate-treated mice. Structural comparison by HPLC peptide mapping. *Biosci. Rep.* **1987**, *7*, 891–896. [CrossRef]
91. Nashed, N.T.; Michaud, D.P.; Levin, W.; Jerina, N.M. Properties of liver microsomal cholesterol 5,6-oxide hydrolase. *Arch. Biochem. Biophys.* **1985**, *241*, 149–162. [CrossRef]
92. Silvente-Poirot, S.; Poirot, M. Cholesterol epoxide hydrolase and cancer. *Curr. Opin. Pharmacol.* **2012**, *12*, 696–703. [CrossRef]
93. Watabe, T.; Ozawa, N.; Ishii, H.; Chiba, K.; Hiratsuka, A. Hepatic microsomal cholesterol epoxide hydrolase: Selective inhibition by detergents and separation from xenobiotic epoxide hydrolase. *Biochem. Biophys. Res. Commun.* **1986**, *140*, 632–637. [CrossRef]
94. Dalenc, F.; Poirot, M.; Silvente-Poirot, S. Dendrogenin A: A Mammalian Metabolite of Cholesterol with Tumor Suppressor and Neurostimulating Properties. *Curr. Med. Chem.* **2015**, *22*, 3533–3549. [CrossRef]

95. De Medina, P.; Paillasse, M.R.; Segala, G.; Voisin, M.; Mhamdi, L.; Dalenc, F.; Lacroix-Triki, M.; Filleron, T.; Pont, F.; Al Saati, T.; et al. Dendrogenin A arises from cholesterol and histamine metabolism and shows cell differentiation and anti-tumour properties. *Nat. Commun.* **2013**, *4*, 1840. [CrossRef]
96. Fransson, A.; De Medina, P.; Paillasse, M.R.; Silvente-Poirot, S.; Poirot, M.; Ulfendahl, M. Dendrogenin A and B two new steroidal alkaloids increasing neural responsiveness in the deafened guinea pig. *Front. Aging Neurosci.* **2015**, *7*, 145. [CrossRef] [PubMed]
97. Poirot, M.; Silvente-Poirot, S. Oxysterols and related sterols: Implications in pharmacology and pathophysiology. *Biochem. Pharmacol.* **2013**, *86*, 1–2. [CrossRef] [PubMed]
98. Poirot, M.; Silvente-Poirot, S. Cholesterol-5,6-epoxides: Chemistry, biochemistry, metabolic fate and cancer. *Biochimie* **2013**, *95*, 622–631. [CrossRef] [PubMed]
99. Poirot, M.; Silvente-Poirot, S. When cholesterol meets histamine, it gives rise to dendrogenin A: A tumour suppressor metabolite1. *Biochem. Soc. Trans.* **2016**, *44*, 631–637. [CrossRef]
100. Segala, G.; David, M.; De Medina, P.; Poirot, M.C.; Serhan, N.; Vergez, F.; Mougel, A.; Saland, E.; Carayon, K.; Leignadier, J.; et al. Dendrogenin A drives LXR to trigger lethal autophagy in cancers. *Nat. Commun.* **2017**, *8*, 1903. [CrossRef]
101. Silvente-Poirot, S.; De Medina, P.; Record, M.; Poirot, M. From tamoxifen to dendrogenin A: The discovery of a mammalian tumor suppressor and cholesterol metabolite. *Biochimie* **2016**, *130*, 109–114. [CrossRef]
102. Windsor, K.; Genaro-Mattos, T.; Miyamoto, S.; Stec, N.F.; Kim, H.-Y.H.; Tallman, K.A.; Porter, N.A. Assay of Protein and Peptide Adducts of Cholesterol Ozonolysis Products by Hydrophobic and Click Enrichment Methods. *Chem. Res. Toxicol.* **2014**, *27*, 1757–1768. [CrossRef]
103. Hock, H.; Lang, S. Autoxydation von Koblen-wasserstoffen, IX. Mitteil: Uber Peroxyde von Benzol-Derivaten. *Chem. Ber.* **1944**, *77*, 257–264. [CrossRef]
104. Lee, J.B.; Uff, B.C. Organic reactions involving electrophilic oxygen. *Q. Rev. Chem. Soc.* **1967**, *21*, 429. [CrossRef]
105. Frimer, A.A. The reaction of singlet oxygen with olefins: The question of mechanism. *Chem. Rev.* **1979**, *79*, 359–387. [CrossRef]
106. Brinkhorst, J.; Nara, S.J.; Pratt, D.A. Hock Cleavage of Cholesterol 5α-Hydroperoxide: An Ozone-Free Pathway to the Cholesterol Ozonolysis Products Identified in Arterial Plaque and Brain Tissue. *J. Am. Chem. Soc.* **2008**, *130*, 12224–12225. [CrossRef] [PubMed]
107. Pryor, W.A. How far does ozone penetrate into the pulmonary air/tissue boundary before it reacts? *Free Radic. Boil. Med.* **1992**, *12*, 83–88. [CrossRef]
108. Pryor, W.A. Mechanisms of radical formation from reactions of ozone with target molecules in the lung. *Free Radic. Boil. Med.* **1994**, *17*, 451–465. [CrossRef]
109. Ursini, F.; Maiorino, M.; Valente, M.; Ferri, L.; Gregolin, C. Purification from pig liver of a protein which protects liposomes and biomembranes from peroxidative degradation and exhibits glutathione peroxidase activity on phosphatidylcholine hydroperoxides. *Biochim. Biophys. Acta* **1982**, *710*, 197–211. [CrossRef]
110. Korytowski, W.; Geiger, P.G.; Girotti, A.W. Enzymatic Reducibility in Relation to Cytotoxicity for Various Cholesterol Hydroperoxides. *Biochemistry* **1996**, *35*, 8670–8679. [CrossRef]
111. Ruan, B.; Wilson, W.K.; Pang, J.; Schroepfer, G.J. Synthesis of [3alpha-3H]cholesta-5,8-dien-3beta-ol and tritium-labeled forms of other sterols of potential importance in the Smith-Lemli-Optiz syndrome. *Steroids* **2000**, *65*, 29–39. [CrossRef]
112. Porter, F.D.; Herman, G.E. Malformation syndromes caused by disorders of cholesterol synthesis. *J. Lipid Res.* **2010**, *52*, 6–34. [CrossRef] [PubMed]
113. Miller, W.L. Genetic disorders of Vitamin D biosynthesis and degradation. *J. Steroid Biochem. Mol. Boil.* **2017**, *165*, 101–108. [CrossRef]
114. Prabhu, A.V.; Luu, W.; Li, D.; Sharpe, L.J.; Brown, A.J. DHCR7: A vital enzyme switch between cholesterol and vitamin D production. *Prog. Lipid Res.* **2016**, *64*, 138–151. [CrossRef] [PubMed]
115. Tanret, C. Sur un Nouveau Principe Immediat de l"Ergot de Seigle. l'Ergosterine. *Ann. Chim. Phys.* **1890**, *6*, 289–293.
116. Tanret, C. Sur l'Ergosterine et la Fongisterine. *Ann. Chim. Phys.* **1908**, *8*, 313–317.
117. Meyer, K. On Catalytic Oxidations V. The Oxidation of Ergosterol. *J. Biol. Chem.* **1933**, *103*, 607–616.

118. Weenen, H.; Porter, N.A. Autoxidation of model membrane systems: Cooxidation of polyunsaturated lecithins with steroids, fatty acids, and alpha.-tocopherol. *J. Am. Chem. Soc.* **1982**, *104*, 5216–5221. [CrossRef]
119. Porter, N.A.; Weber, B.A.; Weenen, H. Autoxidation of polyunsaturated lipids. Factors controlling the stereochemistry of product hydroperoxides. *J. Am. Chem. Soc.* **1980**, *102*, 5597–5601. [CrossRef]
120. Roschek, B.; Tallman, K.A.; Rector, C.L.; Gillmore, J.; Pratt, D.A.; Punta, C.; Porter, N.A. Peroxyl Radical Clocks. *J. Org. Chem.* **2006**, *71*, 3527–3532. [CrossRef]
121. Xu, L.; Porter, N.A. Reactivities and Products of Free Radical Oxidation of Cholestadienols. *J. Am. Chem. Soc.* **2014**, *136*, 5443–5450. [CrossRef]
122. Rajeev, R.; Sunoj, R.B. Mechanism and Stereoselectivity of Biologically Important Oxygenation Reactions of the 7-Dehydrocholesterol Radical. *J. Org. Chem.* **2013**, *78*, 7023–7029. [CrossRef]
123. Korade, Z.; Xu, L.; Mirnics, K.; Porter, N.A. Lipid biomarkers of oxidative stress in a genetic mouse model of Smith-Lemli-Opitz syndrome. *J. Inherit. Metab. Dis.* **2012**, *36*, 113–122. [CrossRef]
124. Korade, Z.; Xu, L.; Shelton, R.; Porter, N.A. Biological activities of 7-dehydrocholesterol-derived oxysterols: Implications for Smith-Lemli-Opitz syndrome. *J. Lipid Res.* **2010**, *51*, 3259–3269. [CrossRef] [PubMed]
125. Pfeffer, B.A.; Xu, L.; Porter, N.A.; Rao, S.R.; Fliesler, S.J. Differential cytotoxic effects of 7-dehydrocholesterol-derived oxysterols on cultured retina-derived cells: Dependence on sterol structure, cell type, and density. *Exp. Eye Res.* **2016**, *145*, 297–316. [CrossRef] [PubMed]
126. Watabe, T.; Kanai, M.; Isobe, M.; Ozawa, N. The hepatic microsomal biotransformation of delta 5-steroids to 5 alpha, 6 beta-glycols via alpha- and beta-epoxides. *J. Boil. Chem.* **1981**, *256*, 2900–2907.
127. Mohana, K.; Achary, A. Human cytosolic glutathione-S-transferases: Quantitative analysis of expression, comparative analysis of structures and inhibition strategies of isozymes involved in drug resistance. *Drug Metab. Rev.* **2017**, *49*, 318–337. [CrossRef]
128. Meyer, D.J.; Ketterer, B. 5 alpha,6 alpha-Epoxy-cholestan-3 beta-ol (cholesterol alpha-oxide): A specific substrate for rat liver glutathione transferase B. *FEBS Lett.* **1982**, *150*, 499–502. [CrossRef]
129. Fliesler, S.J.; Peachey, N.S.; Herron, J.; Hines, K.; Weinstock, N.I.; Rao, S.R.; Xu, L. Prevention of Retinal Degeneration in a Rat Model of Smith-Lemli-Opitz Syndrome. *Sci. Rep.* **2018**, *8*, 1286. [CrossRef]
130. Kapphahn, R.J.; Richards, M.; Ferrington, D.A.; Fliesler, S.J. Lipid-derived and other oxidative modifications of retinal proteins in a rat model of Smith-Lemli-Opitz syndrome. *Exp. Eye Res.* **2019**, *178*, 247–254. [CrossRef]
131. Esterbauer, H. Estimation of peroxidative damage. A critical review. *Pathol. Boil.* **1996**, *44*, 25–28.
132. Doorn, J.A.; Petersen, D.R. Covalent Modification of Amino Acid Nucleophiles by the Lipid Peroxidation Products 4-Hydroxy-2-nonenal and 4-Oxo-2-nonenal. *Chem. Res. Toxicol.* **2002**, *15*, 1445–1450. [CrossRef]
133. Nadkarni, D.V.; Sayre, L.M. Structural Definition of Early Lysine and Histidine Adduction Chemistry of 4-Hydroxynonenal. *Chem. Res. Toxicol.* **1995**, *8*, 284–291. [CrossRef]
134. Uchida, K. 4-Hydroxy-2-nonenal: A product and mediator of oxidative stress. *Prog. Lipid Res.* **2003**, *42*, 318–343. [CrossRef]
135. Uchida, K.; Stadtman, E.R. Modification of histidine residues in proteins by reaction with 4-hydroxynonenal. *Proc. Natl. Acad. Sci. USA* **1992**, *89*, 4544–4548. [CrossRef] [PubMed]
136. West, J.D.; Marnett, L.J. Alterations in Gene Expression Induced by the Lipid Peroxidation Product, 4-Hydroxy-2-nonenal. *Chem. Res. Toxicol.* **2005**, *18*, 1642–1653. [CrossRef] [PubMed]
137. Butterfield, D.A.; Reed, T.; Perluigi, M.; De Marco, C.; Coccia, R.; Cini, C.; Sultana, R. Elevated protein-bound levels of the lipid peroxidation product, 4-hydroxy-2-nonenal, in brain from persons with mild cognitive impairment. *Neurosci. Lett.* **2006**, *397*, 170–173. [CrossRef]
138. Long, E.K.; Picklo, M.J., Sr. Trans-4-hydroxy-2-hexenal, a product of n-3 fatty acid peroxidation: Make some room HNE. *Free. Radic. Boil. Med.* **2010**, *49*, 1–8. [CrossRef]
139. Codreanu, S.G.; Kim, H.-Y.H.; Porter, N.A.; Liebler, D. Biotinylated probes for the analysis of protein modification by electrophiles. *Breast Cancer* **2012**, *803*, 77–95.
140. Codreanu, S.G.; Ullery, J.C.; Zhu, J.; Tallman, K.A.; Beavers, W.N.; Porter, N.A.; Marnett, L.J.; Zhang, B.; Liebler, D. Alkylation damage by lipid electrophiles targets functional protein systems. *Mol. Cell. Proteom.* **2014**, *13*, 849–859. [CrossRef]
141. Kim, H.-Y.H.; Tallman, K.A.; Liebler, D.; Porter, N.A. An azido-biotin reagent for use in the isolation of protein adducts of lipid-derived electrophiles by streptavidin catch and photorelease. *Mol. Cell. Proteom.* **2009**, *8*, 2080–2089. [CrossRef]

142. Sun, R.; Fu, L.; Liu, K.; Tian, C.; Yang, Y.; Tallman, K.A.; Porter, N.A.; Liebler, D.C.; Yang, J. Chemoproteomics Reveals Chemical Diversity and Dynamics of 4-Oxo-2-nonenal Modifications in Cells. *Mol. Cell. Proteom.* **2017**, *16*, 1789–1800. [CrossRef]
143. Tallman, K.A.; Kim, H.-Y.H.; Ji, J.-X.; Szapacs, M.E.; Yin, H.; McIntosh, T.J.; Liebler, D.C.; Porter, N.A.; Yin, H. Phospholipid–Protein Adducts of Lipid Peroxidation: Synthesis and Study of New Biotinylated Phosphatidylcholines. *Chem. Res. Toxicol.* **2007**, *20*, 227–234. [CrossRef]
144. Vila, A.; Tallman, K.A.; Jacobs, A.T.; Liebler, D.; Porter, N.A.; Marnett, L.J. Identification of Protein Targets of 4-Hydroxynonenal Using Click Chemistry for ex Vivo Biotinylation of Azido and Alkynyl Derivatives. *Chem. Res. Toxicol.* **2008**, *21*, 432–444. [CrossRef] [PubMed]
145. Vila, A.; Tallman, K.A.; Porter, N.; Liebler, D.C.; Marnett, L.J. Proteomic analysis of azido-HNE adducted proteins in RKO cells. *Chem. Res. Tox.* **2007**, *20*, 2010.
146. Higdon, A.N.; Dranka, B.; Hill, B.; Oh, J.-Y.; Johnson, M.S.; Landar, A.; Darley-Usmar, V. Methods for imaging and detecting modification of proteins by reactive lipid species. *Free. Radic. Boil. Med.* **2009**, *47*, 201–212. [CrossRef]
147. Sayre, L.M.; Lin, D.; Yuan, Q.; Zhu, X.; Tang, X. Protein Adducts Generated from Products of Lipid Oxidation: Focus on HNE and ONE. *Drug Metab. Rev.* **2006**, *38*, 651–675. [CrossRef]
148. Wentworth, A.D.; Nieva, J.; Takeuchi, C.; Galve, R.; Wentworth, A.D.; Dilley, R.B.; Delaria, G.A.; Saven, A.; Babior, B.M.; Janda, K.D.; et al. Evidence for Ozone Formation in Human Atherosclerotic Arteries. *Science* **2003**, *302*, 1053–1056. [CrossRef]
149. Wentworth, A.D.; Song, B.-D.; Nieva, J.; Shafton, A.; Tripurenani, S.; Wentworth, P. The ratio of cholesterol 5,6-secosterols formed from ozone and singlet oxygen offers insight into the oxidation of cholesterol in vivo. *Chem. Commun.* **2009**, 3098–3100. [CrossRef] [PubMed]
150. Scheinost, J.C.; Witter, D.P.; Boldt, G.E.; Offer, J.; Wentworth, P., Jr. Cholesterol secosterol adduction inhibits the misfolding of a mutant prion protein fragment that induces neurodegeneration. *Angew Chem. Int. Ed. Engl.* **2009**, *48*, 9469–9472. [CrossRef] [PubMed]
151. Nieva, J.; Song, B.-D.; Rogel, J.K.; Kujawara, D.; Altobel, L.; Izharrudin, A.; Boldt, G.E.; Grover, R.K.; Wentworth, A.D.; Wentworth, P. Cholesterol Secosterol Aldehydes Induce Amyloidogenesis and Dysfunction of Wild-Type Tumor Protein p53. *Chem. Boil.* **2011**, *18*, 920–927. [CrossRef]
152. Genaro-Mattos, T.; Appolinário, P.P.; Mugnol, K.C.U.; Bloch, C., Jr.; Nantes-Cardos, I.L.; Di Mascio, P.; Miyamoto, S. Covalent Binding and Anchoring of Cytochrome c to Mitochondrial Mimetic Membranes Promoted by Cholesterol Carboxyaldehyde. *Chem. Res. Toxicol.* **2013**, *26*, 1536–1544. [CrossRef]
153. Rostovtsev, V.; Green, L.G.; Fokin, V.V.; Sharpless, K.B. A Stepwise Huisgen Cycloaddition Process: Copper(I)-Catalyzed Regioselective "Ligation" of Azides and Terminal Alkynes. *Angew. Chem. Int. Ed.* **2002**, *41*, 2596–2599. [CrossRef]
154. Best, M.D. Click Chemistry and Bioorthogonal Reactions: Unprecedented Selectivity in the Labeling of Biological Molecules. *Biochemistry* **2009**, *48*, 6571–6584. [CrossRef] [PubMed]
155. Dantas, L.S.; Filho, A.B.C.; Coelho, F.R.; Genaro-Mattos, T.C.; Tallman, K.A.; Porter, N.A.; Augusto, O.; Miyamoto, S. Cholesterol secosterol aldehyde adduction and aggregation of Cu, Zn-superoxide dismutase: Potential implications in ALS. *Redox Boil.* **2018**, *19*, 105–115. [CrossRef] [PubMed]
156. Pratt, A.J.; Shin, D.S.; Merz, G.E.; Rambo, R.P.; Lancaster, W.A.; Dyer, K.N.; Borbat, P.P.; Poole, F.L.; Adams, M.W.W.; Freed, J.H.; et al. Aggregation propensities of superoxide dismutase G93 hotspot mutants mirror ALS clinical phenotypes. *Proc. Natl. Acad. Sci. USA* **2014**, *111*, E4568–E4576. [CrossRef] [PubMed]
157. Speen, A.M.; Kim, H.-Y.H.; Bauer, R.N.; Meyer, M.; Gowdy, K.M.; Fessler, M.B.; Duncan, K.E.; Liu, W.; Porter, N.A.; Jaspers, I. Ozone-derived Oxysterols Affect Liver X Receptor (LXR) Signaling. *J. Boil. Chem.* **2016**, *291*, 25192–25206. [CrossRef] [PubMed]
158. Pearl, L.; Prodromou, C.; Workman, P. The Hsp90 molecular chaperone: An open and shut case for treatment. *Biochem. J.* **2008**, *410*, 439–453. [CrossRef] [PubMed]
159. Connor, R.E.; Marnett, L.J.; Liebler, D. Protein-Selective Capture to Analyze Electrophile Adduction of Hsp90 by 4-Hydroxynonenal. *Chem. Res. Toxicol.* **2011**, *24*, 1275–1282. [CrossRef]
160. Tallman, K.A.; Kim, H.-Y.H.; Korade, Z.; Genaro-Mattos, T.; Wages, P.; Liu, W.; Porter, N.A. Probes for protein adduction in cholesterol biosynthesis disorders: Alkynyl lanosterol as a viable sterol precursor. *Redox Boil.* **2017**, *12*, 182–190. [CrossRef]

161. Schopfer, F.J.; Cipollina, C.; Freeman, B.A. Formation and Signaling Actions of Electrophilic Lipids. *Chem. Rev.* **2011**, *111*, 5997–6021. [CrossRef]
162. Garcia-Bermudez, J.; Baudrier, L.; Bayraktar, E.C.; Shen, Y.; La, K.; Guarecuco, R.; Yucel, B.; Fiore, D.; Tavora, B.; Freinkman, E.; et al. Squalene accumulation in cholesterol auxotrophic lymphomas prevents oxidative cell death. *Nature* **2019**, *567*, 118–122. [CrossRef]

© 2020 by the authors. Licensee MDPI, Basel, Switzerland. This article is an open access article distributed under the terms and conditions of the Creative Commons Attribution (CC BY) license (http://creativecommons.org/licenses/by/4.0/).

Article

Oxidative Repair of Pyrimidine Cyclobutane Dimers by Nitrate Radicals (NO₃•): A Kinetic and Computational Study

Tomas Haddad, Joses G. Nathanael, Jonathan M. White and Uta Wille *

School of Chemistry, Bio21 Institute, The University of Melbourne, 30 Flemington Road, Parkville, Victoria 3010, Australia; tomashaddad@gmail.com (T.H.); joses.nathanael@unimelb.edu.au (J.G.N.); whitejm@unimelb.edu.au (J.M.W.)
* Correspondence: uwille@unimelb.edu.au

Received: 14 April 2020; Accepted: 6 May 2020; Published: 9 May 2020

Abstract: Pyrimidine cyclobutane dimers are hazardous DNA lesions formed upon exposure of DNA to UV light, which can be repaired through oxidative electron transfer (ET). Laser flash photolysis and computational studies were performed to explore the role of configuration and constitution at the cyclobutane ring on the oxidative repair process, using the nitrate radical (NO₃•) as oxidant. The rate coefficients of $8-280 \times 10^7$ M^{-1} s^{-1} in acetonitrile revealed a very high reactivity of the cyclobutane dimers of *N,N'*-dimethylated uracil (DMU), thymine (DMT), and 6-methyluracil (DMU$^{6\text{-Me}}$) towards NO₃•, which likely proceeds via ET at N(1) as a major pathway. The overall rate of NO₃• consumption was determined by (i) the redox potential, which was lower for the *syn*- than for the *anti*-configured dimers, and (ii) the accessibility of the reaction site for NO₃•. In the *trans* dimers, both N(1) atoms could be approached from above and below the molecular plane, whereas in the *cis* dimers, only the convex side was readily accessible for NO₃•. The higher reactivity of the DMT dimers compared with isomeric DMU dimers was due to the electron-donating methyl groups on the cyclobutane ring, which increased their susceptibility to oxidation. On the other hand, the approach of NO₃• to the dimers of DMU$^{6\text{-Me}}$ was hindered by the methyl substituents adjacent to N(1), making these dimers the least reactive in this series.

Keywords: pyrimidine cyclobutane dimers; nitrate radicals; kinetic studies; DFT calculations; oxidation

1. Introduction

Exposure to UV light from the sun (λ = 180–400 nm) causes hazardous DNA lesions in the human body, which can lead to melanoma, basal cell, and squamous cell skin cancer [1]. The primary photo adducts implicated in skin cancer are pyrimidine cyclobutane dimers, which are formed through [2 + 2] cycloaddition between adjacent thymine nucleobases in the same oligonucleotide strand (Scheme 1, path A). A minor pathway (ca. 25%) proceeds through [2 + 2] photocycloaddition between the alkene and carbonyl moiety to give an oxetane intermediate that undergoes ring-opening to the 6–4 adduct (Scheme 1, path B) [2]. Cytosine forms the same type of dimers as thymine, but deamination occurs within only a few hours to yield the highly mutagenic uracil dimer (not shown) [3]. In general, pyrimidine cyclobutane dimers can exist in four isomeric structures with head-to-head (*syn*, s), head-to-tail (*anti*, a), as well as *cis* (c) and *trans* (t) configuration at the cyclobutane ring. However, because of geometrical constraints, only the *cis,syn* dimer is formed in DNA [4]. Such pyrimidine dimers can prevent DNA polymerases from transcribing DNA, thereby promoting the accumulation of errors [5].

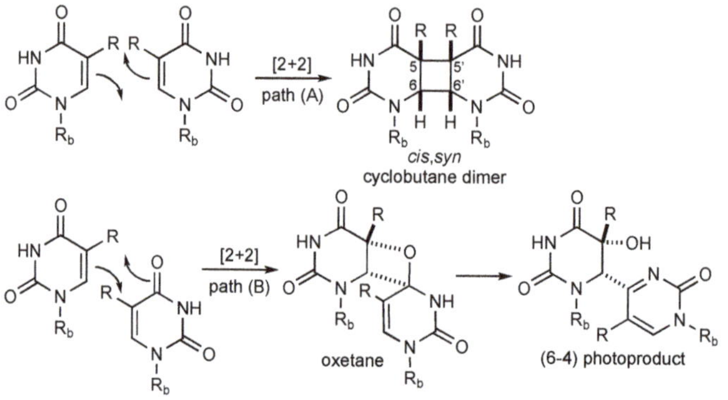

Scheme 1. Light-induced dimerization of pyrimidines. The deoxyribose phosphate backbone of DNA is abbreviated as R_b; thymine (R = Me), uracil (R = H).

The endogenous repair mechanism of damaged DNA nucleotides in placental mammals involves excision and replacement of the affected base or strand cluster that is orchestrated by a large system of reparative enzymes [6,7]. Other species, such as plants or fish, employ the enzyme photolyase and its reducing cofactor FADH (reduced flavine adenine dinucleotide) to repair dimer lesions through a reductive one-electron transfer (ET) process [8,9].

In addition to repair by excision or enzymatic reduction, pyrimidine cyclobutane dimers can also be oxidatively cleaved into their monomers, for example, by oxidizing radicals and radical ions [10]. Heelis et al. showed that the reaction of the thymine cyclobutane dimer with sulfate radical anions ($SO_4^{\bullet-}$) or hydroxyl radicals ($^\bullet OH$, this radical is commonly formed in vivo and can lead to oxidative damage in biomolecules [11]) led to the recovery of the monomeric pyrimidine and suggested a mechanism initiated by the abstraction of a hydrogen atom at C(6) (see below) [12]. In experiments involving the cis,syn-configured N,N'-dimethyluracil cyclobutane dimer (c,s-DMU<>DMU), Yan et al. showed that oxidative repair and formation of N,N'-dimethyluracil (DMU) occurred by reaction with $^\bullet OH$ (possibly through initial hydrogen abstraction) and bromide radical anions ($Br_2^{\bullet-}$, presumably through initial oxidation) but not with the azide radical ($N_3^\bullet$) [13], which could be due to the only moderately high oxidizing ability of this radical (E^0 ($N_3^\bullet/N_3^-$) = 1.35 V vs. NHE [14]). Comparable results were obtained by Ito et al., who found in the reaction of the C(5)–C(5') linked thymine dimer with $^\bullet OH$, $SO_4^{\bullet-}$ and $N_3^\bullet$ that repair only occurred in the case of the former two radicals [15].

In previous published and unpublished work, we performed product studies of the oxidative repair of pyrimidine cyclobutane dimers by nitrate radicals ($NO_3^\bullet$), which are important oxidants in the nighttime troposphere. $NO_3^\bullet$ was generated through photo-induced ET from cerium(IV) ammonium nitrate (CAN) at λ = 350 nm in acetonitrile in the presence of various stereoisomeric N,N'-dimethylated cyclobutane dimers of uracil (DMU), thymine (DMT), and the unnatural pyrimidine 6-methyl uracil (DMU^{6-Me}), according to Reaction (1) [16,17]:

$$(NH_4)_2Ce(NO_3)_6 + h\nu \, (350 \text{ nm}) \rightarrow NO_3^\bullet + (NH_4)_2Ce(NO_3)_5 \quad (1)$$

Table 1 shows the major products formed after a reaction time of two hours with a reactant ratio of [dimer]:[CAN] = 5. The data were obtained by gas chromatographic (GC) analysis of the reaction mixture from the relative peak areas [17] and, therefore, provided only a qualitative picture. In the case of c,s-DMU<>DMU, the major products were the monomer DMU and the partially cleaved dimer **1**, whereas only small amounts of the intact dimer remained (entry 1). Cyclobutane cleavage was considerably less efficient with t,s-DMU<>DMU, where mainly **1** and smaller amounts of DMU were obtained (entry 2). Likewise, repair of c,a-DMU<>DMU, t,a-DMU<>DMU, c,s-DMT<>DMT, and c,a-DMT<>DMT was also incomplete with c,a-DMT<>DMT being the least efficiently cleaved dimer under these conditions (entries 3–6). Repair of the *cis,syn* configured dimer of DMU[6-Me], on the other hand, appeared to be more efficient with only 22% of c,s-DMU[6-Me]<>DMU[6-Me] still present after the reaction (entry 7).

Table 1. Product and competition studies of the reaction of NO$_3^\bullet$ with isomeric pyrimidine cyclobutane dimers of DMU, DMT, and DMU[6-Me].[1]

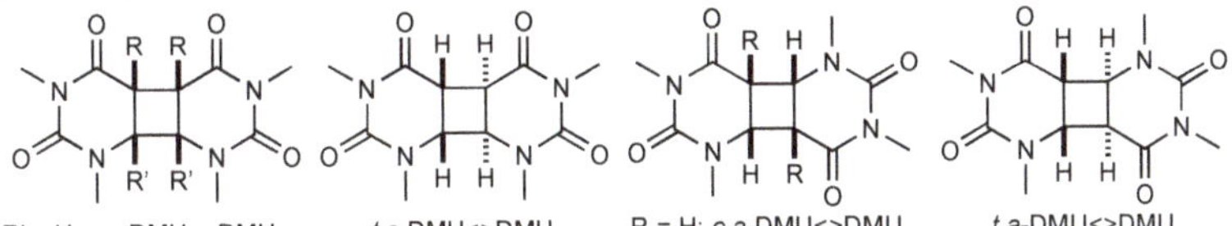

R, R' = H: c,s-DMU<>DMU
R = Me, R' = H: c,s-DMT<>DMT
R = H, R' = Me: c,s-DMU[6-Me]<>DMU[6-Me]

t,s-DMU<>DMU

R = H: c,a-DMU<>DMU
R = Me: c,a-DMT<>DMT

t,a-DMU<>DMU

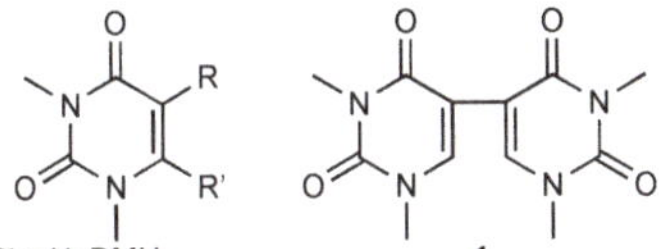

R, R' = H: DMU
R = Me, R' = H: DMT
R = H, R' = Me: DMU[6-Me]

1

Entry	Dimer(s) [1]	E^0 [2]	Products [3]
Product Studies [4]			
1 [5]	c,s-DMU<>DMU	1.825	c,s-DMU<>DMU (7%), DMU (50%), **1** (33%)
2 [5]	t,s-DMU<>DMU	1.850	t,s-DMU<>DMU (37%), DMU (14%), **1** (35%)
3 [5]	c,a-DMU<>DMU	≈2.195	c,a-DMU<>DMU (39%), DMU (33%)
4 [5]	t,a-DMU<>DMU		t,a-DMU<>DMU (30%), DMU (34%)
5	c,s-DMT<>DMT	1.815	c,s-DMT<>DMT (48%), DMT (48%)
6	c,a-DMT<>DMT		c,a-DMT<>DMT (58%), DMT (25%)
7	c,s-DMU[6-Me]<>DMU[6-Me]		c,s-DMU[6-Me]<>DMU[6-Me] (22%), DMU[6-Me] (73%)
Competition Studies [6]			
8	c,s-DMU<>DMU + t,s-DMU<>DMU		c,s-DMU<>DMU (76%), t,s-DMU<>DMU (24%)
9	c,a-DMU<>DMU + t,a-DMU<>DMU		c,a-DMU<>DMU (60%), t,a-DMU<>DMU (40%)
10	c,s-DMT<>DMT + c,a-DMT<>DMT		c,s-DMT<>DMT (19%), c,a-DMT<>DMT (81%)
11	c,s-DMU<>DMU + c,s-DMT<>DMT		c,s-DMU<>DMU (29%), c,s-DMT<>DMT (71%)
12	c,s-DMU<>DMU + c,s-DMU[6-Me]<>DMU[6-Me]		c,s-DMU<>DMU (8%), c,s-DMU[6-Me]<>DMU[6-Me] (92%)
13	c,s-DMT<>DMT + c,s-DMU[6-Me]<>DMU[6-Me]		c,s-DMT<>DMT (31%), c,s-DMU[6-Me]<>DMU[6-Me] (69%)

[1] DMU = 1,3-dimethyluracil, DMT = 1,3-dimethylthymine, DMU[6-Me] = 1,3,6-trimethyluracil; c = cis, t = trans, s = syn, a = anti. [2] E^0(dimer$^{\bullet+}$/dimer) in V vs. SCE (in MeCN; error ±0.025 V) determined by cyclic voltammetry [18]. [3] GC analysis, relative peak area of unreacted dimer and major products (>10%). [4] Reaction time 2 h, in acetonitrile; [dimer]:[CAN] = 5, unless stated otherwise. [5] Ref [17]. [6] Reaction time 2 h, in acetonitrile; [dimer 1]:[dimer 2]:[CAN] = 1:1:0.5; only the ratio of unreacted starting material was given, products arising from the reaction were not included.

Since NO$_3^\bullet$ is a strong one-electron oxidant (E^0 (NO$_3^\bullet$/NO$_3^-$) = 2.3–2.5 V vs. NHE [19]), these data suggested that NO$_3^\bullet$ initiated the cleavage process by oxidative ET to form the dimer radical cation, as shown in Scheme 2a for c,s-DMU<>DMU. Stepwise scission of the C(6)–C(6') bond followed by the C(5)–C(5') bond gives the repaired monomer DMU and its radical cation DMU$^{\bullet+}$ [17,18]. Similarly, oxidative repair of c,a-DMU<>DMU should proceed through the first scission of the C(6)–C(5') and then the C(5)–C(6') bond (Scheme 2b). However, according to the standard potentials E^0 available for some of the dimers [18], the rate of oxidation should not be significantly different for dimers with a *cis* or a *trans* configuration at the cyclobutane ring and whether or not a methyl group at the C(5) and C(5') position is present (entries 1, 2, and 5) [13], which contradicts some of the experimental findings from the product studies. On the other hand, the observed lower efficiency of the oxidative repair of the *anti*-configured dimers by NO$_3^\bullet$ confirmed previous findings with other oxidizing species [20] and could be rationalized by their higher E^0 value (see entry 3 as an example) [18,21]. It should be noted that in many cases, the extent of dimer consumption was larger than would be expected on the basis of the reactant ratio, suggesting that the monomer radical cation, such as DMU$^{\bullet+}$, could subsequently oxidize another dimer molecule, thereby propagating a radical chain process [17].

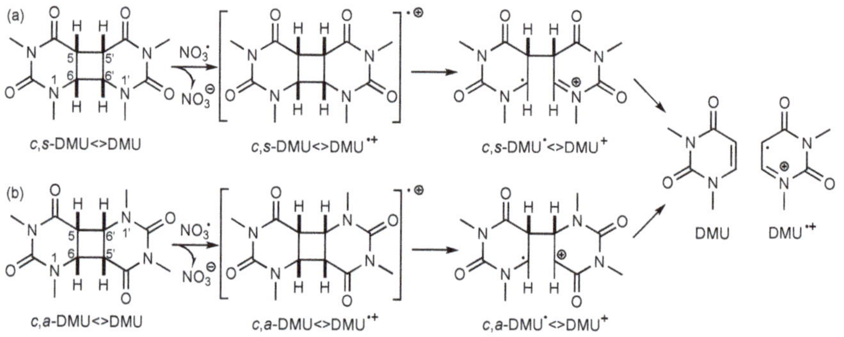

Scheme 2. Suggested cleavage mechanism of (a) c,s-DMU<>DMU and (b) c,a-DMU<>DMU following one-electron oxidation by NO$_3^\bullet$.

The discrepancies become even more apparent in competition studies, where equal amounts of two isomeric dimers were irradiated in the presence of one equivalent of CAN (entries 8–13). For example, a mixture of c,s- and t,s-DMU<>DMU revealed a much more efficient cleavage of the *trans,syn*-configured isomer (entry 8), whereas in a mixture of c,s-DMU<>DMU and c,s-DMT<>DMT, preferential cleavage of the former dimer occurred (entry 11) [22]. However, as mentioned above, it is highly likely that, apart from NO$_3^\bullet$-induced dimer cleavage, additional oxidation of both dimers could also occur by the monomer radical cations formed. Such competition experiments can, therefore, only provide qualitative information about the susceptibility of particular dimers for radical-induced oxidative repair. To obtain quantitative data whether and how the configuration and constitution at the cyclobutane ring affect the rate of the initial step of the oxidative repair, we performed kinetic studies to determine the absolute rate coefficients for the reaction of a series of stereo- and regioisomeric pyrimidine cyclobutane dimers with NO$_3^\bullet$, using laser flash photolysis techniques. Apart from being an important environmental oxidant, a major advantage of NO$_3^\bullet$ is that this radical allows investigating irreversible oxidation processes in biological molecules, since the back-electron transfer, which often hampers studies where the oxidant is produced through photo-induced ET in donor-acceptor pairs, cannot occur. The experimental studies were augmented with density functional theory (DFT) calculations to provide further mechanistic insight.

2. Materials and Methods

2.1. Synthesis of the Pyrimidine Cyclobutane Dimers

2.1.1. General

The starting materials for the synthesis were purchased from commercial suppliers (Sigma-Aldrich, Castle Hill, Australia) and used without purification. Thin-layer chromatography (TLC) was performed to monitor the reactions using aluminum plates coated with silica gel 60 F_{254} (Merck, ASIS Scientific, Hindmarsh, Australia). UV light at $\lambda = 254$ nm, potassium permanganate ($KMnO_4$) stain followed by heating, or iodine ground in Davisil Chromatography Silica Gel LC60A (40–63 microns, 230–400 mesh) were used to visualize TLC plates. The crude products were purified by recrystallization from hot solvent or silica column chromatography with approximately 30–50 g of dry silica (Davisil Chromatography Silica Gel LC60A, 40–63 microns, 230–400 mesh) per 1 g of the crude product mixture. The eluting solvent consisted of a mixture of chloroform and acetone. Solvents were removed under reduced pressure and elevated temperature using a rotary evaporator (Büchi, In Vitro Technologies, Melbourne, Australia). The purity was assessed by an Agilent 1100 reversed-phase HPLC on a Phenomenex Aeris XB-C18 column 250 mm × 4.5 mm × 3.6 µm (Gradient: 100% water buffered with 0.1% trifluoroacetic acid (TFA) to 100% acetonitrile buffered with 0.1% TFA over 25 min, 4% min^{-1}, flow rate: 1 mL min^{-1}).

^1H and ^{13}C-NMR spectra were recorded on an Agilent MR 400 MHz NMR spectrometer or an Agilent DD2 500 MHz NMR spectrometer in deuterated dimethyl sulfoxide (DMSO-d_6) or in deuterated chloroform (CDCl$_3$) at 25 °C. Chemical shifts were expressed in parts per million (ppm, δ) relative to either DMSO-d_6 (^1H δ = 2.50 ppm, ^{13}C δ = 39.5 ppm) or CDCl$_3$ (^1H δ = 7.26 ppm, ^{13}C δ = 77.2 ppm). High-resolution mass spectrometry (HRMS) was performed by ionizing the sample using electrospray ionization (ESI) into a Thermo Scientific Exactive Plus Orbitrap mass spectrometer (Thermo Scientific Australia, Scoresby, Australia). The detected molecular ions were formed in the positive ion mode and were expressed as $[M + H]^+$. The ^1H and ^{13}C-NMR spectra of the pyrimidines and pyrimidine cyclobutane dimers prepared in this study are provided in the Supplementary Material.

2.1.2. Synthesis of N-Methylated Pyrimidines

Method 1. The pyrimidine was stirred in a solution of powdered potassium hydroxide (8 equiv.) and dimethyl sulfoxide (1 mL per mmol of KOH) at 0 °C for 10 min under an inert atmosphere. Methyl iodide (16 equiv.) was then added dropwise, and the mixture was stirred overnight at room temperature. The solution was diluted with cold water and extracted three times with dichloromethane. The combined organic layers were washed with 2 M aqueous sodium hydroxide solution and brine, dried over magnesium sulfate, and the solvent was removed in vacuo. The crude product was purified by recrystallization from ethanol.

Method 2. The pyrimidine and potassium carbonate (3.5 equiv.) were refluxed in acetone (2 mL per mmol of pyrimidine) for 10 min under an inert atmosphere. Methyl iodide (8 equiv.) was then added dropwise, and the reaction mixture was stirred overnight. The reaction was quenched with water (2 mL per mL of acetone) and extracted three times with dichloromethane. The combined organic layers were washed with 2 M aqueous sodium hydroxide solution and brine, dried over magnesium sulfate, and the solvent was removed in vacuo. The crude solid was purified by recrystallization from ethanol.

1,3-Dimethyluracil (DMU)

Prepared from uracil using both methods 1 and 2. ^1H-NMR (400 MHz, DMSO-d_6): δ 7.66 (d, J = 7.8 Hz, 1H), 5.66 (d, J = 7.8 Hz, 1H), 3.29 (s, 3H), 3.15 ppm (s, 3H). ^{13}C {^1H}-NMR (101 MHz, DMSO-d_6): δ 162.8, 151.5, 144.6, 99.6, 36.4, 27.2 ppm. HRMS (ESI) m/z calcd. for $[C_6H_9N_2O_2]^+$: 141.0664 $[M + H]^+$, found 141.0660.

1,3-Dimethylthymine (DMT)

Prepared from thymine using both methods 1 and 2. ^{1}H-NMR (400 MHz, DMSO-d_6): δ 7.56 (s, 1H), 3.26 (s, 3H), 3.16 (s, 3H), 1.79 ppm (s, 3H). ^{13}C {^{1}H}-NMR (101 MHz, DMSO-d_6): δ 163.4, 151.3, 140.7, 107.1, 36.1, 27.4, 12.5 ppm. HRMS (ESI) m/z calcd. for [C$_7$H$_{11}$N$_2$O$_2$]$^+$: 155.0821 [M + H]$^+$, found 155.0815.

1,3,6-Trimethyluracil (DMU$^{6\text{-Me}}$)

N,N'-Dimethylurea (8.81 g, 100 mmol), 4-dimethylamino pyridine (12.2 g, 100 mmol), and acetic anhydride (31.2 mL, 330 mmol) were stirred in anhydrous pyridine (150 mL) under an inert atmosphere at room temperature for 2 h. The resulting red solution was concentrated in vacuo; the residue was diluted with dichloromethane (700 mL) and washed sequentially with 2 M aqueous hydrochloric acid (100 mL), 2 M aqueous sodium bicarbonate solution (100 mL), and 2 M aqueous copper(II) sulfate solution (100 mL). The organic layer was dried over magnesium sulfate. The solvent was removed in vacuo. The crude pale-yellow solid was recrystallized from ethyl acetate and washed repeatedly with cold ethyl acetate until the yellow color had disappeared. White crystals (10.9 g, 70%). ^{1}H-NMR (400 MHz, CDCl$_3$): δ 5.61 (s, 1H), 3.39 (s, 3H), 3.32 (s, 3H), 2.23 ppm (s, 3H). ^{13}C {^{1}H}-NMR (101 MHz, CDCl$_3$): δ 162.4, 152.6, 151.4, 101.2, 31.7, 27.9, 20.2 ppm. HRMS (ESI) m/z calcd. for [C$_7$H$_{11}$N$_2$O$_2$]$^+$: 155.0821 [M + H]$^+$, found 155.0817.

2.1.3. Dimerization of N-Methylated Pyrimidines

The pyrimidine was dissolved in UV-grade acetone and degassed under an inert atmosphere with sonication for 30 min. The solution was then added to a Pyrex reactor and diluted to 300 mL with UV-grade acetone. Under an inert atmosphere, the solution was stirred for 10 min before exposure to a medium-pressure mercury lamp. Reaction progress was monitored using TLC (KMnO$_4$ stain) until complete consumption of starting material was observed. The solvent was removed in vacuo to afford a mixture of the dimerized substrates as a white solid, which were separated by column chromatography (SiO$_2$, a mixture of CHCl$_3$/acetone). The separated dimers were further purified by recrystallization from ethyl acetate or ethanol and identified by X-ray crystallography. The X-ray structures are given in the Supplementary Material.

Synthesis of DMU<>DMU Isomers

DMU (14.00 g, 100.00 mmol) was dimerized with an overall yield of 25% (3.57 g, 12.70 mmol).

(a) t,a-DMU<>DMU: White crystals (0.35 g, 1.25 mmol, 3%, R_f = 0.26 (EtOAc)). ^{1}H-NMR (400 MHz, CDCl$_3$): δ 4.10 (dd, J = 9.4, 4.9 Hz, 2H), 3.53 (dd, J = 9.4, 4.9 Hz, 2H), 3.24 (s, 6H), 3.08 ppm (s, 6H). ^{13}C {^{1}H}-NMR (101 MHz, CDCl$_3$): δ 167.1 (2C), 151.6 (2C), 53.7 (2C), 44.5 (2C), 33.8 (2C), 28.1 ppm (2C). HRMS (ESI) m/z calcd. for [C$_{12}$H$_{17}$N$_4$O$_4$]$^+$: 281.1250 [M + H]$^+$, found 281.1245.

(b) t,s-DMU<>DMU: White crystals (1.12 g, 4.00 mmol, 8%, R_f = 0.18 (EtOAc)). ^{1}H-NMR (400 MHz, CDCl$_3$): δ 3.87 (d, J = 8.0 Hz, 2H), 3.62 (d, J = 8.0 Hz, 2H), 3.25 (s, 6H), 3.06 ppm (s, 6H). ^{13}C {^{1}H}-NMR (101 MHz, CDCl$_3$): δ 168.1 (2C), 151.8 (2C), 59.3 (2C), 39.1 (2C), 35.1 (2C), 28.3 ppm (2C). HRMS (ESI) m/z calcd. for [C$_{12}$H$_{17}$N$_4$O$_4$]$^+$: 281.1250 [M + H]$^+$, found 281.1245.

(c) c,a-DMU<>DMU: White crystals (0.56 g, 2.00 mmol, 4%, R_f = 0.08 (EtOAc)). ^{1}H-NMR (400 MHz, CDCl$_3$): δ 4.09 (t, J = 8.5 Hz, 2H), 3.77 (t, J = 8.6 Hz, 2H), 3.12 (s, 6H), 3.11 ppm (s, 6H). ^{13}C {^{1}H}-NMR (101 MHz, CDCl$_3$): δ 166.1 (2C), 152.2 (2C), 49.3 (2C), 45.4 (2C), 35.7 (2C), 27.8 ppm (2C). HRMS (ESI) m/z calcd. for [C$_{12}$H$_{17}$N$_4$O$_4$]$^+$: 281.1250 [M + H]$^+$, found 281.1245.

(d) c,s-DMU<>DMU: White crystals (1.54 g, 5.49 mmol, 11%, R_f = 0.04 (EtOAc)). ^{1}H-NMR (400 MHz, CDCl$_3$): δ 4.06 (dd, J = 6.1, 3.9 Hz, 2H), 3.78 (dd, J = 6.1, 3.9 Hz, 2H), 3.16 (s, 6H), 3.00 ppm (s, 6H). ^{13}C {^{1}H}-NMR (101 MHz, CDCl$_3$): δ 165.8 (2C), 152.8 (2C), 55.5 (2C), 39.6 (2C), 35.7 (2C), 28.0 ppm (2C). HRMS (ESI) m/z calcd. for [C$_{12}$H$_{17}$N$_4$O$_4$]$^+$: 281.1250 [M + H]$^+$, found 281.1245.

Synthesis of DMT<>DMT Isomers

DMT (9.30 g, 60.30 mmol) was dimerized with an overall yield of 29% (2.73 g, 8.85 mmol).

(a) c,a-DMT<>DMT: White crystals (0.70 g, 2.27 mmol, 7%, R_f = 0.04 (1:1 n-pentane/EtOAc)). ^{1}H-NMR (400 MHz, CDCl$_3$): δ 3.27 (s, 2H), 3.16 (s, 6H), 3.08 (s, 6H), 1.58 ppm (s, 6H). ^{13}C {^{1}H}-NMR (101 MHz, CDCl$_3$): δ 169.6 (2C), 151.9 (2C), 64.4 (2C), 48.8 (2C), 36.3 (2C), 27.9 (2C), 25.0 ppm (2C). HRMS (ESI) m/z calcd. for [C$_{14}$H$_{21}$N$_4$O$_4$]$^+$: 309.1563 [M + H]$^+$, found 309.1557.

(b) c,s-DMT<>DMT: White crystals (2.03 g, 6.58 mmol, 22%, R_f = 0.10 (EtOAc)). ^{1}H-NMR (500 MHz, CDCl$_3$): δ 3.72 (s, 2H), 3.14 (s, 6H), 3.00 (s, 6H), 1.50 ppm (s, 6H). ^{13}C {^{1}H}-NMR (126 MHz, CDCl$_3$): δ 169.5 (2C), 152.5 (2C), 60.6 (2C), 47.6 (2C), 35.9 (2C), 28.3 (2C), 19.4 ppm (2C). HRMS (ESI) m/z calcd. for [C$_{14}$H$_{21}$N$_4$O$_4$]$^+$: 309.1563 [M + H]$^+$, found 309.1582.

Synthesis of DMU$^{6\text{-Me}}$<>DMU$^{6\text{-Me}}$ Isomers

DMU$^{6\text{-Me}}$ (7.71 g, 50.00 mmol) was dimerized with an overall yield of 67% (5.16 g, 16.70 mmol). The crystal structure of t,s-DMU$^{6\text{-Me}}$<>DMU$^{6\text{-Me}}$ could not be solved, and this compound was, therefore, characterized by comparing the NMR data with ref. [22].

(a) t,s-DMU$^{6\text{-Me}}$<>DMU$^{6\text{-Me}}$: White crystals (0.21 g, 0.67 mmol, 3%, R_f = 0.05 (1:1 n-pentane/EtOAc)). ^{1}H-NMR (400 MHz, CDCl$_3$): δ 3.27 (s, 6H), 3.23 (s, 2H), 3.00 (s, 6H), 1.42 ppm (s, 6H). ^{13}C {^{1}H}-NMR (101 MHz, CDCl$_3$): δ 168.1 (2C), 152.4 (2C), 62.2 (2C), 45.1 (2C), 31.2 (2C), 28.5 (2C), 22.3 ppm (2C). HRMS (ESI) m/z calcd. for [C$_{14}$H$_{21}$N$_4$O$_4$]$^+$: 309.1563 [M + H]$^+$, found 309.1556.

(b) c,a-DMU$^{6\text{-Me}}$<>DMU$^{6\text{-Me}}$: White crystals (4.11 g, 13.30 mmol, 53%, R_f = 0.19 (EtOAc)). ^{1}H-NMR (400 MHz, CDCl$_3$): δ 3.21 (s, 6H), 3.16 (s, 2H), 2.82 (s, 6H), 1.65 ppm (s, 6H). ^{13}C {^{1}H}-NMR (101 MHz, CDCl$_3$): δ 165.3 (2C), 152.0 (2C), 58.0 (2C), 53.6 (2C), 31.2 (2C), 28.2 (2C), 27.8 ppm (2C). HRMS (ESI) m/z calcd. for [C$_{14}$H$_{21}$N$_4$O$_4$]$^+$: 309.1563 [M + H]$^+$, found 309.1557.

(c) c,s-DMU$^{6\text{-Me}}$<>DMU$^{6\text{-Me}}$: White crystals (0.82 g, 2.67 mmol, 11%, R_f = 0.03 (2:1 n-pentane/EtOAc)). ^{1}H-NMR (400 MHz, CDCl$_3$): δ 3.41 (s, 2H), 3.16 (s, 6H), 2.92 (s, 6H), 1.52 ppm (s, 6H). ^{13}C {^{1}H}-NMR (101 MHz, CDCl$_3$): δ 165.6 (2C), 152.8 (2C), 62.5 (2C), 44.9 (2C), 32.2 (2C), 28.0 (2C), 21.6 ppm (2C). HRMS (ESI) m/z calcd. for [C$_{14}$H$_{21}$N$_4$O$_4$]$^+$: 309.1563 [M + H]$^+$, found 309.1557.

2.2. X-ray Data

Intensity data were collected on a Rigaku XtalLAB Synergy at 100.0(1) K diffractometer using either Cu-Kα or Mo-Kα radiation at 100.0(1) K. The temperature was maintained using an Oxford Cryostream cooling device. The structures were solved by direct methods and difference Fourier synthesis [23]. The thermal ellipsoid plot was generated using the program Mercury-3 [24] integrated within the WINGX [25] suite of programs. Detailed information on the crystal structures is provided in the Supplementary Material. X-ray crystal structures have been deposited at the Cambridge Structural Database and assigned the CCDC codes 1995805–1995812.

2.3. Laser Flash Photolysis Studies

Kinetic measurements were performed using an Edinburgh Instruments LP920 spectrometer using the third harmonic of a Quantel Brilliant B Nd:YAG laser (6 ns pulse; 10–20 mJ/pulse, λ = 355 nm) to generate the radical transient. A Hamamatsu R2856 photomultiplier tube (PMT) interfaced with a Tektronix TDS 3012C Digital Phosphor Oscilloscope was used for the detection system. Samples were exposed to laser light in Starna Spectrosil Quartz fluorometer cells (10 × 10 × 48 mm).

All kinetic experiments were carried out under pseudo-first-order conditions following the established procedure described in refs. [26–28], with the pyrimidine monomer and cyclobutane dimer as the excess component ([substrate] = 0.2–11 mM, [CAN] = 0.33 mM; using the molar extinction coefficient of ε = 1350 M^{-1} cm^{-1} at λ = 630 nm in acetonitrile [29], [NO$_3$$^{\bullet}$] in the range of 108–135 µM was generated). Each data point was determined from the average of three measurements. Experimental

2.4. DFT Calculations

Density functional theory (DFT) calculations were carried out with the Gaussian suite of programs [30] using the M062X method [31–33] in combination with the 6-31+G* basis set, which has been employed previously to investigate reactions of $NO_3^\bullet$ with biomolecules [26,28]. Calculations in acetonitrile were performed using the conductor-like polarizable continuum model (CPCM) [34]. All equilibrium geometries and transition structures were verified by vibrational frequency analysis at the same level of theory, and all identified transition structures showed only one imaginary frequency. The spin expectation value, $<s^2>$, was very close to 0.75 after spin annihilation. The Gaussian archive entries for all optimized geometries, including free energy data and imaginary frequencies of the transition structures, are given in the Supplementary Material.

3. Results

The pyrimidine cyclobutane dimers studied in this work are shown in Figure 1. These compounds were prepared by photochemical dimerization of the respective N,N'-dimethylated pyrimidine using acetone as triplet sensitizer, as described previously (see Materials and Methods section) [17]. Under such conditions, dimerization was not stereospecific, and formation of *syn*- and *anti*-isomers with both *cis* and *trans* stereochemistry at the cyclobutane ring could principally occur. However, only in the case of DMU, all four possible stereoisomers were formed in synthetically useful yield, whereas, for DMT, only the *cis*-configured isomers, and, in the case of $DMU^{6\text{-Me}}$, all isomers, except for the *trans,anti*-isomer, were obtained.

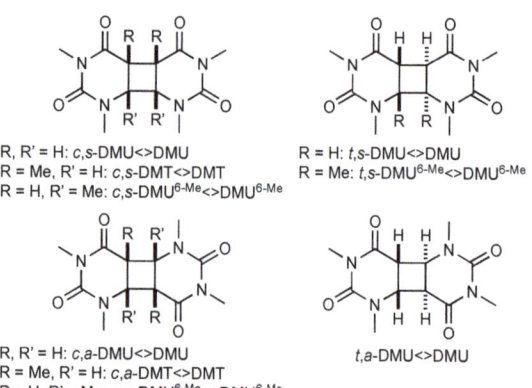

Figure 1. Pyrimidine cyclobutane dimers studied in this work. DMU = 1,3-dimethyluracil, DMT = 1,3-dimethylthymine, $DMU^{6\text{-Me}}$ = 1,3,6-trimethyluracil; c = cis, t = trans, s = syn, a = anti.

The laser flash photolysis experiments were performed in acetonitrile under pseudo-first-order conditions with the dimer as an excess component by monitoring the decay of the $NO_3^\bullet$ signal at λ = 630 nm. In previous work, we found that purging the solutions to remove oxygen was detrimental to the $NO_3^\bullet$ signal for reasons not yet understood [26]. However, since the lifetime of $NO_3^\bullet$ is independent of the presence of oxygen [29], the rate data were measured without degassing the reaction mixtures.

The decay profiles depicted in Figure 2a for the exemplary reaction of $NO_3^\bullet$ with different excess concentrations of c,s-DMT<>DMT clearly showed that the rate of $NO_3^\bullet$ consumption increased with increasing [c,s-DMT<>DMT]. Determination of the second-order rate coefficient of the reaction, k, was obtained from the slope of the plot of the pseudo-first-order rate coefficient k_{obs} vs. [c,s-DMT<>DMT] (Figure 2b).

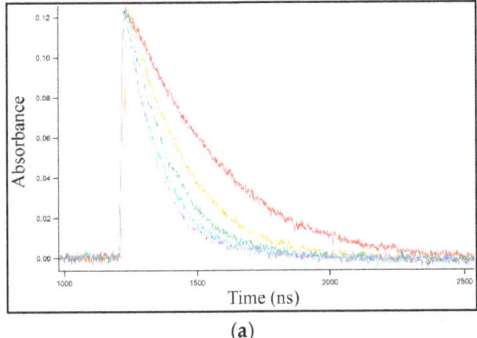
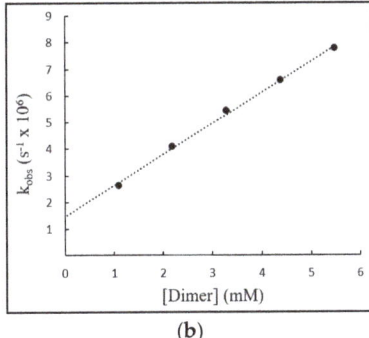

Figure 2. The reaction of $NO_3^{\bullet}$ with c,s-DMT<>DMT. (**a**) Concentration-time profiles for $NO_3^{\bullet}$ recorded at λ = 630 nm with different excess concentrations of c,s-DMT<>DMT: red 1.1 mM, yellow 2.2 mM, green 3.3 mM, blue 4.4 mM, purple 5.5 mM. (**b**) The plot of the pseudo-first-order rate coefficient k_{obs} vs. [dimer] to determine the second-order rate coefficient k from the slope.

The intercept in Figure 2b could be attributed to the background reaction of $NO_3^{\bullet}$ with the solvent acetonitrile, which proceeded via hydrogen abstraction (HAT) with a rate coefficient of 9×10^2 M^{-1} s^{-1} and led to the depletion of the $NO_3^{\bullet}$ signal after about 300 µs [26]. However, this decay was much slower than the consumption of $NO_3^{\bullet}$ through reaction with the cyclobutane dimers, which occurred on the nanosecond timescale and could, therefore, be neglected.

The second-order rate coefficients k for the reactions of $NO_3^{\bullet}$ with the pyrimidine cyclobutane dimers and with the respective monomers are compiled in Table 2.

Table 2. Absolute second-order rate coefficients k for the reaction of $NO_3^{\bullet}$ with pyrimidine cyclobutane dimers and the corresponding monomers using nanosecond laser flash photolysis [1,2].

Entry	Dimer	k/M^{-1} s^{-1}	Monomer: k/M^{-1} s^{-1}
1	t,s-DMU<>DMU	2.8×10^9	
2	c,s-DMU<>DMU	9.0×10^8	DMU: 1.0×10^9
3	t,a-DMU<>DMU	5.5×10^8	
4	c,a-DMU<>DMU	3.1×10^8	
5	c,s-DMT<>DMT	1.2×10^9	DMT: 6.3×10^9
6	c,a-DMT<>DMT	5.2×10^8	
7	t,s-DMU$^{6\text{-Me}}$<>DMU$^{6\text{-Me}}$	5.7×10^8	
8	c,s-DMU$^{6\text{-Me}}$<>DMU$^{6\text{-Me}}$	3.9×10^8	DMU$^{6\text{-Me}}$: 4.8×10^9
9	c,a-DMU$^{6\text{-Me}}$<>DMU$^{6\text{-Me}}$	8.0×10^7	

[1] In acetonitrile, at 25 ± 1 °C. [2] Experimental error ±15%.

The rate data ranged from $8-280 \times 10^7$ M^{-1} s^{-1}, revealing a very high susceptibility of all pyrimidine cyclobutane dimers towards reaction with $NO_3^{\bullet}$. This high reactivity was further illustrated by comparison with the rate coefficients for the reaction of $NO_3^{\bullet}$ with amino acids and short peptides, which were in the range of 10^6-10^8 M^{-1} s^{-1}, with the fastest reactions occurring with aromatic amino acids and proline [26–28,35]. In general, for a particular pyrimidine, the *syn* isomers reacted faster than the *anti*-configured isomers by a factor of about 2–4 (entries 1 vs. 3, 2 vs. 4, 5 vs. 6, and 8 vs. 9), and the *trans* isomers reacted faster than the *cis* isomers by a factor of about 1.5–3 (entries 1 vs. 2, 3 vs. 4, and 7 vs. 8). This finding largely supported the observations from the previous competition studies (see Table 1; for example, entries 8 vs. 9 and entry 10). Overall, the reactivity increased in the order DMU$^{6\text{-Me}}$<>DMU$^{6\text{-Me}}$ → DMU<>DMU → DMT<>DMT for dimers with similar geometry and configuration at the cyclobutane ring. For example, the ratio of the rate coefficients for c,s-DMU$^{6\text{-Me}}$<>DMU$^{6\text{-Me}}$ and c,s-DMT<>DMT was about 0.3 (entry 8 vs. 5) and only about 0.15 for the pair c,a-DMU$^{6\text{-Me}}$<>DMU$^{6\text{-Me}}$ and c,a-DMT<>DMT (entry 9 vs. 6). This finding was remarkable since the only difference between the latter two dimers was the site of the methyl groups at the cyclobutane ring.

Rate coefficients were also determined for the reaction of NO$_3$• with the monomers formed upon dimer cleavage (see Scheme 2). The data, which are included in Table 2, revealed also a very high reactivity with these pyrimidines. According to previous work by us, the initial step in the reaction of NO$_3$• with pyrimidines likely proceeded through oxidative ET, leading to different products depending on the nature of the pyrimidine [36,37]. However, since the kinetic measurements were performed with a large excess of the dimers to ensure pseudo-first-order conditions, errors arising from the second-order consumption of NO$_3$• through reaction with the respective monomers should be negligible.

4. Discussion

The overall trend that for the same stereochemistry and configuration at the cyclobutane ring (where available), DMT<>DMT is the most reactive and DMU$^{6\text{-Me}}$<>DMU$^{6\text{-Me}}$ the least reactive dimer system is remarkable: it indicates that not only the absence or presence of a methyl group at the cyclobutane ring but also the site of methyl substitution significantly impacts the reaction rate. For example, the rate coefficients for the *cis,syn*-configured dimers, which would be expected to react with NO$_3$• through ET with similar rates on the basis of the available E^0 data (see Table 1), varied by a factor of 3 (entries 2, 5, vs. 8). This finding raised the question of whether these reactions proceeded indeed via initial ET, in particular since reactions of NO$_3$• with biomolecules, such as peptides, commonly also occur through HAT [26,28]. Therefore, we explored possible alternative mechanisms in the reaction of NO$_3$• with selected pyrimidine cyclobutane dimers using density functional theory (DFT) calculations (see Materials and Methods section).

Using c,a-DMU<>DMU as a model system, potential HAT pathways were examined by calculating the energies associated for hydrogen abstraction from the four different positions in the dimer, i.e., from C(6), C(5), and from the methyl groups at N(1) and N(3), respectively (Scheme 3a).

The computations predicted the formation of a charge-transfer (CT) complex between the dimer and NO$_3$• prior to any actual reaction. The optimized geometry of the CT complex, which is included in Scheme 3, showed that NO$_3$• coordinated via two of its oxygen atoms to N(1) and the hydrogen at C(6) of the dimer, but the structure of c,a-DMU<>DMU itself was very similar to that of the 'intact' dimer (not shown). However, analysis of the spin density in this reactant complex showed a partition of the spin between NO$_3$• and N(1). Furthermore, the Mulliken charges showed a value of −0.39e on NO$_3$•, indicating an accumulation of negative charge on this moiety. The calculated energy of a reactant complex without such CT interactions was higher by 27.4 kJ mol^{-1} (not shown), and this complex was used as a reference point in Scheme 3a with its energy set to 0 kJ mol^{-1}.

NO$_3$•-induced HAT from all four positions in c,a-DMU<>DMU was thermodynamically highly favorable with reaction energies exceeding −40 kJ mol^{-1}. Of the four possible HAT pathways, abstraction from the methyl group at N(1) via *TS3* is predicted to be the kinetically most preferred pathway, which is only about 8 kJ mol^{-1} above the energy of the CT complex. Hydrogen abstraction from C(6) is slightly less favorable, with *TS1* being about 10 kJ mol^{-1} higher in energy than *TS3*. In contrast to this, the barriers for HAT from C(5) via *TS2* and from the methyl group at N(3) via *TS4* are considerably higher, and both pathways are, therefore, not kinetically competitive compared with the reaction channels via *TS1* and *TS3* [38].

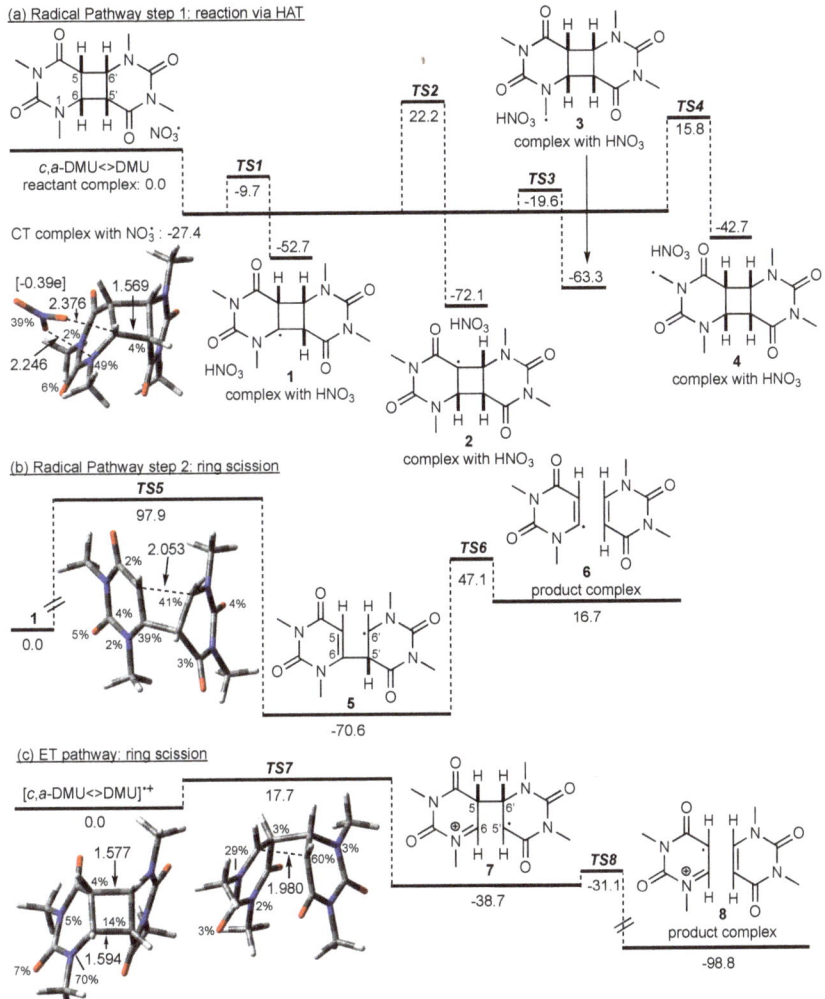

Scheme 3. Possible pathways in the reaction of NO$_3$• with *c,a*-DMU<>DMU. M062X/6-31+G* geometries and free energies in kJ mol^{-1} (in acetonitrile). Spin density in percentage, distances in Å, Mulliken charge on the NO$_3$ fragment in square brackets. For the HAT pathway (**a**), the energy of the reactant complex without CT interactions was used as a reference point; the energy for the radical ring scission in (**b**) was given relative to the dimer radical **1**; for the ET pathway (**c**), the energy of the dimer radical cation was used as a reference point (see text).

It is highly unlikely that the dimer radical **3** resulting from the kinetically most favorable HAT from the N(1)-Me group could lead to cleavage of the cyclobutane ring in a straightforward fashion. On the other hand, to explore whether the formation of a radical site at the cyclobutane ring itself could principally lead to ring scission, which has been suggested for the reaction of pyrimidine cyclobutane dimers with other O-centered radicals [12], homolytic fragmentation in the C(6) radical **1** was exemplary calculated. The energy profile for this reaction is given in Scheme 3b, which showed that the stepwise bond scissions were associated with high barriers **TS5** and **TS6**. The geometry of **TS5** was characterized by a similar spin density at C(6) and C(6′) on both pyrimidine moieties and a considerably large distance between C(5) and C(6') of 2.053 Å, suggesting a relatively late transition state. Furthermore, while the formation of **5** through scission of the C(5)–C(6') bond was exothermic, the subsequent cleavage of the C(6)–C(5') bond and formation of the product complex consisting of the repaired monomer and the monomer vinyl radical was endothermic, rendering repair through a radical-only mechanism as highly unlikely. Based on these data, $NO_3^{\bullet}$-induced HAT in *c,a*-DMU<>DMU could, therefore, not be considered as a 'productive' pathway with regards to cleavage of the cyclobutane ring under these conditions.

In contrast to this, oxidation of *c,a*-DMU<>DMU by $NO_3^{\bullet}$ led to the dimer radical cation [*c,a*-DMU<>DMU]$^{\bullet+}$, which still had an intact cyclobutane ring where most of the spin was localized on N(1). The energy diagram for the cleavage of *c,a*-DMU<>DMU]$^{\bullet+}$ is shown in Scheme 3c. Thus, fragmentation of the C(6)–C(5') bond was associated with a barrier **TS7** of only about 18 kJ mol^{-1} and gave the singly bonded dimer **7** in an exothermic process. The distance between C(6) and C(5') in **TS7** was considerably shorter than that of the breaking bond in **TS5**, which was in line with an earlier, energetically more favorable process. Subsequent homolytic scission of the C(5)–C(6') bond via **TS8** was only about 8 kJ mol^{-1} higher in energy than **7** and should occur rapidly to give the product complex **8** in a highly exothermic reaction.

Scheme 4a shows the calculated energy profile for the reaction of $NO_3^{\bullet}$ with the isomeric dimer *c,s*-DMU<>DMU. Interestingly, the formation of a complex with a broken C(6)–C(6') bond occurred, as revealed by the distance of 1.955 Å (Scheme 4a). A complex with an intact cyclobutane ring could not be located, which suggested that *c,s*-DMU<>DMU underwent barrierless ring scission upon encountering $NO_3^{\bullet}$. Analysis of the spin density and Mulliken charges in this complex showed complete loss of radical character and accumulation of negative charge on the NO_3 fragment. The spin was distributed over the dimer with 27% located on each of C(6) and C(6') and about 20% on each of N(1) and N(1'). These data suggested that the complex should, in fact, be regarded as a nitrate-dimer radical cation pair resulting from ET, i.e., [NO_3^-] [*c,s*-DMU<>DMU]$^{\bullet+}$. The energy of [NO_3^-] [*c,s*-DMU<>DMU]$^{\bullet+}$ was about 43 kJ mol^{-1} lower than the sum of the energies of the free reactants, which were used as a reference point for the energy profile. A reactant complex without any CT contribution (similar to *c,a*-DMU<>DMU) could not be located computationally for *c,s*-DMU<>DMU.

To explore a potential homolytic fragmentation of the cyclobutane ring in *c,s*-DMU<>DMU, the first step through $NO_3^{\bullet}$-induced HAT from C(6) was calculated. This process was associated with a barrier (**TS9**) of about 28 kJ mol^{-1} relative to [NO_3^-] [*c,s*-DMU<>DMU]$^{\bullet+}$ and led to the product complex of the C(6) dimer radical **9** and HNO_3 in an overall exothermic reaction with respect to the free reactants. Inspection of the spin density in **TS9** revealed that only 18% of the spin was remaining on the NO_3 fragment, indicating a relatively late transition state that was also supported by the comparably long C(6)–H distance of 1.255 Å compared with 1.087 Å in the dimer (not shown). Interestingly, in **TS9**, the C(6)–C(6') distance was only 1.630 Å, indicating a nearly restored cyclobutane ring (1.568 Å in the intact dimer, not shown). Intrinsic reaction coordinate (IRC) calculations confirmed that the complex and **TS9** were connected and that with increasing C(6)–H distance, the C(6)–C(6') bond snapped close. Together with the reduced negative charge on the NO_3 moiety in **TS9** (Mulliken charge −0.52e), these findings suggested that if a radical HAT from C(6) starting from [NO_3^-] [*c,s*-DMU<>DMU]$^{\bullet+}$ were to occur, a reverse charge transfer from the dimer to the NO_3 moiety would be required to traverse **TS9**.

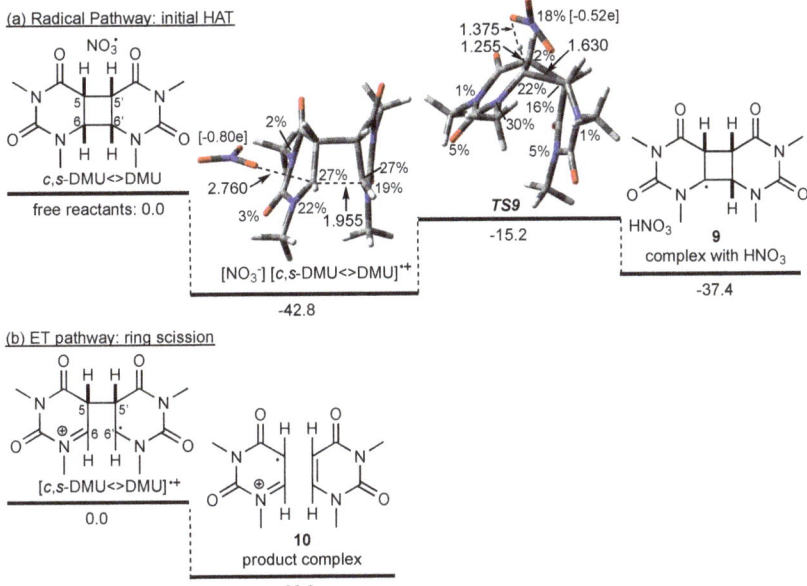

Scheme 4. Possible pathways in the reaction of NO$_3^\bullet$ with c,s-DMU<>DMU. M062X/6-31+G* geometry and free energies in kJ mol^{-1} (in acetonitrile). Spin density in percentage, distances in Å, Mulliken charge on NO$_3$ fragment in square brackets. For the HAT pathway (a), the sum of the energies of the free reactants is used as a reference point; for the ET pathway (b), the energy of the dimer radical cation was used as a reference point (see text).

On the other hand, scission of the remaining C(5)–C(5′) bond in the dimer radical cation [c,s-DMU<>DMU]$^{\bullet+}$, which according to the calculations should be immediately formed when NO$_3^\bullet$ encounters c,s-DMU<>DMU, was both barrier-less (i.e., a TS could not be located) and led to the product complex **10** in an exothermic reaction, relative to [c,s-DMU<>DMU]$^{\bullet+}$ (Scheme 4b). From these data, it could, therefore, be concluded that repair of c,s-DMU<>DMU was initiated by NO$_3^\bullet$-induced ET, whereas a HAT pathway was not competitive.

Calculations of the reactant complexes with other pyrimidine cyclobutane dimers used in the kinetic studies clearly revealed that all *syn*-configured dimers formed similar nitrate-dimer radical cation pairs, where the dimer had an already cleaved C(6)–C(6′) bond and should undergo barrierless scission of the remaining C(5)–C(5′) bond. This behavior was not limited to acetonitrile as a solvent but was similarly also found in water (data not shown). It should be noted that it was not possible to locate a geometry for any *syn*-configured dimer radical cation in which the cyclobutane ring was not broken.

The reactant complexes between NO$_3^\bullet$ and *anti*-configured dimers generally constituted CT complexes with a largely intact cyclobutane ring, in which the spin density was distributed over both NO$_3^\bullet$ and dimer and negative charge accumulating on the NO$_3$ moiety. These data suggested a less favorable ET process, which was in line with the higher E^0 value determined for *anti*-configured dimers (see Table 1). The scission of the cyclobutane ring was associated with a barrier, which slowed the overall repair process down. Because ET was less favorable, it should be noted that it could not be excluded that a competing HAT reaction at the methyl group at N(1) or at C(6) could occur in *anti*-configured dimers to some extent. However, the rate coefficient determined experimentally for HAT from N$_{amide}$-alkyl groups of about 6–10 × 10^7 M^{-1} s^{-1} [35] suggested that such a reaction should only be a minor contributor to the overall reaction of NO$_3^\bullet$ with the dimer.

The finding that the *trans*-configured dimers reacted faster with NO$_3^\bullet$ than the isomeric *cis*-configured dimers of the same pyrimidine could be rationalized by steric effects. Analysis of the spin densities in the NO$_3^\bullet$ complexes with the dimers (see Schemes 3 and 4) suggested that oxidation occurred at the amide nitrogen N(1) in both pyrimidine moieties. Therefore, accessibility of this site for NO$_3^\bullet$ should be a factor that modulates the rate of oxidation—a proposition that is supported by recent findings that steric hindrance at tertiary amides slows down the rate of oxidation by NO$_3^\bullet$ [35]. Figure 3 shows in the top row the X-ray structures of *t,s*-DMU<>DMU and *c,s*-DMU<>DMU. *t,s*-DMU<>DMU had a largely linear shape, where N(1) in both pyrimidine rings could be accessed by NO$_3^\bullet$ from both above and below the molecular plane, as indicated by the green arrows. In contrast to this, *c,s*-DMU<>DMU clearly had a freely accessible convex side, but the approach of NO$_3^\bullet$ from the concave side was sterically hindered, which should lower the rate of oxidation.

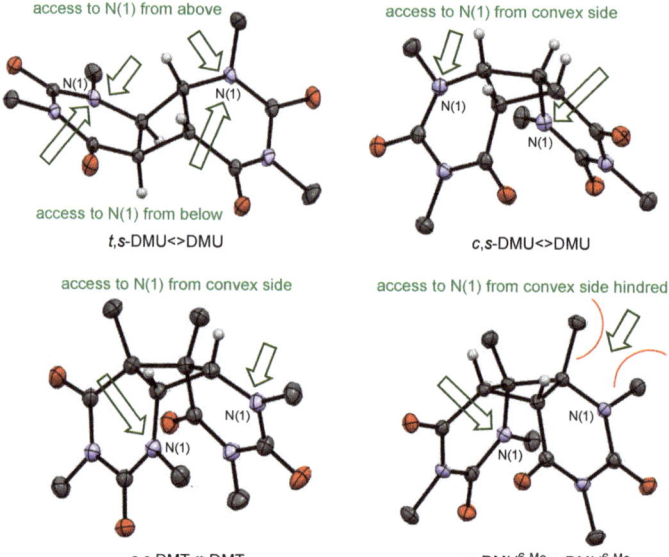

Figure 3. X-ray structures for selected dimers and likely approach by NO$_3^\bullet$ (green arrow).

Similarly, steric hindrance could also explain the slower reaction of the dimers of DMU$^{6\text{-Me}}$, compared with those of DMU and DMT with similar configurations. Figure 3 presents in the bottom row the X-ray structures of *c,s*-DMT<>DMT and *c,s*-DMU$^{6\text{-Me}}$<>DMU$^{6\text{-Me}}$. It would be expected that the higher degree of alkyl substitution at the cyclobutane ring in both dimers, compared with the isomeric *c,s*-DMU<>DMU, should facilitate oxidation. Indeed, the rate coefficient, shown in Table 2, for the reaction with *c,s*-DMT<>DMT was about 30% higher than with *c,s*-DMU<>DMU. On the other hand, the considerably lower reactivity of *c,s*-DMU$^{6\text{-Me}}$<>DMU$^{6\text{-Me}}$ suggested that steric hindrance caused by the methyl substituents at C(6) was impeding NO$_3^\bullet$ access to the adjacent amide N(1) in both pyrimidine moieties, which slowed down the rate of oxidation.

5. Conclusions

NO$_3^\bullet$ reacted with pyrimidine cyclobutane dimers via a rapid ET at N(1), which triggered cleavage of the cyclobutane ring and regeneration of a pyrimidine monomer and a monomer radical cation.

The four possible geometric isomers of the pyrimidine cyclobutane dimers differed considerably in their reactivity towards NO$_3^\bullet$, with the *syn*-configured dimers being more reactive than their *anti*-configured counterparts. The higher reactivity of the former was manifested by the formation of a nitrate/dimer radical cation pair, where the negative charge had accumulated on the nitrate moiety, and the C(6)–C(6′) bond in the dimer was already broken. The scission of the remaining C(5)–C(5)′ bond was barrierless and exothermic. In the case of the *anti*-configured dimers, ET was less progressed, which was in line with the higher E^0 value, and a CT complex with $NO_3^\bullet$ was formed. The cyclobutane ring in this complex, as well as in the corresponding dimer radical cation, was still intact. Therefore, ring-opening in the dimer radical cation required sequential cleavage of two bonds, of which the first step was associated with a moderate barrier. However, because of the slower ET in the *anti*-configured dimers, it could not be excluded that $NO_3^\bullet$-induced HAT at the methyl group of N(1) and from C(6) at the cyclobutane ring might also occur to some extent. However, subsequent fragmentation of the cyclobutane ring through radical-only pathways was not competitive with the cleavage process in a dimer radical cation.

Steric effects were likely responsible for the higher reactivity of *trans*-configured dimers compared with their *cis* isomers. Thus, the geometry of the *trans* dimers allowed the approach of $NO_3^\bullet$ to the N(1) atoms in the pyrimidines from both sides of the molecular plane, whereas, in the *cis* dimers, only the convex side was readily accessible for $NO_3^\bullet$. Likewise, the order of reactivity of DMU^{6-Me}<>DMU^{6-Me} → DMU<>DMU → DMT<>DMT towards $NO_3^\bullet$ was a result of steric and electronic effects. The dimers of DMT generally showed the highest reactivity due to the electron-donating methyl groups at C(5,5′) on the cyclobutane ring that increased their susceptibility to oxidation by $NO_3^\bullet$. Remarkably, methyl substituents at C(6,6′) in the DMU^{6-Me} dimers lowered their reactivity, even below that of the DMU dimers. This finding could be rationalized by the difficulty for $NO_3^\bullet$ to approach N(1) in DMU^{6-Me}<>DMU^{6-Me} due to steric hindrance caused by the adjacent methyl groups.

Overall, this study clearly revealed that the rate of the $NO_3^\bullet$ reaction was not only determined by the redox potential of the dimers but also to a considerable extent by the accessibility of the reaction site for $NO_3^\bullet$. In future work, we will further explore the role of steric effects on the radical-induced oxidative damage of DNA constituents, such as pyrimidine deoxy nucleosides and their cyclobutane dimers.

Supplementary Materials: The following are available online at http://www.mdpi.com/2624-8549/2/2/453\T1\textendash469/s1, ^{1}H and ^{13}C-NMR spectra of substrates used in the laser flash photolysis study, crystallographic data for pyrimidine cyclobutane dimers, details of laser flash photolysis experiments and kinetic plots, Gaussian archive entries.

Author Contributions: Conceptualization, U.W.; methodology, T.H., J.G.N., and U.W.; validation, T.H. and J.G.N.; formal analysis, T.H., J.M.W., and J.G.N.; investigation, T.H., J.M.W, J.G.N., and U.W.; resources, T.H., J.G.N., and U.W.; data curation, T.H. and J.G.N.; writing—original draft preparation, J.G.N. and U.W.; writing—review and editing, T.H., J.M.W., J.G.N., and U.W.; visualization, J.G.N. and U.W.; supervision, J.G.N. and U.W.; project administration, U.W.; funding acquisition, U.W. All authors have read and agreed to the published version of the manuscript.

Funding: This work was funded by The University of Melbourne and the Australian Research Council (LE0989197 and LE170100065).

Acknowledgments: We acknowledge the support of this work by Oliver Krüger.

Conflicts of Interest: The authors declare no conflict of interest.

References and Notes

1. D'Orazio, J.; Jarrett, S.; Amaro-Ortiz, A.; Scott, T. UV Radiation and the Skin. *Int. J. Mol. Sci.* **2013**, *14*, 12222–12248. [CrossRef] [PubMed]
2. Douki, T.; Sauvaigo, S.; Odin, F.; Cadet, J. Formation of the Main UV-induced Thymine Dimeric Lesions within Isolated and Cellular DNA as Measured by High Performance Liquid Chromatography-Tandem Mass Spectrometry. *J. Biol. Chem.* **2000**, *275*, 11678–11685. [CrossRef] [PubMed]
3. Barak, Y.; Cohen-Fix, O.; Livneh, Z. Deamination of Cytosine-containing Pyrimidine Photodimers in UV-irradiated DNA. Significance for UV Light Mutagenesis. *J. Biol. Chem.* **1995**, *270*, 24174–24179. [CrossRef] [PubMed]
4. Ravanat, J.-L.; Douki, T.; Cadet, J. Direct and indirect effects of UV radiation on DNA and its components. *J. Photochem. Photobiol. B* **2001**, *63*, 88–102. [CrossRef]

5. Thomas, D.C.; Kunkel, T.A. Replication of UV-irradiated DNA in human cell extracts: Evidence for mutagenic bypass of pyrimidine dimers. *Proc. Natl. Acad. Sci. USA* **1993**, *90*, 7744–7748. [CrossRef] [PubMed]
6. Burroughs, A.M.; Aravind, L. RNA damage in biological conflicts and the diversity of responding RNA repair systems. *Nucleic Acids Res.* **2016**, *44*, 8525–8555. [CrossRef]
7. Krokan, H.E.; Standal, R.; Slupphaug, G. DNA glycosylases in the base excision repair of DNA. *Biochem. J.* **1997**, *325*, 1–16. [CrossRef]
8. Brettel, K.; Byrdin, M. Reaction mechanisms of DNA photolyase. *Curr. Opin. Struct. Biol.* **2010**, *20*, 693–701. [CrossRef]
9. Friedel, M.G.; Gierlich, J.; Carell, T. Cyclobutane Pyrimidine Dimers as UV-Induced DNA Lesions. In *PATAI'S Chemistry of Functional Groups*; Rappoport, Z., Ed.; John Wiley & Sons, Ltd.: Hoboken, NJ, USA, 2009. [CrossRef]
10. Huntley, J.J.; Nieman, R.A.; Rose, S.D. Development and Investigation of a Novel Oxidative Pyrimidine Dimer Splitting Model. *Photochem. Photobiol.* **1999**, *69*, 1–7. [CrossRef]
11. Treml, J.; Smejkal, K. Flavonoids as Potent Scavengers of Hydroxyl Radicals. *Compr. Rev. Food Sci. F.* **2016**, *15*, 720–738. [CrossRef]
12. Heelis, P.; Deeble, D.; Kim, S.-T.; Sancar, A. Splitting of Cis-syn Cyclobutane Thymine-thymine Dimers by Radiolysis and Its Relevance to Enzymatic Photoreactivation. *Int. J. Radiat. Biol.* **1992**, *62*, 137–143. [CrossRef] [PubMed]
13. Yan, L.-Q.; Song, Q.-H.; Hei, X.-M.; Guo, Q.-X.; Lin, W.-Z. Oxidative Splitting of a Pyrimidine Cyclobutane Dimer: A Pulse Radiolysis Study. *Chin. J. Chem.* **2003**, *21*, 16–19. [CrossRef]
14. Alfassi, Z.B.; Harriman, A.; Huie, R.E.; Mosseri, S.; Neta, P. The redox potential of the azide/azidyl couple. *J. Phys. Chem.* **1987**, *91*, 2120–2122. [CrossRef]
15. Ito, T.; Shinohara, H.; Hatta, H.; Nishimoto, S.-I.; Fujita, S.-I. Radiation-Induced and Photosensitized Splitting of C5–C5′-Linked Dihydrothymine Dimers: Product and Laser Flash Photolysis Studies on the Oxidative Splitting Mechanism. *J. Phys. Chem. A* **1999**, *103*, 8413–8420. [CrossRef]
16. Baciocchi, E.; Del Giacco, T.; Murgia, S.M.; Sebastiani, G.V. Rate and mechanism for the reaction of the nitrate radical with aromatic and alkylaromatic compounds in acetonitrile. *J. Chem. Soc. Chem. Commun.* **1987**, 1246–1248. [CrossRef]
17. Krüger, O.; Wille, U. Oxidative Cleavage of a Cyclobutane Pyrimidine Dimer by Photochemically Generated Nitrate Radicals ($NO_3^{\bullet}$). *Org. Lett.* **2001**, *3*, 1455–1458. [CrossRef]
18. Boussicault, F.; Krüger, O.; Robert, M.; Wille, U. Dissociative electron transfer to and from pyrimidine cyclobutane dimers: An electrochemical study. *Org. Biomol. Chem.* **2004**, *2*, 2742–2750. [CrossRef]
19. Neta, P.; Huie, R.E.; Ross, A.B. Rate Constants for Reactions of Inorganic Radicals in Aqueous Solution. *J. Phys. Chem. Ref. Data* **1988**, *17*, 1027–1284. [CrossRef]
20. Wenska, G.; Paszyc, S. Electron-acceptor-sensitized splitting of cyclobutane-type thymine dimers. *J. Photochem. Photobiol. B* **1990**, *8*, 27–37. [CrossRef]
21. Pac, C.; Kubo, J.; Majima, T.; Sakurai, H. Structure-reactivity relationships in redox-photosensitized splitting of pyrimidine dimers and unusual enhancing effect of molecular oxygen. *Photochem. Photobiol.* **1982**, *36*, 273–282. [CrossRef]
22. Krüger, O. Oxidative Schädigung und Reparatur von Nucleobasen und Nucleosiden durch NO_3 Radikale. Ph.D. Thesis, Christian-Albrechts University, Kiel, Germany, 2002.
23. Sheldrick, G. SHELXT—Integrated space-group and crystal-structure determination. *Acta Crystallogr. Sect. C* **2015**, *71*, 3–8. [CrossRef] [PubMed]
24. Macrae, C.F.; Bruno, I.J.; Chisholm, J.A.; Edlington, P.R.; McCabe, P.; Pidcock, E.; Rodriguez-Monge, L.; Taylor, R.; van de Streek, J.; Wood, P.A. Mercury CSD 2.0—New features for the visualization and investigation of crystal structures. *J. Appl. Cryst.* **2008**, *41*, 466–470. [CrossRef]
25. Farrugia, L.J. WinGX suite for small-molecule single-crystal crystallography. *J. Appl. Cryst.* **1999**, *32*, 837–838. [CrossRef]
26. Nathanael, J.G.; Hancock, A.N.; Wille, U. Reaction of Amino Acids, Di- and Tripeptides with the Environmental Oxidant $NO_3^{\bullet}$: A Laser Flash Photolysis and Computational Study. *Chem. Asian J.* **2016**, *11*, 3188–3195. [CrossRef]
27. Nathanael, J.G.; Gamon, L.F.; Cordes, M.; Rablen, P.R.; Bally, T.; Fromm, K.F.; Giese, B.; Wille, U. Amide Neighbouring-Group Effects in Peptides: Phenylalanine as Relay Amino Acid in Long-Distance Electron Transfer. *ChemBioChem* **2018**, *19*, 922–926. [CrossRef]

28. Nathanael, J.G.; Wille, U. Oxidative Damage in Aliphatic Amino Acids and Di- and Tripeptides by the Environmental Free Radical Oxidant $NO_3^\bullet$: The Role of the Amide Bond Revealed by Kinetic and Computational Studies. *J. Org. Chem.* **2019**, *84*, 3405–3418. [CrossRef]
29. Del Giacco, T.; Baciocchi, E.; Steenken, S. One-Electron Oxidation of Alkylbenzenes in Acetonitrile by Photochemically Produced Nitrate Radical: Evidence for an Inner-Sphere Mechanism. *J. Phys. Chem.* **1993**, *97*, 5451–5456. [CrossRef]
30. Frisch, M.J.; Trucks, G.W.; Schlegel, H.B.; Scuseria, G.E.; Robb, M.A.; Cheeseman, J.R.; Scalmani, G.; Barone, V.; Petersson, G.A.; Nakatsuji, H.; et al. *Gaussian 16, Revision B.01*; Gaussian, Inc.: Wallingford, CT, USA, 2016.
31. Zhao, Y.; Truhlar, D.G. The M06 suite of density functionals for main group thermochemistry, thermochemical kinetics, noncovalent interactions, excited states, and transition elements: Two new functionals and systematic testing of four M06-class functionals and 12 other functionals. *Theor. Chem. Acc.* **2008**, *120*, 215–241. [CrossRef]
32. Goerigk, L.; Hansen, A.; Bauer, C.; Ehrlich, S.; Najibi, A.; Grimme, S. A look at the density functional theory zoo with the advanced GMTKN55 database for general main group thermochemistry, kinetics and noncovalent interactions. *Phys. Chem. Chem. Phys.* **2017**, *19*, 32184–32215. [CrossRef]
33. Goerigk, L.; Grimme, S. A thorough benchmark of density functional methods for general main group thermochemistry, kinetics, and noncovalent interactions. *Phys. Chem. Chem. Phys.* **2011**, *13*, 6670–6688. [CrossRef]
34. Takano, Y.; Houk, K.N. Benchmarking the Conductor-like Polarizable Continuum Model (CPCM) for Aqueous Solvation Free Energies of Neutral and Ionic Organic Molecules. *J. Chem. Theory Comput.* **2005**, *1*, 70–77. [CrossRef] [PubMed]
35. Nathanael, J.G.; White, J.M.; Richter, A.; Nuske, M.; Wille, U. Radical-induced oxidative damage of proline is influenced by its position in a peptide: A kinetic and product study. manuscript in preparation.
36. Goeschen, C.; White, J.M.; Gable, R.W.; Wille, U. Oxidative Damage of Pyrimidine Nucleosides by the Environmental Free Radical Oxidant $NO_3^\bullet$ in the Absence and Presence of $NO_2^\bullet$ and Other Radical and Non-Radical Oxidants. *Aust. J. Chem.* **2012**, *65*, 427–437. [CrossRef]
37. The higher reactivity of DMT and $DMU^{6\text{-}Me}$ could be due to the additional methyl group at C(5) and C(6), respectively, which should lower the redox potential compared with that of DMU.
38. Quantum mechanical tunneling was not considered in the HAT calculations, as these were aimed to qualitatively assess the most reactive sites and not to predict rate coefficients for these processes. Experimental rate data have been determined previously (see ref. [35]).

© 2020 by the authors. Licensee MDPI, Basel, Switzerland. This article is an open access article distributed under the terms and conditions of the Creative Commons Attribution (CC BY) license (http://creativecommons.org/licenses/by/4.0/).

 *chemistry*

Article

Size and Surface Charge Dependent Impregnation of Nanoparticles in Soft- and Hardwood [†]

David Bossert [1,‡], Christoph Geers [1,*,‡], Maria Inés Placencia Peña [2], Thomas Volkmer [2], Barbara Rothen-Rutishauser [1] and Alke Petri-Fink [1,3,*]

1. Adolphe Merkle Institute, University of Fribourg, Chemin des Verdiers 4, 1700 Fribourg, Switzerland; david.bossert@unifr.ch (D.B.); barbara.rothen@unifr.ch (B.R.-R.)
2. Bern University of Applied Sciences, Architecture, Wood and Civil Engineering, Solothurnstrasse 102, 2500 Biel, Switzerland; ines.placencia@gmail.com (M.I.P.P.); thomas.volkmer@bfh.ch (T.V.)
3. Chemistry Department, University of Fribourg, Chemin du Musée 9, 1700 Fribourg, Switzerland
* Correspondence: christoph.geers@unifr.ch (C.G.); alke.fink@unifr.ch (A.P.-F.)
† In Honor of Professor Bernd Giese on the Occasion of His 80th Birthday.
‡ These authors contributed equally.

Received: 2 April 2020; Accepted: 24 April 2020; Published: 2 May 2020

Abstract: Recent progress in wood preservative research has led to the use of insoluble copper carbonate in the form of nano- to micron-sized particles in combination with known triazole fungicides to combat fungal decay and thus decrease physical material properties. Evidently, particle-based agents could lead to issues regarding impregnation of a micro-structured material like wood. In this study, we analyzed these limitations via silicon dioxide particles in impregnation experiments of pine and beech wood. In our experiments, we showed that limitations already existed prior to assumed particle size thresholds of 400–600 nm. In pine wood, 70 nm sized particles were efficiently impregnated, in contrast to 170 nm particles. Further we showed that surface functionalized silica nanoparticles have a major impact on the impregnation efficiency. Silica surfaces bearing amino groups were shown to have strong interactions with the wood cell surface, whereas pentyl chains on the SiO_2 surfaces tended to lower the particle–wood interaction. The acquired results illustrate an important extension of the currently limited knowledge of nanoparticles and wood impregnation and contribute to future improvements in the field of particle-based wood preservatives.

Keywords: wood; nanoparticles; surface modification; localization; distribution; impregnation; scanning electron microscopy

---

1. Introduction

Wood has unique properties and is a renewable but also degradable natural material. However, its longevity is affected by microorganisms and environmental factors, which can decompose the wood structure, leading to altered appearance and lowered stability. To counteract these, wood can be protected by a variety of methods, amongst them, for example, treatment with organic solvent preservatives, creosote or waterborne wood preservatives [1]. Among the latter, copper-containing, and in particular chromated copper arsenate (CCA), preservatives have been used in the past decades [2,3]. In 2002, approximately 30 million cubic meters of wood were treated with an estimated consumption of 500,000 tons of CCA worldwide [4]. Since 2004, arsenate-containing wood preservatives were restricted for industrial use only, and thus their demand has been declining [5]. Copper-containing formulations still prevail as the most used class of wood preserving agents due to their excellent activity against many wood-destroying organisms [6,7]. Recently, formulations using insolubilized copper species for wood preservation have been investigated [8–14]. These formulations rely on micronized copper (MC) particles, which are obtained from ball-milling solid copper carbonate into the size range of 1 nm–25 µm particles, which are then pressure impregnated into wood [15–17]. Studies have demonstrated that micronized copper azole and micronized copper quaternary comprising wood preservatives showed better performance compared to formulations using soluble copper [8,9,18]. Soil pH and composition have drastic effects on the protective ability of MC [9,19], but overall, MC particles show much lower leaching compared to soluble copper formulations due to the reservoir effect [20]. The same effect was confirmed for other particle-based wood preservatives. For example, zinc oxide nanoparticles (NPs) showed reduced leaching compared to zinc sulfate solutions [21–23]. Other NPs used for wood protection include, for example, silver NPs, which exhibit a broad range of antimicrobial activity. In the presence of moisture, metallic silver NPs oxidize, which results in the release of silver ions. Because silver oxidation is a slow reaction, the size of the silver NPs is critical to achieve microorganism growth control. In general, the smaller the particle size, the higher the surface area, and the larger the area available for oxidation. In addition, NPs are increasingly investigated for transparent wood coatings. Such nanocomposite coatings have the potential of not only preserving the natural color of the wood, but also stabilizing the wood surface against the combined degradative effects of sunlight and moisture.

Wood preservatives are expected to impregnate via the water conducting system with membrane openings of 400–600 nm in diameter. These elements pose a structural limitation, and therefore larger particles are not expected to penetrate the wood structure due to the clogging of conducting pathways [15,17]. The latter is in line with recent findings of MC-based wood impregnation; it was shown that most of the copper particles were deposited on the wood surface since the larger particles (the bulk of the material consists of larger micron size particles [24]) could not penetrate the wood [13,24].

Particle surface charge is another important parameter to be considered. For example, positively charged NPs form stronger interactions with wood than negatively charged NPs, since cellulose fibers (i.e., cell wall constituents) are negatively charged due to presence of acidic groups (e.g., carboxyl, sulphonic acid, or hydroxyl groups). This was previously confirmed with, e.g., cationic polymers and silicon–aluminum oxide nanocomposites bearing a positive surface charge; both form strong non-covalent interactions with the anionic surface of cellulose and wood, respectively [25,26]. Furthermore, cationic NPs were used as nanocarriers to deliver wood preservatives into the wood structure [27,28].

Although the results of previous studies on the use of NPs for wood preservation are promising, our understanding of the new properties of the applied (nano)particles and their mode of action is still limited [29], and wood scientists have encouraged broader engagement of nanoscientists in this field [30]. In our study, we investigated the influence of particle size and surface charge during pressure impregnation of soft- and hardwood. We used spherical silicon dioxide model particles with low polydispersity to evaluate impregnation depth and efficiency as a function of particle size and surface charge.

2. Materials and Methods

2.1. Synthesis of Silica Nanoparticles

Silica NPs were synthesized following a modified Stöber method [31]. The synthesis was standardized for all particle sizes. Briefly, a solution of ethanol (absolute, Honeywell, Charlotte, NC, USA), ammonia (25% aqueous solution, Merck, Darmstadt, Germany) and water (MilliQ, Arium 611DI, Sartorius Stedim Biotech, Göttingen, Germany) was heated to temperature (T) and equilibrated for 1 h before adding tetraethyl orthosilicate (TEOS, 98%, Sigma-Aldrich, St. Louis, MO, USA). The mixture was stirred overnight at the given temperature and allowed to cool down to room temperature (RT). The formed particles were washed by three centrifugation (Heraeus Multifuge X1R, equipped with an F15-8 × 50cy fixed-angle rotor, Thermo Scientific, Waltham, MA, USA)–redispersion (MilliQ) cycles and finally dispersed in water. If the particles had to be dispersed in EtOH, three additional centrifugation–redispersion (EtOH) cycles were applied. The quantitative reaction parameters for each particle size were as follows: SiO$_2$-NP 70 nm: water (24.5 mL), EtOH (200 mL), ammonia (7.8 mL), TEOS (13.94 mL), T = 60 °C, centrifugation (15,000× g, 20 min). SiO$_2$-NPs 170 nm: MilliQ (58.5 mL), EtOH (162 mL), ammonia (7.8 mL), TEOS (22 mL), T = 65 °C, centrifugation (10,000× g, 10 min). SiO$_2$-NPs 350 nm: MilliQ (13.5 mL), EtOH (174 mL), ammonia (40.9 mL), TEOS (21 mL), T = 60 °C, centrifugation (5000× g, 10 min).

2.2. Functionalization of Silica Nanoparticles

SiO$_2$-NPs 70 nm (100 mL of a 7.2 mg mL^{-1} suspension in EtOH, 720 mg SiO$_2$, 12 mmol), ammonia (7.35 mL) and functionalized triethoxysilane ((3-aminopropyl)triethoxysilane (APTES, 252 µL, 321 µmol, Sigma-Aldrich) for APTES-SiO$_2$ 70 nm; pentyltriethoxysilane (PETES, 281 µL, 321 µmol, Gelest Inc., Morrisville, PA, USA) for PETES-SiO$_2$ 70 nm) were mixed and stirred at reflux overnight. The reaction mixture was cooled to RT; the particles were washed by three centrifugation (10 min, 10,000× g) and redispersion (EtOH) cycles and finally dispersed in EtOH. To transfer the particles into water, three additional centrifugation–redispersion (water) cycles were applied. Note: The surface chemistry of amorphous silica and more specifically the hydroxylation of the silica surface was estimated by the Zhuravlev model. Equal stoichiometry was applied to ensure equivalent surface functionalization.

2.3. Characterization of Nanoparticles

To prepare a NP sample suitable for transmission electron microscopy (TEM), samples were prepared by diluting the particle suspension (1 µL) with ethanol (5 µL, absolute, Honeywell) directly on the TEM grids (carbon film, 300 mesh on Cu, Electron Microscopy Sciences, Hatfield, PA, USA) and removing the remaining liquid using a fuzz-free tissue (precision wipes, Kimtech Science, Kimberly Clark, Dallas, TX, USA). For difficult to disperse samples, the particle suspensions were treated according to a previously published protocol [32] In brief, the particle suspension was mixed with an aqueous solution of bovine serum albumin (BSA); 5 µL of this mixture was drop cast onto a carbon-film square mesh copper grid (Electron Microscopy Sciences, CF-300-Cu) and dried in ambient air in a dust-free environment. The images were recorded with 2048 × 2048 pixels (Veleta CCD camera, Olympus, Shinjuku, Tokio, Japan) on a Tecnai Spirit transmission electron microscope (FEI), operating at an acceleration voltage of 120 kV. The size distribution of the NPs (>300 particle count) was determined using particle analysis software (ImageJ, National Institutes of Health, Bethesda, MD, USA). Dynamic light scattering (DLS) data was collected at constant temperature (25 °C) on a commercial goniometer instrument (3D LS Spectrometer, LS Instruments AG, Fribourg, Switzerland). Zeta-potential measurements were performed using a Brookhaven Plus90 particle size analyzer (USA) measuring at a scattering angle of 90° for 1 min and 10 repetitions. Fourier-transform infrared spectroscopy (FTIR) was measured (Spectrum 65, Perkin Elmer, Waltham, MA, USA) with lyophilized SiO$_2$-NPs as background.

2.4. Wood Samples

Beech (*Fagus sylvatica*) and pine (*Pinus sylvestris*) wood were used for all experiments. The samples were selected using the EN113 [33] standard. Wood samples of 30 × 10 × 10 mm^3 (longitudinal × radial × tangential) in size were used in all impregnation studies. The cross sections of the wood pieces were covered with solvent-free glue (picodent twinsil®, Wipperfürth, Germany).

Impregnation: For impregnation, dry wood samples were placed in polypropylene flasks, the NP suspension was added, and the samples were placed in an autoclave. A vacuum of −95 mbar was applied for 20 min followed by a pressure cycle of 5 bars for 1 h. These parameters are typically used in laboratory conditions for impregnation of small wood samples and can be compared to the industry scale in terms of material uptake of the wood samples. A ballasted plastic grid was used to ensure that the wood samples stayed entirely submerged in the liquid. After impregnation, the wood samples were dried in an oven at 60 °C for 48 h.

Sample preparation: The wood samples were carefully split in radial or tangential directions, respectively, using a scalpel. The split in the tangential direction was positioned in the transition between early and latewood. The exposed wood surface was analyzed by scanning electron microscopy (SEM) and energy-dispersive X-ray spectroscopy (EDX) analysis.

2.5. Scanning Electron Microscopy and EDX-Mapping

The split wood samples were glued to graphite plates using conductive carbon paste, and the exposed wood surface was coated with a conductive layer to ensure conductivity. For SEM maps, samples were sputtered with Au or Pd/Pt, and the sputtered surface was connected with a silver stripe to the sample holder to ensure high conductivity for high magnifications. For low background EDX measurements, the samples were coated with carbon. The samples were analyzed on a Mira3 LM (Tescan) FE SEM scanning electron microscope (high voltage: 5 kV, working distance: 15 mm) using a secondary electron (SE) detector or an Inbeam detector. The samples were mapped using an energy dispersive X-ray spectrometer from EDAX equipped with a lithium-doped silicon detector and analyzed using EDAX Genesis software (version 5.2).

3. Results and Discussion

3.1. Particle Characterization

SiO_2-NPs were synthesized as model particles to investigate the impregnation of NPs in beech and pine wood. TEM micrographs and particle size analyses are summarized in Figure 1. All particles, i.e., SiO_2 70 nm, SiO_2 170 nm and SiO_2 350 nm, showed a low polydispersity and no size overlap between the three different groups. The diameters were determined as 70.3 ± 7.0 nm, 172.8 ± 14.9 nm and 351.4 ± 21.5 nm, respectively. All particles were spherical and colloidally stable (~−40 mV at pH 7).

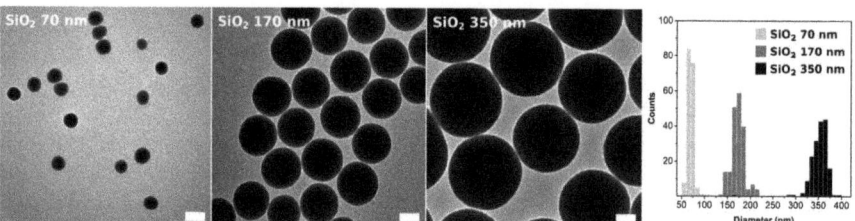

Figure 1. SiO_2-NPs transmission electron micrographs and particle size distributions. The obtained mean particle diameters are 70.3 ± 7.0 nm (SiO_2 70 nm), 173 ± 15 nm (SiO_2 170 nm) and 351 ± 22 nm (SiO_2 350 nm), respectively. Particle size distributions do not overlap (see histogram), which is important to determine impregnation size thresholds for soft- and hardwood samples. Scale bar represents 100 nm.

SiO_2 70 nm NPs were surface-modified to investigate the interaction of wood with different particle surfaces. The results are summarized in Table 1. Successful functionalization with APTES was confirmed by a significant change of zeta-potential from ~−38 mV (SiO_2-NPs) to +42mV (APTES–SiO_2). Despite the high zeta-potential, DLS showed particle aggregation at the concentration used for wood impregnation. The same particle concentration in EtOH yielded more stable particles (Table 1), which can be explained by the stabilizing interactions between EtOH and the aminopropyl chain.

The pentyl group of PETES–SiO_2 impacted the zeta potential insignificantly compared to unfunctionalized SiO_2-NPs. Additionally, the applied protocol for PETES-functionalization resulted in particles that were still dispersible in water but showed better dispersibility in EtOH. Successful functionalization was confirmed by Fourier-transform infrared spectroscopy (FTIR) of lyophilized particles (Supplementary Materials Figure S1), showing weak but characteristic peaks for C–H bonds (2913 cm^{-1}) and Si–CH$_2$ bonds (2983 cm^{-1}, 785 cm^{-1}), in accordance with cited literature [34,35].

Table 1. Size and zeta potential of pristine and functionalized SiO$_2$ 70 nm particles prior to impregnation.

Particle Type	TEM (nm)	DLS (nm)		Zeta Potential (mV)	
		Water	EtOH	Water	EtOH
APTES–SiO$_2$	68.2 ± 10.6	935.2	110	41.6	46.3
SiO$_2$	67.8 ± 10.2	88.8	68.8	−37.5	−33.1
PETES–SiO$_2$	69.8 ± 6.6	89.8	99.8	−34	−30.1

The particles were measured at 1 mg mL^{-1} as used during impregnation.

3.2. Impregnation of Beech Wood

Wood impregnation was analyzed by SEM-EDX mapping of wood samples (Figure 2). Figure 3 displays the SEM images and the corresponding silicon X-ray map (silicon in magenta) up to a depth of 5 mm (half width of sample), corresponding to full impregnation of the wood sample. The maps were recorded with the axial direction aligned vertically, from left (sample edge) to right (sample core). The particles were localized in the main components of the water conducting system, apparent from the bright colored spots in Figure 3. The strongest signal resulted from the beech vessels, indicating the water conducting system as the main entrance and distribution route. It was assumed that the large vessel diameter of around 50 µm interconnected with unobstructed perforation plates allowed free movement of SiO$_2$-NPs in suspension [36]. It was previously reported that vessel to vessel contact was not given over growth ring boundaries and, therefore, we argue that vessel–ray intersections were mainly responsible for the radial distribution of particle suspensions [37,38]. These results are in line with previous studies, where highly viscous tannin solutions were successfully impregnated in beech wood [39]. No signal in wood rays was detected, although wood rays are responsible for radial transport of liquid. In comparable studies [40], EDX measurements were usually acquired by preparation of flat sample surfaces (e.g., microtome cut, polished surfaces), and high topography samples suffered from absorption of emitted X-rays before reaching the detector.

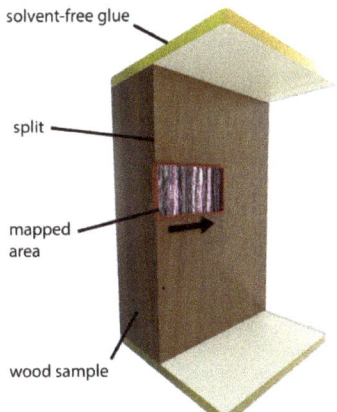

Figure 2. Illustration of wood sample preparation for SEM–EDX mapping. The cross sections were closed with a solvent-free glue to force radial and tangential particle penetration. The arrow represents the direction of impregnation from sample edge towards the core.

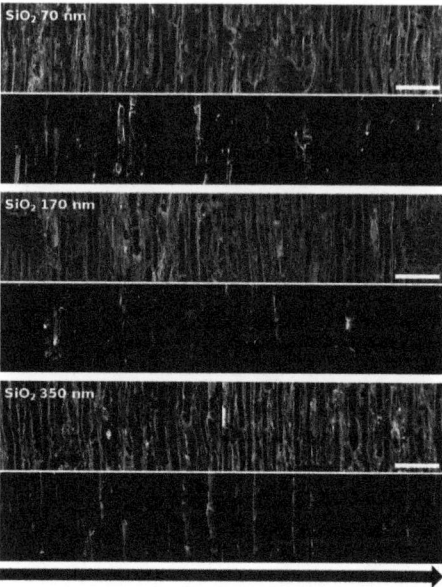

Figure 3. SEM–EDX analysis on split beech wood samples. The EDX map, with the silicon signal pseudo-colored in pink, is shown below each SEM map. Particles were detected mostly in large structures, e.g., vessel elements. All three investigated SiO_2 particle sizes penetrated beech wood samples down to the maximal depth of 5 mm. The arrow indicates the direction of impregnation, and the scale bar represents 500 µm.

Additionally, we tested the impregnation of SiO_2 70 nm suspensions in EtOH. As depicted in Figure 4, SiO_2-NPs were distributed uniformly on the vessel walls. The particles were colloidally stable in both water and EtOH prior to impregnation. Faster evaporation of EtOH compared to water could have contributed substantially to the observed effect when drying the wood samples at 60 °C.

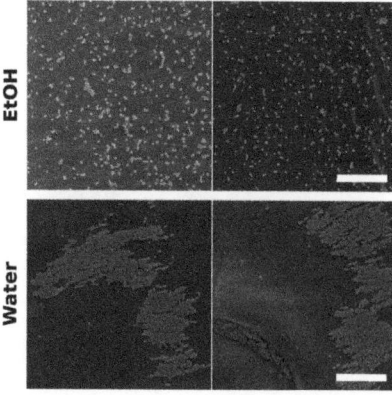

Figure 4. SiO_2 70 nm were transferred to either water or ethanol and impregnated into beech wood samples. The particles are detected primarily in the vessel elements, and particles suspended in ethanol were well dispersed over the vessel surface, whereas particles suspended in water aggregated locally. Neither suspensions were aggregated prior to impregnation (Table 1). The scale bar represents 2 µm.

Overall the results showed complete penetration of all particle sizes, and no evident threshold could be detected for beech wood in the size range investigated. Although the connecting pit structures are the dimensionally most restricting structural elements of hardwood, their diameter of 4–10 µm [36] (which depends on the type of pit and hardwood species) clearly is above the requirements for physical penetration of the here investigated particle sizes.

3.3. Impregnation of Pine Wood

SiO$_2$-NPs were impregnated in pine wood samples with closed cross-section and split in the radial direction (Figure 2). Figure 5 displays the SEM images and the corresponding silicon X-ray map (silicon signal in magenta). The samples were measured again down to 5 mm, which corresponds to the half width of the wood sample. For SiO$_2$ 70 nm, a signal was observed over the whole range, and particles were detected mainly in the water conducting system of softwood, i.e., in the tracheids. We observed particle accumulation in the ray cells and in areas of bordered pits, visible by the bright spots in the EDX-map of SiO$_2$ 70 nm. This indicated that the particles entered via ray parenchyma and subsequently distributed via half-bordered pits into tracheids.

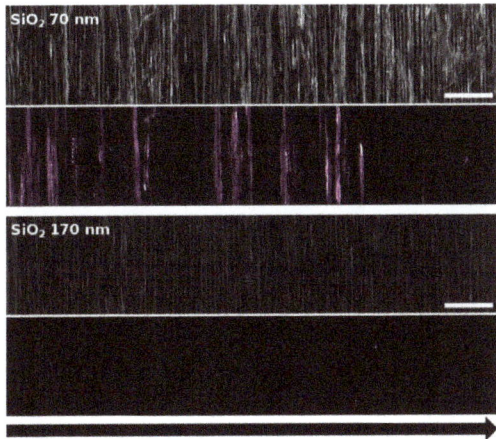

Figure 5. SEM–EDX analysis on split pine wood samples impregnated with the same particles used for beech wood. The corresponding EDX map below each SEM map shows thorough impregnation of SiO$_2$ 70 nm, but only background noise in the case of pine samples impregnated with SiO$_2$ 170 nm. The arrow indicates the direction of impregnation and the scale bar represents 500 µm.

EDX maps of pine samples impregnated with SiO$_2$ 170 nm (Figure 5, SiO$_2$ 170 nm) showed only marginal impregnation with NPs. Similar to beech wood, the connecting features were the bottleneck for impregnation since the diameter of tracheids is much larger (20–40 µm) than the diameter of the applied SiO$_2$ NPs [41,42]. The bordered pits and half-bordered pits, which are responsible for the intertracheid and the tracheid to ray parenchyma transport of liquids, respectively, are permeable for particles with a diameter of 400–600 nm with their pit membrane intact [17,43]. The pit membranes were not aspirated as this would completely seal off any liquid transport between the lumen and would therefore hinder the impregnation drastically. Comparison to previously published studies is difficult, since either particles with different particle sizes [27,44,45] or very broad particle size distributions [14,46,47] are reported, which makes it impossible to unambiguously define particle size thresholds. In our study, we observed a much lower size limit for particle impregnation in pine wood impregnation than the reported 400–600 nm. Arguably, the fraction of smaller particles mostly contributed to impregnation in the reported studies, and the larger particles were just deposited on the surface of the wood sample.

3.4. Influence of the Particle Surface on Impregnation

In the next step, we investigated the influence of surface properties on impregnation (efficiency) of SiO_2 70 nm. The impregnated samples were analyzed by SEM and the results are presented in Figures 6–8. Pristine SiO_2 70 nm and PETES–SiO_2 70 nm showed stronger impregnation compared to APTES–SiO_2. Pristine SiO_2 70 nm particles were detected in pine wood down to 5 mm (Figure 6), confirming the results of a previous study [27]. In some spots (Figure 6, pictures 4, 6, 8 and 10), the particles were aggregated on the cell wall. This could be explained by interferences of SiO_2 with possible extractives during the impregnation process. Extractives have shielding effects, which lower the charge and thus the electrostatic repulsion between the particles. Samples impregnated with APTES–SiO_2 70 nm did not penetrate the sample, as shown by SEM (Figure 7). Although the TEM size of APTES–SiO_2 was comparable to the pristine particles (Table 1), DLS showed that the particles flocculated at the concentrations used for impregnation, i.e., above the determined threshold of 400–600 nm. APTES–SiO_2 were more stable in EtOH than in water (Table 1). However, impregnation of pine samples with APTES–SiO_2 in EtOH only marginally improved wood infiltration (Supplementary Materials Figure S2), indicating that particle size (alone) is not the determining parameter, as shown previously by Renneckar et al. [25]. APTES–SiO_2 NPs can interact with the wood surface through the aminopropyl functional groups, since amine groups act as hydrogen-bond donors and acceptors, thus facilitating the interactions with the cellulose surface of the cell wall.

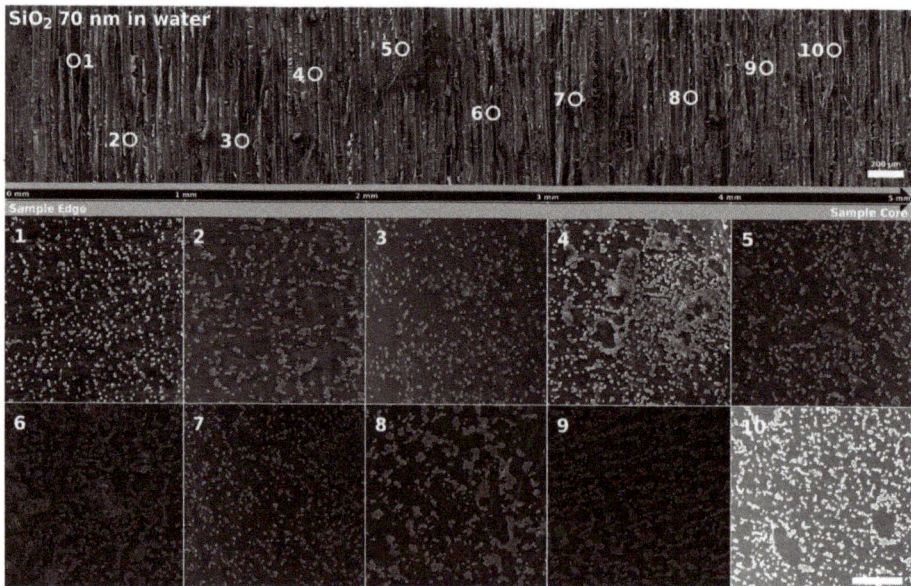

Figure 6. SEM analysis on split pine wood samples impregnated with pristine SiO_2 70 nm particles. Particles were detected until the maximum invested depth of 5 mm and were located in the water conducting system and on the cell wall. Some degree of aggregation was observed after drying, although the NPs were stable prior to and after the impregnation.

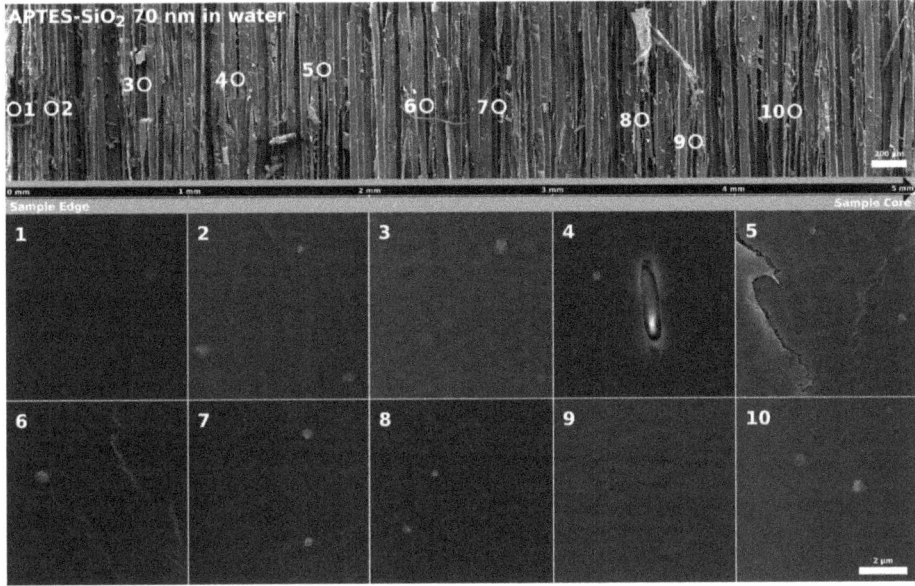

Figure 7. SEM analysis of split wood samples pressure-treated with APTES–SiO$_2$ 70 nm does not show any impregnation. No particles can be detected even in the tracheids closest to the sample edge. Suspensions of APTES–SiO$_2$ 70 nm in water were not sufficiently stable during the impregnation and large aggregates were probably filtered by the wood structure.

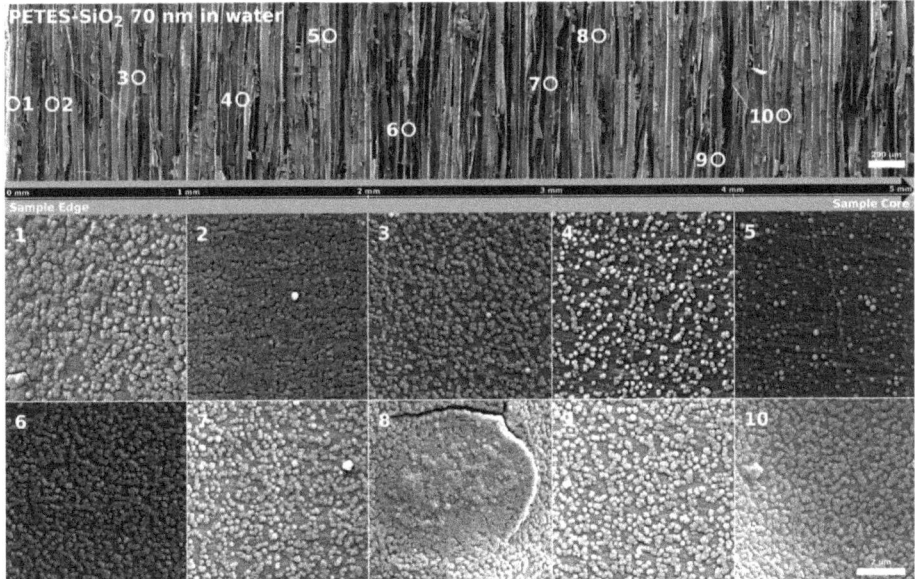

Figure 8. SEM analysis of wood samples impregnated with PETES–SiO$_2$ 70 nm. Particles are easily detectable and form almost confluent sheets. Functionalization with pentyl groups allows for deeper penetration of high amounts of particles.

As shown in Figure 8, the impregnation of PETES–SiO$_2$ was similar to what we observed with pristine SiO$_2$-NPs. PETES–SiO$_2$ were detected down to 5 mm in the lumens of tracheids. However, in comparison to pristine SiO$_2$-NPs, PETES–SiO$_2$ were found in almost confluent sheets even at the maximal depth, indicating that the introduction of pentyl groups on the SiO$_2$ surface could partially shield any interaction between pristine SiO$_2$ and the cellulose cell wall surface. This may explain the deeper penetration and almost full coverage of tracheids with PETES–SiO$_2$ (Figure 6). The ability of PETES–SiO$_2$ to form a confluent layer of particles could be applied to further investigations on similar particles with sufficient hydrophobicity [48] to introduce a hydrophobic character to wood, which hinders water uptake and hence such adverse effects as fungal decay or dimensional instability [49,50] However, SiO$_2$-NPs with a much more hydrophobic surface would not be stable in aqueous suspension [51], which is why we also dispersed and impregnated PETES–SiO$_2$ in EtOH (Supplementary Materials Figure S3). Although SiO$_2$-NPs and PETES–SiO$_2$ were found in the wood core, it was observed that impregnation of particles in EtOH allowed a larger bulk of PETES–SiO$_2$ to penetrate to the core of the samples. No aggregation in pine wood was observed for formulations, similar to the described phenomena in beech vessels.

4. Conclusions

Pine and beech wood samples with closed cross sections were pressure-treated with SiO$_2$-NPs in the size range from 70 to 350 nm. Particles were detected on the cell walls in the water conducting elements in beech and pine wood. For beech wood, all sizes were successfully impregnated, but for pine wood, the size threshold was found to be lower than previously reported [17]; 70 nm NPs were successfully incorporated, whereas 170nm NPs were no longer impregnated in pine wood. These results are in line with common findings of high loads of superficial particulate preserving agents filtered by the wood structure. Particle-bearing formulations with defined size distributions below the physical size restrictions of a particular wood species are expected to have improved structural penetration and hence better resistance towards leaching.

The impact of surface functionalization was investigated on SiO$_2$ 70 nm. Our investigations showed that colloidal stability (i.e., aggregation behavior) and surface charge have a significant impact on particle impregnation. Compared to unfunctionalized SiO$_2$ particles, PETES-functionalization enhanced particle infiltration. APTES–SiO$_2$ particles, featuring amine groups on the surface, were not impregnated, neither in water nor in ethanol. In water, particles were colloidally instable and the resulting aggregates were "filtered" by the wood structure. The same particles were colloidally more stable in ethanol, but showed only minor improvement with respect to penetration depth, probably due to strong interactions of the positive amine groups with the (negatively charged) cellulose/hemicellulose cell wall surface.

Supplementary Materials: The following are available online at http://www.mdpi.com/2624-8549/2/2/361\T1\textendash373/s1, Figure S1: Pine wood impregnation of pristine SiO$_2$ 70 nm nanoparticles dispersed in EtOH, Figure S2: Impregnation of APTES–SiO$_2$ 70 nm in EtOH, Figure S3: Impregnation of PETES–SiO$_2$ 70 nm in EtOH.

Author Contributions: D.B. and C.G. contributed equally to this work and have prepared and characterized all particles, recorded all SEM images and performed wood impregnation together with M.I.P.P. and T.V. The manuscript was written by D.B., C.G. and A.P.-F. with input from B.R.-R. and T.V. All authors have read and agreed to the published version of the manuscript.

Funding: This work was financially supported by the Swiss National Foundation (136976), the Adolphe Merkle Foundation and the University of Fribourg.

Acknowledgments: We would like to thank Miguel Spuch for preparing Figure 2.

Conflicts of Interest: The authors declare no conflict of interest.

References

1. European Industry Trade Association Representing the Pressure Treated Wood Industry. Available online: http://www.wei-ieo.org/woodpreservation.html (accessed on 17 March 2018).
2. Mohajerani, A.; Vajna, J.; Ellcock, R. Chromated copper arsenate timber: A review of products, leachate studies and recycling. *J. Clean. Prod.* **2018**, *179*, 292–307. [CrossRef]
3. Borazjani, H.; Ferguson, B.J.; McFarland, L.K.; McGinnis, G.D.; Pope, D.F.; Strobel, D.A.; Wagner, J.L. Evaluation of Wood-Treating Plant Sites for Land Treatment of Creosote- and Pentachlorophenol-Contaminated Soils. *ACS Sym. Ser.* **1990**, *422*, 252–266.
4. Humphrey, D.G. The chemistry of chromated copper arsenate wood preservatives. *Rev. Inorg. Chem.* **2002**, *22*, 1–40. [CrossRef]
5. Regulation (EC) No 1907/2006 of the European Parliament and of the Council. Available online: https://eur-lex.europa.eu/legal-content/EN/TXT/?uri=CELEX%3A02006R1907-20140410 (accessed on 14 March 2018).
6. Humar, M.; Lesar, B. Fungicidal properties of individual components of copper–ethanolamine-based wood preservatives. *Int. Biodeterior. Biodegrad.* **2008**, *62*, 46–50. [CrossRef]
7. Gadd, G.M. Interactions of fungip with toxic metals. *New Phytol.* **1993**, *124*, 25–60. [CrossRef]
8. Nicholas, D. Performance of waterborne copper/organic wood preservatives in an AWPA E14 soft-rot laboratory soil bed test using modified soil. *Holzforschung* **2017**, *71*, 759–763. [CrossRef]
9. Xue, W.; Ruddick, J.N.R.; Kennepohl, P. Solubilisation and chemical fixation of copper(ii) in micronized copper treated wood. *Dalton Trans.* **2016**, *45*, 3679–3686. [CrossRef]
10. Xue, W.; Kennepohl, P.; Ruddick, J.N. Reacted copper(II) concentrations in earlywood and latewood of micronized copper-treated Canadian softwood species. *Holzforschung* **2015**, *69*, 509–512. [CrossRef]
11. Schmitt, S.; Zhang, J.; Shields, S.; Schultz, T.P. Copper-Based Wood Preservative Systems Used for Residential Applications in North America and Europe. *ACS Symp. Ser.* **2014**, *1158*, 217–225.
12. Platten, W.E.; Luxton, P.T.; Gerke, T.; Harmon, S.; Sylvest, N.; Bradham, K.; Rogers, K. *Release of Micronized Copper Particles from Pressure-Treated Wood Products*; EPA Report; United States Environmental Protection Agency: Washington, DC, USA, 2014. Available online: https://cfpub.epa.gov/si/si_public_record_report.cfm?dirEntryId=307040&Lab=NRMRL (accessed on 14 March 2018).
13. Civardi, C.; Bulcke, J.V.D.; Schubert, M.; Michel, E.; Butron, M.I.; Boone, M.; Dierick, M.; Van Acker, J.; Wick, P.; Schwarze, F.W.M. Penetration and Effectiveness of Micronized Copper in Refractory Wood Species. *PLoS ONE* **2016**, *11*, e0163124. [CrossRef]
14. Civardi, C.; Schubert, M.; Fey, A.; Wick, P.; Schwarze, F.W.M. Micronized Copper Wood Preservatives: Efficacy of Ion, Nano, and Bulk Copper against the Brown Rot Fungus Rhodonia placenta. *PLoS ONE* **2015**, *10*, e0142578. [CrossRef] [PubMed]
15. Zhang, J.; Zhang, W. *Micronized Wood Preservative Formulations Comprising Copper and Zinc*; Osmose, Inc.: Atlanta, GA, USA, 2009.
16. Leach, R.M.; Zhang, J. Micronized Wood Preservative Compositions. U.S. Patent Application 20060288904, 28 December 2006.
17. Freeman, M.H.; McIntyre, C.R. A Comprehensive Review of Copper-Based Wood Preservatives. *For. Prod. J.* **2008**, *58*, 6–27.
18. McCallan, S.E.A. The nature of the fungicidal action of copper and sulfur. *Bot. Rev.* **1949**, *15*, 629–643. [CrossRef]
19. Pankras, S.; Cooper, P.A. Effect of ammonia addition to alkaline copper quaternary wood preservative solution on the distribution of copper complexes and leaching. *Holzforschung* **2012**, *66*. [CrossRef]
20. Freeman, M.H.; McIntyre, C.R. Micronized Copper Wood Preservatives: Strong indications of the Reservoir Effect. In Proceedings of the IRG Annual Meeting, Stockholm, Sweden, 16–20 June 2013.
21. Clausen, C.A.; Kartal, S.N.; Arango, R.A.; Green, F. Erratum to: The role of particle size of particulate nano-zinc oxide wood preservatives on termite mortality and leach resistance. *Nanoscale Res. Lett.* **2011**, *6*, 465. [CrossRef]
22. Kartal, S.N.; Green, F.; Clausen, C. Do the unique properties of nanometals affect leachability or efficacy against fungi and termites? *Int. Biodeterior. Biodegradation* **2009**, *63*, 490–495. [CrossRef]

23. Ghorbani, M.; Taghiyari, H.R.; Siahposht, H. Effects of heat treatment and impregnation with zinc-oxide nanoparticles on physical, mechanical, and biological properties of beech wood. *Wood Sci. Technol.* **2014**, *48*, 727–736. [CrossRef]
24. Clar, J.; Platten, W.E.; Baumann, E.J.; Remsen, A.; Harmon, S.M.; Bennett-Stamper, C.L.; Thomas, T.A.; Luxton, T.P. Dermal transfer and environmental release of CeO_2 nanoparticles used as UV inhibitors on outdoor surfaces: Implications for human and environmental health. *Sci. Total. Environ.* **2018**, *613*, 714–723. [CrossRef]
25. Renneckar, S.; Zhou, Y. Nanoscale Coatings on Wood: Polyelectrolyte Adsorption and Layer-by-Layer Assembled Film Formation. *ACS Appl. Mater. Interfaces* **2009**, *1*, 559–566. [CrossRef]
26. Nechyporchuk, O.; Bordes, R.; Köhnke, T. Wet Spinning of Flame-Retardant Cellulosic Fibers Supported by Interfacial Complexation of Cellulose Nanofibrils with Silica Nanoparticles. *ACS Appl. Mater. Interfaces* **2017**, *9*, 39069–39077. [CrossRef]
27. Geers, C.; Rodriguez-Lorenzo, L.; Peña, M.I.P.; Brodard, P.; Volkmer, T.; Rothen-Rutishauser, B.; Petri-Fink, A. Distribution of Silica-Coated Silver/Gold Nanostars in Soft- and Hardwood Applying SERS-Based Imaging. *Langmuir* **2015**, *32*, 274–283. [CrossRef] [PubMed]
28. Liu, Y.; Laks, P.; Heiden, P. Nanoparticles for the Controlled Release of Fungicides in Wood: Soil Jar Studies Using G. Trabeum and T. Versicolor Wood Decay Fungi. *Holzforschung* **2003**, *57*, 135–139. [CrossRef]
29. Evans, P.; Matsunaga, H.; Kiguchi, M. Large-scale application of nanotechnology for wood protection. *Nat. Nanotechnol.* **2008**, *3*, 577. [CrossRef]
30. Beecher, J.F. Wood, trees and nanotechnology. *Nat. Nanotechnol.* **2007**, *2*, 466–467. [CrossRef] [PubMed]
31. Stöber, W.; Fink, A.; Bohn, E. Controlled growth of monodisperse silica spheres in the micron size range. *J. Colloid Interface Sci.* **1968**, *26*, 62–69. [CrossRef]
32. Michen, B.; Geers, C.; Vanhecke, D.; Endes, C.; Rothen-Rutishauser, B.; Balog, S.; Petri-Fink, A. Avoiding drying-artifacts in transmission electron microscopy: Characterizing the size and colloidal state of nanoparticles. *Sci. Rep.* **2015**, *5*, 9793. [CrossRef]
33. British Standards Institution. Wood Preservatives. Test Method for Determining the Protective Effectiveness against Wood Destroying Basidiomycetes. Determination of the Toxic Values. 1997. Available online: https://shop.bsigroup.com/ProductDetail?pid=000000000030112176 (accessed on 14 March 2018).
34. Rahman, I.; Jafarzadeh, M.; Sipaut, C.S. Synthesis of organo-functionalized nanosilica via a co-condensation modification using γ-aminopropyltriethoxysilane (APTES). *Ceram. Int.* **2009**, *35*, 1883–1888. [CrossRef]
35. Sriramulu, D.; Reed, E.L.; Annamalai, M.; Venkatesan, T.V.; Valiyaveettil, S. Synthesis and Characterization of Superhydrophobic, Self-cleaning NIR-reflective Silica Nanoparticles. *Sci. Rep.* **2016**, *6*, 35993. [CrossRef]
36. Wagenführ, R. *Anatomie des Holzes: Struktur, Identifizierung, Nomenklatur, Mikrotechnologie*, 5th ed.; DRW Verlag: Leinfelden-Echterdingen, Germany, 1999.
37. Hass, P.; Wittel, F.K.; McDonald, S.A.; Marone, F.; Stampanoni, M.; Herrmann, H.J.; Niemz, P. Pore space analysis of beech wood: The vessel network. *Holzforschung* **2010**, *64*, 639–644. [CrossRef]
38. Kucera, L. Die dreidimensionale Strukturanalyse des Holzes. *Holz Roh Werkst.* **1975**, *33*, 276–282. [CrossRef]
39. Tondi, G.; Thevenon, M.F.; Mies, B.; Standfest, G.; Petutschnigg, A.; Wieland, S. Impregnation of Scots pine and beech with tannin solutions: Effect of viscosity and wood anatomy in wood infiltration. *Wood Sci. Technol.* **2013**, *47*, 615–626. [CrossRef] [PubMed]
40. Matsunaga, H.; Kiguchi, M.; Roth, B.; Evans, P. Visualisation of Metals in Pine Treated with Preservative Containing Copper and Iron Nanoparticles. *IAWA J.* **2008**, *29*, 387–396. [CrossRef]
41. Mencuccini, M.; Grace, J.; Fioravanti, M. Biomechanical and hydraulic determinants of tree structure in Scots pine: Anatomical characteristics. *Tree Physiol.* **1997**, *17*, 105–113. [CrossRef] [PubMed]
42. Kilpeläinen, A.; Gerendiain, A.Z.; Luostarinen, K.; Peltola, H.; Kellomäki, S. Elevated temperature and CO(2) concentration effects on xylem anatomy of Scots pine. *Tree Physiol.* **2007**, *27*, 1329–1338. [CrossRef]
43. Clausen, C.A.; Green, F.; Kartal, S.N. Weatherability and Leach Resistance of Wood Impregnated with Nano-Zinc Oxide. *Nanoscale Res. Lett.* **2010**, *5*, 1464–1467. [CrossRef]
44. Dong, Y.; Yan, Y.; Zhang, S.; Li, J.; Wang, J. Flammability and physical–mechanical properties assessment of wood treated with furfuryl alcohol and nano-SiO_2. *Holz Roh Werkst.* **2015**, *73*, 457–464. [CrossRef]
45. Soltani, M.; Najafi, A.; Yousefian, S.; Naji, H.; Bakar, E.S. Water Repellent Effect and Dimension Stability of Beech Wood Impregnated with Nano-Zinc Oxide. *Bioresources* **2013**, *8*, 6280–6287. [CrossRef]

46. Alex, T. Short Communication: The clay nanoparticle impregnation for increasing the strength and quality of sengon (*Paraserianthes falcataria*) and white meranti (*Shorea bracteolata*) timber. *Nusant. Biosci.* **2017**, *9*, 107–110. [CrossRef]
47. Matsunaga, H.; Kiguchi, M.; Evans, P.D. Microdistribution of copper-carbonate and iron oxide nanoparticles in treated wood. *J. Nanoparticle Res.* **2008**, *11*, 1087–1098. [CrossRef]
48. Mahltig, B.; Bottcher, H. Modified Silica Sol Coatings for Water-Repellent Textiles. *J. Sol.-Gel Sci. Technol.* **2003**, *27*, 43–52. [CrossRef]
49. Lourençon, T.V.; Mattos, B.; Cademartori, P.H.; Magalhães, W.L.E. Bio-oil from a fast pyrolysis pilot plant as antifungal and hydrophobic agent for wood preservation. *J. Anal. Appl. Pyrolysis* **2016**, *122*, 1–6. [CrossRef]
50. Wang, X.; Chai, Y.; Liu, J. Formation of highly hydrophobic wood surfaces using silica nanoparticles modified with long-chain alkylsilane. *Holzforschung* **2013**, *67*, 667–672. [CrossRef]
51. Pellegrino, T.; Manna, L.; Kudera, S.; Liedl, T.; Koktysh, D.; Rogach, A.L.; Keller, S.; Rädler, J.; Natile, G.; Parak, W.J. Hydrophobic Nanocrystals Coated with an Amphiphilic Polymer Shell: A General Route to Water Soluble Nanocrystals. *Nano Lett.* **2004**, *4*, 703–707. [CrossRef]

© 2020 by the authors. Licensee MDPI, Basel, Switzerland. This article is an open access article distributed under the terms and conditions of the Creative Commons Attribution (CC BY) license (http://creativecommons.org/licenses/by/4.0/).

Review

Recent Studies on the Antimicrobial Activity of Transition Metal Complexes of Groups 6–12

Sara Nasiri Sovari and Fabio Zobi *

Department of Chemistry, Fribourg University, Chemin du Musée 9, 1700 Fribourg, Switzerland; sara.nasirisovari@unifr.ch
* Correspondence: fabio.zobi@unifr.ch

Received: 6 April 2020; Accepted: 6 May 2020; Published: 9 May 2020

Abstract: Antimicrobial resistance is an increasingly serious threat to global public health that requires innovative solutions to counteract new resistance mechanisms emerging and spreading globally in infectious pathogens. Classic organic antibiotics are rapidly exhausting the structural variations available for an effective antimicrobial drug and new compounds emerging from the industrial pharmaceutical pipeline will likely have a short-term and limited impact before the pathogens can adapt. Inorganic and organometallic complexes offer the opportunity to discover and develop new active antimicrobial agents by exploiting their wide range of three-dimensional geometries and virtually infinite design possibilities that can affect their substitution kinetics, charge, lipophilicity, biological targets and modes of action. This review describes recent studies on the antimicrobial activity of transition metal complexes of groups 6–12. It focuses on the effectiveness of the metal complexes in relation to the rich structural chemical variations of the same. The aim is to provide a short *vade mecum* for the readers interested in the subject that can complement other reviews.

Keywords: antimicrobial; transition metals; complex; organometallic; drug-resistant

---

1. Introduction

Antimicrobial resistance has become a global concern ultimately affecting humans' ability to prevent and treat an increasing number of infections caused by bacteria, parasites, viruses and fungi and the success of surgery and cancer chemotherapy. It occurs naturally over time, usually through genetic changes of the pathogens when exposed to antimicrobial drugs. One of the causes for the emergence of the problem is the overuse and misuse of existing antibiotics, which fueled the evolution of pathogens resistant to the current library of antimicrobial agents [1,2]. As a result, available medicines become ineffective, infections persist in the body, increasing the risk to patients' health, spreading and health care costs. Multidrug resistant bacteria, such as *Enterococcus faecium, Staphylococcus aureus, Klebsiella pneumoniae, Acetinobacter baumanii, Pseudomonas aeruginosa,* and *Enterobacteriaceae* ("ESKAPE") species, are a major concern of the World Health Organization (WHO) and health authorities. These pathogens cause a large number of victims worldwide [3–5]. As an example, methicillin-resistant *Staphylococcus aureus* (MRSA) is one of the most critical causes of healthcare-related or community-related infections, because of the multiple resistances to antibiotics and the toxins produced [6]. It is, therefore, evident that there is an urgent need for the development of new antimicrobial agents with more effective mechanisms of action [7].

While the problem is escalating, major pharmaceutical companies have interrupted their antibiotic drug discovery programs, leaving academia at the forefront of the discovery of new classes of active compounds, especially for Gram-negative bacteria [8,9]. The classical approach of medicinal chemists based exclusively on organic molecules is poised to have a short-term limited impact because pathogens will adapt and develop resistance to new drugs. Furthermore, as recently pointed out by Frei [10], only ~25% of the compounds currently in clinical development represent entirely new structural classes, with the remaining 75% of drugs being merely derivatives and modifications of already approved antibiotics. Thus, there is not just an urgent need for new antibiotics, but also a need for entirely new classes of molecules for the purpose. Transition metal complexes offer this possibility. They possess a wide range of three-dimensional geometries, virtually infinite possibilities to design their coordination sphere in order to affect their substitution kinetics, charge, lipophilicity biological targets and modes of action. Such complexes, however, are still (prejudicially) ignored by pharmaceutical companies despite the fact that several of them are used in hospitals worldwide. For example, arsphenamine, also known as Salvarsan or compound 606, is an effective drug for the treatment for syphilis; cisplatin is a chemotherapeutic agent, administered intravenously, and used to treat a number of cancers (e.g., testicular, ovarian, cervical, breast, bladder, head and neck, esophageal cancer); auranofin is a gold salt approved by the WHO as an antirheumatic agent; technetium sestamibi (trade name *Cardiolite*) is a pharmaceutical agent used in nuclear medicine imaging to visualize the myocardium.

In the last ten years, inorganic and organometallic transition metal medicinal chemists have begun to develop new antimicrobial agents with great promise and remarkable success. Complexes of virtually all ions of the transition periods have been tested. In this article, we present an overview of antimicrobial transition metal (groups 6–12) complexes published in the scientific literature in the last five years. We only describe inorganic and organometallic complexes of group 6–12 with a few exceptions. Thus, e.g., silver and iron complexes are not included. Antimicrobial iron and silver complexes and (nano)materials have been recently reviewed elsewhere [11–25]. Due to the growing interest in the field, recent reviews and prospective on the antibacterial applications of transition metal complexes have appeared [10,26–29]. This work aims at being complementary to those, including some important common seminal examples but mostly species not included by the other authors.

2. Group 6

2.1. Chromium Complexes

Schiff base complexes of chromium are most commonly studied for their antimicrobial efficacy, but the species have seldom shown high potency. Kumar et al. synthesized a new class of tetradentate Schiff bases as ligands and their corresponding chromium(III) complexes (**1**, Figure 1) by using $CrCl_3$ as the metal ion source [30]. The antimicrobial activities of the chromium(III) complexes were tested against *S. aureus* (Gram-positive), *E. coli* and *P. aeruginosa* (Gram-negative) bacterial strains, but their efficacies were lower than the standard drug, i.e., ampicillin. Rathi et al. reported the preparation of thiophene based macrocyclic Schiff base complexes from the reaction of succinohydrazide and thiophene-2,5-dicarbaldehyde in the presence of chromium(III) and iron(III) salts of chloride, nitrate and acetate (**2**, Figure 1) [31]. The antimicrobial activities of all synthesized complexes were tested against bacterial strains, such as *B. subtilis* and *E. coli*, and fungal strains, such as *S. cerevisiae* and *C. albicans*. The data showed good activity of compounds against all tested microbial strains with the MIC values in the range of 8–128 µg/mL. In 2017, Shaabani et al. prepared bridged chromium(III) complexes, of hydrazine Schiff bases tridentate ligands and azide (**3**, Figure 1) [32]. The complexes, however, were not particularly effective against tested pathogens with MIC values (~1250 µg/mL) higher than standard drugs (MIC = 8–28 µg/mL). Kafi-Ahmadi and coworkers synthesized thiourea derivatives as Schiff base ligands (**4**, Figure 1) and their chromium(III) complexes [33]. The complexes were tested for their antibacterial activities against clinically important bacteria, such as *E. coli*, *S. aureus*, and *B. subtilis*, and they showed good activities against all strains, comparable to that of streptomycin as the standard. The mechanism of action of these complexes is unknown. The authors suggested that chelation theory might help explain the biological activities of the metal chelates. This phenomenon relates to a decrease in the polarity of the metal ion due to overlap metal and ligand orbitals, resulting in partial sharing of the positive charge of the metal ion with donor groups and possibly electron delocalization over the whole molecule [33–36].

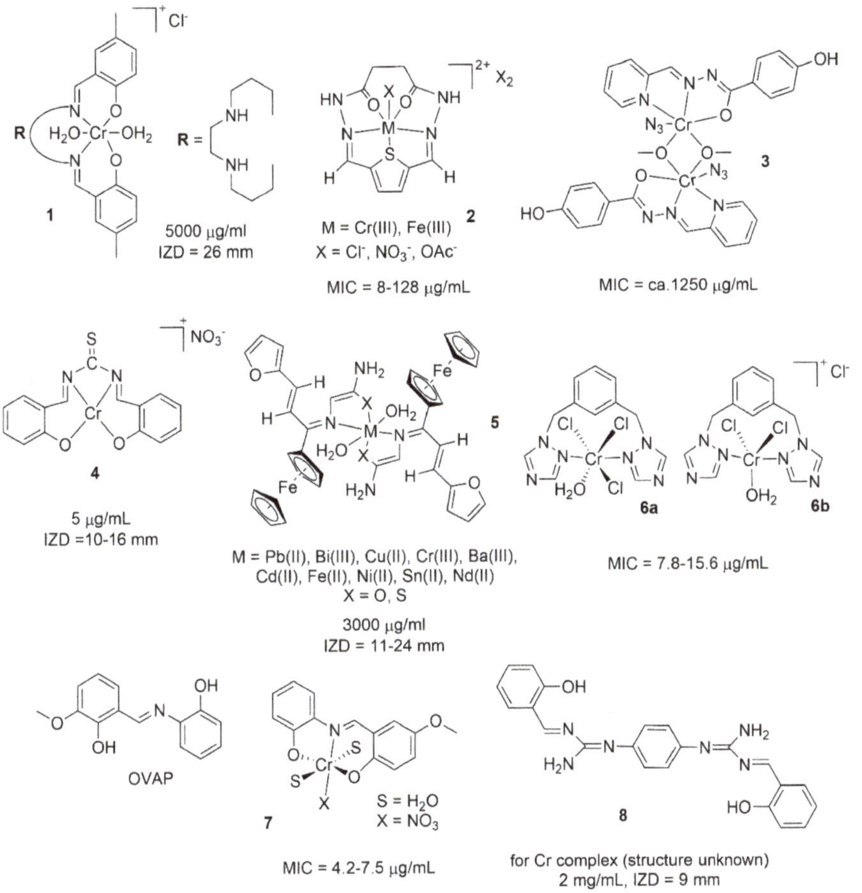

Figure 1. Structural formula of selected antimicrobial chromium(III) complexes and corresponding ligands. MIC = minimal inhibitory concentration; IZD = inhibition zone diameter at given concentration.

In 2018, Liu et al. introduced a Schiff base ligand, 1-ferrocenyl-3-(2-furyl) propenone diamino (thio) urea, and coordinated it to a range of metal ions (e.g., Pb(II), Bi(III), Cu(II), Cr(III), Ba(III), Cd(II), Fe(II), Ni(II), Sn(II) and Nd(II), **5**, Figure 1) [37]. All compounds were screened for their antimicrobial activities against bacteria, such as *E. coli*, *S. aureus* and MRSA, also fungi, such as *C. albicans* and *A. flavus*. The complexes were not particularly effective. The zones of inhibition (mm) were found in a range between 11 and 24 mm (3 mg/mL concentrations) with the chromium(III) complex being amongst the least effective compounds. In 2018, tridentate triazole based ligands of chromium(III) complexes were reported by Murcia et al. (**6a** and **6b**, Figure 1) [38]. The antimicrobial activities of both ligands and complexes were tested against a wide range of bacterial and fungal strains of clinical relevance. The results indicated that the chromium(III) complexes were more potent than free ligands and more effective against fungi than bacteria. The complexes **6a** and **6b** showed MIC values in the range of 7.8 to 15.6 µg/mL. A study on azomethine chelates of Cu(II), Pd(II), Zn(II) and Cr(III) with tridentate dianionic azomethine OVAP ligand (where OVAP = 2-[(2-hydroxyphenylimino)methyl]-6-methoxyphenol), was carried out by Abu-Dief et al. (**7**, Figure 1) [39]. All OVAP metal complexes were screened against a broad-spectrum of antimicrobial strains (bacterial strains: *M. luteus*, *E. coli* and *S. marcescence*; fungal strains: *A. flavus*, *G. candidum* and *F. oxysporum*) and showed MIC values between 4.25 and 7.50 µg/mL.

The investigated azomethine metal chelates revealed significantly enhanced antimicrobial activities in comparison to the free ligand (MIC = 9.0–10.75 μg/mL) and showed comparable activities to ofloxacin and fluconazol. Very recently, copper(II), nickel(II), cobalt(II), manganese(II), iron(III), chromium(III), bismuth(III), and zinc(II) complexes of the guanidine Schiff bases **8** (Figure 1) were reported [40]. Compounds were screened against *S. aureus* (Gram-positive), *P. aeruginosa* (Gram-negative) and the fungi strains, such as *C. albicans* and *A. niger*. In general, metal complexes were found to be more toxic than ligand **8** and showed greater activity than neomycin and the naturally occurring fungicide cycloheximide. Several other chromium metal complexes have been recently tested for their antimicrobial efficacy but were not found to be active [41–45].

2.2. Molybdenum Complexes

In comparison to chromium complexes, only a few studies on the antimicrobial potential of molybdenum metal complexes have appeared over the last decade in the literature, while we are not aware of tungsten species having been reported lately. A series of *cis*-dichloro/dibromodioxidobis (2-amino-6-substitutedbenzothiazole) molybdenum(VI) complexes (**9**, Figure 2) were reported by Saraswat et al. in 2013 [46]. The complexes of dihalodioxidomolybdenum(VI) have played a special role in the higher valent molybdenum enzymes such as sulfite oxidase, nitrate reductase, xanthine oxidase and xanthine dehydrogenase during biological processes [47,48]. The authors reported the antibacterial activities of the species against *P. aeruginosa*, *S aureus* and *K. pneumoniae* and antifungal activities against *A. flavus* and *A. niger*. The results showed that derivatives of **9** were generally as active as ampicillin against bacteria strains, and most effective against *P. aeruginosa* and *K. pneumonia*, while their antifungal MIC values were in the range of 10–20 μg/mL.

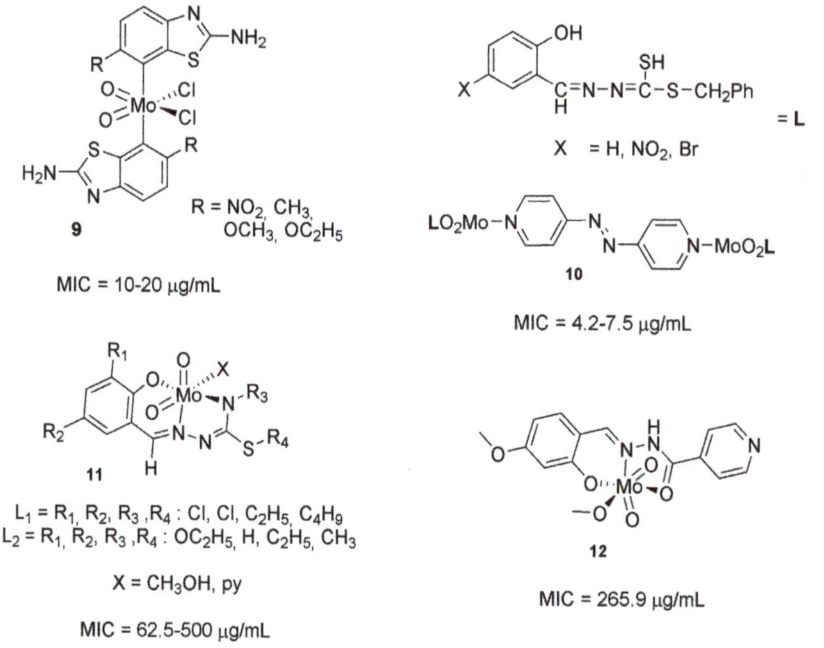

Figure 2. Structural formula of selected antimicrobial molybdenum(VI) complexes and corresponding ligands.

In 2015, Biswal et al. reported isostructural 4,4'-azopyridine (4,4'-azpy) pillared binuclear dioxomolybdenum(VI) complexes of formula [(MoO$_2$L$_1$)$_2$(4,4'-azpy)], [(MoO$_2$L$_2$)$_2$(4,4'-azpy)] and

[(MoO$_2$L$_3$)$_2$(4,4′-azpy)] (where L$_\#$ = Schiff base ligand, **10**, Figure 2) [49]. The ligands and molybdenum(VI) complexes were tested against *B. cerus* and *L. monocytogenes* (Gram-positive), *E. coli* and *S. aureus* (Gram-negative) bacteria by the disc diffusion method. The compounds exhibited different degrees of antimicrobial activities at a concentration of 10 µg per disc against the pathogens with antimicrobial activities comparable with those of common antibiotics ampicillin and tetracycline (as standard drugs). Schiff base ligands also showed moderate to good antimicrobial activities against all test microorganisms and were more active than their corresponding complexes. In 2019, Çelen et al. synthesized a group of thiosemicarbazonato-based ligands (2-hydroxy-3-methoxy/3,5-dibromo benzaldehyde 4-phenyl/ethyl-S-methyl/butyl thiosemicarbazones), and then coordinated the ligands through the ONN set to the molybdenum(VI) ion center to prepare *cis*-dioxomolybdenum(VI) complexes (**11**, Figure 2) [50]. All ligands and the complexes were tested (10 mg/mL) against *C. albicans*, *E. coli*, *P. aeruginosa* and *S. aureus*. The results confirmed the antimicrobial activities of all thiosemicarbazones and their dioxomolybdenum(VI) complexes and MIC values were in the range of 62.5–500 µg/mL. In 2020, Sang et al. reported the synthesis and the antimicrobial properties of a dioxidomolybdenum(VI) complex of N′-(2-hydroxy-4-methoxybenzylidene)isonicotinohydrazide (**12**, Figure 2) [51]. The free ligand showed modest antibacterial activity against *S. aureus* and *E. coli* (MIC = ~5 mmol/L), however, the molybdenum complex showed higher antibacterial activity against *E. coli* with MIC value of 0.62 ± 0.04 mmol/L. Finally, a report on two cationic cluster complexes based on the {Mo$_6$I$_8$}$^{4+}$ core with (4-carboxybutyl)triphenylphosphonium and 4-carboxy-1-methylpyridinium as apical ligands, indicated no antimicrobial activities of the species [52].

3. Group 7

3.1. Manganese Complexes

Several manganese complexes have been reported in the field, including photoactivatable CO-releasing molecules, which are described separately in the following section. In 2013, Zampakou et al. prepared [KMn(oxo)$_3$(MeOH)$_3$] and [Mn(erx)$_2$(phen)] complexes by reacting MnCl$_2$ with the quinolone antibacterial drug oxolinic acid (Hoxo), enrofloxacin (Herx) and the N,N′-donor heterocyclic ligand 1,10-phenanthroline (phen), respectively (**13**, Figure 3) [53]. Complexes were found significantly active against three Gram-positive (*B. subtilis*, *B. cereus* and *S. aureus*) and two Gram-negative (*X. campestris* and *E. coli*) bacterial strains with half-minimum inhibitory concentration (MIC) between 1.2 and 44 µg/mL. In 2018, Barmpa et al. reported similar types of manganese(II) complexes by using the quinolone antimicrobial agent sparfloxacin (Hsf) and flumequine (Hflmq) with or without nitrogen-donor heterocyclic ligands 1,10–phenanthroline (phen), 2,2′–bipyridine (bipy), 2,2′–bipyridylamine (bipyam) or pyridine (py) (**14**, Figure 3) [54]. The in vitro antimicrobial tests gave MIC values for the complexes in the range of, or slightly better than free Hsf. Against bacterial strains, such as *E. coli*, *B. subtilis*, and *S. aureus*, MICs were significantly low ranging from 0.0625–1.000 and 0.5–19 µg/mL. In 2015, P. Arthi and coworkers reported a series of pendant-armed Schiff base hexaaza macrocycles manganese(II) complexes by the condensation of equimolar amounts of terephthalaldehyde and N,N-bis(2-aminoethyl)benzamide derivatives in the presence of Mn(ClO$_4$)$_2$·6H$_2$O as a templating agent (**15**, Figure 3) [55]. In comparison to the standard drug, ciprofloxacin, the complexes showed good activities against both Gram-negative (*K. pneumoniae*, *P. aeruginosa*, *V. alginolyticus*, *V. cholerae* and *V. harveyi*) and Gram-positive (*S. aureus* and *S. mutans*) bacterial strains. The mean zone of inhibition values of the complexes and the standard were in the range of 4–21 and 20–25 mm (100 µg/mL), respectively. Also in 2015, Simpson et al. described the antibacterial and antiparasitic activities of manganese(I) tricarbonyl complexes with ketoconazole, miconazole, and clotrimazole ligands [56]. The molecules were tested against eight different bacterial strains: Gram-positive, such as *S. aureus*, *S. epidermidis*, *E. faekalis*, and *E. faecium*, and Gram-negative, such as *E. coli*, *P. aeruginosa*, *Y. pseudotuberculosa*, and *Y. pestis*. Only the miconazole complex (MIC values of 10–20 µM on *E. coli*, *Y. pseudotuberculosa*, and *Y. pestis*) was active against Gram-negative bacteria and showed higher activity than miconazole alone. Conversely, all species were active against Gram-positive bacteria at submicromolar concentrations (MIC = 0.625 to 2.5 µM), particularly against *staphylococci*. The complexation of luteolin to manganese(II) was carried out to prepare manganese(IV) complex **16**, (Figure 3) [57]. The ligand and complex were screened against different microbial strains (e.g., *E. coli*, *S. aureus*, *L. monocytogenes* and *P. aeruginosa*) and **16** was found ~x1.5 more active than the ligand alone. A study on a series of manganese(I) tricarbonyl complexes bearing bis(2-pyridinylmethyl)(2-quinolinylmethyl)amine, bis(2-quinolinylmethyl)(2-pyridinylmethyl)amine, tris(2-quinolinylmethyl)amine, and tris(2-pyridinylmethyl)amine ligands (**17**, Figure 3), was reported recently by Güntzel and coworkers [58]. The compounds were examined against 14 different multidrug-resistant clinical isolates of *A. baumannii* and *P. aeruginosa* showing MIC values in the range of 0.2–0.8 mM. Finally, Kottelat et al. described a series of carbonyl complexes of manganese bearing isocyanide ligands of formula *fac*-[Mn(CO)$_3$(CNR)$_2$Br] and found that for CNR = (1-isocyanoethyl)benzene, the complex showed a MIC of 128 µg/mL against *E. coli* [59]. Several other manganese metal complexes have been recently tested for their antimicrobial efficacy but were not found to be active [60–63].

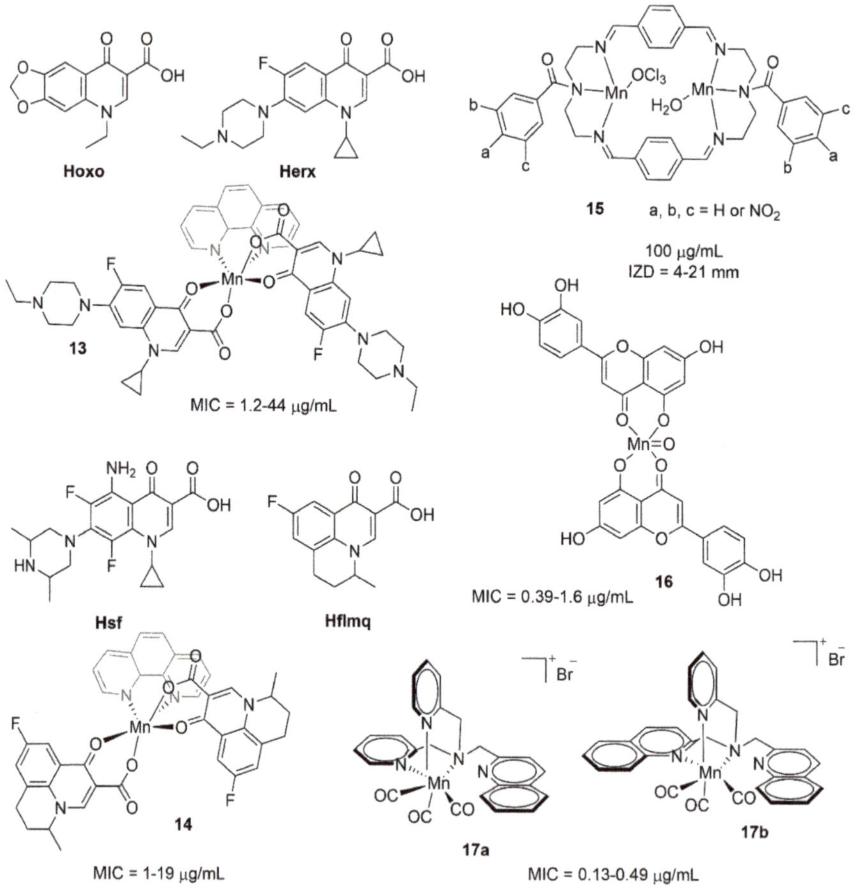

Figure 3. Structural formula of selected manganese(I), (II) and (IV) complexes and antibacterial drugs oxalinic acid (Hoxo), enrofloxacin (Herox), sparfloxacin (Hsf) and flumequine (Hflmq).

3.2. Manganese Photoactivatable CO-Releasing Molecules (PhotoCORMs)

PhotoCORMs are a special class of manganese-based antimicrobial complexes (Figure 4). The molecules are able to release carbon monoxide when activated with light. Carbon monoxide then acts in concert with the metal fragment to impart antimicrobial efficacy to the species. The [Mn(CO)$_3$(tpa-k^{3}N)]Br complex (**18**, Figure 4) was the first one reported in the literature and it remains the most extensively studied [64–67]. It was active against several *E. coli* strains (K12 derivative MG1655, EC958, APEC), if photo-activated and perturbs the growth of multidrug-resistant isolates of Avian Pathogenic *E. coli* (APEC) (both in vitro and in vivo) without the need of light activation. In vivo (*G. mellonella* wax moth model), **18** showed no toxicity at double the concentration required in the treatment assay. The complex **19** (known as Trypto-CORM), was described in 2014 by Ward et al. [68,69]. The compound was not toxic to eukaryotic RAW264.7 cells but showed a strong antibacterial effect against *E. coli* strain W3110, *N. gonorrhoeae* and *S. aureus*. It completely inhibited *E. coli* growth following irradiation, leading to a loss of >99.9% of cell viability. Trypto-CORM was similarly toxic to *N. gonorrhoeae*, in the dark resulting in a loss of >99% cellular viability (half maximal inhibitory concentration (IC$_{50}$) value of 22 µM). Furthermore, complex **19** exhibited a cytostatic effect in the dark and cytotoxic effect if exposed to light against *S. aureus*. Mann et al. introduced molecule **20** and studied the broad-spectrum antimicrobial potential of the molecules [70,71]. The complex **20** inhibited growth of *E. coli* and several antibiotic resistant clinical isolates of pathogenic bacteria in a concentration-dependent manner. It extensively concentrated in *E. coli* cells, reaching concentrations of ~3.5 mM after 80 min of incubation. Significantly, **20** was effective against several pathogens isolated from clinical infections and causes in vitro a complete growth arrest of the multidrug-resistant *E. coli* EC958 clinical pathogen, *K. pneumoniae*, *S. flexneri*, *S. kedougou* and *E. hormaechei*, but it was ineffective against growth of *P. aeruginosa*, *C. koseri*, and *A. baumannii*.

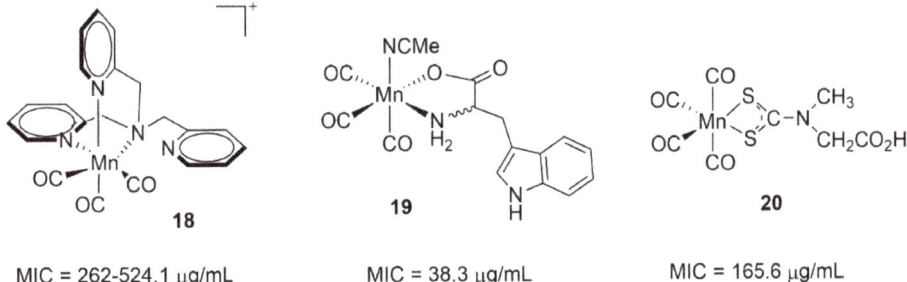

Figure 4. Structural formula of selected antimicrobial manganese(I) photoactivatable CO-releasing molecules.

3.3. Rhenium Complexes

In 2014, Noor et al. described a family of bioconjugated tridentate pyridyl-1,2,3-triazole macrocycles and the corresponding rhenium(I) complexes (**21**, Figure 5), which were screened for antimicrobial activities in vitro against both Gram-positive (*S. aureus*) and Gram-negative (*E. coli*) bacterial strains [72]. The minimum inhibitory concentrations for the compounds, however, showed values >256 µg/mL. At the same time Partra and coworkers introduced an interesting trimetallic complex (**22**, Figure 5) containing a ferrocenyl (Fc), a cymantrene and a [(dpa)Re(CO)$_s$] residue (dpa = N,N-bis(pyridine-2-ylmethyl)prop-2-yn-1-amine) as the main biological active moiety of the construct [73]. A systematic structure–activity relationship (SAR) study against various Gram-positive pathogenic bacteria, including methicillin-resistant *S. aureus* (MRSA) strains proved that [(dpa)Re(CO)$_3$] moiety was the essential part for the antibacterial activity of the trimetallic complex. The other two metallic units (Fc and cymantrene) could be replaced by organic compounds without affecting the antibacterial activities of the construct. The MIC values of the compounds against Gram-positive bacterial strains, such as *B. subtilis*, *S. aureus* DSM 20231 and *S. aureus* ATCC43300 (MRSA) were found in the range of 1.4–21 µM. In 2016, Kumar et al. [74] reported a group of mono- and bis-*fac*-rhenium tricarbonyl 2-pyridyl-1,2,3-triazole complexes with different aliphatic and aromatic substituents (**23**, Figure 5) which were tested for antimicrobial activities in vitro against both Gram-positive (*S. aureus*) and Gram-negative (*E. coli*) bacterial strains. The MICs for all the complexes were measured between 16 and 1024 µg/mL. In 2017, a family of N-heterocyclic carbene (NHC) *fac*-[Re(I)(CO)$_3$] complexes containing unsubstituted benzimidazol-2-ylidene and bisimine ligands (N^N) ligands (**24**, Figure 5), were reported by Siegmund et al. [75]. The antimicrobial tests gave MIC values of the complexes between 0.7–2 µg/mL against Gram-positive strains, such as *B. subtilits* and *S. aureus*. However, the same complexes were inactive against Gram-negative strains, such as *E. coli*, *A. baumannii* and *P. aeruginosa*. Recently, Frei et al. reported the synthesis and antibacterial profiling of three rhenium bisquinoline complexes (**25**, Figure 5) [76]. The complexes displayed light-induced activities against drug-resistant *S. aureus* and *E. coli* showed MICs under photo-irradiation between 4- to 16-fold lower than in the dark. Other rhenium metal complexes have been tested for their antimicrobial efficacy but were inactive [77–80].

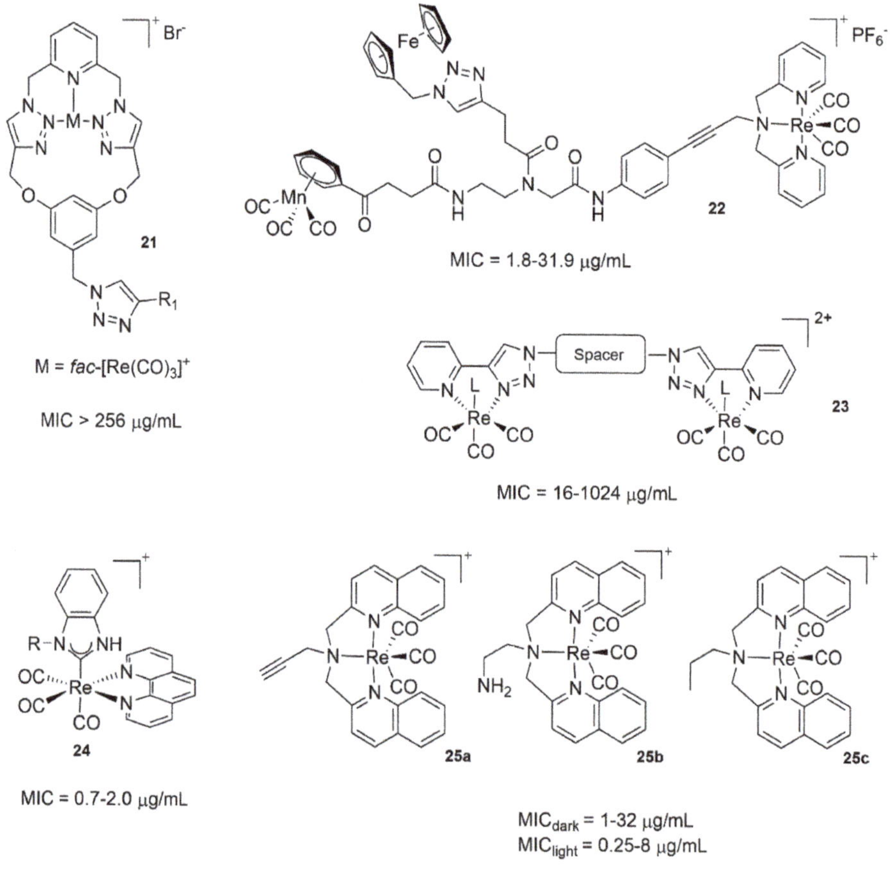

Figure 5. Structural formula of selected antimicrobial rhenium(I) complexes.

4. Group 8

4.1. Ruthenium Complexes

As stated in the introduction, iron complexes are not treated in this review; however, before discussing Ru species, the helicates-chiral assemblies reported by Howson et al. in 2012 deserves a special mention for their unique structure [81]. The species (**26**, Figure 6) were prepared by alkylation of 2 equiv. of (*R*)-2-phenylglycinol with 1 equiv. of α,α'-dibromo-p-xylene followed by reaction with 2-pyridinecarboxaldehyde and Fe(ClO$_4$)$_2$·6H$_2$O in the proportions 3:6:2. Single bimetallic diastereomerically pure flexicates Δ_{Fe},R_C-[Fe$_2$L$_3$][ClO$_4$]$_4$ were isolated following heating of the mixture at 85 °C for 24 h. These flexicates showed good antibacterial activities against MRSA and *E. coli* with MIC values in the 4–8 µg/mL range.

Ruthenium, as the second member of group 8 transition metals, has appeared in many reports as the central metal ion of new potential antimicrobial agents [82–93]. Here we describe the latest examples, but a recent perspective offers more details on the subject [10]. In 2016, Kumar and coworkers reported a series of tris(homoleptic) ruthenium(II) complexes with 2-(1-*R*-1*H*-1,2,3-triazol-4-yl)pyridine ligands (R-pytri) containing different aliphatic and aromatic substituents (**27**, Figure 7) [94]. The in vitro antimicrobial activities of R-pytri ligands and their *mer*- and *fac*-[Ru(R-pytri)$_3$]$^{2+}$ complexes were screened against both Gram-positive (*S. aureus*, *S. pyogenes* and MRSA) and Gram-negative (*A. calcoaceticus*) bacterial strains. The experiments resulted in the good activity of two [Ru(R-pytri)$_3$]$^{2+}$ complexes (where R = hexyl or octyl) against Gram-positive bacteria with MIC values between 1 and 8 µg/mL (depending on the strain), but lower activity was seen against Gram-negative *A. calcoaceticus* (MIC = 16–128 µg/mL). More importantly, both complexes showed stronger antibacterial effects (MIC = 4–8 µg/mL) than gentamicin as the control (MIC = 16 µg/mL) against two strains of MRSA (MR 4393 and MR 4549). Liao et al. have reported a study involving octahedral ruthenium(II) complexes as antimicrobial agents against the mycobacterium *M. smegmatis* [95]. The complex **28** (Figure 7) selectively inhibited *M. smegmatis* growth with MIC of 2 µg/mL comparable to those of norfloxacin and rifampicin (MIC of 2 and 1 µg/mL, respectively). All complexes, however, were found to be inactive against *S. aureus* (MSSA), *P. aeruginosa*, *E. coli*, *C. albicans* and *C. neoformans*.

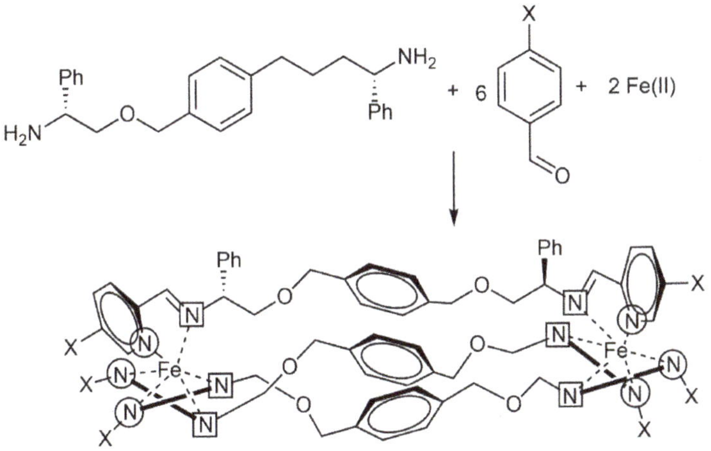

Figure 6. Structural formula of antimicrobial iron(II) flexicates.

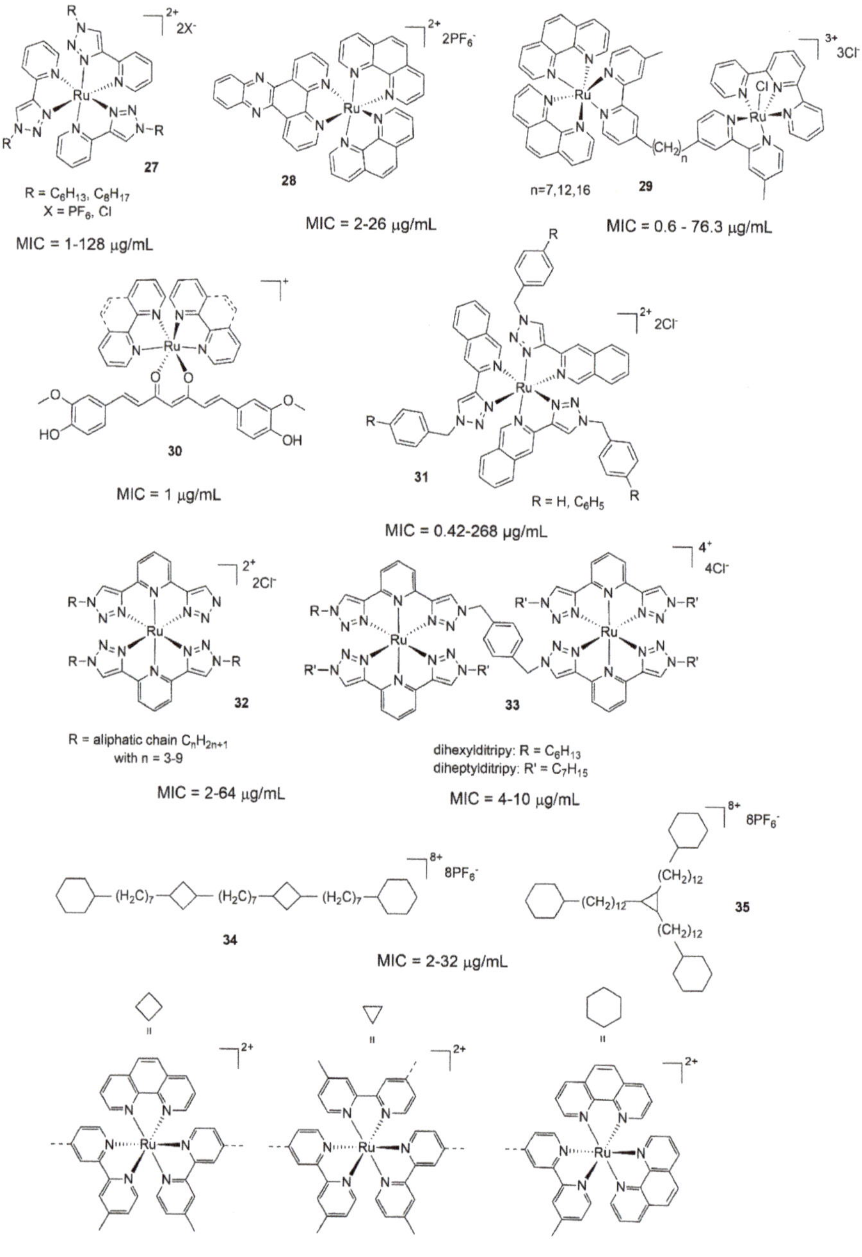

Figure 7. Structural formula of selected antimicrobial ruthenium(II) complexes.

In a study published in 2016 [96], Li et al. introduced a series of non-symmetric dinuclear polypyridylruthenium(II) complexes (**29**, Figure 7), and tested the same as antimicrobial agents. These complexes contained one inert metal center and one coordinatively-labile metal center, linked via the bis[4(4′-methyl-2,2′-bipyridyl)]-1,n-alkane ligand. The ruthenium(II) complexes were tested against four strains of bacteria, *S. aureus* and MRSA (Gram-positive), and *E. coli* and *P. aeruginosa*

(Gram-negative). In most cases, the compounds showed good MIC values (0.6–0.7 µM, comparable to gentamicin) against MRSA, they were less effective against E. coli and nearly inactive against P. aeruginosa. More recently, Srivastava et al. reported ruthenium(II) polypyridyl complexes [97] coordinated to curcumin, [Ru(NN)$_2$(cur)](PF$_6$) [NN = bpy, phen], (30, Figure 7) and tested them against a panel of ESKAPE pathogens, including the drug resistant S. aureus ATCC. The results revealed a good inhibitory effect of the complexes against the latter pathogen and a remarkably high selectivity index (MIC = 1 µg/mL vs 0.25 for levofloxacin, SI = 80). Also in 2019, ruthenium(II) complexes of bidentate chelators 1-(1-benzyl-1,2,3-triazol-4-yl)isoquinoline and 3-(1-benzyl-1,2,3-triazol-4-yl)isoquinoline (31, Figure 7) were reported by Kreofsky et al. [98]. The complexes were screened against Grampositive bacteria (e.g., B. subtilis and S. epidermidis), and revealed a very low MIC value of 0.4 µM. In the same year, van Hilst et al. described mono and dinuclear ruthenium(II) complexes of 2,6-bis(1-R-1,2,3-triazol-4-yl)pyridine ligands (32 and 33, Figure 7), bearing aliphatic substituents [99]. The antibacterial activities of the complexes were evaluated by in vitro tests against S. aureus, and E. coli strains. The MIC values for the most active mononuclear complex, [Ru(hexyltripy)(heptyltripy)]$^{2+}$ (i.e., 33 with n = 7 in Figure 7), were 2 µg/mL and 8 µg/mL, against S. aureus and E. coli, respectively. [Ru(hexyltripy)(heptyltripy)]$^{2+}$ and [Ru$_2$(dihexylditripy)(hexyltripy)$_2$]$^{4+}$ also showed good activities against the Gram positive and Gram negative methicillin resistant S. aureus strains (MICs = 4–8 µg/mL and 8–16 µg/mL, respectively). Finally, linear (34) and non-linear (35) tetranuclear ruthenium(II) complexes were reported by Sun and coworkers [100], as having MIC values against six strains of bacteria (Gram-positive S. aureus and MRSA; Gram-negative, E. coli strains MG1655, APEC, UPEC and P. aeruginosa) in the range between 2 and 32 µg/mL.

4.2. Osmium Complexes

There are only a few reports that have appeared lately detailing antimicrobial studies of osmium complexes. In 2015, a series of enantiopure (S,S)-iPr-pybox and {(S,S)-iPr-pybox = 2,6-bis[4(S)-isopropyloxazolin-2-yl]pyridine} osmium(II) complexes (36, Figure 8), were prepared by Menéndez-Pedregal and coworkers [101]. The complexes were screened against M. luteus, B. subtilis, E. coli, S. coelicolor, S. antibioticus, and P. aeruginosa bacteria. The results showed inhibition halos (mm) of the complexes in the range of 6–20 mm at concentrations between 99 and 500 µg/mL. Gichumbi and coworkers reported a class of osmium(II)-arene complexes with bidentate N,N'-ligands (37, Figure 8) [102]. A panel of antimicrobial-susceptible and -resistant Gram-negative (E. coli, K. pneumonia and P. aeruginosa) and Gram-positive (B. subtilis, E. faecalis, S. aureus, S. aureus, S. saprophyticus and M. smegmatis) bacterial strains were used to examine the antimicrobial activities of the synthesized complexes. The results showed promising anti-mycobacterial activity against M. smegmatis, and bactericidal activity against drug-resistant E. faecalis and methicillin-resistant S. aureus ATCC 43300.

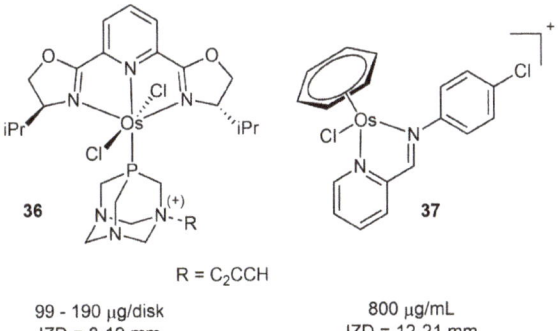

Figure 8. Structural formula of selected antimicrobial osmium(II) complexes.

5. Group 9

5.1. Cobalt Complexes

In 2010, Zhu et al. reported azide bridged Schiff bases 5-methoxy-2-[(2-morpholin-4-ylethylimino) methyl]phenol and 2-ethoxy-6-[(2-isopropylaminoethylimino)methyl]phenol cobalt(III) complexes (**38** and **39**, Figure 9) [103], and tested them against a panel of pathogens. The complexes **38** and **39** were active against *B. subtilis*, *E. coli* and *S. aureus* with the MIC values in the range of 4–18.5 µg/mL, but less active against *P. fluorescens* with MIC values of 21.7 and 37.3 µg/mL, respectively. The authors explained the bactericidal mechanism of action of the metal complexes by Overtone's concept [104] and Tweedy's chelation theory [105]. Coordination to the metal ion of the chelating Schiff bases results in the overlap of the ligand orbital and partial sharing of the positive charge of the ion with donor groups, which gives rise to a decrease in the polarity of the metal ion. As a result, the delocalization of π-electrons over the whole chelate ring increases and, consequently, enhances the lipophilicity of the complex. The main effect of increased lipophilicity is that of improving complexes penetration through lipid membranes, and finally, deactivation of the binding sites on enzymes of microorganisms.

Irgi et al. reported in 2015 a study of the antimicrobial potential of cobalt(II) complexes featuring coordination to the quinolone oxolinic acid drug (Hoxo), and 2,2′-bipyridine (bipy), 2,2′-bipyridylamine (bipyam), 1,10-phenanthroline (phen), pyridine (py) or 4-benzylpyridine (4bzpy) ligands (**40**, Figure 9) [106]. The antimicrobial activities of Hoxo and its complexes were screened against Gram-negative (*E. coli* NCTC 29212 and *X. campestris* ATCC 1395), and Gram-positive (*S. aureus* ATCC 6538 and *B. subtilis* ATCC 6633) bacterial species. Oxolinic acid and its cobalt(II) complexes showed inhibitory action against all the microorganisms tested, with MIC values in the 1–2 µg/mL range for most of the complexes. Similar cobalt(II) complexes, based on a series of coordinated quinolone sparfloxacin and nitrogen-donor heterocyclic ligands bipy, phen or 2,2′-bipyridylamine (bipyam) (**41**, Figure 9), were introduced in 2016 by Kouris et al. [107]. The ligand and complexes showed remarkable antimicrobial activities against bacteria strains, such as *X. campestris*, *S. aureus*, *B. subtilis* and *E. coli* with MICs of 0.031–0.500 µg/mL. The authors suggested that the chelate effect and the presence of sparfloxacinato and N-donor ligands, as well as the generation of the quinolone ligand, could be the prevailing factors contributing to the antimicrobial activities of the complexes.

A class of [CoCl$_2$(dap)$_2$]Cl (dap = 1,3-diaminopropane) and [CoCl$_2$(en)$_2$]Cl (en = ethylenediamine) were recently reported by Turecka et al. [108], and tested against a broad spectrum of reference and clinical fungal strains of *Candida*. The complexes showed MICs of ~16 µg/mL on the selected species (e.g., *C. glabrata* ATCC 2001) but were not as effective as amphotericin B and ketoconazole. A series of zinc(II), copper(II) and cobalt(II) metallophthalocyanine (Pc) compounds derivatized with four 2-methoxy-4-{(Z)-[(4-morpholin-4-ylphenyl)imino]methyl}phenol at the peripheral positions (**42**, Figure 9) were reported by Unluer et al. [109]. According to the in vitro studies, cobalt(II)Pc and copper(II)Pc complexes, in particular, showed antibacterial activities against *S. typhimurium* and *E. coli*. Recently, a series of 2-formylpyridine 4-allyl-S-methylisothiosemicarbazone of zinc(II), copper(II), nickel(II) and cobalt(III) complexes were reported [110]. The in vitro tests showed that cobalt(III) complexes (**43**, Figure 9) were more active against Gram-positive bacteria (e.g., *S. aureus*) and fungal strains (*C. albicans*) with MIC values of 0.7–3 and 7–250 µg/mL, respectively, and less active against Gram-negative strains, such as *E. coli* and *K. pneumoniae*. Several other types of cobalt metal complexes have been tested for their antimicrobial efficacy, however, their activities were not found remarkably high [111–121].

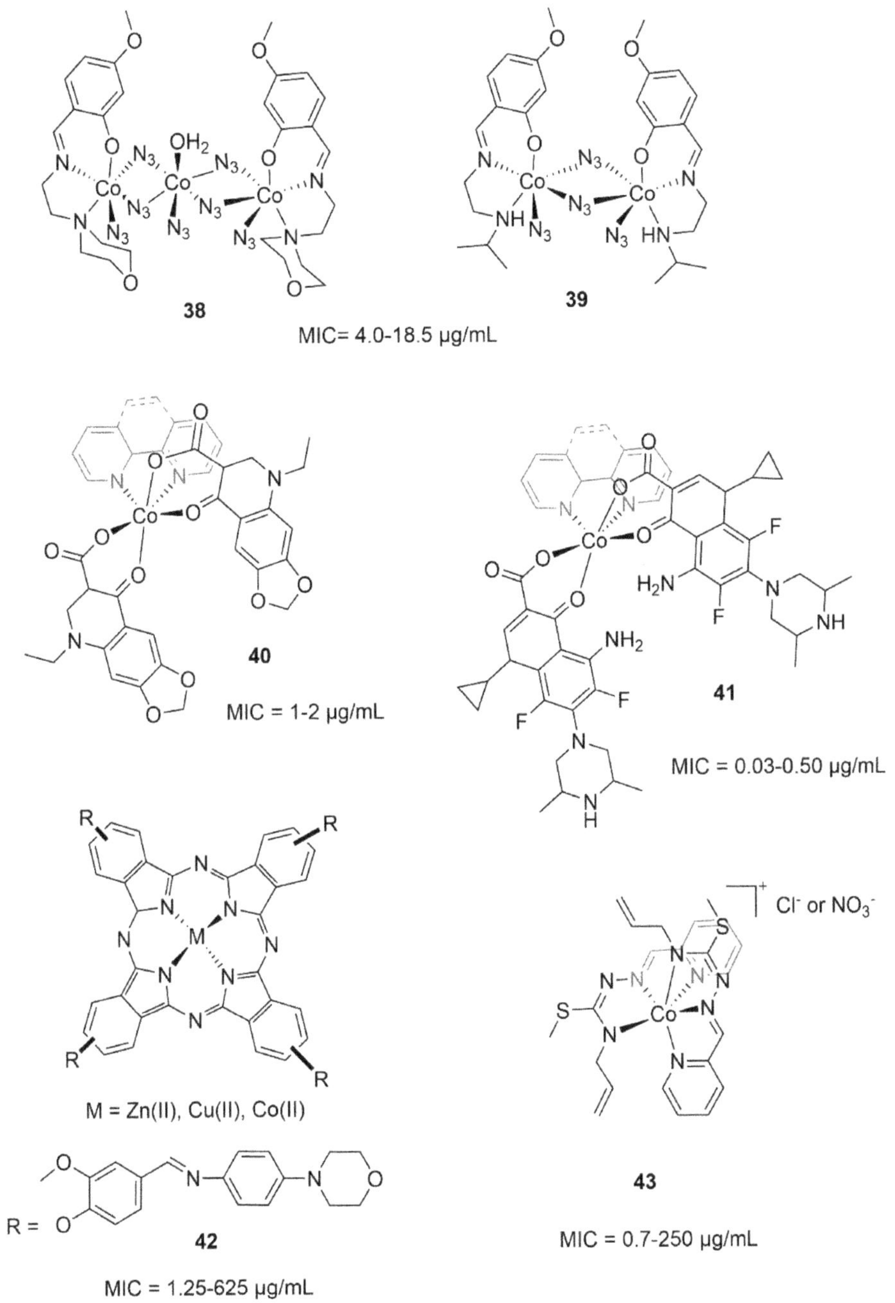

Figure 9. Structural formula of selected antimicrobial cobalt(II) and cobalt(III) complexes and corresponding ligands.

5.2. Rhodium and Iridium Complexes

Rhodium and iridium complexes hold great potential as metal-based antimicrobial agents. In 2015, Lu et al. tested a series of cyclometallated rhodium(III) and iridium(III) complexes for their antimicrobial activities [122]. The in vitro tests revealed that complex **44** (Figure 10) had a selective inhibitory effect against *S. aureus* growth with MIC and MBC values of 3.60 and 7.19 µM, respectively. The complex was the first example of a substitutionally-inert, group 9 organometallic compound utilized as a direct inhibitor of *S. aureus*. In 2017, Fiorini et al. [123] reported methylation of iridium(III) tetrazolato complexes as an effective route to modulate the emission outputs and to switch the antimicrobial properties of the species. Transformation of neutral iridium(III) tetrazolato complexes **45** to the equivalent methylated cations **46** (Figure 10), was accompanied by a remarkable change in the antimicrobial activities of the complexes. Compounds of general structure **45** were inactive against Gram-negative (*E. coli*) and Gram-positive (*D. radiodurans*) microorganisms. However, by converting them to methylated cationic derivatives **46**, the MIC values of the latter dropped to 1–4 µg/mL against the *D. radiodurans* bacterial strain. The same year, Kumar et al. prepared an iridium(III) complex of formula [Ir(cod)(dmtu)$_2$]Cl (where cod = 1,5-cyclooctadiene and dmtu = N,N'-dimethylthiourea, **47** in Figure 10), from the reaction of dmtu with the [Ir(cod)(Cl)]$_2$ dimer [124]. The antimicrobial activity of the complex was investigated against *E. coli*, *S. aureus* and *P. aeruginosa*, and it showed good activity against the two latter strains. In 2018, DuChane and coworkers reported a series of ~40 rhodium(III) and iridium(III) half-sandwich complexes of formula [(η^5-Cp*R)M(β-diketonato)Cl] (M = Rh(III), Ir(III), **48** in Figure 10) [125] and tested them against *M. smegmatis*. The rhodium(III) complexes were found consistently more active than the iridium analogs with MIC values in the range of 2–16 µM and 15–69 µM for the two ions, respectively. The most active rhodium(III) complex was the one bearing pentamethylcyclopentadiene (η^5-Cp*R where R = -CH$_3$) and dipivaloylmethane as the β-diketonato chelate.

Recently, Lapasam and coworkers have reported a family of mononuclear metal complexes containing hydrazone ligands (L) of the type [(arene)MLCl]$^+$ (M = Ru(II), Rh(III) and Ir(III), **49** in Figure 10) [126]. The antibacterial efficacies of the complexes were evaluated against four pathogenic bacteria, such as *S. aureus*, *E. coli*, *B. thuringiensis* and *P. aeruginosa*. All the complexes behaved selectively against *P. aeruginosa* and *B. thuringiensis* with comparable activities to gentamycin but were inactive against *E. coli* and *S. aureus*. In a report in 2019, the same author described related ruthenium(II), rhodium(III) and iridium(III) arene complexes bearing pyridyl azine Schiff base ligands (**50** and **51**, Figure 10) showing potent antibacterial activities against *S. aureus*, *E. coli* and *K. pneumonia* with the zone of inhibition (at conc. 2.0 mg/mL) greater than that of ciprofloxacin [127]. A class of neutral heteroleptic cyclometalated iridium(III) complexes linked to boron dipyrromethene (BODIPY) substituted *N*-heterocyclic carbene (NHC) ligands was characterized by Liu et al. in 2019 [128]. The antimicrobial photo-biological properties of **52** and **53** (Figure 10) were evaluated against *S. aureus* bacteria growing as planktonic cultures. The results revealed good activity of **53** against the pathogen upon visible light activation, with a phototherapeutic index >15 and the half-maximal effective concentration (EC$_{50}$) value of 6.67 µM.

In 2018, a series of organoiridium(III) antimicrobial complexes containing biguanides derivatives as chelated ligands were reported by Chen et al. (**54**, Figure 10) [129]. The compounds have remarkable activities against both Gram-negative and Gram-positive bacteria, including MRSA with MICs as low as 0.125 µg/mL. The complexes also exhibited a high fungicidal effect toward *C. albicans* and *C. neoformans* with MIC values of 0.25 µg/mL (0.34 µM), and generally, low cytotoxicity toward mammalian cells. In 2019, DuChane et al. evaluated a series of piano-stool iridium complexes with 1,2-diaminoethane ligands against bacterial strains of *S. aureus*, including various isolates of methicillin-resistant strains (MRSA) [130]. The in vitro tests indicated an interesting difference between stereoisomers of the species with complex **55** (*cis* isomer, Figure 10) being the most effective compound with MIC values of 5 and 7.5 µg/mL against *S. aureus* and MRSA, respectively. Recently, Lapasam et al. introduced a series of ruthenium(II), rhodium(III) and iridium(III) complexes with 4-phenyl-1-(pyridin-4yl)methylene thiosemicarbazide and 4-phenyl-1-(pyridin-4yl)ethylidene thiosemicarbazide ligands (**56**, Figure 10) with comparable antibacterial properties to that of ciprofloxacin [131]. The MIC values of the complexes were as low as 0.015 mg/mL (MIC of ciprofloxacin = 0.031–0.062 mg/mL) against *S. aureus*, *E. Coli* and *K. pneumonia*.

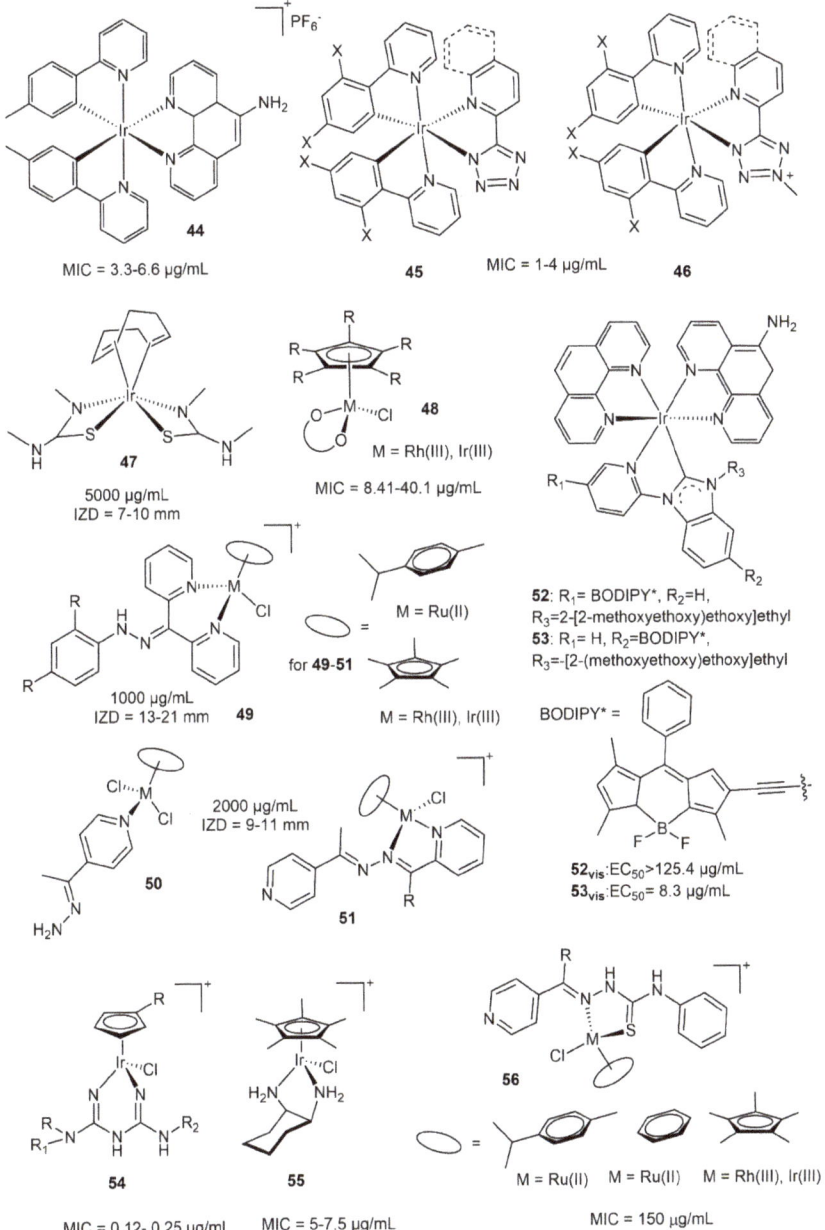

Figure 10. Structural formula of selected antimicrobial rhodium(III) and iridium(III) complexes. EC_{50} = half-maximal effective concentration.

6. Group 10

6.1. Nickel Complexes

Group 10 antimicrobial complexes are not as active as metal complexes of other groups and generally show relatively high MICs when compared to other transition metal species. Nickel is no exception. Therefore, only two selected cases will be given in the current section. In 2017, Raj et al. described Schiff base (57, Figure 11) nickel(II) complexes with MIC values against *S. aureus* (15–30 µg/mL) comparable to the standard drug, ciprofloxacin [132]. The complexes, whose structures remained undefined, also showed good MICs against methicillin resistant *S. aureus* (MRSA, 20–50 µg/mL), but were inactive against other tested pathogens (e.g., *S. flexneri* MTCC-1457, *P. aeruginosa* MTCC-741, and *E. coli* MTCC-119) and several fungal strains. The complexes exert their antimicrobial action by disintegrating the bacterial cell membrane. Recently, Ibrahim et al. [133] presented nickel(II) complexes of NNS tridentate thiosemicarbazone based ligands (58, Figure 11) and evaluated them against several bacterial (e.g., *E. coli*, *P. aeruginosa*, *B. cereus*, *S. aureus*, *M. luteus* and *S. marcescens*) and fungal (e.g., *F. oxysporum*, *C. albicans*, *G. candidum*, *A. flavus*, *S. brevicaulis* and *T. rubrum*) strains. The complexes all showed similar and comparable effects as the standard antibacterial chloramphenicol drug. The results varied in terms of the antifungal potency of complexes 58, but the active ones showed greater inhibition than clotrimazole (the standard drug).

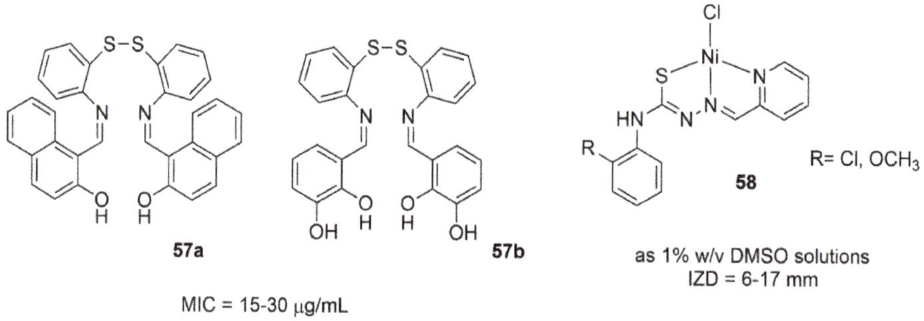

Figure 11. Structural formula of Schiff base ligands 57a and 57b and nickel(II) antimicrobial complex 58.

6.2. Palladium and Platinum Complexes

Several palladium and platinum complexes have been tested for their antimicrobial potencies and a few species showed significant effects. In general, the reported complexes of the two metal ions were not as effective as those of other metals and palladium compounds were more active than the platinum ones. It is, however, instructive to also overview some of the latest reported examples not showing antimicrobial potential. By varying reaction conditions and stoichiometry of reagents, Juribašić et al. [134] prepared a series of quinolinylaminophosphonate palladium(II) halide complexes (59–61, Figure 12) and tested them on a wide spectrum of bacterial and fungal strains. None of the species was active. Similarly, the methylpyrazole-4-carboxaldehyde thiosemicarbazone and the 2-((6-allylidene-2-hydroxycyclohexa-1,3-dienylmethylene)amino)benzoic acid complexes (62 and 63, Figure 12) were inactive [135]. Radić et al. introduced S-alkyl thiosalicylic acid derivatives of palladium(II) (64, Figure 12) and investigated the antimicrobial potential of the ligands and complexes on a wide panel of 26 microorganism species [136]. The palladium(II) complexes were inactive against nearly all pathogens with the exception of fungal strains (e.g., *A. fumigatus* and *A. flavus*) with MICs <7.8 µg/mL.

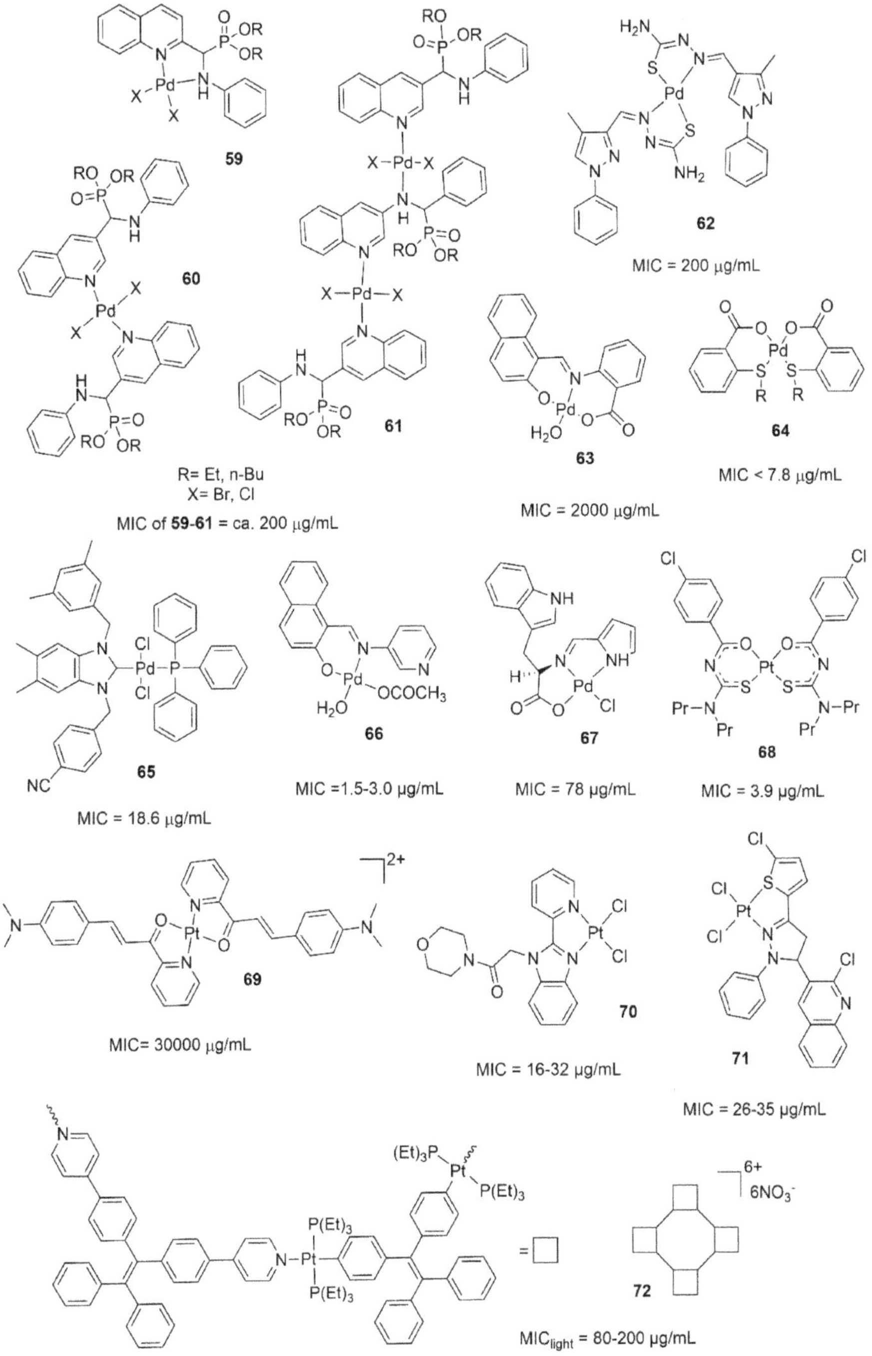

Figure 12. Structural formula of selected antimicrobial palladium(II) and platinum(II) complexes.

In 2018, Boubakri et al., reported the synthesis and antibacterial properties of triphenylphosphine (PPh$_3$) N-heterocycle carbene (NHC) complexes of palladium(II) (**65**, Figure 12) [137]. The complexes were prepared by combining the NHC benzimidazolium salts with, PdCl$_2$, K$_2$CO$_3$ in pyridine at 80 °C, followed by reaction with triphenylphosphine. The in vitro tests of palladium(II)-NHC-PPh$_3$ complexes against Gram-positive (*M. luteus* LB 14110, *S. aureus* ATCC 6538 and *L. monocytogenes* ATCC 19117) and Gram-negative (*S. typhimurium* ATCC14028 and *P. aeruginosa* ATCC 49189) pathogens showed moderate to significant activities of the complexes against the different bacterial strains. The MIC values against *M. luteus*, *L. monocytogenes* and *S. typhimurium* were in the range of 0.0197–0.625, 0.078–1.25, and 1.25–5 mg/mL, respectively. A remarkable example of active palladium(II) complexes was obtained by Abu-Dief et al. [138]. The authors prepared a series of metal complexes bearing the 1-(pyridin-3-yliminomethyl)-naphthalen-2-ol ligand and tested the silver(I), palladium(II) and vanadium(II) oxide derivatives against different strains of bacteria and fungi (*S. Marcescens*, *E. coli*, *M. Luteus F. oxysporum*, *G. candidum* and *A. flavus*). The palladium(II) complex (**66**, Figure 12) showed MIC values against all tested strains between 1.50 and 3.00 µg/mL, close to the standard drugs (ofloxacin and fluconazole). Recently, Nyawade et al. reported new 2-pyrral amino acid Schiff base palladium(II) complexes [139] and investigated their antibacterial effects against six species (Gram-positive, such as *S. aureus*, MRSA, *S. epidermidis*, *S. pyogenes*, and Gram-negative, such as *P. aeruginosa* and *K. pneumonia*). Of the series of compounds, complex **67** (Figure 12) was the most active showing comparable antimicrobial potency to ampicillin against MRSA, *S. epidermidis* and *S. pyogenes*.

Solmaz and coworkers synthesized N,N-Di-(R)-N'-(4-chlorobenzoyl)thiourea platinum(II) complexes (**68**, Figure 12) and carried out antimicrobial tests against *S. aureus*, *S. pneumonia*, *E. coli*, *P. aeruginosa*, *A. baumannii*, *C. albicans* and *C. glabrata* [140]. The compounds were particularly effective against *S. pneumonia*, *P. aeruginosa*, and *A. baumannii* (MIC value of 3.90 µg/mL) and moderately active against *S. aureus*, *E. coli* and *C. albicans* (MIC value of 15.62 µg/mL). More recently, Gaber et al. reported palladium(II) and platinum(II) chalcone complexes of the bidentate ligand, (E)-3-(4-(dimethylamino)phenyl)-1-(pyridin-2-yl)prop-2-en-1-one (**69**, Figure 12) [141]. The platinum(II) complex showed low IC$_{50}$ values but virtually no antimicrobial potency (MIC value of ~30 mg/mL) against *C. albicans*, *A. flavus*, *E. coli* or *S. aureus*. Palladium(II) and platinum(II) complexes with good antifungal activities against *C. albicans* and *C. neoformans* (MIC values of 32 and 16 µg/mL, respectively for the two species) were those bearing a derivatized N,N-bidentate pyridyl benzimidazole ligand (**70**, Figure 12) reported by Mansour et al. [142]. In 2018, Lunagariya et al. tested square planar mononuclear platinum(II) complexes bearing 5-quinoline 1,3,5-tri-substituted pyrazole scaffolds against *S. Aureus*, *B. subtilis*, *S. marcescens*, *P. aeruginosa* and *E. coli* [143]. Within the series, compound **71** (Figure 12) showed good activity against the pathogens with MIC values between 25 and 35 µg/mL. Finally, in 2019, Gao and coworkers published a bacterial membrane intercalation-enhanced photodynamic inactivation (PDI) system, of discrete organoplatinum(II) metallacycles (**72**, Figure 12) [144]. The compound acted as a photosensitizer with aggregation-induced emission. It self-assembled with a transacting activator of the transduction (TAT) peptide-decorated virus coat protein. The resulting aggregate intercalated in the bacterial cell membrane and decreased the survival rate of Gram-negative *E. coli* to nearly zero and that of Gram-positive *S. aureus* to ~30% upon light irradiation. Several other complexes of these ions have been tested for their antimicrobial efficacy, however, their activities were not found remarkably high [145–169].

7. Group 11

7.1. Copper Complexes

In the last five years, hundreds of scientific publications have reported antimicrobial properties of copper complexes. As for iron and the other members of this group, the complexes of the metal ion would be best reviewed alone, but for completeness, a few recent selected examples will be mentioned in this section. In 2019, Kaushal et al. described the synthesis and characterization of several 2-acetylpyridine-*N*-substituted thiosemicarbazonates of copper(II) species (**73**, Figure 13) with remarkable antimicrobial activities against methicillin resistant *S. aureus* (MRSA), *K. pneumoniae* and *C. albicans* [170]. The complexes showed MICs values between 0.5 and 5 µg/mL and often equated the potency of amphotericin and gentamicin. The authors attempted a structure–activity relationship of the variation of antimicrobial bioactivity with variations of R substituents and halogens (X).

In general, for all pathogens, the halogens did not provide any preferential trend but variations occurred due to the substituents R, with ethyl/methyl substituents showing high activity. Oladipo et al. reported a synthetic and structural study of copper(II) *N,N'*-diarylformamidine dithiocarbamate complexes (**74**, Figure 13), showing excellent antibacterial activities against Gram-negative, *S. typhimurium*, *P. aeruginosa*, *E. coli* and *K. pneumoniae* and Gram-positive, *S. aureus* bacteria, including MRSA [171]. The MIC values of complexes were in the order of 6.25 ng/mL to 0.8 µg/mL, surpassing in many cases the potency of ciprofloxacin. Krishnegowda and coworkers prepared 1-phenyl-1,3-butanedione copper(II) complexes (**75**, Figure 13), showing activity against *B. cereus*, *Bacillus substilis*, methicillin-resistant *S. aureus*, *E. coli*, *P. aerogenes* and *K. pneumonia* (MICs in the range of 10.4–16.5 µg/mL) similar to ampicillin [172].

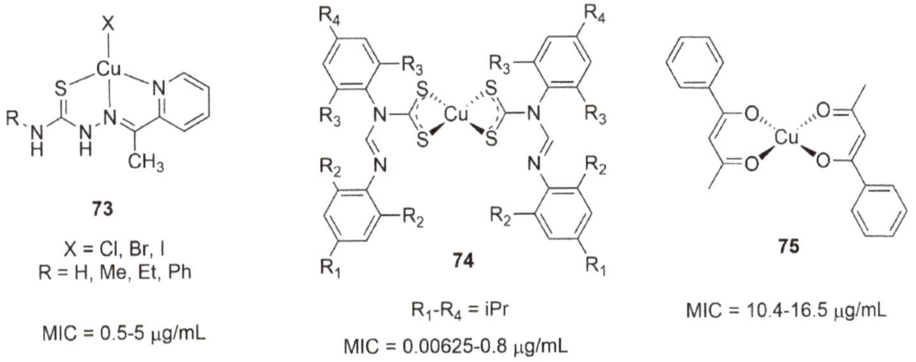

Figure 13. Structural formula of selected antimicrobial copper(II) complexes.

7.2. Gold Complexes

Gold complexes have been investigated in a wide range of therapeutic applications (e.g., as antiarthritic agents for the treatment of rheumatoid arthritis and a variety of rheumatic diseases, including psoriatic arthritis, juvenile arthritis, palindromic rheumatism and discoid lupus erythematosus [173]), and continue to attract the attention of many organometallic chemists [174]. They also have great potential as antimicrobial agents. In this section, we have selected only a few examples, but a recent perspective offers more details on the subject [10].

In 2016, Savić et al. reported a series of aromatic nitrogen-containing heterocycles gold(III) species (**76**, Figure 14) in a comparative antimicrobial and toxicological study of gold and silver complexes of the same [175]. All square-planar gold complexes were evaluated in vitro against *P. aeruginosa*, *E. coli*, *S. aureus*, *L. monocytogenes* and *C. albicans*. They revealed good antibacterial activity with the MIC values in the 2.5 to 100 µg/mL range but were not as effective as the silver analogues. Hikisz et al. studied the antibacterial activities of the gold(I) alkynyl chromone complexes (**77**, Figure 14) against *E. coli* and Gram-positive methicillin-sensitive (MSSA) and methicillin-resistant (MRSA) *S. aureus* including clinical isolates [176]. In vitro tests of the complexes showed high activities against *S. aureus* pathogens with MICs between 2 and 32 µg/mL, but they were not active against *E. coli*. Glišić et al. prepared dinuclear gold(III) complexes with bridging aromatic nitrogen-containing heterocyclic ligands (**78**, Figure 14) and studied their antimicrobial activities in relation to the complex nuclearity [174]. In most cases, complexes showed higher antibacterial activity than K[AuCl$_4$] with MICs in the range of 3.9–62.5 µg/mL. The complexes **78** were particularly effective against *M. luteus* being ~3x more potent than kanamycin. In 2017, Schmidt and co-workers evaluated a series of gold(I) bis-*N*-heterocyclic carbene complexes (**79**, Figure 14) [177] for their effects against pathogenic bacteria *E. faecium, E. coli, P. aeruginosa, A. baumannii, K. pneumonia* and methicillin-resistant *S. aureus* strains (MRSA). The complexes showed good activity against MRSA (for R = Phe, MIC = 1.7–2.3 µM) but were not as effective as auranofin or standard antibiotics. These biscarbene gold complexes act by inhibiting bacterial thioredoxin reductase (TrxRs) with moderate potency. Finally, Pöthig et al. recently described structurally interesting gold pillarplexes [178]. The compounds (**80**, Figure 14), however, showed little or no activity against *B. subtilis, S. aureus, E. coli, P. aeruginosa* or *C. albicans*.

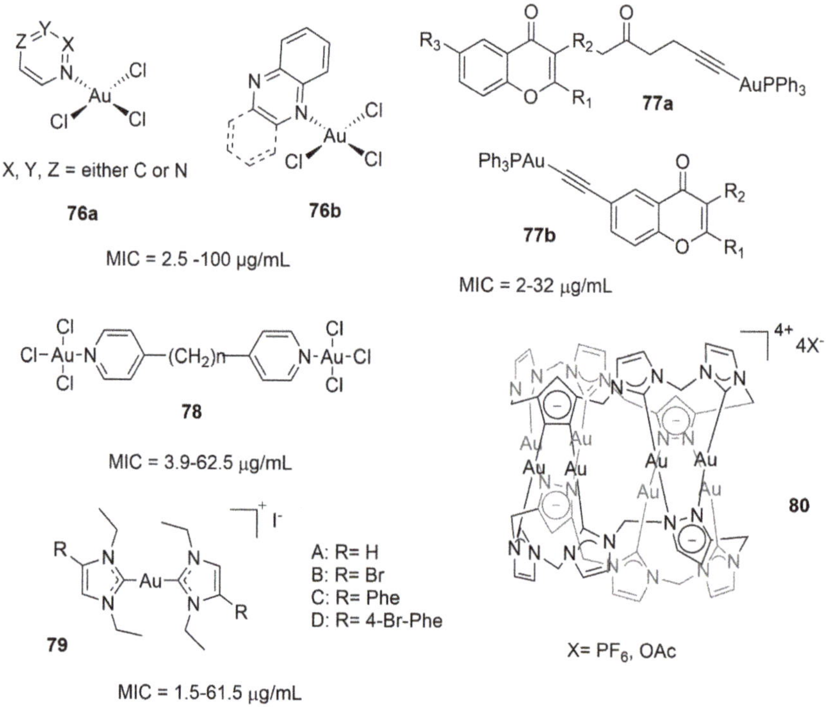

Figure 14. Structural formula of selected antimicrobial gold(I) and gold(III) complexes.

8. Group 12

8.1. Zinc Complexes

Zinc, the first element in group 12, is the only metal that appears in all enzyme classes [179–182]. Complexes of the element have been the topic of many studies, including antibacterial and antiviral activities [183]. In the last five years, more than 100 scientific publications on the antimicrobial properties of zinc complexes have been reported. As mentioned in Section 7, it is beyond the scope of this short review to detail all these studies. We have selected, therefore, only a few cases as interesting examples from the structural and chemical point of view.

In 2015, Zaltariov et al. reported zinc(II) complexes of trimethylsilyl-propyl-p-aminobenzoate (**81**, Figure 15) with remarkable antimicrobial properties [184]. The compounds showed the MIC values as low as 16 ng/mL against *A. fumigatus, P. chrysogenum* and *Fusarium*, and 0.38 µg/mL against *Bacillus* sp. and *Pseudomonas* sp., being more active than the standards, i.e., caspofungin and kanamycin. Abu Ali et al. investigated ibuprofen zinc(II) complexes in combination with mono and bidentate ligands such as 2-aminopyridine, 2-aminomethylpyridine and 2,2'-bipyridine (**82** and **83**,

Figure 15) [185]. Compounds were screened against three Gram-positive (*M. luteus*, *S. aureus* and *B. subtilis*) and three Gram-negative (*E. coli*, *K. pneumonia* and *P. mirabilis*) bacterial strains. The complex containing ibuprofen and 2,2'-bipy (**83**, Figure 15) was the most potent compound against all bacteria with MICs of ~1.5–3 mg/mL. In 2018, Boughougal et al. reported a series of zinc(II) complexes coordinated to sulfadiazine and enrofloxacin (**84**, Figure 15) [186].

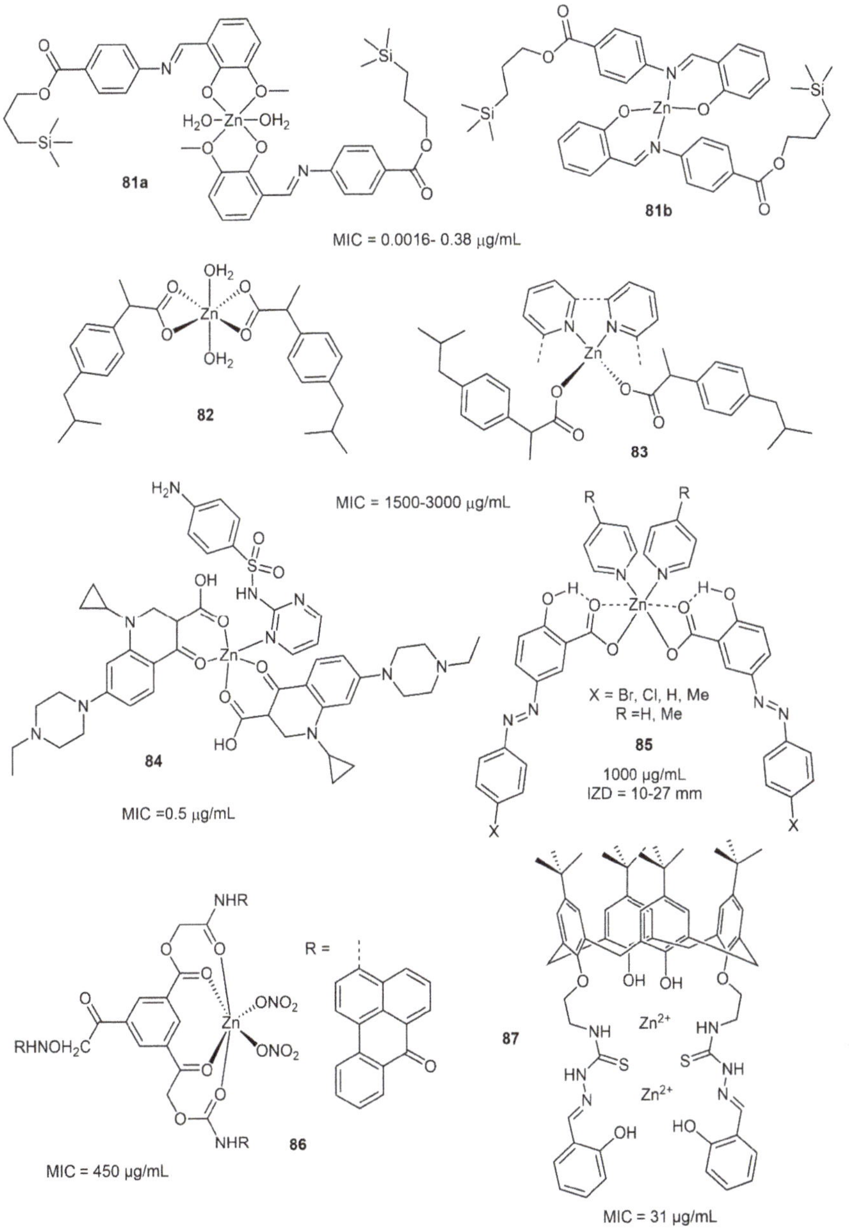

Figure 15. Structural formula of selected antimicrobial zinc(II) complexes.

In all complexes, enrofloxacin acted as a bidentate ligand via the pyridinone and carboxylate oxygens. Free ligands and complexes showed good antibacterial activity against E. Coli, S. Aureus and E. Faecalis with MICs lower than 0.5 mg/L. A series of zinc(II) compounds of aryl-substituted diazosalicylato- and pyridine ligands, was recently described by Basu Baul et al. (**85**, Figure 15) and tested along with copper and cadmium analogs against B. subtilis, S. aureus and K. pneumonia and C. albicans [187]. The zinc(II) complexes showed comparable activity to the standard chloramphenicol and fluconazole antimicrobial drugs. In 2019, Stataneva et al. described a new bioactive zinc(II) complex with a fluorescent symmetrical benzanthrone tripod for applications in antibacterial textiles (**86**, Figure 15) [188]. Tested against different pathogens, the complex showed the highest activity against B. cereus with a MIC of 450 µg/mL. Recently, Noruzi et al. reported the biological activities of metal complexes of a multidentate calix[4]arene ligand doubly functionalized by 2-hydroxybenzeledene-thiosemicarbazone (**87**, Figure 15) [189]. Both the calix[4]arene ligand and its zinc(II) complex showed activity against B. subtilis, E. coli and P. aeruginosa with MICs of 31 µg/mL.

8.2. Cadmium and Mercury Complexes

Several studies have described the antimicrobial properties of cadmium and mercury complexes since 2015. Despite the harmful nature of the metal ions and their complexes, they can still be remarkably useful for antimicrobial applications and they should not be neglected. However, given the inherent high toxicity associated with the metal ions, we decided to select only studies of complexes showing MICs in the low µg/mL/µM range and (where possible) with activities comparable to tested standard drugs. These stringent requirements considerably reduced the number of studies that we could consider here. Montazerozohori et al. have reported cadmium(II) and mercury(II) complexes of the bidentate Schiff base ligand 4-(3-(2-(4-(dimethyl aminophenyl alylidene aminopropylimino)prop-1-ethyl)-N,N-dimethyl benzene (**88**, Figure 16) and tested the molecules against two Gram-positive (B. substilis and S. aureus), and two Gram-negative (P. aeruginosa and E. coli) bacterial strains [190]. Mercury complexes with X = I and SCN showed minimum bactericidal concentration (MBC) of 3.7 and 7.5 µg/mL, respectively, against S. aureus and P. aeruginosa (SCN complex only). The cadmium complexes were less toxic, with the most active species (X = SCN) showing a MIC of 25 µg/mL against P. aeruginosa. In 2016, Agertt et al. evaluated sulfonamide metal complexes of Au, Ag, Cd, Cu and Hg for their antimycobacterial activities against M. abscessus, M. fortuitum and M. massiliense [191]. Cadmium and mercury complexes showed MICs of 4.9 µg/mL against M. fortuitum and M. massiliense and of 19.5 and 9.8 µg/mL, respectively, against M. abscessus. It should be noted that the study did not report a full characterization of cadmium(II) and mercury(II) complexes and their structures are unknown. In a study published in 2019, Matiadis et al. investigated the antimicrobial properties of cadmium(II) metal complexes of the N-acetyl-3-acetyl-5-benzylidenetetramic acid (**89**, Figure 16) [192]. The in vitro tests against five key "ESKAPE" pathogens (E. coli, MRSA, K. pneumoniae, A. baumannii and P. aeruginosa) and two fungi (C. neoformans and C. albicans) revealed that **89** was active only against C. neoformans (MIC = 8 µg/mL).

In 2017 and 2018, Mandal et al. reported the synthesis, characterization and antimicrobial activities of cadmium(II) and mercury(II) complexes of 5-methyl pyrazole-3yl-N-(2′-methylthiophenyl) methyleneimine [193] and pyrazol-3-yl-N-(2-methoxyphenyl) methanimine [194] (**90**–**93**, Figure 16) against a panel of pathogens. In comparison to amoxicillin, cadmium(II) and mercury(II) complexes **90** and **92** showed very good antimicrobial activity against P. vulgaris and S. aureus with MICs of 35 and 25 µg/mL and 5 and 2 µg/mL, respectively (MICs of amoxicillin = 129 and 85 µg/mL, respectively). Furthermore, **92** was 8-fold more effective than amoxicillin against E. aerogenes (MIC of **92** = 35 µg/mL) [193]. The complex **91** was inactive while **93** showed a MIC value of 10 µg/mL against different V. cholerae strains, P. aeruginosa and M. luteus [194]. Lam et al. have reported a series of bis-(alkynyl)mercury(II) complexes with oligothiophene and bithiazole linking units (**94** and **95**, Figure 16) with remarkable antimicrobial activity against MRSA and C. albicans [195]. Complex **94** showed the strongest bactericidal activity against MRSA with MIC and MBC values 0.2 µg/mL, and fungicidal effect against C. albicans with MIC and MBC values 0.4 µg/mL. Finally, Weng et al.

reported cadmium(II) supramolecular Kandinsky circles (**96**, Figure 16) with high antibacterial activity against Gram-positive methicillin-resistant *S. aureus* (MRSA) [196]. The MIC values of the different supramolecular were between 0.5 and 3 µg/mL. The compounds **96a–96c** were not active against *E. coli* and showed negligible toxicity to eukaryotic cells.

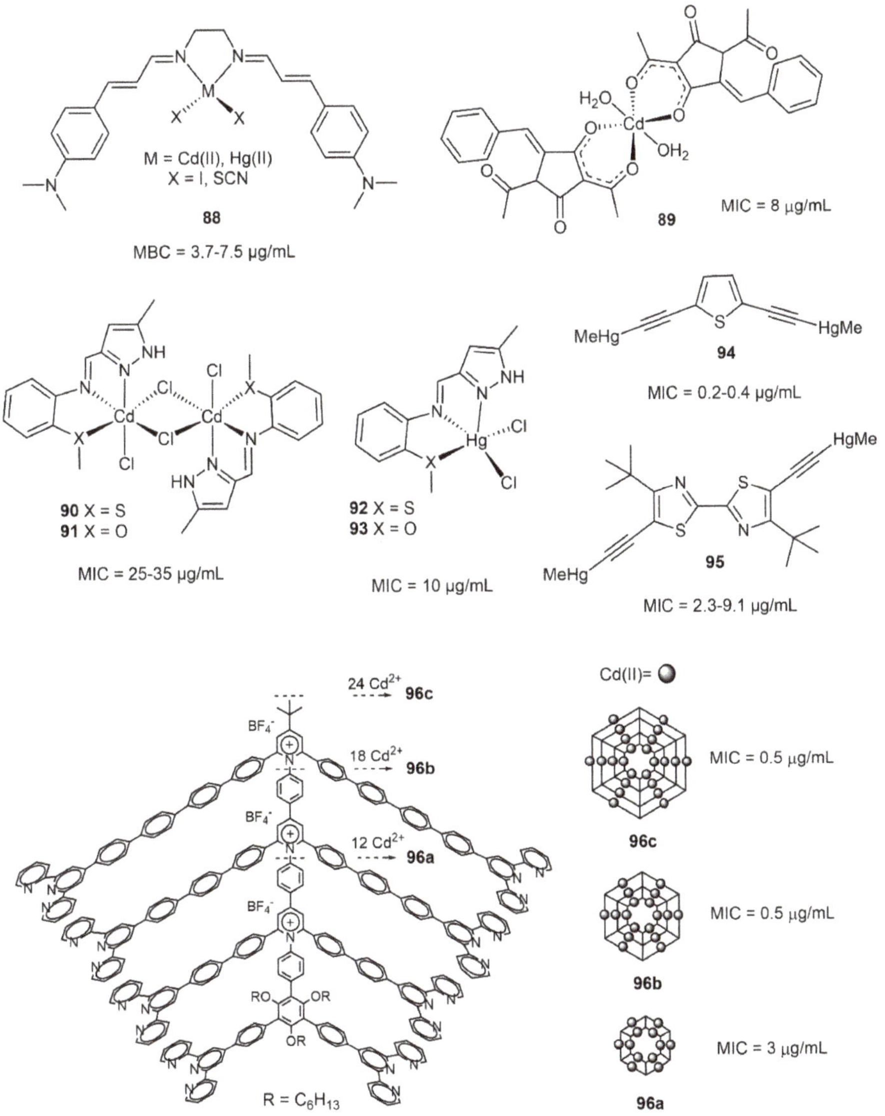

Figure 16. Structural formula of selected antimicrobial cadmium(II) and mercury(II) complexes.

9. Conclusions

In the last ten years, inorganic and organometallic transition metal medicinal chemists have begun to develop new antimicrobial agents with great promise and noteworthy success. Complexes of virtually all ions of the transition periods have been tested. In this review, we have detailed in particular recent studies on the antimicrobial activities and potential of transition metal complexes of groups 6–12. Several species show remarkable prospective as candidates for the development of new classes of highly active antimicrobial agents. The majority of compounds still need validation in vivo but the unique properties of the complexes offer the possibility of fine-tuning in the future their properties, reactivity and toxicological profiles. Metal complexes operate via specific modes of actions unknown to carbon-based drugs and yet unexperienced by infectious pathogens. This will likely translate into long-term new strategies in this urgent global fight.

Author Contributions: Both authors equally contributed to the work. All authors have read and agreed to the published version of the manuscript.

Funding: This research received no external funding.

Conflicts of Interest: The authors declare no conflict of interest.

References

1. Bush, K.; Courvalin, P.; Dantas, G.; Davies, J.; Eisenstein, B.; Huovinen, P.; Jacoby, G.A.; Kishony, R.; Kreiswirth, B.N.; Kutter, E.; et al. Tackling antibiotic resistance. *Nat. Rev. Microbiol.* **2011**, *9*, 894–896. [CrossRef]
2. Davies, J.; Davies, D. Origins and evolution of antibiotic resistance. *Microbiol. Mol. Biol. Rev.* **2010**, *74*, 417–433. [CrossRef]
3. Bouley, R.; Ding, D.; Peng, Z.; Bastian, M.; Lastochkin, E.; Song, W.; Suckow, M.A.; Schroeder, V.A.; Wolter, W.R.; Mobashery, S.; et al. Structure-Activity Relationship for the 4(3*H*)-Quinazolinone Antibacterials. *J. Med. Chem.* **2016**, *59*, 5011–5021. [CrossRef]
4. Ng, N.S.; Leverett, P.; Hibbs, D.E.; Yang, Q.; Bulanadi, J.C.; Wu, M.J.; Aldrich-Wright, J.R. The antimicrobial properties of some copper(II) and platinum(II) 1,10-phenanthroline complexes. *Dalton Trans.* **2013**, *42*, 3196–3209. [CrossRef]
5. Zhao, Y.; Chen, Z.; Chen, Y.; Xu, J.; Li, J.; Jiang, X. Synergy of non-antibiotic drugs and pyrimidinethiol on gold nanoparticles against superbugs. *J. Am. Chem. Soc.* **2013**, *135*, 12940–12943. [CrossRef]
6. Xia, J.; Gao, J.; Kokudo, N.; Hasegawa, K.; Tang, W. Methicillin-resistant Staphylococcus aureus antibiotic resistance and virulence. *Biosci. Trends* **2013**, *7*, 113–121.
7. Von Nussbaum, F.; Brands, M.; Hinzen, B.; Weigand, S.; Habich, D. Antibacterial natural products in medicinal chemistry-exodus or revival? *Angew. Chem. Int. Ed. Engl.* **2006**, *45*, 5072–5129. [CrossRef]
8. Pawlowski, A.C.; Johnson, J.W.; Wright, G.D. Evolving medicinal chemistry strategies in antibiotic discovery. *Curr. Opin. Biotech.* **2016**, *42*, 108–117. [CrossRef]
9. Talbot, G.H.; Jezek, A.; Murray, B.E.; Jones, R.N.; Ebright, R.H.; Nau, G.J.; Rodvold, K.A.; Newland, J.G.; Boucher, H.W.; Infectious Diseases Society of America. The Infectious Diseases Society of America's 10 × '20 Initiative (10 New Systemic Antibacterial Agents US Food and Drug Administration Approved by 2020): Is 20 × '20 a Possibility? *Arch. Clin. Infect. Dis.* **2019**, *69*, 1–11. [CrossRef]
10. Frei, A. Metal Complexes, an Untapped Source of Antibiotic Potential? *Antibiotics* **2020**, *9*, 90. [CrossRef]
11. Arias, L.S.; Pessan, J.P.; Vieira, A.P.M.; Lima, T.M.T.; Delbem, A.C.B.; Monteiro, D.R. Iron Oxide Nanoparticles for Biomedical Applications: A Perspective on Synthesis, Drugs, Antimicrobial Activity, and Toxicity. *Antibiotics* **2018**, *7*, 46. [CrossRef]
12. De Toledo, L.d.A.S.; Rosseto, H.C.; Bruschi, M.L. Iron oxide magnetic nanoparticles as antimicrobials for therapeutics. *Pharm. Dev. Technol.* **2018**, *23*, 316–323. [CrossRef]
13. Dutta, P.; Wang, B. Zeolite-supported silver as antimicrobial agents. *Coord. Chem. Rev.* **2019**, *383*, 1–29. [CrossRef]
14. Duval, R.E.; Gouyau, J.; Lamouroux, E. Limitations of Recent Studies Dealing with the Antibacterial Properties of Silver Nanoparticles: Fact and Opinion. *Nanomaterials* **2019**, *9*, 1775. [CrossRef]
15. Fahmy, H.M.; Mosleh, A.M.; Elghany, A.A.; Shams-Eldin, E.; Abu Serea, E.S.; Ali, S.A.; Shalan, A.E. Coated silver nanoparticles: Synthesis, cytotoxicity, and optical properties. *RSC Adv.* **2019**, *9*, 20118–20136. [CrossRef]
16. Gunawan, C.; Marquis, C.P.; Amal, R.; Sotiriou, G.A.; Rice, S.A.; Harry, E.J. Widespread and Indiscriminate Nanosilver Use: Genuine Potential for Microbial Resistance. *ACS Nano* **2017**, *11*, 3438–3445. [CrossRef]
17. Hashim, A.; Agool, I.R.; Kadhim, K.J. Modern Developments in Polymer Nanocomposites For Antibacterial and Antimicrobial Applications: A Review. *J. Bionanosci.* **2018**, *12*, 608–613. [CrossRef]

18. Hussaini, S.Y.; Haque, R.A.; Razali, M.R. Recent progress in silver(I)-, gold(I)/(III)- and palladium(II)-N-heterocyclic carbene complexes: A review towards biological perspectives. *J. Organomet. Chem.* **2019**, *882*, 96–111. [CrossRef]
19. Leitao, J.H.; Sousa, S.A.; Leite, S.A.; Carvalho, M. Silver Camphor Imine Complexes: Novel Antibacterial Compounds from Old Medicines. *Antibiotics* **2018**, *7*, 65. [CrossRef]
20. Medici, S.; Peana, M.; Nurchi, V.M.; Zoroddu, M.A. Medical Uses of Silver: History, Myths, and Scientific Evidence. *J. Med. Chem.* **2019**, *62*, 5923–5943. [CrossRef]
21. Möhler, J.S.; Sim, W.; Blaskovich, M.A.T.; Cooper, M.A.; Ziora, Z.M. Silver bullets: A new lustre on an old antimicrobial agent. *Biotech. Adv.* **2018**, *36*, 1391–1411. [CrossRef] [PubMed]
22. Rusu, A.; Hancu, G.; Cristina Munteanu, A.; Uivarosi, V. Development perspectives of silver complexes with antibacterial quinolones: Successful or not? *J. Organomet. Chem.* **2017**, *839*, 19–30. [CrossRef]
23. Singh, R.; Shedbalkar, U.U.; Wadhwani, S.A.; Chopade, B.A. Bacteriagenic silver nanoparticles: Synthesis, mechanism, and applications. *Appl. Microbiol. Biotechnol.* **2015**, *99*, 4579–4593. [CrossRef] [PubMed]
24. Zheng, K.; Setyawati, M.I.; Leong, D.T.; Xie, J. Antimicrobial silver nanomaterials. *Coord. Chem. Rev.* **2018**, *357*, 1–17. [CrossRef]
25. Patil, S.A.; Patil, S.A.; Patil, R.; Keri, R.S.; Budagumpi, S.; Balakrishna, G.R.; Tacke, M. N-heterocyclic carbene metal complexes as bio-organometallic antimicrobial and anticancer drugs. *Future Med. Chem.* **2015**, *7*, 1305–1333. [CrossRef]
26. Li, F.; Collins, J.G.; Keene, F.R. Ruthenium complexes as antimicrobial agents. *Chem. Soc. Rev.* **2015**, *44*, 2529–2542. [CrossRef]
27. Patra, M.; Gasser, G.; Metzler-Nolte, N. Small organometallic compounds as antibacterial agents. *Dalton Trans.* **2012**, *41*, 6350–6358. [CrossRef]
28. Sierra, M.; Casarrubios, L.; De la Torre, M. Bio-Organometallic Derivatives of Antibacterial Drugs. *Chemistry* **2019**, *25*. [CrossRef]
29. Frei, A.; Zuegg, J.; Elliott, A.G.; Baker, M.; Braese, S.; Brown, C.; Chen, F.; C, G.D.; Dujardin, G.; Jung, N.; et al. Metal complexes as a promising source for new antibiotics. *Chem. Sci.* **2020**, *11*, 2627–2639. [CrossRef]
30. Kumar, S.P.; Suresh, R.; Giribabu, K.; Manigandan, R.; Munusamy, S.; Muthamizh, S.; Narayanan, V. Synthesis and characterization of chromium(III) Schiff base complexes: Antimicrobial activity and its electrocatalytic sensing ability of catechol. *Spectrochim. Acta A* **2015**, *139*, 431–441. [CrossRef]
31. Rathi, P.; Singh, D.P. Synthesis, antimicrobial, antioxidant and molecular docking studies of thiophene based macrocyclic Schiff base complexes. *J. Mol. Struct.* **2015**, *1100*, 208–214. [CrossRef]
32. Shaabani, B.; Khandar, A.A.; Ramazani, N.; Fleck, M.; Mobaiyen, H.; Cunha-Silva, L. Chromium(III), manganese(II) and iron(III) complexes based on hydrazone Schiff-base and azide ligands: Synthesis, crystal structure and antimicrobial activity. *J. Coord. Chem.* **2017**, *70*, 696–708. [CrossRef]
33. Kafi-Ahmadi, L.; Shirmohammadzadeh, L. Synthesis of Co(II) and Cr(III) salicylidenic Schiff base complexes derived from thiourea as precursors for nano-sized Co_3O_4 and Cr_2O_3 and their catalytic, antibacterial properties. *J. Nanostruct. Chem.* **2017**, *7*, 179–190. [CrossRef]
34. Abdel Rahman, L.H.; Abu-Dief, A.M.; Hamdan, S.K.; Seleem, A.A. Nano Structure Iron(II) and Copper(II) Schiff-Base Complexes of a NNO Tridentate Ligand as New Antibiotic Agents. Spectral Thermal Behaviors and Dann Binding Ability. *Int. J. Nanomater. Chem.* **2015**, *1*, 65–77. [CrossRef]
35. Abdel-Rahman, L.H.; El-Khatib, R.M.; Nassr, L.A.; Abu-Dief, A.M.; Ismael, M.; Seleem, A.A. Metal based pharmacologically active agents: Synthesis, structural characterization, molecular modeling, CT-DNA binding studies and in vitro antimicrobial screening of iron(II) bromosalicylidene amino acid chelates. *Spectrochim. Acta Part A Mol. Biomol. Spectrosc.* **2014**, *117*, 366–378. [CrossRef]
36. Abu-Dief, A.M.; Nassr, L.A.E. Tailoring, physicochemical characterization, antibacterial and DNA binding mode studies of Cu(II) Schiff bases amino acid bioactive agents incorporating 5-bromo-2-hydroxybenzaldehyde. *J. Iran. Chem. Soc.* **2015**, *12*, 943–955. [CrossRef]
37. Liu, Y.-T.; Sheng, J.; Yin, D.-W.; Xin, H.; Yang, X.-M.; Qiao, Q.-Y.; Yang, Z.-J. Ferrocenyl chalcone-based Schiff bases and their metal complexes: Highly efficient, solvent-free synthesis, characterization, biological research. *J. Organomet. Chem.* **2018**, *856*, 27–33. [CrossRef]
38. Murcia, R.A.; Leal, S.M.; Roa, M.V.; Nagles, E.; Munoz-Castro, A.; Hurtado, J.J. Development of Antibacterial and Antifungal Triazole Chromium(III) and Cobalt(II) Complexes: Synthesis and Biological Activity Evaluations. *Molecules* **2018**, *23*, 2013. [CrossRef]

39. Abu-Dief, A.M.; El-Sagher, H.M.; Shehata, M.R. Fabrication, spectroscopic characterization, calf thymus DNA binding investigation, antioxidant and anticancer activities of some antibiotic azomethine Cu(II), Pd(II), Zn(II) and Cr(III) complexes. *Appl. Organomet. Chem.* **2019**, *33*, e4943. [CrossRef]
40. El-Razek, S.E.A.; El-Gamasy, S.M.; Hassan, M.; Abdel-Aziz, M.S.; Nasr, S.M. Transition metal complexes of a multidentate Schiff base ligand containing guanidine moiety: Synthesis, characterization, anti-cancer effect, and anti-microbial activity. *J. Mol. Struct.* **2020**, *1203*, 127381. [CrossRef]
41. Abdel-Rahman, L.H.; Abu-Dief, A.M.; Newair, E.F.; Hamdan, S.K. Some new nano-sized Cr(III), Fe(II), Co(II), and Ni(II) complexes incorporating 2-((E)-(pyridine-2-ylimino)methyl)napthalen-1-ol ligand: Structural characterization, electrochemical, antioxidant, antimicrobial, antiviral assessment and DNA interaction. *J. Photochem. Photobiol.* **2016**, *160*, 18–31. [CrossRef] [PubMed]
42. Drzeżdżon, J.; Piotrowska-Kirschling, A.; Malinowski, J.; Kloska, A.; Gawdzik, B.; Chmurzyński, L.; Jacewicz, D. Antimicrobial, cytotoxic, and antioxidant activities and physicochemical characteristics of chromium(III) complexes with picolinate, dipicolinate, oxalate, 2,2′-bipyridine, and 4,4′-dimethoxy-2,2′-bipyridine as ligands in aqueous solutions. *J. Mol. Liq.* **2019**, *282*, 441–447. [CrossRef]
43. Mahmoud, W.H.; Deghadi, R.G.; Mohamed, G.G. Metal complexes of novel Schiff base derived from iron sandwiched organometallic and 4-nitro-1,2-phenylenediamine: Synthesis, characterization, DFT studies, antimicrobial activities and molecular docking. *Appl. Organomet. Chem.* **2018**, *32*, e4289. [CrossRef]
44. Pahontu, E.; Usataia, I.; Graur, V.; Chumakov, Y.; Petrenko, P.; Gudumac, V.; Gulea, A. Synthesis, characterization, crystal structure of novel Cu(II), Co(III), Fe(III) and Cr(III) complexes with 2-hydroxybenzaldehyde-4-allyl-S-methylisothiosemicarbazone: Antimicrobial, antioxidant and in vitro antiproliferative activity. *Appl. Organomet. Chem.* **2018**, *32*, e4544. [CrossRef]
45. Srivastva, A.N.; Singh, N.P.; Shriwastaw, C.K. Physicochemical studies on bioactive Cr(III) coordination compounds with esters of hydrazine carboxylic acid as hetero donor ligands. *Res. Chem. Intermed.* **2017**, *43*, 5453–5465. [CrossRef]
46. Saraswat, K.; Kant, R. Synthesis characterization and biological activity of some molybdenum(VI) complexes. *Der Pharma. Chem.* **2013**, *5*, 347–356.
47. Wedd, A.G. Sulfido-Complexes of Molybdenum and Tungsten: Synthetic Aspects. In *Studies in Inorganic Chemistry*; Müller, A., Krebs, B., Eds.; Elsevier: Amsterdam, The Netherlands, 1984; Volume 5, pp. 181–193.
48. Zhang, L.; Nelson, K.J.; Rajagopalan, K.V.; George, G.N. Structure of the Molybdenum Site of Escherichia coli Trimethylamine N-Oxide Reductase. *Inorg. Chem.* **2008**, *47*, 1074–1078. [CrossRef]
49. Biswal, D.; Pramanik, N.R.; Chakrabarti, S.; Drew, M.G.B.; Mitra, P.; Acharya, K.; Biswas, S.; Mondal, T.K. Supramolecular frameworks of binuclear dioxomolybdenum(VI) complexes with ONS donor ligands using 4,4′-azopyridine as a pillar: Crystal structure, DFT calculations and biological study. *N. J. Chem.* **2015**, *39*, 8681–8694. [CrossRef]
50. Çelen, Ş.; Eğlence-Bakır, S.; Şahin, M.; Deniz, I.; Celik, H.; Kizilcikli, I. Synthesis and characterization of new thiosemicarbazonato molybdenum(VI) complexes and their in vitro antimicrobial activities. *J. Coord. Chem.* **2019**, *72*, 1747–1758. [CrossRef]
51. Sang, Y.-L.; Zhang, X.-H.; Lin, X.-S.; Liu, Y.-H.; Liu, X.-Y. Syntheses, crystal structures, and antibacterial activity of oxidovanadium(V) and dioxidomolybdenum(VI) complexes derived from N′-(2-hydroxy-4-methoxybenzylidene)isonicotinohydrazide. *J. Coord. Chem.* **2020**, *73*, 164–174. [CrossRef]
52. Kirakci, K.; Zelenka, J.; Rumlová, M.; Cvačka, J.; Ruml, T.; Lang, K. Cationic octahedral molybdenum cluster complexes functionalized with mitochondria-targeting ligands: Photodynamic anticancer and antibacterial activities. *Biomater. Sci.* **2019**, *7*, 1386–1392. [CrossRef]
53. Zampakou, M.; Akrivou, M.; Andreadou, E.G.; Raptopoulou, C.P.; Psycharis, V.; Pantazaki, A.A.; Psomas, G. Structure, antimicrobial activity, DNA- and albumin-binding of manganese(II) complexes with the quinolone antimicrobial agents oxolinic acid and enrofloxacin. *J. Inorg. Biochem.* **2013**, *121*, 88–99. [CrossRef]
54. Barmpa, A.; Frousiou, O.; Kalogiannis, S.; Perdih, F.; Turel, I.; Psomas, G. Manganese(II) complexes of the quinolone family member flumequine: Structure, antimicrobial activity and affinity for albumins and calf-thymus DNA. *Polyhedron* **2018**, *145*, 166–175. [CrossRef]
55. Arthi, P.; Shobana, S.; Srinivasan, P.; Prabhu, D.; Arulvasu, C.; Kalilur Rahiman, A. Dinuclear manganese(II) complexes of hexaazamacrocycles bearing N-benzoylated pendant separated by aromatic spacers: Antibacterial, DNA interaction, cytotoxic and molecular docking studies. *J. Photoch. Photobio. B* **2015**, *153*, 247–260. [CrossRef]

56. Simpson, P.V.; Nagel, C.; Bruhn, H.; Schatzschneider, U. Antibacterial and Antiparasitic Activity of Manganese(I) Tricarbonyl Complexes with Ketoconazole, Miconazole, and Clotrimazole Ligands. *Organometallics* **2015**, *34*, 3809–3815. [CrossRef]
57. Dong, H.; Yang, X.; He, J.; Cai, S.; Xiao, K.; Zhu, L. Enhanced antioxidant activity, antibacterial activity and hypoglycemic effect of luteolin by complexation with manganese(II) and its inhibition kinetics on xanthine oxidase. *RSC Adv.* **2017**, *7*, 53385–53395. [CrossRef]
58. Güntzel, P.; Nagel, C.; Weigelt, J.; Betts, J.W.; Pattrick, C.A.; Southam, H.M.; La Ragione, R.M.; Poole, R.K.; Schatzschneider, U. Biological activity of manganese(I) tricarbonyl complexes on multidrug-resistant Gram-negative bacteria: From functional studies to in vivo activity in Galleria mellonella. *Metallomics* **2019**, *11*, 2033–2042. [CrossRef]
59. Kottelat, E.; Chabert, V.; Crochet, A.; Fromm, K.M.; Zobi, F. Towards Cardiolite-Inspired Carbon Monoxide Releasing Molecules–Reactivity of d^4, d^5 Rhenium and d^6 Manganese Carb-onyl Complexes with Isocyanide Ligands. *Eur. J. Inorg. Chem.* **2015**, *2015*, 5628–5638. [CrossRef]
60. Devi, P.P.; Chipem, F.A.S.; Singh, C.B.; Lonibala, R.K. Complexation of 2-amino-3-(4-hydroxyphenyl)-N'-[(2-hydroxyphenyl) methylene] propane hydrazide with Mn(II), Co(II), Ni(II), Cu(II) and Zn(II) ions: Structural characterization, DFT, DNA binding and antimicrobial studies. *J. Mol. Struct.* **2019**, *1176*, 7–18. [CrossRef]
61. Jayakumar, S.; Mahendiran, D.; Rahiman, A.K. Theoretical, antimicrobial, antioxidant, in vitro cytotoxicity, and cyclin-dependent kinase 2 inhibitor studies of metal(II) complexes with bis(imidazol-1-yl)methane-based heteroscorpionate ligands. *J. Coord. Chem.* **2019**, *72*, 2015–2034. [CrossRef]
62. Özgür, M.; Yılmaz, M.; Nishino, H.; Çinar Avar, E.; Dal, H.; Pekel, A.T.; Hökelek, T. Efficient syntheses and antimicrobial activities of new thiophene containing pyranone and quinolinone derivatives using manganese(III) acetate: The effect of thiophene on ring closure–opening reactions. *N. J. Chem.* **2019**, *43*, 5737–5751. [CrossRef]
63. Pulina, N.A.; Kuznetsov, A.S.; Krasnova, A.I.; Novikova, V.V. Synthesis, Antimicrobial Activity, and Behavioral Response Effects of N-Substituted 4-Aryl-2-Hydroxy-4-Oxobut-2-Enoic Acid Hydrazides and Their Metal Complexes. *Pharm. Chem. J.* **2019**, *53*, 220–224. [CrossRef]
64. Betts, J.; Nagel, C.; Schatzschneider, U.; Poole, R.; La Ragione, R.M. Antimicrobial activity of carbon monoxide-releasing molecule [Mn(CO)$_3$(tpa-k^{3}N)]Br versus multidrug-resistant isolates of Avian Pathogenic Escherichia coli and its synergy with colistin. *PLoS ONE* **2017**, *12*, e0186359. [CrossRef]
65. Nagel, C.; McLean, S.; Poole, R.K.; Braunschweig, H.; Kramer, T.; Schatzschneider, U. Introducing [Mn(CO)$_3$(tpa-k^{3}N)]$^+$ as a novel photoactivatable CO-releasing molecule with well-defined iCORM intermediates–synthesis, spectroscopy, and antibacterial activity. *Dalton Trans.* **2014**, *43*, 9986–9997. [CrossRef]
66. Rana, N.; Jesse, H.E.; Tinajero-Trejo, M.; Butler, J.A.; Tarlit, J.D.; von Und Zur Muhlen, M.L.; Nagel, C.; Schatzschneider, U.; Poole, R.K. A manganese photosensitive tricarbonyl molecule [Mn(CO)$_3$(tpa-k^{3}N)]Br enhances antibiotic efficacy in a multi-drug-resistant Escherichia coli. *Microbiology* **2017**, *163*, 1477–1489. [CrossRef]
67. Tinajero-Trejo, M.; Rana, N.; Nagel, C.; Jesse, H.E.; Smith, T.W.; Wareham, L.K.; Poole, R.K. Antimicrobial Activity of the Manganese Photoactivated Carbon Monoxide-Releasing Molecule [Mn(CO)$_3$(tpa-k^{3}N)]$^+$ Against a Pathogenic Escherichia coli that Causes Urinary Infections. *Antioxid. Redox. Sign.* **2016**, *24*, 765–780. [CrossRef]
68. Ward, J.S.; Lynam, J.M.; Moir, J.; Fairlamb, I.J.S. Visible-Light-Induced CO Release from a Therapeutically Viable Tryptophan-Derived Manganese(I) Carbonyl (TryptoCORM) Exhibiting Potent Inhibition against E. coli. *Chem. Eur. J.* **2014**, *20*, 15061–15068. [CrossRef]
69. Ward, J.S.; Morgan, R.; Lynam, J.M.; Fairlamb, I.J.S.; Moir, J.W.B. Toxicity of tryptophan manganese(I) carbonyl (Trypto-CORM), against Neisseria gonorrhoeae. *Med. Chem. Commun.* **2017**, *8*, 346–352. [CrossRef]
70. Crook, S.H.; Mann, B.E.; Meijer, A.J.H.M.; Adams, H.; Sawle, P.; Scapens, D.; Motterlini, R. [Mn(CO)$_4$(S$_2$CNMe(CH$_2$CO$_2$H))], a new water-soluble CO-releasing molecule. *Dalton Trans.* **2011**, *40*, 4230–4235. [CrossRef]
71. Wareham, L.K.; McLean, S.; Begg, R.; Rana, N.; Ali, S.; Kendall, J.J.; Sanguinetti, G.; Mann, B.E.; Poole, R.K. The Broad-Spectrum Antimicrobial Potential of [Mn(CO)$_4$(S$_2$CNMe(CH$_2$CO$_2$H))], a Water-Soluble CO-Releasing Molecule (CORM-401): Intracellular Accumulation, Transcriptomic and Statistical Analyses, and Membrane Polarization. *Antioxid. Redox Signal.* **2018**, *28*, 1286–1308. [CrossRef]

72. Noor, A.; Huff, G.S.; Kumar, S.V.; Lewis, J.E.M.; Paterson, B.M.; Schieber, C.; Donnelly, P.S.; Brooks, H.J.L.; Gordon, K.C.; Moratti, S.C.; et al. [Re(CO)$_3$]$^+$ Complexes of exo-Functionalized Tridentate "Click" Macrocycles: Synthesis, Stability, Photophysical Properties, Bioconjugation, and Antibacterial Activity. *Organometallics* **2014**, *33*, 7031–7043. [CrossRef]
73. Patra, M.; Wenzel, M.; Prochnow, P.; Pierroz, V.; Gasser, G.; Bandow, J.E.; Metzler-Nolte, N. An organometallic structure-activity relationship study reveals the essential role of a Re(CO)$_3$ moiety in the activity against gram-positive pathogens including MRSA. *Chem. Sci.* **2015**, *6*, 214–224. [CrossRef]
74. Kumar, S.V.; Lo, W.K.C.; Brooks, H.J.L.; Hanton, L.R.; Crowley, J.D. Antimicrobial Properties of Mono- and Di-*fac*-rhenium Tricarbonyl 2-Pyridyl-1,2,3-triazole Complexes. *Aust. J. Chem.* **2016**, *69*, 489–498. [CrossRef]
75. Siegmund, D.; Lorenz, N.; Gothe, Y.; Spies, C.; Geissler, B.; Prochnow, P.; Nuernberger, P.; Bandow, J.E.; Metzler-Nolte, N. Benzannulated Re(I)–NHC complexes: Synthesis, photophysical properties and antimicrobial activity. *Dalton Trans.* **2017**, *46*, 15269–15279. [CrossRef]
76. Frei, A.; Amado, M.; Cooper, M.A.; Blaskovich, M.A.T. Light-Activated Rhenium Complexes with Dual Mode of Action against Bacteria. *Chem. Eur. J.* **2020**, *26*, 2852–2858. [CrossRef]
77. Carreño, A.; Solís-Céspedes, E.; Zúñiga, C.; Nevermann, J.; Rivera-Zaldívar, M.M.; Gacitúa, M.; Ramírez-Osorio, A.; Páez-Hernández, D.; Arratia-Pérez, R.; Fuentes, J.A. Cyclic voltammetry, relativistic DFT calculations and biological test of cytotoxicity in walled-cell models of two classical rhenium(I) tricarbonyl complexes with 5-amine-1,10-phenanthroline. *Chem. Phys. Lett.* **2019**, *715*, 231–238. [CrossRef]
78. Kydonaki, T.E.; Tsoukas, E.; Mendes, F.; Hatzidimitriou, A.G.; Paulo, A.; Papadopoulou, L.C.; Papagiannopoulou, D.; Psomas, G. Synthesis, characterization and biological evaluation of 99mTc/Re–tricarbonyl quinolone complexes. *J. Inorg. Biochem.* **2016**, *160*, 94–105. [CrossRef]
79. Miller, R.G.; Vázquez-Hernández, M.; Prochnow, P.; Bandow, J.E.; Metzler-Nolte, N. A CuAAC Click Approach for the Introduction of Bidentate Metal Complexes to a Sulfanilamide-Derived Antibiotic Fragment. *Inorg. Chem.* **2019**, *58*, 9404–9413. [CrossRef]
80. Pagoni, C.-C.; Xylouri, V.-S.; Kaiafas, G.C.; Lazou, M.; Bompola, G.; Tsoukas, E.; Papadopoulou, L.C.; Psomas, G.; Papagiannopoulou, D. Organometallic rhenium tricarbonyl–enrofloxacin and –levofloxacin complexes: Synthesis, albumin-binding, DNA-interaction and cell viability studies. *J. Biol. Inorg. Chem.* **2019**, *24*, 609–619. [CrossRef]
81. Howson, S.E.; Bolhuis, A.; Brabec, V.; Clarkson, G.J.; Malina, J.; Rodger, A.; Scott, P. Optically pure, water-stable metallo-helical 'flexicate' assemblies with antibiotic activity. *Nat. Chem.* **2012**, *4*, 31–36. [CrossRef]
82. Allardyce, C.S.; Dyson, P.J. Ruthenium in Medicine_ Current Clinical Uses and Future Prospects. *Platin. Met. Rev.* **2001**, *45*, 62–69.
83. Atton, J.G.D.M.; Gillard, R.D. Equilibria in complexes of N-heterocyclic ligands. Part 33. Ruthenium(II) complex ions with chelating pyridyl-imidazoles. *Transit. Met. Chem.* **1981**, *6*, 351–355. [CrossRef]
84. Bolhuis, A.; Hand, L.; Marshall, J.E.; Richards, A.D.; Rodger, A.; Aldrich-Wright, J. Antimicrobial activity of ruthenium-based intercalators. *Eur. J. Pharm. Sci.* **2011**, *42*, 313–317. [CrossRef]
85. Gill, M.R.; Garcia-Lara, J.; Foster, S.J.; Smythe, C.; Battaglia, G.; Thomas, J.A. A ruthenium(II) polypyridyl complex for direct imaging of DNA structure in living cells. *Nat. Chem.* **2009**, *1*, 662–667. [CrossRef]
86. Gill, M.R.; Thomas, J.A. Ruthenium(II) polypyridyl complexes and DNA—from structural probes to cellular imaging and therapeutics. *Chem. Soc. Rev.* **2012**, *41*, 3179–3192. [CrossRef]
87. Keene, F.R.; Smith, J.A.; Collins, J.G. Metal complexes as structure-selective binding agents for nucleic acids. *Coord. Chem. Rev.* **2009**, *253*, 2021–2035. [CrossRef]
88. Luedtke, N.W.; Hwang, J.S.; Nava, E.; Gut, D.; Kol, M.; Tor, Y. The DNA and RNA specificity of eilatin Ru(II) complexes as compared to eilatin and ethidium bromide. *Nucleic Acids Res.* **2003**, *31*, 5732–5740. [CrossRef]
89. Matson, M.; Svensson, F.R.; Nordén, B.; Lincoln, P. Correlation Between Cellular Localization and Binding Preference to RNA, DNA, and Phospholipid Membrane for Luminescent Ruthenium(II) Complexes. *J. Phys. Chem. B* **2011**, *115*, 1706–1711. [CrossRef]
90. Metcalfe, C.; Thomas, J.A. Kinetically inert transition metal complexes that reversibly bind to DNA. *Chem. Soc. Rev.* **2003**, *32*, 215–224. [CrossRef]
91. Norden, B.; Lincoln, P.; Akerman, B.; Tuite, E. DNA interactions with substitution-inert transition metal ion complexes. *Met. Ions Biol. Syst.* **1996**, *33*, 177–252.

92. Puckett, C.A.; Barton, J.K. Mechanism of Cellular Uptake of a Ruthenium Polypyridyl Complex. *Biochemistry* **2008**, *47*, 11711–11716. [CrossRef]
93. Zeglis, B.M.; Pierre, V.C.; Barton, J.K. Metallo-intercalators and metallo-insertors. *Chem. Commun.* **2007**, *44*, 4565–4579. [CrossRef]
94. Kumar, S.V.; Scottwell, S.Ø.; Waugh, E.; McAdam, C.J.; Hanton, L.R.; Brooks, H.J.L.; Crowley, J.D. Antimicrobial Properties of Tris(homoleptic) Ruthenium(II) 2-Pyridyl-1,2,3-triazole "Click" Complexes against Pathogenic Bacteria, Including Methicillin-Resistant Staphylococcus aureus (MRSA). *Inorg. Chem.* **2016**, *55*, 9767–9777. [CrossRef]
95. Liao, G.; Ye, Z.; Liu, Y.; Fu, B.; Fu, C. Octahedral ruthenium(II) polypyridyl complexes as antimicrobial agents against mycobacterium. *PeerJ.* **2017**, *5*, e3252. [CrossRef]
96. Li, X.; Heimann, K.; Li, F.; Warner, J.M.; Richard Keene, F.; Grant Collins, J. Dinuclear ruthenium(II) complexes containing one inert metal centre and one coordinatively-labile metal centre: Syntheses and biological activities. *Dalton Trans.* **2016**, *45*, 4017–4029. [CrossRef]
97. Srivastava, P.; Shukla, M.; Kaul, G.; Chopra, S.; Patra, A.K. Rationally designed curcumin based ruthenium(II) antimicrobials effective against drug-resistant Staphylococcus aureus. *Dalton Trans.* **2019**, *48*, 11822–11828. [CrossRef]
98. Kreofsky, N.W.; Dillenburg, M.D.; Villa, E.M.; Fletcher, J.T. Ru(II) coordination compounds of NN bidentate chelators with 1,2,3 triazole and isoquinoline subunits: Synthesis, spectroscopy and antimicrobial properties. *Polyhedron* **2020**, *177*, 114259. [CrossRef]
99. Van Hilst, Q.V.C.; Vasdev, R.A.S.; Preston, D.; Findlay, J.A.; Scottwell, S.Ø.; Giles, G.I.; Brooks, H.J.L.; Crowley, J.D. Synthesis, Characterisation and Antimicrobial Studies of some 2,6-bis(1,2,3-Triazol-4-yl)Pyridine Ruthenium(II) "Click" Complexes. *Asian J. Org. Chem.* **2019**, *8*, 496–505. [CrossRef]
100. Sun, B.; Sundaraneedi, M.K.; Southam, H.M.; Poole, R.K.; Musgrave, I.F.; Keene, F.R.; Collins, J.G. Synthesis and biological properties of tetranuclear ruthenium complexes containing the bis[4(4′-methyl-2,2′-bipyridyl)]-1,7-heptane ligand. *Dalton Trans.* **2019**, *48*, 14505–14515. [CrossRef]
101. Menéndez-Pedregal, E.; Manteca, Á.; Sánchez, J.; Díez, J.; Gamasa, M.P.; Lastra, E. Antimicrobial and Antitumor Activity of Enantiopure Pybox–Osmium Complexes. *Eur. J. Inorg. Chem.* **2015**, *2015*, 1424–1432. [CrossRef]
102. Gichumbi, J.M.; Friedrich, H.B.; Omondi, B.; Naicker, K.; Singh, M.; Chenia, H.Y. Synthesis, characterization, antiproliferative, and antimicrobial activity of osmium(II) half-sandwich complexes. *J. Coord. Chem.* **2018**, *71*, 342–354. [CrossRef]
103. Zhu, Y.; Li, W.-H. Syntheses, crystal structures and antibacterial activities of azido-bridged cobalt(III) complexes with Schiff bases. *Transition Met. Chem.* **2010**, *35*, 745–749. [CrossRef]
104. Anjaneyulu, Y.; Rao, R.P. Preparation, Characterization and Antimicrobial Activity Studies on Some Ternary Complexes of Cu(II) with Acetylacetone and Various Salicylic Acids. *Synth. React. Inorg. Met. Org. Chem.* **1986**, *16*, 257–272. [CrossRef]
105. Dharmaraj, N.; Viswanathamurthi, P.; Natarajan, K. Ruthenium(II) complexes containing bidentate Schiff bases and their antifungal activity. *Transit. Met. Chem.* **2001**, *26*, 105–109. [CrossRef]
106. Irgi, E.P.; Geromichalos, G.D.; Balala, S.; Kljun, J.; Kalogiannis, S.; Papadopoulos, A.; Turel, I.; Psomas, G. Cobalt(II) complexes with the quinolone antimicrobial drug oxolinic acid: Structure and biological perspectives. *RSC Adv.* **2015**, *5*, 36353–36367. [CrossRef]
107. Kouris, E.; Kalogiannis, S.; Perdih, F.; Turel, I.; Psomas, G. Cobalt(II) complexes of sparfloxacin: Characterization, structure, antimicrobial activity and interaction with DNA and albumins. *J. Inorg. Biochem.* **2016**, *163*, 18–27. [CrossRef]
108. Turecka, K.; Chylewska, A.; Kawiak, A.; Waleron, K.F. Antifungal Activity and Mechanism of Action of the Co(III) Coordination Complexes With Diamine Chelate Ligands Against Reference and Clinical Strains of *Candida* spp. *Front. Microbiol.* **2018**, *9*. [CrossRef]
109. Unluer, D.; Aktas Kamiloglu, A.; Direkel, S.; Bektas, E.; Kantekin, H.; Sancak, K. Synthesis and characterization of metallophthalocyanine with morpholine containing Schiff base and determination of their antimicrobial and antioxidant activities. *J. Organomet. Chem.* **2019**, *900*, 120936. [CrossRef]

110. Balan, G.; Burduniuc, O.; Usataia, I.; Graur, V.; Chumakov, Y.; Petrenko, P.; Gudumac, V.; Gulea, A.; Pahontu, E. Novel 2-formylpyridine 4-allyl-S-methylisothiosemicarbazone and Zn(II), Cu(II), Ni(II) and Co(III) complexes: Synthesis, characterization, crystal structure, antioxidant, antimicrobial and antiproliferative activity. *Appl. Organomet. Chem.* **2020**, *34*, e5423. [CrossRef]
111. Casanova, I.; Durán, M.L.; Viqueira, J.; Sousa-Pedrares, A.; Zani, F.; Real, J.A.; García-Vázquez, J.A. Metal complexes of a novel heterocyclic benzimidazole ligand formed by rearrangement-cyclization of the corresponding Schiff base. Electrosynthesis, structural characterization and antimicrobial activity. *Dalton Trans.* **2018**, *47*, 4325–4340. [CrossRef]
112. Chandrasekhar, V.R.; Mookkandi Palsamy, K.; Lokesh, R.; Jegathalaprathaban, R.; Gurusamy, R. Biomolecular docking, antimicrobial and cytotoxic studies on new bidentate schiff base ligand derived metal(II) complexes. *Appl. Organomet. Chem.* **2019**, *33*, e4753. [CrossRef]
113. Ebrahimipour, S.Y.; Machura, B.; Mohamadi, M.; Khaleghi, M. A novel cationic cobalt(III) Schiff base complex: Preparation, crystal structure, Hirshfeld surface analysis, antimicrobial activities and molecular docking. *Microb. Pathog.* **2017**, *113*, 160–167. [CrossRef] [PubMed]
114. Gałczyńska, K.; Ciepluch, K.; Madej, Ł.; Kurdziel, K.; Maciejewska, B.; Drulis-Kawa, Z.; Węgierek-Ciuk, A.; Lankoff, A.; Arabski, M. Selective cytotoxicity and antifungal properties of copper(II) and cobalt(II) complexes with imidazole-4-acetate anion or 1-allylimidazole. *Sci. Rep.* **2019**, *9*, 9777. [CrossRef] [PubMed]
115. Orojloo, M.; Zolgharnein, P.; Solimannejad, M.; Amani, S. Synthesis and characterization of cobalt(II), nickel(II), copper(II) and zinc(II) complexes derived from two Schiff base ligands: Spectroscopic, thermal, magnetic moment, electrochemical and antimicrobial studies. *Inorg. Chim. Acta* **2017**, *467*, 227–237. [CrossRef]
116. Palmucci, J.; Mahmudov, K.T.; Guedes da Silva, M.F.C.; Marchetti, F.; Pettinari, C.; Petrelli, D.; Vitali, L.A.; Quassinti, L.; Bramucci, M.; Lupidi, G.; et al. DNA and BSA binding, anticancer and antimicrobial properties of Co(II), Co(II/III), Cu(II) and Ag(I) complexes of arylhydrazones of barbituric acid. *RSC Adv.* **2016**, *6*, 4237–4249. [CrossRef]
117. Sadhu, M.H.; Kumar, S.B.; Saini, J.K.; Purani, S.S.; Khanna, T.R. Mononuclear copper(II) and binuclear cobalt(II) complexes with halides and tetradentate nitrogen coordinate ligand: Synthesis, structures and bioactivities. *Inorg. Chim. Acta* **2017**, *466*, 219–227. [CrossRef]
118. Saha, M.; Biswas, J.K.; Mondal, M.; Ghosh, D.; Mandal, S.; Cordes, D.B.; Slawin, A.M.Z.; Mandal, T.K.; Saha, N.C. Synthesis, characterization and antimicrobial activities of Co(III) and Ni(II) complexes with 5-methyl-3-formylpyrazole-N(4)-dihexylthiosemicarbazone (HMPzNHex$_2$): X-ray crystallography and DFT calculations of [Co(MPzNHex$_2$)$_2$]ClO$_4$·1.5H$_2$O (I) and [Ni(HMPzNHex$_2$)$_2$]Cl$_2$·2H$_2$O (II). *Inorg. Chim. Acta* **2018**, *483*, 271–283. [CrossRef]
119. Saha, S.; Sasmal, A.; Roy Choudhury, C.; Pilet, G.; Bauzá, A.; Frontera, A.; Chakraborty, S.; Mitra, S. Synthesis, crystal structure, antimicrobial screening and density functional theory calculation of nickel(II), cobalt(II) and zinc(II) mononuclear Schiff base complexes. *Inorg. Chim. Acta* **2015**, *425*, 211–220. [CrossRef]
120. Vasdev, R.A.; Preston, D.; Scottwell, S.O.; Brooks, H.J.; Crowley, J.D.; Schramm, M.P. Oxidatively Locked [Co$_2$L$_3$]$^{6+}$ Cylinders Derived from Bis(bidentate) 2-Pyridyl-1,2,3-triazole "Click" Ligands: Synthesis, Stability, and Antimicrobial Studies. *Molecules* **2016**, *21*, 1548. [CrossRef]
121. Woźniczka, M.; Świątek, M.; Pająk, M.; Gądek-Sobczyńska, J.; Chmiela, M.; Gonciarz, W.; Lisiecki, P.; Pasternak, B.; Kufelnicki, A. Complexes in aqueous cobalt(II)–2-picolinehydroxamic acid system: Formation equilibria, DNA-binding ability, antimicrobial and cytotoxic properties. *J. Inorg. Biochem.* **2018**, *187*, 62–72. [CrossRef]
122. Lu, L.; Liu, L.J.; Chao, W.C.; Zhong, H.J.; Wang, M.; Chen, X.P.; Lu, J.J.; Li, R.N.; Ma, D.L.; Leung, C.H. Identification of an iridium(III) complex with anti-bacterial and anti-cancer activity. *Sci. Rep.* **2015**, *5*, e14544. [CrossRef]
123. Fiorini, V.; Zanoni, I.; Zacchini, S.; Costa, A.L.; Hochkoeppler, A.; Zanotti, V.; Ranieri, A.M.; Massi, M.; Stefan, A.; Stagni, S. Methylation of Ir(III)-tetrazolato complexes: An effective route to modulate the emission outputs and to switch to antimicrobial properties. *Dalton Trans.* **2017**, *46*, 12328–12338. [CrossRef]
124. Kumar, S.; Purcell, W.; Conradie, J.; Bragg, R.R.; Langner, E.H.G. Synthesis, characterization, computational and antimicrobial activities of a novel iridium thiourea complex. *New J. Chem.* **2017**, *41*, 10919–10928. [CrossRef]
125. DuChane, C.M.; Brown, L.C.; Dozier, V.S.; Merola, J.S. Synthesis, Characterization, and Antimicrobial Activity of Rh(III) and Ir(III) β-Diketonato Piano-Stool Compounds. *Organometallics* **2018**, *37*, 530–538. [CrossRef]

126. Lapasam, A.; Dkhar, L.; Joshi, N.; Poluri, K.M.; Kollipara, M.R. Antimicrobial selectivity of ruthenium, rhodium, and iridium half sandwich complexes containing phenyl hydrazone Schiff base ligands towards B. thuringiensis and P. aeruginosa bacteria. *Inorg. Chim. Acta* **2019**, *484*, 255–263. [CrossRef]
127. Lapasam, A.; Banothu, V.; Uma, A.; Kollipara, M. Synthesis, structural and antimicrobial studies of half-sandwich ruthenium, rhodium and iridium complexes containing nitrogen donor Schiff-base ligands. *J. Mol. Struct.* **2019**, *1191*. [CrossRef]
128. Liu, B.; Monro, S.; Jabed, M.A.; Cameron, C.G.; Colón, K.L.; Xu, W.; Kilina, S.; McFarland, S.A.; Sun, W. Neutral iridium(III) complexes bearing BODIPY-substituted N-heterocyclic carbene (NHC) ligands: Synthesis, photophysics, in vitro theranostic photodynamic therapy, and antimicrobial activity. *Photochem. Photobiol. Sci.* **2019**, *18*, 2381–2396. [CrossRef] [PubMed]
129. Chen, F.; Moat, J.; McFeely, D.; Clarkson, G.; Hands-Portman, I.J.; Furner-Pardoe, J.P.; Harrison, F.; Dowson, C.G.; Sadler, P.J. Biguanide Iridium(III) Complexes with Potent Antimicrobial Activity. *J. Med. Chem.* **2018**, *61*, 7330–7344. [CrossRef] [PubMed]
130. DuChane, C.M.; Karpin, G.W.; Ehrich, M.; Falkinham, J.O.; Merola, J.S. Iridium piano stool complexes with activity against S. aureus and MRSA: It is past time to truly think outside of the box. *Med. Chem. Commun.* **2019**, *10*, 1391–1398. [CrossRef]
131. Lapasam, A.; Banothu, V.; Addepally, U.; Kollipara, M.R. Half-sandwich arene ruthenium, rhodium and iridium thiosemicarbazone complexes: Synthesis, characterization and biological evaluation. *J. Chem. Sci.* **2020**, *132*, 34. [CrossRef]
132. Raj, P.; Singh, A.; Singh, A.; Singh, N. Syntheses and Photophysical Properties of Schiff Base Ni(II) Complexes: Application for Sustainable Antibacterial Activity and Cytotoxicity. *ACS Sustain. Chem. Eng.* **2017**, *5*, 6070–6080. [CrossRef]
133. Ibrahim, A.B.M.; Farh, M.K.; El-Gyar, S.A.; El-Gahami, M.A.; Fouad, D.M.; Silva, F.; Santos, I.C.; Paulo, A. Synthesis, structural studies and antimicrobial activities of manganese, nickel and copper complexes of two new tridentate 2-formylpyridine thiosemicarbazone ligands. *Inorg. Chem. Commun.* **2018**, *96*, 194–201. [CrossRef]
134. Juribašić, M.; Molčanov, K.; Kojić-Prodić, B.; Bellotto, L.; Kralj, M.; Zani, F.; Tušek-Božić, L. Palladium(II) complexes of quinolinylaminophosphonates: Synthesis, structural characterization, antitumor and antimicrobial activity. *J. Inorg. Biochem.* **2011**, *105*, 867–879. [CrossRef] [PubMed]
135. Tamayo, L.V.; Burgos, A.E.; Brandão, P.F.B. Synthesis, Characterization, and Antimicrobial Activity of the Ligand 3-Methylpyrazole- 4-Carboxaldehyde Thiosemicarbazone and Its Pd(II) Complex. *Phosphorus Sulfur Relat. Elem.* **2014**, *189*, 52–59. [CrossRef]
136. Radić, G.P.; Glođović, V.V.; Radojević, I.D.; Stefanović, O.D.; Čomić, L.R.; Ratković, Z.R.; Valkonen, A.; Rissanen, K.; Trifunović, S.R. Synthesis, characterization and antimicrobial activity of palladium(II) complexes with some alkyl derivates of thiosalicylic acids: Crystal structure of the bis(S-benzyl-thiosalicylate)–palladium(II) complex, [Pd(S-bz-thiosal)$_2$]. *Polyhedron* **2012**, *31*, 69–76. [CrossRef]
137. Boubakri, L.; Mansour, L.; Harrath, A.H.; Özdemir, I.; Yaşar, S.; Hamdi, N. N-Heterocyclic carbene-Pd(II)-PPh$_3$ complexes as a new highly efficient catalyst system for the Sonogashira cross-coupling reaction: Synthesis, characterization and biological activities. *J. Coord. Chem.* **2018**, *71*, 183–199. [CrossRef]
138. Abu-Dief, A.M.; Abdel-Rahman, L.H.; Abdel-Mawgoud, A.A.H. A robust in vitro Anticancer, Antioxidant and Antimicrobial Agents Based on New Metal-Azomethine Chelates Incorporating Ag(I), Pd(II) and VO(II) Cations: Probing the Aspects of DNA Interaction. *Appl. Organomet. Chem.* **2020**, *34*, e5373. [CrossRef]
139. Nyawade, E.A.; Onani, M.O.; Meyer, S.; Dube, P. Synthesis, characterization and antibacterial activity studies of new 2-pyrral-L-amino acid Schiff base palladium (II) complexes. *Chem. Pap.* **2020**, 1–11. [CrossRef]
140. Solmaz, U.; Gumus, I.; Binzet, G.; Celik, O.; Balci, G.K.; Dogen, A.; Arslan, H. Synthesis, characterization, crystal structure, and antimicrobial studies of novel thiourea derivative ligands and their platinum complexes. *J. Coord. Chem.* **2018**, *71*, 200–218. [CrossRef]
141. Gaber, M.; El-Ghamry, H.A.; Mansour, M.A. Pd(II) and Pt(II) chalcone complexes. Synthesis, spectral characterization, molecular modeling, biomolecular docking, antimicrobial and antitumor activities. *J. Photochem. Photobiol.* **2018**, *354*, 163–174. [CrossRef]
142. Mansour, A.M. Antifungal activity, DNA and lysozyme binding affinity of Pd(II) and Pt(II) complexes bearing N,N-pyridylbenzimidazole ligand. *J. Coord. Chem.* **2018**, *71*, 3381–3391. [CrossRef]

143. Lunagariya, M.V.; Thakor, K.P.; Varma, R.R.; Waghela, B.N.; Pathak, C.; Patel, M.N. Synthesis, characterization and biological application of 5-quinoline 1,3,5-trisubstituted pyrazole based platinum(II) complexes. *Med. Chem. Commun.* **2018**, *9*, 282–298. [CrossRef]

144. Gao, S.; Yan, X.; Xie, G.; Zhu, M.; Ju, X.; Stang, P.J.; Tian, Y.; Niu, Z. Membrane intercalation-enhanced photodynamic inactivation of bacteria by a metallacycle and TAT-decorated virus coat protein. *Proc. Natl. Acad. Sci. USA* **2019**, *116*, 23437–23443. [CrossRef]

145. Alam, M.N.; Yu, J.Q.; Beale, P.; Turner, P.; Proschogo, N.; Huq, F. Crystal Structure, Antitumour and Antibacterial Activity of Imidazo[1, 2–α]pyridine Ligand Containing Palladium Complexes. *ChemistrySelect* **2020**, *5*, 668–673. [CrossRef]

146. Basto, A.P.; Muller, J.; Rubbiani, R.; Stibal, D.; Giannini, F.; Suss-Fink, G.; Balmer, V.; Hemphill, A.; Gasser, G.; Furrer, J. Characterization of the Activities of Dinuclear Thiolato-Bridged Arene Ruthenium Complexes against Toxoplasma gondii. *Antimicrob. Agents Chemother.* **2017**, *61*, e01031-17. [CrossRef] [PubMed]

147. Bugarčić, Z.M.; Divac, V.M.; Kostić, M.D.; Janković, N.Ž.; Heinemann, F.W.; Radulović, N.S.; Stojanović-Radić, Z.Z. Synthesis, crystal and solution structures and antimicrobial screening of palladium(II) complexes with 2-(phenylselanylmethyl)oxolane and 2-(phenylselanylmethyl)oxane as ligands. *J. Inorg. Biochem.* **2015**, *143*, 9–19. [CrossRef] [PubMed]

148. Choo, K.B.; Mah, W.L.; Lee, S.M.; Lee, W.L.; Cheow, Y.L. Palladium complexes of bidentate pyridine N-heterocyclic carbenes: Optical resolution, antimicrobial and cytotoxicity studies. *Appl. Organomet. Chem.* **2018**, *32*, e4377. [CrossRef]

149. Farkasová, V.; Drweesh, S.A.; Lüköová, A.; Sabolová, D.; Radojević, I.D.; Čomić, L.R.; Vasić, S.M.; Paulíková, H.; Fečko, S.; Balašková, T.; et al. Low-dimensional compounds containing bioactive ligands. Part VIII: DNA interaction, antimicrobial and antitumor activities of ionic 5,7-dihalo-8-quinolinolato palladium(II) complexes with K^+ and Cs^+ cations. *J. Inorg. Biochem.* **2017**, *167*, 80–88. [CrossRef]

150. Gorle, A.K.; Li, X.; Primrose, S.; Li, F.; Feterl, M.; Kinobe, R.T.; Heimann, K.; Warner, J.M.; Keene, F.R.; Collins, J.G. Oligonuclear polypyridylruthenium(II) complexes: Selectivity between bacteria and eukaryotic cells. *J. Antimicrob. Chemother.* **2016**, *71*, 1547–1555. [CrossRef]

151. Laurent, Q.; Batchelor, L.K.; Dyson, P.J. Applying a Trojan Horse Strategy to Ruthenium Complexes in the Pursuit of Novel Antibacterial Agents. *Organometallics* **2018**, *37*, 915–923. [CrossRef]

152. Li, F.; Feterl, M.; Mulyana, Y.; Warner, J.M.; Collins, J.G.; Keene, F.R. In vitro susceptibility and cellular uptake for a new class of antimicrobial agents: Dinuclear ruthenium(II) complexes. *J. Antimicrob. Chemother.* **2012**, *67*, 2686–2695. [CrossRef] [PubMed]

153. Liu, X.; Sun, B.; Kell, R.E.M.; Southam, H.M.; Butler, J.A.; Li, X.; Poole, R.K.; Keene, F.R.; Collins, J.G. The Antimicrobial Activity of Mononuclear Ruthenium(II) Complexes Containing the dppz Ligand. *ChemPlusChem* **2018**, *83*, 643–650. [CrossRef]

154. Onar, G.; Gürses, C.; Karataş, M.O.; Balcıoğlu, S.; Akbay, N.; Özdemir, N.; Ateş, B.; Alıcı, B. Palladium(II) and ruthenium(II) complexes of benzotriazole functionalized N-heterocyclic carbenes: Cytotoxicity, antimicrobial, and DNA interaction studies. *J. Organomet. Chem.* **2019**, *886*, 48–56. [CrossRef]

155. Pahontu, E.; Paraschivescu, C.; Ilies, D.C.; Poirier, D.; Oprean, C.; Paunescu, V.; Gulea, A.; Rosu, T.; Bratu, O. Synthesis and Characterization of Novel Cu(II), Pd(II) and Pt(II) Complexes with 8-Ethyl-2-hydroxytricyclo(7.3.1.0(2,7))tridecan-13-one-thiosemicarbazone: Antimicrobial and in Vitro Antiproliferative Activity. *Molecules* **2016**, *21*, 674. [CrossRef] [PubMed]

156. Patra, S.C.; Saha Roy, A.; Banerjee, S.; Banerjee, A.; Das Saha, K.; Bhadra, R.; Pramanik, K.; Ghosh, P. Palladium(II) and platinum(II) complexes of glyoxalbis(N-aryl)osazone: Molecular and electronic structures, anti-microbial activities and DNA-binding study. *New J. Chem.* **2019**, *43*, 9891–9901. [CrossRef]

157. Satheesh, C.E.; Raghavendra Kumar, P.; Sharma, P.; Lingaraju, K.; Palakshamurthy, B.S.; Raja Naika, H. Synthesis, characterisation and antimicrobial activity of new palladium and nickel complexes containing Schiff bases. *Inorg. Chim. Acta* **2016**, *442*, 1–9. [CrossRef]

158. Southam, H.M.; Butler, J.A.; Chapman, J.A.; Poole, R.K. Chapter One—The Microbiology of Ruthenium Complexes. *Adv. Microb. Physiol.* **2017**, *71*, 1–96.

159. Turel, I.; Kljun, J.; Perdih, F.; Morozova, E.; Bakulev, V.; Kasyanenko, N.; Byl, J.A.; Osheroff, N. First ruthenium organometallic complex of antibacterial agent ofloxacin. Crystal structure and interactions with DNA. *Inorg. Chem.* **2010**, *49*, 10750–10752. [CrossRef]

160. Yang, Y.; Liao, G.; Fu, C. Recent Advances on Octahedral Polypyridyl Ruthenium(II) Complexes as Antimicrobial Agents. *Polymers* **2018**, *10*, 650. [CrossRef]
161. Jević, V.V.; Radić, G.P.; Stefanović, O.D.; Radojević, I.D.; Vasić, S.; Čomić, L.R.; Đinović, V.M.; Trifunović, S.R. Part XXIII. Synthesis and characterization of platinum(IV) complexes with O,O'-dialkyl esters of (S,S)-ethylenediamine-N,N'-di-2-(3-methyl)butanoic acid and bromido ligands. Antimicrobial, antibiofilm and antioxidant screening. *Inorg. Chim. Acta* **2016**, *442*, 105–110. [CrossRef]
162. Dkhar, L.; Banothu, V.; Kaminsky, W.; Kollipara, M.R. Synthesis of half sandwich platinum group metal complexes containing pyridyl benzothiazole hydrazones: Study of bonding modes and antimicrobial activity. *J. Organomet. Chem.* **2020**, *914*, 121225. [CrossRef]
163. Sankarganesh, M.; Vijay Solomon, R.; Dhaveethu Raja, J. Platinum complex with pyrimidine- and morpholine-based ligand: Synthesis, spectroscopic, DFT, TDDFT, catalytic reduction, in vitro anticancer, antioxidant, antimicrobial, DNA binding and molecular modeling studies. *J. Biomol. Struct. Dyn.* **2020**, 1–13. [CrossRef] [PubMed]
164. Choo, K.B.; Lee, S.M.; Lee, W.L.; Cheow, Y.L. Synthesis, characterization, in vitro antimicrobial and anticancer studies of new platinum N-heterocyclic carbene (NHC) complexes and unexpected nickel complexes. *J. Organomet. Chem.* **2019**, *898*, 120868. [CrossRef]
165. Musa, T.M.; Al-jibouri, M.N.; Al-bayati, R.I.H. Synthesis, Characterization and Antimicrobial Study of nickel(II), palladium(II), platinum(II), rhodium(III), cadmium(II) and zirconium(IV) complexes with (E)-1-(benzo[d]thiazol-2-yl)-4-(hydroxy(2-hydroxyphenyl)methylene)-3-methyl-1H-pyrazol-5(4H)-one. *J. Phys. Conf. Ser.* **2018**, *1032*, 012057. [CrossRef]
166. Bobinihi, F.F.; Onwudiwe, D.C.; Ekennia, A.C.; Okpareke, O.C.; Arderne, C.; Lane, J.R. Group 10 metal complexes of dithiocarbamates derived from primary anilines: Synthesis, characterization, computational and antimicrobial studies. *Polyhedron* **2019**, *158*, 296–310. [CrossRef]
167. Stojković, D.L.; Jevtić, V.V.; Radić, G.P.; Đukić, M.B.; Jelić, R.M.; Zarić, M.M.; Anđelković, M.V.; Mišić, M.S.; Baskić, D.D.; Trifunović, S.R. Stereospecific ligands and their complexes. XXIV. Synthesis, characterization and some biological properties of Pd(II) and Pt(II) complexes with R_2-S,S-eddtyr. *New J. Chem.* **2018**, *42*, 3924–3935. [CrossRef]
168. Ahmad, S.; Seerat ur, R.; Rüffer, T.; Khalid, T.; Isab, A.A.; Al-Arfaj, A.R.; Saleem, M.; Khan, I.U.; Choudhary, M.A. Crystal structure and antimicrobial activity of a transplatin adduct of N,N'-dimethylthiourea, trans-[Pt(NH$_3$)$_2$(dmtu)$_2$]Cl$_2$. *Monatsh. Chem.* **2017**, *148*, 669–674. [CrossRef]
169. Onwudiwe, D.C.; Ekennia, A.C.; Mogwase, B.M.S.; Olubiyi, O.O.; Hosten, E. Palladium(II) and platinum(II) complexes of N-butyl-N-phenyldithiocarbamate: Synthesis, characterization, biological activities and molecular docking studies. *Inorg. Chim. Acta* **2016**, *450*, 69–80. [CrossRef]
170. Kaushal, M.; Lobana, T.S.; Nim, L.; Bala, R.; Arora, D.S.; Garcia-Santos, I.; Duff, C.E.; Jasinski, J.P. Synthesis of 2-acetylpyridine-N-substituted thiosemicarbazonates of copper(II) with high antimicrobial activity against methicillin resistant S. aureus, K. pneumoniae 1 and C. albicans. *New J. Chem.* **2019**, *43*, 11727–11742. [CrossRef]
171. Oladipo, S.D.; Omondi, B.; Mocktar, C. Synthesis and structural studies of nickel(II)- and copper(II)-N,N'-diarylformamidine dithiocarbamate complexes as antimicrobial and antioxidant agents. *Polyhedron* **2019**, *170*, 712–722. [CrossRef]
172. Krishnegowda, H.M.; Karthik, C.S.; Marichannegowda, M.H.; Kumara, K.; Kudigana, P.J.; Lingappa, M.; Mallu, P.; Neratur, L.K. Synthesis and structural studies of 1-phenyl-1,3-butanedione copper(II) complexes as an excellent antimicrobial agent against methicillin-resistant Staphylococcus aureus. *Inorg. Chim. Acta* **2019**, *484*, 227–236. [CrossRef]
173. Tiekink, E.R.T. Gold compounds in medicine: Potential anti-tumour agents. *Gold Bull.* **2003**, *36*, 117–124. [CrossRef]
174. Glišić, B.Đ.; Djuran, M.I. Gold complexes as antimicrobial agents: An overview of different biological activities in relation to the oxidation state of the gold ion and the ligand structure. *Dalton Trans.* **2014**, *43*, 5950–5969. [CrossRef] [PubMed]
175. Savić, N.D.; Milivojevic, D.R.; Glišić, B.Đ.; Ilic-Tomic, T.; Veselinovic, J.; Pavic, A.; Vasiljevic, B.; Nikodinovic-Runic, J.; Djuran, M.I. A comparative antimicrobial and toxicological study of gold(III) and silver(I) complexes with aromatic nitrogen-containing heterocycles: Synergistic activity and improved selectivity index of Au(III)/Ag(I) complexes mixture. *RSC Adv.* **2016**, *6*, 13193–13206. [CrossRef]

176. Hikisz, P.; Szczupak, L.; Koceva-Chyla, A.; Gu Spiel, A.; Oehninger, L.; Ott, I.; Therrien, B.; Solecka, J.; Kowalski, K. Anticancer and Antibacterial Activity Studies of Gold(I)-Alkynyl Chromones. *Molecules* **2015**, *20*, 19699–19718. [CrossRef]
177. Schmidt, C.; Karge, B.; Misgeld, R.; Prokop, A.; Brönstrup, M.; Ott, I. Biscarbene gold(I) complexes: Structure–activity-relationships regarding antibacterial effects, cytotoxicity, TrxR inhibition and cellular bioavailability. *Med. Chem. Commun.* **2017**, *8*, 1681–1689. [CrossRef]
178. Pöthig, A.; Ahmed, S.; Winther-Larsen, H.C.; Guan, S.; Altmann, P.J.; Kudermann, J.; Santos Andresen, A.M.; Gjøen, T.; Høgmoen Åstrand, O.A. Antimicrobial Activity and Cytotoxicity of Ag(I) and Au(I) Pillarplexes. *Front. Chem.* **2018**, *6*, 584. [CrossRef]
179. Crichton, R. *Biological Inorganic Chemistry*, 3rd ed.; Elsevier: London, UK, 2019; pp. 339–362. ISBN 978-0-12-811741-5.
180. Cuevas, L.; Koyanagi, A. Zinc and infection: A review. *Ann. Trop. Paediatr.* **2005**, *25*, 149–160. [CrossRef]
181. Harrison, P.M.; Hoare, R.J. *Metals in Biochemistry*; Chapman and Hall: London, UK; New York, NY, USA, 1980.
182. Osredkar, J. Copper and Zinc, Biological Role and Significance of Copper/Zinc Imbalance. *J. Clin. Toxicol.* **2011**, *3*, 495. [CrossRef]
183. Zelenak, V.; Gyoryova, K.; Mlynarcik, D. Antibacterial and Antifungal Activity of Zinc(II) Carboxylates with/without N-Donor Organic LIigands. *Met. Based Drugs* **2002**, *8*, 269–274. [CrossRef]
184. Zaltariov, M.-F.; Cazacu, M.; Avadanei, M.; Shova, S.; Balan, M.; Vornicu, N.; Vlad, A.; Dobrov, A.; Varganici, C.-D. Synthesis, characterization and antimicrobial activity of new Cu(II) and Zn(II) complexes with Schiff bases derived from trimethylsilyl-propyl-p-aminobenzoate. *Polyhedron* **2015**, *100*, 121–131. [CrossRef]
185. Abu Ali, H.; Omar, S.N.; Darawsheh, M.D.; Fares, H. Synthesis, characterization and antimicrobial activity of zinc(II) ibuprofen complexes with nitrogen-based ligands. *J. Coord. Chem.* **2016**, *69*, 1110–1122. [CrossRef]
186. Boughougal, A.; Cherchali, F.Z.; Messai, A.; Attik, N.; Decoret, D.; Hologne, M.; Sanglar, C.; Pilet, G.; Tommasino, J.B.; Luneau, D. New model of metalloantibiotic: Synthesis, structure and biological activity of a zinc(II) mononuclear complex carrying two enrofloxacin and sulfadiazine antibiotics. *N. J. Chem.* **2018**, *42*, 15346–15352. [CrossRef]
187. Basu Baul, T.; Nongsiej, K.; Ka-Ot, A.; Joshi, S.R.; León, I.; Höpfl, H. Tweaking the affinity of aryl-substituted diazosalicylato- and pyridine ligands towards Zn(II) and its neighbors in the periodic system of the elements, Cu(II) and Cd(II), and their antimicrobial activity. *Appl. Organomet. Chem.* **2019**, *33*, e4905. [CrossRef]
188. Staneva, D.; Vasileva-Tonkova, E.; Grabchev, I. A New Bioactive Complex between Zn(II) and a Fluorescent Symmetrical Benzanthrone Tripod for an Antibacterial Textile. *Materials* **2019**, *12*, 3473. [CrossRef]
189. Bahojb Noruzi, E.; Shaabani, B.; Geremia, S.; Hickey, N.; Nitti, P.; Kafil, H.S. Synthesis, Crystal Structure, and Biological Activity of a Multidentate Calix[4]arene Ligand Doubly Functionalized by 2-Hydroxybenzeledene-Thiosemicarbazone. *Molecules* **2020**, *25*, 370. [CrossRef]
190. Montazerozohori, M.; Nasr-Esfahani, M.; Hoseinpour, M.; Naghiha, A.; Zahedi, S. Synthesis of some new antibacterial active cadmium and mercury complexes of 4-(3-(2-(4-(dimethyl aminophenyl allylidene aminopropyl-imino)prop-1-ethyl)-N,N-dimethyl benzene amine. *Chem. Spec. Bioavailab.* **2014**, *26*, 240–248. [CrossRef]
191. Agertt, V.A.; Bonez, P.C.; Rossi, G.G.; Flores, V.d.C.; Siqueira, F.d.S.; Mizdal, C.R.; Marques, L.L.; de Oliveira, G.N.M.; de Campos, M.M.A. Identification of antimicrobial activity among new sulfonamide metal complexes for combating rapidly growing mycobacteria. *BioMetals* **2016**, *29*, 807–816. [CrossRef]
192. Matiadis, D.; Tsironis, D.; Stefanou, V.; Elliott, A.G.; Kordatos, K.; Zahariou, G.; Ioannidis, N.; McKee, V.; Panagiotopoulou, A.; Igglessi-Markopoulou, O.; et al. Synthesis, characterization and antimicrobial activity of N-acetyl-3-acetyl-5-benzylidene tetramic acid-metal complexes. X-ray analysis and identification of the Cd(II) complex as a potent antifungal agent. *J. Inorg. Biochem.* **2019**, *194*, 65–73. [CrossRef]
193. Mandal, S.; Mondal, M.; Biswas, J.K.; Cordes, D.B.; Slawin, A.M.Z.; Butcher, R.J.; Saha, M.; Chandra Saha, N. Synthesis, characterization and antimicrobial activity of some nickel, cadmium and mercury complexes of 5-methyl pyrazole-3yl-N-(2′-methylthiophenyl) methyleneimine, (MPzOATA) ligand. *J. Mol. Struct.* **2018**, *1152*, 189–198. [CrossRef]

194. Mandal, S.; Das, M.; Das, P.; Samanta, A.; Butcher, R.J.; Saha, M.; Alswaidan, I.A.; Rhyman, L.; Ramasami, P.; Saha, N.C. Synthesis, characterization, DFT and antimicrobial studies of transition metal ion complexes of a new schiff base ligand, 5-methylpyrazole-3yl-N-(2′-hydroxyphenylamine)methyleneimine, (MPzOAP). *J. Mol. Struct.* **2019**, *1178*, 100–111. [CrossRef]
195. Lam, P.L.; Lu, G.L.; Choi, K.H.; Lin, Z.; Kok, S.H.L.; Lee, K.K.H.; Lam, K.H.; Li, H.; Gambari, R.; Bian, Z.X.; et al. Antimicrobial and toxicological evaluations of binuclear mercury(II) bis(alkynyl) complexes containing oligothiophenes and bithiazoles. *RSC Adv.* **2016**, *6*, 16736–16744. [CrossRef]
196. Wang, H.; Qian, X.; Wang, K.; Su, M.; Haoyang, W.-W.; Jiang, X.; Brzozowski, R.; Wang, M.; Gao, X.; Li, Y.; et al. Supramolecular Kandinsky circles with high antibacterial activity. *Nat. Commun.* **2018**, *9*, 1815. [CrossRef] [PubMed]

© 2020 by the authors. Licensee MDPI, Basel, Switzerland. This article is an open access article distributed under the terms and conditions of the Creative Commons Attribution (CC BY) license (http://creativecommons.org/licenses/by/4.0/).

Article

Topological Dynamics of a Radical Ion Pair: Experimental and Computational Assessment at the Relevant Nanosecond Timescale [†]

Helmut Quast [1,‡], Georg Gescheidt [2,*] and Martin Spichty [3,*,§]

1. Institute of Organic Chemistry, University of Würzburg, Am Hubland, 97074 Würzburg, Germany; hquast@uni-osnabrueck.de
2. Institute of Physical and Theoretical Chemistry, Technical University of Graz, NAWI Graz, Stremayrgasse 9, 8010 Graz, Austria
3. Laboratoire de Biologie et Modélisation de la Cellule, Ecole Normale Supérieure de Lyon, CNRS, Université Lyon 1, Université de Lyon, 46 allée d'Italie, 69364 Lyon CEDEX 07, France
* Correspondence: g.gescheidt-demner@tugraz.at (G.G.); martin.spichty@uha.fr (M.S.); Tel.: +43-316-873 32220 (G.G.); +33-389-336862 (M.S.)
† Dedicated to Professor Bernd Giese on behalf of his 80th birthday.
‡ Current address: Hoetgerstrasse 10, 49080 Osnabrück, Germany
§ Current address: Laboratoire d'Innovation Moléculaire et Applications, Site de Mulhouse – IRJBD, 3 bis rue Alfred Werner, 68057 Mulhouse Cedex, France

Received: 28 February 2020; Accepted: 27 March 2020; Published: 31 March 2020

Abstract: Chemical processes mostly happen in fluid environments where reaction partners encounter via diffusion. The bimolecular encounters take place at a nanosecond time scale. The chemical environment (e.g., solvent molecules, (counter)ions) has a decisive influence on the reactivity as it determines the contact time between two molecules and affects the energetics. For understanding reactivity at an atomic level and at the appropriate dynamic time scale, it is crucial to combine matching experimental and theoretical data. Here, we have utilized all-atom molecular-dynamics simulations for accessing the key time scale (nanoseconds) using a QM/MM-Hamiltonian. Ion pairs consisting of a radical ion and its counterion are ideal systems to assess the theoretical predictions because they reflect dynamics at an appropriate time scale when studied by temperature-dependent EPR spectroscopy. We have investigated a diketone radical anion with its tetra-ethylammonium counterion. We have established a funnel-like transition path connecting two (equivalent) complexation sites. The agreement between the molecular-dynamics simulation and the experimental data presents a new paradigm for ion–ion interactions. This study exemplarily demonstrates the impact of the molecular environment on the topological states of reaction intermediates and how these states can be consistently elucidated through the combination of theory and experiment. We anticipate that our findings will contribute to the prediction of bimolecular transformations in the condensed phase with relevance to chemical synthesis, polymers, and biological activity.

Keywords: ion pairing; radical anion; kinetics; thermodynamics; molecular dynamics; QM/MM; EPR

1. Introduction

Photo-induced and organic electrochemical procedures have experienced a substantial push in recent times. These methodologies generally involve the formation of charged species. Here, ion pairing plays an important role in solution, as indicated in recent publications [1–4]. The important role of ion pairing extends in fields like catalysis [5], organic synthesis [6–8], and the chemistry of flavonoids [9,10] and radical-induced DNA-strand breaks [11]. These charge-separated reaction intermediates are often decisive for the character and efficiency of the follow-up reactions and product formation. The question is, how the corresponding ion-pairing phenomena impact chemical reactivity [12]. Ion pairs possess highly dynamic topologies in the fluid phase. The association between charged species alters between ps and µs, reflecting decisive time scales of chemical reactivity. Especially photo-induced reactions involve intermediates with particularly short lifetimes that overlap with the time scale of topological dynamics.

The aim of our work is to validate an approach that has the potential to serve as a predictive tool for assessing the time-dependent topologies of ion pairs at the atomic level. This should provide unique insights, which can be translated into strategies for the optimization of reactions involving charge-separated states. The requirement for such an approach is providing reliable parameters for the structural, electronic, and kinetic properties of ion pairs. Beside the thermodynamics (energies) of its constituents, the solvation sphere and the time profile of the interacting species have to be determined.

All-atom molecular-dynamics simulations provide the required structural and kinetic resolution, which has to be linked to experimental data. Here, the dynamic hyperfine time scale of electron paramagnetic resonance (EPR) ranges from MHz to a few GHz (µs–ns), matching the required kinetic regime. For testing our approach, we chose diketone 1 ((3aS,6aS)-3a,6a-dimethyl-3a,6a-dihydropentalene-1,4-dione, Scheme 1), possessing two identical non-conjugated 2-en-1-one moieties. It forms sufficiently persistent radical anions (upon electrolytic reduction). With a counterion such as tetraethylammonium (Et$_4$N$^+$) encounters $1^{\bullet-}$, it can bind to either of the two equivalent oxygen atoms (Scheme 1). There is a dynamic equilibrium between these two states, leading to fluctuations of the spin distribution, which can be conveniently followed by variable-temperature EPR. The radical nature of the ion pair requires time-consuming quantum-mechanical (QM) calculations. In addition, the solvent and the dynamic nature of the entire system have to be regarded at the nanosecond time scale. These requirements are matched by QM/MM-based molecular dynamics simulations [13–15] (Figure 1). To access the nanosecond time-scale, we use a semi-empirical method for the quantum mechanical (QM) part and a force field method for the molecular mechanical (MM) part. As both methods rely on parameterization, it is essential to verify the validity of the chosen QM/MM-Hamiltonian. In these terms, the relatively small size of $1^{\bullet-}$ makes it an ideal paradigm for linking the EPR experiment and all-atom molecular-dynamics simulations. This enables us to validate atomistic predictions from the theory versus macroscopic data from the experiment.

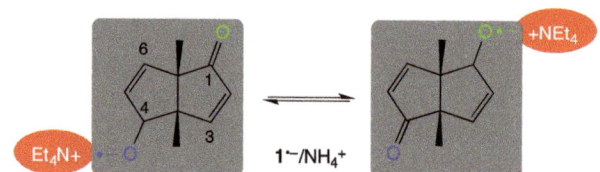

Scheme 1. Electron delocalization after electrolytic reduction of diketone **1**.

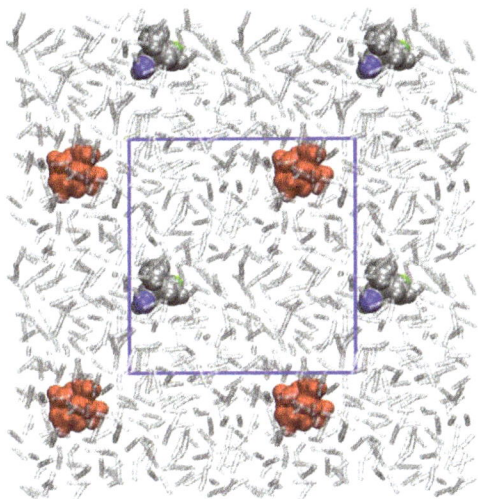

Figure 1. Molecular setup for the QM/MM-simulations: **1**$^{\bullet-}$ (grey space-filling model, oxygen atoms in green and blue) forms the QM part, Et$_4$N$^+$ (red) and acetonitrile (512 molecules, transparent) forms the MM part. The molality of the counterion is 0.01 m. Periodic boundary conditions apply to the simulation cell (length: 32 Å). The borders of the cell are shown (blue lines) as well as parts of the neighboring replicas.

2. Materials and Methods

2.1. EPR Experiments

EPR spectra were performed on a Varian E9 spectrometer (T$_{102}$ cavity) equipped with a variable temperature controller. The radical anions were generated on an amalgamated helical Au electrode (counterelectrode; Pt wire), according to the design by Allendoerfer et al. [16]. The voltage was gradually (200 mV steps) increased until an EPR spectrum was detected. Electrolysis was performed in acetonitrile as the solvent and tetraethylammonium perchlorate was used as the supporting salt (0.01 m). The simulations were performed using EasySpin [17,18]. Life-time broadening [19–21] owing to dynamic chemical exchange is taken into account with a two-jump model, as described in the literature [18,22]. Here, the slow-exchange limit is characterized by the two protons H–C(3) and H–C(6) having distinctly differing isotropic hyperfine coupling constants (hfcs). Whereas one hfc is set to 0 mT, the other has an hfc of 1.14 mT. Accordingly, in the case of fast exchange, both of these protons become formally equivalent displaying the averaged hfc of 0.57 mT (2 equiv. H, see "Results").

2.2. Computer Simulations

2.2.1. Software

The QM/MM-based molecular dynamics simulations were carried out with the program CHARMM [23]. The analysis of the data was performed with the program WORDOM [24] and the python tool set Pynomarix [25]. Molecular visualizations were done with VMD [26].

2.2.2. Molecular Dynamics Simulations

The simulation setup is displayed in Figure 1. The QM/MM-Hamiltonian treated the radical anion with the semi-empirical SCC-DFTB method [27,28] and the remaining environment with CHARMM's generalized force field CGenFF36 [29]. A cutoff value of 12 Ang was used for the short-range non-bonded interactions, and the long-range interactions were treated by Particle-Mesh Ewald. For attaining computational predictions, two independent simulations at 288 K and 203 K (according to the EPR experiments) were performed. The equation of motion was integrated with a time step of 0.5 fs (no SHAKE constraints were applied to bonds). The non-boned list was updated when needed (using CHARMM's heuristic updating algorithm). Constant pressure (1 atm) and temperature were maintained with Langevin piston masses of 500 ps^{-1} and 1000 ps^{-1}, respectively. To achieve the effects based on diffusive encounters, the simulation was run for 150 ns. Snapshots were saved along the trajectory at every 100 fs. The subsequent quantitative analyses were performed only for the simulation at the higher temperature on the last 145 ns of the simulation (5 ns served as equilibration). The simulation at the lower temperature served for qualitatively confirming the predictions from the high temperature molecular dynamics (MD) simulations.

2.2.3. Thermodynamic Analysis

The 1,450,000 snapshots from the MD trajectory at 288 K were overlaid to minimize the root-mean-square deviation between the coordinates of $1^{\bullet-}$. The resulting distribution of counterion positions (nitrogen atoms only) was then symmetrized and binned with a small grid size of 0.25 Å into two dimensions (d_1 and d_2). The population of the bins was translated into the free-energy surface using the usual Boltzmann weighting. Lowering the bin size to 0.125 Ang leads to the same free energy barriers for the relevant transition state. A converged free energy profile is obtained for a simulation length of at least 100 ns (Supplementary Table S1).

2.2.4. Kinetic Analysis

A kinetic network analysis was performed for the MD trajectory at 288 K using a Markov clustering (MCL) algorithm [30]; for a detailed description of this approach, see Gfeller et al. [31]. Here, only the essence of the analysis is given; the accessible topological configuration space of the system is discretized into cells based on measurable order parameters. Every snapshot of the trajectory is assigned to one of the cells. Each cell is considered as a microstate of a Markov model. Transition probabilities for going from one microstate to another for a specific lag time are calculated from the sequence of microstates on the MD trajectory. The lag time should be sufficiently large to ensure Markovian conditions, but small enough to resolve the relevant time scale. The M groups microstates into nodes of a network depending on the chosen granularity parameter. Mean first passage times for the connection between two nodes can then be extracted for this reduced network using a Brownian walker.

For our system, we discretized the configurational space with three order parameters. Beside the two obvious order parameters d_1 and d_2, we used the dihedral angle N(Et$_4$N$^+$)-O(green)-O(blue)-M, where M is the midpoint of the two methyl carbon atoms of the radical anion. With these three order parameters the position of the counterion center relative to the radical anion is unambiguously defined. For d_1 and d_2, we used a cutoff of 15 Ang, that is, all distances beyond this value were considered the same as the cutoff value. This was done to reduce the number of microstates describing the unbound states. The distances were discretized into 10 bins, and the dihedral angle into 36 bins so that the 1,450,000 snapshots were distributed into 3600 cells (microstates). Transition probabilities were calculated for a lag time of 1 ps. We also tested values of 0.1 ps and 10 ps, yielding essentially the same result for the granularity level of interest (see below). Using the MCL, the number of nodes was monitored as a function of the granularity parameter (Supplementary Figure S1A). We see the formation of two plateaus at granularity values of 1.2 and 1.3 with three and seven nodes, respectively. Mean first passage times (mfpt) were extracted using a Brownian walker simulation of 10^7 steps. The smallest mpft of the lower-resolved network (granularity value of 1.2) is already significantly smaller than the time resolution of the EPR experiments, so we present in the main text only the result for the network with three nodes. Supplementary Figure S1B presents the properties of the higher-resolved network with seven nodes.

The rotational correlation time of the radical anion was determined by calculating the time correlation functions for three vectors corresponding to the carbon bonds tert-C–C(3), tert-C–C(4), and tert-C-C(Methyl) using CHARMM's CORFUN module. Fitting the correlation functions with a bi-exponential function yields a fast and slow correlation time for each vector. The mean value of the slow correlation time is given as rotational correlation time in the section "Results".

3. Results

3.1. Temperature-Dependent EPR Experiments

The electrolytic generation of **1**$^{\bullet-}$ allows precisely controlling the character and the concentration of counterions for EPR experiments. With respect to the radical species, the counterions are in large excess, thus bimolecular encounters between radical species and counterions follow pseudo-first order kinetics [32]. The temperature-dependent dynamic exchange (two-jump model) in solution between the two chemically identical topologies forms the ground of this study. It is analyzed in terms of line-shape analysis based on Redfield theory [33].

Cathodic reduction of **1** in acetonitrile afforded EPR spectra that are sufficiently resolved to allow a clear cut-distinction of the corresponding species. At high temperatures, the dominating pattern is triplet-like (somehow deviating from the 1:2:1 ratio), which is further split into multiplets (Figure 2A, right). This EPR signal is produced by the coupling with the unpaired electron with three sets of ^{1}H nuclei. The largest hfc of 0.57 mT is attributed to the two protons at the 3 and 6 position, whereas the substantially smaller hfcs of 0.04 and 0.02 mT are assigned to the protons at C(2)/C(5) and the 3a/6a methyl groups, respectively.

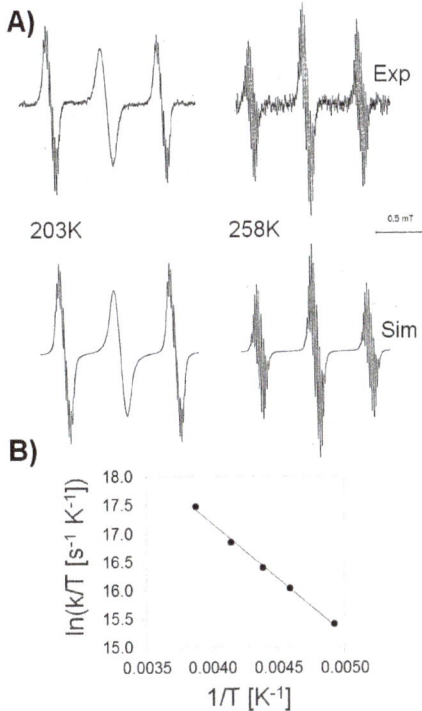

Figure 2. (**A**) EPR spectra obtained upon electrolytic reduction of **1** taken at 203 and 258 K (for additional temperatures, see Supplementary Material). (**B**) Eyring–Polanyi plot for the following values of k (in MHz): 1000 (203 K), 2000 (218 K), 3000 (228 K), 5000 (242 K), and 10,000 (258 K). Linear regression yields a slope of −1939 K and an intercept of 24.92.

The intensity distribution of the EPR lines is strongly dependent on the temperature. The triplet almost reaches a 1:2:1 intensity ratio at 258 K. However, it reversibly alters with lowering the temperature. At 203 K (before the solution freezes), the central line becomes markedly broadened (Figure 2A, left). These alternating line widths can be traced back to electron localization/delocalization phenomena between the two equivalent, non-conjugated 2-en-1-one moieties. At high temperatures, the exchange between the two topologies of Scheme 1 is faster than the time resolution of the experiment, thus the two 2-en-1-one moieties feature identical spectral properties (electron delocalization). At low temperatures, however, the exchange rate reaches the EPR time scale and the electron localization starts being detectable. Using a two-jump model for electron delocalization (see Methods), the experimental EPR spectra can be simulated. An Eyring–Polanyi plot (Figure 2B) of $\ln(k/T)$ (with $k = 1/\tau$, where τ is the lifetime of the localized state) versus $1/T$ yields an activation enthalpy of 3.8 kcal mol^{-1} and an activation entropy of 2.3 cal mol^{-1} K^{-1}, matching those for similar previously-studied systems [32]. In that study, electron delocalization was explained by a mechanism of counterion decomplexation from one side and recomplexation at the other side. The positive activation entropy was attributed to the increased flexibility of the two ions in the intermediate, unbound state. Such an interpretation neglects, however, the decrease of the solvent entropy upon dissociation of ionic particles.

The calculated rotational correlation times at 288 K and 203 K (as determined from the MD simulation, see below) are 2.5 and 19.6 ps, respectively, and thereby 100 times faster than the chemical exchange (see below) and significantly faster than the EPR time scale. This is in line with the observation that inhomogeneous line broadening owing to the lowering of the rotational motion is hardly visible in the experimental EPR spectra.

3.2. QM/MM-Based Molecular Dynamics Simulations

Our computational model treats **1**$^{•-}$ semi-empirically (QM part), while the counterion Et4N$^+$ and the solvent molecules (acetonitrile) are controlled by a force field (MM part, Figure 1). We performed the molecular dynamic simulations at two temperatures. The simulation at 203 K matches the lowest experimentally achievable temperature where exchange phenomena could be observed. At this low temperature, only a few exchange events can be expected in an affordable simulation time (i.e., 150 ns with our QM/MM-approach), which limits any quantitative analyses. Therefore, we decided to contrast this setup with a simulation temperature where sufficient statistics can be obtained for exchange events. From the experimental spectra at 258 K, we conclude that this temperature is still not high enough to reach the fast-exchange limit where both molecular moieties should appear as being equivalent at the EPR hyperfine time scale, representing a 1:2:1 ratio in the dominating triplet (see above). To reach the fast-exchange limit (not achievable experimentally), we thus set the high temperature to 288 K.

We first analyzed the dynamics of the charge distribution in radical anion **1**$^{•-}$ by monitoring the fluctuations of the atomic charge at the two oxygen atoms (Figure 3). The time scale of these charge fluctuations is distinctly different at 203 K and 288 K. This is clearly reflected by the thick lines in Figure 3A representing the time evolution of the charges at the two oxygen atoms when averaged over a window of 20 ns ($\approx$ time resolution limit of the EPR experiment). At 203 K, charge exchange can well be detected (i.e., the thick blue and green lines form mirrored minima and maxima). This is in agreement with the EPR experiment that displayed line broadening at 203 K.

At 288 K, the charges are equally distributed at the 20 ns time scale (i.e., the two thick lines are flat), well in line with the assumption that, at this temperature, the fast exchange limit is reached. Figure 3B shows the oxygen charges and the oxygen–counterion distance at ten times higher time resolution. Here, it can be clearly seen that the charge fluctuations correlate with the binding and unbinding of the counter ion to **1**$^{•-}$. When the counterion is not bound to the radical anion (both oxygen/counterion distances are larger than 10 Å, see snapshot a), the atomic charge is identical for both oxygen atoms. Binding of the counterion to one of the oxygen atoms (snapshots b and c) polarizes the charge distribution; the oxygen atom in contact with the counterion lowers its charge by about 0.04 e, and the distant oxygen atom slightly increases its charge by 0.01 e.

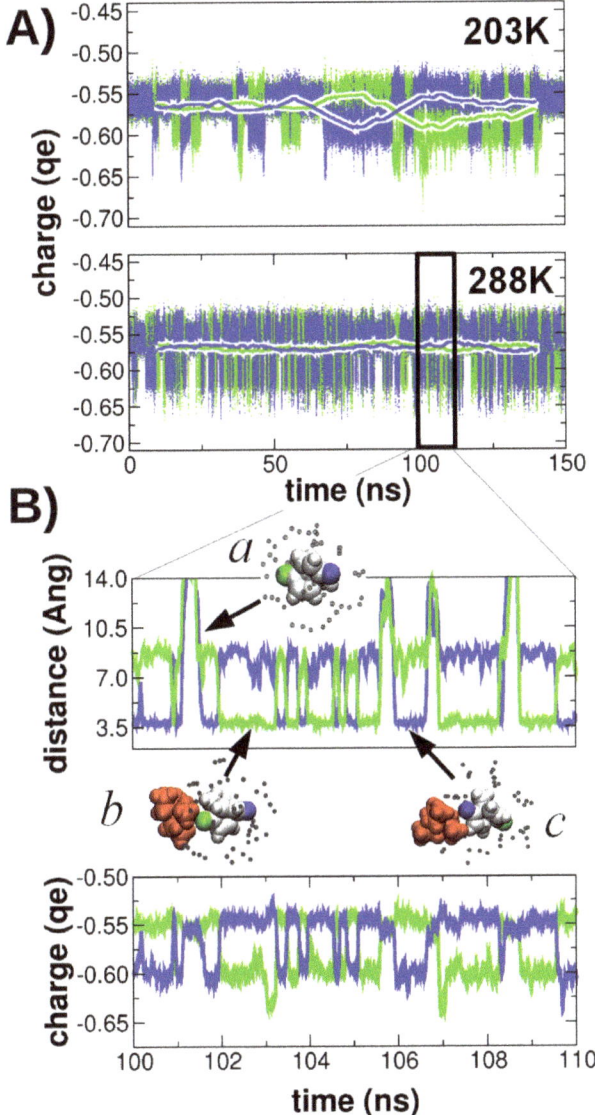

Figure 3. (**A**) Time evolution of the charge on the two oxygen atoms (shown in blue and green) of **1**[•−] at 203 K and 288 K. Running averages are shown for windows of 1 ps (dots) and 20 ns (thick lines). (**B**) Enlargement of the cutout region (as indicated in subfigure (**A**)) for the evolution of the charge as well as the distance between the nitrogen atom of the cation and the oxygen atoms. Structures are depicted (counterion in red, solvent nitrogen in grey, radical anion in white, with the two oxygens colored in blue and green) for selective snapshots (*a*,*b*,*c*).

It is remarkable that the transition between snapshots *b* and *c* (with the counterion bound once to each of the two "distinct" oxygen atoms) occurs without the unbinding of the counterion from the solvation shell (i.e., the counterion "slides" on the surface of **1**[•−], see Supplementary Video S1). It even appears that the solvent-caged transitions are more probable than the unbinding and re-binding pathway involving penetration of the solvation shell.

Figure 4 visualizes the probability distribution of the counterion position (central nitrogen atom) relative to **1**$^{\bullet-}$, revealing a considerable topological flexibility. There are six topological states; three for each complexation site of the radical anion (1a–c and 2a–c). The transitions between the three states at the same complexation site (i.e., within 1a–c or within 2a–c) are very rapid (<100 ps), so we can consider them as a single state on the EPR time scale (nanoseconds); we group them into states 1 and 2, respectively. Snapshots b and c (Figure 3B) illustrate members of states 1 and 2. There is a funnel-like solvent-caged transition path for the counterion to pass from one complexation site to the other one; the transition state is marked by an asterisk in Figure 4A. This transition state is 3.2 kcal/mol higher in free energy than the basins 1 and 2 (Figure 4B). It is significantly lower in free energy than the transition state of the decomplexation path (4.8 kcal mol^{-1}).

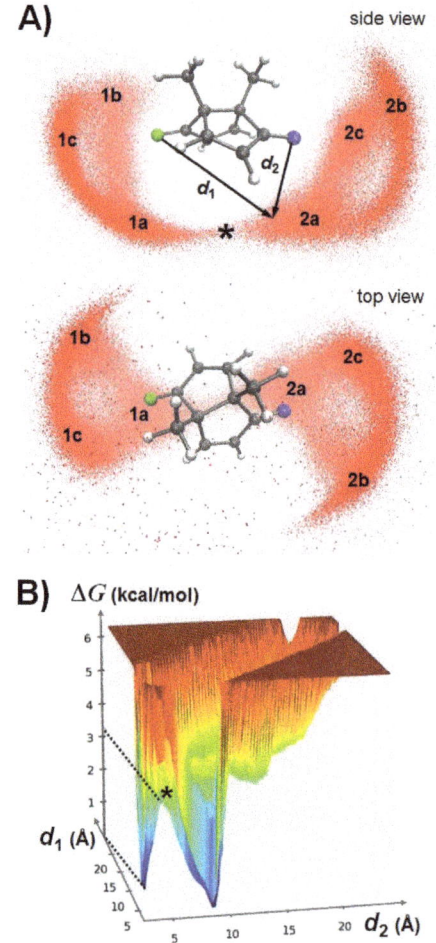

Figure 4. (**A**) Counterion distribution around **1**$^{\bullet-}$. The image shows **1**$^{\bullet-}$ in a ball-and-stick representation and the relative position of the counterion (nitrogen atom) as red dots. States 1a–c and 2a–c are indicated as well as the the location of the transition state (*) for the counterion exchange between the two complexation sites (oxygen atoms) of **1**$^{\bullet-}$. For better visibility, a rotating movie is provided as Supplementary Video S2. (**B**) Free-energy surface for the complexation of NEt$_4^+$ to **1**$^{\bullet-}$. The relative free energy is projected on the distance order parameters d_1 and d_2 (as indicated in (**A**)).

The kinetics connecting the topological states was extracted from a kinetic network analysis of the MD simulation at 288 K using a Markov clustering algorithm (see Methods for more details). Accordingly, the kinetics can be described with a network of three nodes (Figure 5A): two nodes correspond to the topological states 1 and 2 (with complexed counterion), and the remaining cluster describes the unbound state 0. The unbound state is about five times less populated than the two complexed states. Thus, at the simulated salt concentration (which matches the experimental setup, 0.01 m), the EPR spectrum is dominated by states 1 and 2. The exchange between these two states can occur either through a fast or a slow route: the solvent-caged direct path features a mean first passage time (mfpt) of only 270 ps. The indirect path via decomplexation (mfpt: 3.6 ns) and recomplexation (160 ps) is more than ten times slower. This reflects the free energy difference of the transition states (3.2 vs. 4.8 kcal mol^{-1}).

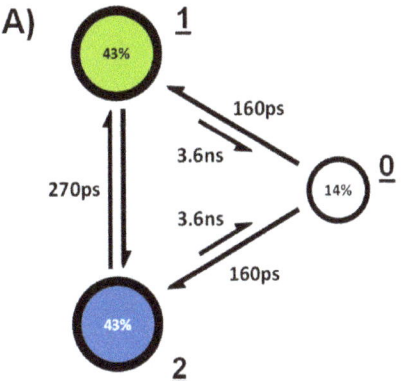

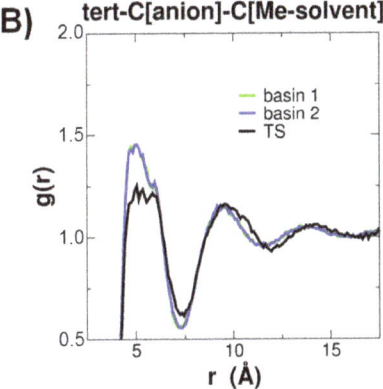

Figure 5. (A) Kinetic network extracted from the analysis of the molecular dynamics (MD) simulation at 288 K. The population of the nodes (in %), the mean first passage times (in ps and ns), and the corresponding topological states (underlined numbers) are indicated. (B) Radial distribution functions of the solvent–radical anion interaction for nodes 1 and 2 and the transition state (TS) (as indicated in Figure 4).

4. Discussion

From the molecular dynamics simulations, it can be observed that, for time scales larger than 100 ps, there are three topological states for the complexation between EtN$_4$$^+$ and 1$^{•-}$. The most populated states are those where the counterion is directly bound to one of the two oxygen atoms. The fastest transition between those two states occurs via a solvent-caged transition state. The observation that mainly only two states (1 and 2) are populated and that the exchange between these two states occurs predominantly by a single path justifies the use of a two-jump model for the analyses of the experimental spectra (see above). More importantly, the extracted thermodynamic data from the experiment can be directly associated to the solvent-caged exchange path as determined by the simulation. With the experimental values of the activation enthalpy and entropy ($\Delta H^{\#} = 3.8$ kcal mol^{-1}, $\Delta S^{\#} = 2.3$ cal mol^{-1} K^{-1}), a free energy of activation of 3.1 kcal mol^{-1} can be extrapolated to the simulation temperature at 288 K, which is in excellent agreement with the observed value (3.2 kcal mol^{-1}). The positive value of the activation entropy might seem counterintuitive for a funnel-like transition path (which lowers the degrees of freedom of the counterion). We argue that the positive sign of the activation entropy is a result of the increased flexibility of the solvent: in the transition state, the ion pair features a smaller surface and thus has fewer solvent molecules bound than in states 1 and 2, especially the counterion and the side opposite to the methyl groups of the radical anion (i.e., close to the tert-carbon atoms) are lesser solvated in the transition state (see Figure 5B).

With Eyring's equation [34], we find that the decrease of the temperature from 288 K to 203 K decreases the exchange rate by a factor of about 15. Thus, the passage time for the solvent-caged transition should increase from 270 ps at 288 K to about 4 ns at 203 K. Indeed, the simulation at 203 K reveals transition times on this time scale (Figure 3A) in agreement with the corresponding EPR line broadening upon lowering the temperature.

It is tempting to discuss the formal charge exchange between the two 2-en-1-one moieties of radical anion $1^{\bullet-}$ in terms of an intramolecular electron transfer, especially owing to laureate's milestone contributions in this field [35,36]. In our case, the transfer is between a donor and acceptor state with identical free energy. The rate-limiting step is the reorganization of the chemical environment (counterion). Marcus' theory [37] would yield a reorganization energy λ of about 12.5 kcal mol^{-1} ($\Delta G^{\#} = 3.1$ kcal mol$^{-1} = \lambda/4$ because $\Delta G = 0$). However, electron transfer theories rely on the fact that the transfer rate is controlled by vibronically coupled events, and that there are now slow gating processes that are rate-limiting. The latter applies to our case though. The negative charge is gated by the solvent-caged exchange of the positively charged counter ion. Thus, the formal charge transfer between the two 2-en-1-one moieties is probably best described by conventional transition state theory (Eyring's equation), as is done above.

In conclusion, we have shown a proof of the concept that all-atom molecular dynamics simulations, when validated against experimental data, provide unique insights into the structural, thermodynamic, and kinetic characteristics of topological processes in ion pairs. Our results indicate that even a simple system like $1^{\bullet-}$/Et$_4$N$^+$ shows an unexpectedly rich variety of dynamically interchanging molecular topologies with distinct electronic properties. Their lifetimes vary between picosecond and nanoseconds. In reactions involving ion pairs, various reaction pathways become imaginable—it depends on the kinetics of follow-up reactions. Their course can potentially be controlled by the concentrations of substrates. For more complex charged molecules, the structures of ion/counterion arrangements very likely become more complex, but, more importantly, they may achieve substantially longer lifetimes. This enhances the reaction energy hypersurface and increases the palette of feasible reaction pathways. In these terms, our results suggest that just changing the concentrations of reactants in reactions involving ion-paired species may lead to the observation of unexpected products.

Supplementary Materials: The following are available online at http://www.mdpi.com/2624-8549/2/2/219\T1\textendash230/s1, Table S1: Dependence of the free-energy profile on the simulation time; Figure S1: Kinetic network with seven nodes; Figure S2–S6: EPR spectra and their simulation at five different temperatures; Video S1: Transition path for counterion exchange; Video S2: Rotating 3D counterion distribution.

Author Contributions: Conceptualization, G.G. and M.S.; experiments, H.Q.; computer simulations, M.S.; analysis of experiments, G.G.; analyses of computer simulations, M.S., writing, G.G. and M.S.; visualization M.S.; funding acquisition, G.G and M.S. All authors have read and agreed to the published version of the manuscript.

Funding: This research was funded by the Grand Equipement National de Calcul Intensif (GENCI), DARI grant number x2015077214, and the interuniversity program in natural sciences, NAWI Graz.

Acknowledgments: M.S. thanks the Pôle Scientifique de Modélisation Numérique (PSMN) of the ENS Lyon for computer time and Francesco Rao for help in using the software Pynomarix. G.G. thanks NAWI Graz for support.

Conflicts of Interest: The authors declare no conflict of interest. The funders had no role in the design of the study; in the collection, analyses, or interpretation of data; in the writing of the manuscript, or in the decision to publish the results.

References

1. Rockl, J.L.; Pollok, D.; Franke, R.; Waldvogel, S.R. A Decade of Electrochemical Dehydrogenative C,C-Coupling of Aryls. *Acc. Chem. Res.* **2020**, *53*, 45–61. [CrossRef] [PubMed]
2. Farney, E.P.; Chapman, S.J.; Swords, W.B.; Torelli, M.D.; Hamers, R.J.; Yoon, T.P. Discovery and Elucidation of Counteranion Dependence in Photoredox Catalysis. *J. Am. Chem. Soc.* **2019**, *141*, 6385–6391. [CrossRef] [PubMed]
3. Chiarotto, I.; Mattiello, L.; Feroci, M. The Electrogenerated Cyanomethyl Anion: An Old Base Still Smart. *Acc. Chem. Res.* **2019**, *52*, 3297–3308. [CrossRef] [PubMed]
4. Li, G.C.; Brady, M.D.; Meyer, G.J. Visible Light Driven Bromide Oxidation and Ligand Substitution Photochemistry of a Ru Diimine Complex. *J. Am. Chem. Soc.* **2018**, *140*, 5447–5456. [CrossRef] [PubMed]
5. Jiang, C.; Chen, W.; Zheng, W.-H.; Lu, H. Advances in asymmetric visible-light photocatalysis, 2015–2019. *Org. Biomol. Chem.* **2019**, *17*, 8673–8689. [CrossRef] [PubMed]
6. Morack, T.; Muck-Lichtenfeld, C.; Gilmour, R. Bioinspired Radical Stetter Reaction: Radical Umpolung Enabled by Ion-Pair Photocatalysis. *Angew. Chem.-Int. Ed.* **2019**, *58*, 1208–1212. [CrossRef]
7. Ravelli, D.; Protti, S.; Fagnoni, M. Carbon-Carbon Bond Forming Reactions via Photogenerated Intermediates. *Chem. Rev.* **2016**, *116*, 9850–9913. [CrossRef]
8. Lim, C.H.; Ryan, M.D.; McCarthy, B.G.; Theriot, J.C.; Sartor, S.M.; Damrauer, N.H.; Musgrave, C.B.; Miyake, G.M. Intramolecular Charge Transfer and Ion Pairing in N,N-Diaryl Dihydrophenazine Photoredox Catalysts for Efficient Organocatalyzed Atom Transfer Radical Polymerization. *J. Am. Chem. Soc.* **2017**, *139*, 348–355. [CrossRef]
9. Zollitsch, T.M.; Jarocha, L.E.; Bialas, C.; Henbest, K.B.; Kodali, G.; Dutton, P.L.; Moser, C.C.; Timmel, C.R.; Hore, P.J.; Mackenzie, S.R. Magnetically Sensitive Radical Photochemistry of Non-natural Flavoproteins. *J. Am. Chem. Soc.* **2018**, *140*, 8705–8713. [CrossRef]
10. Zelenka, J.; Cibulka, R.; Roithova, J. Flavinium Catalysed Photooxidation: Detection and Characterization of Elusive Peroxyflavinium Intermediates. *Angew. Chem.-Int. Ed.* **2019**, *58*, 15412–15420. [CrossRef]
11. Glatthar, R.; Spichty, M.; Gugger, A.; Batra, R.; Damm, W.; Mohr, M.; Zipse, H.; Giese, B. Mechanistic studies in the radical induced DNA strand cleavage-Formation and reactivity of the radical cation intermediate. *Tetrahedron* **2000**, *56*, 4117–4128. [CrossRef]
12. Horibe, T.; Ohmura, S.; Ishihara, K. Structure and Reactivity of Aromatic Radical Cations Generated by FeCl3. *J. Am. Chem. Soc.* **2019**, *141*, 1877–1881. [CrossRef] [PubMed]
13. Karplus, M. Development of Multiscale Models for Complex Chemical Systems: From H+H2 to Biomolecules (Nobel Lecture). *Angew. Chem. Int. Ed.* **2014**, *53*, 9992–10005. [CrossRef] [PubMed]
14. Levitt, M. Birth and Future of Multiscale Modeling for Macromolecular Systems (Nobel Lecture). *Angew. Chem. Int. Ed.* **2014**, *53*, 10006–10018. [CrossRef] [PubMed]
15. Warshel, A. Multiscale Modeling of Biological Functions: From Enzymes to Molecular Machines (Nobel Lecture). *Angew. Chem. Int. Ed.* **2014**, *53*, 10020–10031. [CrossRef]
16. Allendoerfer, R.D.; Martinchek, G.A.; Bruckenstein, S. Simultaneous electrochemical-electron spin resonance measurements with a coaxial microwave cavity. *Anal. Chem.* **1975**, *47*, 890–894. [CrossRef]
17. Stoll, S.; Schweiger, A. EasySpin, a comprehensive software package for spectral simulation and analysis in EPR. *J. Magn. Reson. San Diego Calif 1997* **2006**, *178*, 42–55. [CrossRef]
18. Zalibera, M.; Jalilov, A.S.; Stoll, S.; Guzei, I.A.; Gescheidt, G.; Nelsen, S.F. Monotrimethylene-bridged bis-p-phenylenediamine radical cations and dications: Spin states, conformations, and dynamics. *J. Phys. Chem. A* **2013**, *117*, 1439–1448. [CrossRef]
19. Binsch, G. Unified theory of exchange effects on nuclear magnetic resonance line shapes. *J. Am. Chem. Soc.* **1969**, *91*, 1304–1309. [CrossRef]

20. Heinzer, J. Fast computation of exchange-broadened isotropic E.S.R. spectra. *Mol. Phys.* **1971**, *22*, 167–177. [CrossRef]
21. Gerson, F.; Huber, W. *Electron Spin Resonance Spectroscopy of Organic Radicals*; John Wiley & Sons: Weinheim, Germany, 2003.
22. Spichty, M.; Giese, B.; Matsumoto, A.; Fischer, H.; Gescheidt, G. Conformational dynamics in a methacrylate-derived radical: A computational and EPR study. *Macromolecules* **2001**, *34*, 723–726. [CrossRef]
23. Brooks, B.R.; Brooks III, C.L.; Mackerell, A.D.; Nilsson, L.; Petrella, R.J.; Roux, B.; Won, Y.; Archontis, G.; Bartels, C.; Boresch, S.; et al. CHARMM: The biomolecular simulation program. *J. Comput. Chem.* **2009**, *30*, 1545–1614. [CrossRef] [PubMed]
24. Seeber, M.; Cecchini, M.; Rao, F.; Settanni, G.; Caflisch, A. Wordom: A program for efficient analysis of molecular dynamics simulations. *Bioinformatics* **2007**, *23*, 2625–2627. [CrossRef] [PubMed]
25. Rao, F. PYNORAMIX Is a Fast and Efficient Python Library to Analyze Water Structure and Dynamics. Available online: https://github.com/ruvido/Pynoramix (accessed on 14 May 2016).
26. Humphrey, W.; Dalke, A.; Schulten, K. VMD–Visual Molecular Dynamics. *J. Mol. Graph.* **1996**, *14*, 33–38. [CrossRef]
27. Elstner, M.; Porezag, D.; Jungnickel, G.; Elsner, J.; Haugk, M.; Frauenheim, T.; Suhai, S.; Seifert, G. Self-consistent-charge density-functional tight-binding method for simulations of complex materials properties. *Phys. Rev. B* **1998**, *58*, 7260–7268. [CrossRef]
28. Cui, Q.; Elstner, M.; Kaxiras, E.; Frauenheim, T.; Karplus, M. A QM/MM implementation of the self-consistent charge density functional tight binding (SCC-DFTB) method. *J. Phys. Chem. B* **2001**, *105*, 569–585. [CrossRef]
29. Vanommeslaeghe, K.; Hatcher, E.; Acharya, C.; Kundu, S.; Zhong, S.; Shim, J.; Darian, E.; Guvench, O.; Lopes, P.; Vorobyov, I.; et al. CHARMM General Force Field (CGenFF): A force field for drug-like molecules compatible with the CHARMM all-atom additive biological force fields. *J. Comput. Chem.* **2010**, *31*, 671–690.
30. Van Dongen, S. Graph. Clustering by Flow Simulation. Ph.D. Thesis, University of Utrecht, Utrecht, The Netherlands, 2000.
31. Gfeller, D.; Rios, P.D.L.; Caflisch, A.; Rao, F. Complex network analysis of free-energy landscapes. *Proc. Natl. Acad. Sci. USA* **2007**, *104*, 1817–1822. [CrossRef]
32. Furderer, P.; Gerson, F.; Heinzer, J.; Mazur, S.; Ohyanishiguchi, H.; Schroeder, A.H. Esr and Endor Studies of a Radical-Anion with Formally Nonconjugated Keto Groups - Cis-10,11-Dimethyldiphensuccindan-9,12-Dione. *J. Am. Chem. Soc.* **1979**, *101*, 2275–2281. [CrossRef]
33. Redfield, A.G. On the Theory of Relaxation Processes. *IBM J. Res. Dev.* **1957**, *1*, 19–31. [CrossRef]
34. Eyring, H. The Activated Complex in Chemical Reactions. *J. Chem. Phys.* **1935**, *3*, 107–115. [CrossRef]
35. Giese, B. Electron transfer in DNA. *Curr. Opin. Chem. Biol.* **2002**, *6*, 612–618. [CrossRef]
36. Cordes, M.; Giese, B. Electron transfer in peptides and proteins. *Chem. Soc. Rev.* **2009**, *38*, 892–901. [CrossRef] [PubMed]
37. Marcus, R.A. Electron Transfer Reactions in Chemistry: Theory and Experiment (Nobel Lecture). *Angew. Chem. Intl. Ed.* **1993**, *32*, 1111–1121. [CrossRef]

© 2020 by the authors. Licensee MDPI, Basel, Switzerland. This article is an open access article distributed under the terms and conditions of the Creative Commons Attribution (CC BY) license (http://creativecommons.org/licenses/by/4.0/).

Article

A Procedure for Computing Hydrocarbon Strain Energies Using Computational Group Equivalents, with Application to 66 Molecules [†]

Paul R. Rablen

Department of Chemistry and Biochemistry, Swarthmore College, Swarthmore, PA 19081, USA; prablen1@swarthmore.edu

† Dedication: This paper is dedicated to Professor Bernd Giese on the occasion of his 80th birthday. It was a pleasure and an honor to work with you on the question of electron conduction in peptides a few years ago.

Received: 4 April 2020; Accepted: 27 April 2020; Published: 30 April 2020

Abstract: A method is presented for the direct computation of hydrocarbon strain energies using computational group equivalents. Parameters are provided at several high levels of electronic structure theory: W1BD, G-4, CBS-APNO, CBS-QB3, and M062X/6-31+G(2df,p). As an illustration of the procedure, strain energies are computed for 66 hydrocarbons, most of them highly strained.

Keywords: strain; strain energy; group equivalents; strained hydrocarbons; calculated strain; quadricyclane; cubane; prismane; fenestranes; propellanes; spiroalkanes

1. Introduction

The concept of strain has long held interest for organic chemists, going back all the way to Baeyer [1–5]. Strain refers to the amount by which the energy of a molecule exceeds that which one would expect if all bond lengths, bond angles, and dihedral angles could simultaneously hold their ideal values, and if no repulsive nonbonded interactions (steric repulsions) were present. As such, strain is generally assumed to be absent in molecules such as straight-chain alkanes in which the bond lengths, angles, and dihedral angles are not geometrically constrained, and in which the extended conformation avoids repulsive nonbonded interactions. Small rings, on the other hand, force bond angles to be smaller than ideal, and lead to other nonidealities (such as torsional strain) as well. In highly strained molecules, bond angle and steric strain are almost always the main contributors to the overall strain energy [5]. Syntheses of a wide variety of highly strained compounds have been carried out in ingenious ways, allowing the experimental study of these elusive species. It is frequently of interest to quantify the strain, generally as an energy of some sort, and many approaches exist for doing so [2,3,5–15]. Most of these approaches rely, either explicitly or implicitly, on comparison of the molecular energy to that of a "strain-free" reference system. It is only in describing such a procedure that the somewhat fuzzy concept of strain becomes precisely, if also somewhat arbitrarily, defined.

One straightforward approach is to use isodesmic [16], homodesmotic [14], or group equivalent reactions [17], in which the reactants and products of a hypothetical reaction are paired so to isolate the source of strain from other contributing factors. Thus, for instance, one can design a reaction in which the reactant and product sides have equal numbers of bonds of a given type, and all compounds but the single compound of interest can reasonably be assumed to be free of strain. The energy of the reaction, obtained either by experimental or computational means, can then be associated with the strain. Of course, different levels of exactitude are possible regarding what is meant by "bond type". Wheeler et al. have provided careful and elegant definitions of different orders of homodesmotic reactions that provide progressively more complete definitions of "bond types" and "atom types", and thus, in principle, more precisely defined strain energies [18].

Another approach involves comparing the experimental heat of formation of a given compound to a hypothetical "strain free" value derived from a more general model. These models rely on the additivity of the energies of molecular fragments, a phenomenon that has long been recognized and often used, and that holds remarkably accurately for even quite generic fragments [19–27]. For instance, both Franklin [24] and Benson [19–21] pioneered the notion of "group equivalents" that could be used to estimate the heat of formation for a novel structure, based on patterns in the known experimental data. A simple approach is to use the number of methyl, methylene, methane, and quaternary carbon groups, plus the number of alkene functional groups, to estimate an enthalpy of formation for a hydrocarbon. As the increments are based on data for unstrained compounds, one can define as the strain energy the difference between the actual, experimental enthalpy of formation and the estimate obtained by summing the unstrained increments. Along similar lines, Benson defined a far more extensive set of group equivalents, permitting a more precise prediction of strain-free enthalpy. This procedure is, for instance, presented in a leading advanced organic chemistry textbook [28].

Computational methods for assessing strain follow the same patterns as experimental methods. A common and versatile approach is simply to compute energies for the components of an isodesmic or homodesmotic reaction using electronic structure theory. Another approach is to use computational methods to obtain an enthalpy of formation, which can then be compared to strain-free estimates generated by Franklin's or Benson's methods [29]. A still more direct approach, however, is to use computational group equivalents: that is, to develop group increments that permit the estimation of a strain-free electronic energy, that can then be directly compared to the result of an actual electronic structure calculation for the compound of interest. The intermediate step of predicting an enthalpy of formation is thus avoided. Wiberg [30,31] first used such an approach in 1984, when HF/6-31G(d) represented a fairly high level of calculation, and Schleyer [32] further elaborated the scheme.

This direct computational technique offers several advantages. First, once the group increments for a given calculational level are available, only one electronic structure calculation is required to obtain a strain energy for a new molecule of interest: a calculation of that molecule. That stands in contrast to the isodesmic/homodesmotic approach, in which all components of the reaction must be computed. Perhaps more importantly, the approach is conceptually more direct; it removes the unnecessary intermediate step of estimating an experimental heat of formation from an electronic structure calculation, as well as the additional labor and potential sources of error thereby introduced. Finally, one might argue that chemists are most interested, conceptually speaking, in the strain as defined in the pure essence of an electronic energy, without the complications of thermodynamic factors that affect enthalpies at 298K. In such a sense, the ability to define strain energies in terms of energy/enthalpy at absolute zero (with only the zero-point energy as a thermodynamic correction), and in the absence of medium effects, is perhaps a conceptual advantage.

The computational group equivalent approach first explored by Wiberg is thus a valuable one. However, the original version involves electronic structure methods that are suboptimal by today's standards (HF/6-31G(d)), as well as a very simple and thus somewhat limited definition of the strain-free reference. Here, the approach is updated and expanded in two ways. First, a much wider variety of group equivalents is used, following the approach of Benson rather than of Franklin, permitting both a wider variety of hydrocarbons to be considered, and also providing a somewhat more precisely calibrated definition of the strain-free reference than is possible using more limited definitions. Second, the approach is modernized by using highly accurate compound procedures of the type available and routinely used today: W1BD [33], G-4 [34], CBS-APNO [35], and CBS-QB3 [36,37], as well as a modern density functional method, M062X/6-31+G(2df,p) [38], that was found in a previous study to offer results in generally good accord with the aforementioned multi-component procedures [39].

2. Materials and Methods

All calculations were carried out using either G09 [40] or G16 [41]. For geometry optimization, force constants were calculated analytically and tight convergence criteria were used (fopt = (calcfc, tight)).

Structures were verified as minima on the potential energy surface via calculation of second derivatives (frequency calculation). Thermodynamic corrections for enthalpy at 0 and 298 K were obtained using the frequency calculations, without empirical scaling. The compound methods (W1BD [33], G-4 [34], CBS-APNO [35], and CBS-QB3 [36,37]) were carried out using the corresponding keywords. The latter methods were chosen as they represent some of the most accurate, reliable, and extensively validated electronic structure methods available for calculating the energies of small- to medium-sized organic molecules. The DFT approach using M062X/6-31G(2df,p) [38], on the other hand, represents a much more economical but also popular approach, that was found previously to compare well to the more expensive compound methods [39].

Calculations were carried out on the molecules shown in Figures 1–3 using all methods, with the exception of a few of the largest molecules for which W1BD was impractical. The structures in Figure 1 were used to define the group equivalents. They were chosen for this purpose because they are the smallest and simplest structures that contain the requisite atom types, and because they are expected to be free of strain, or at least as free of strain as possible while having the necessary structural characteristics.

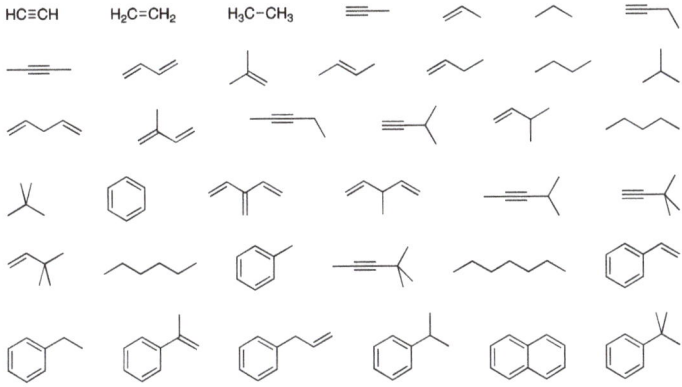

Figure 1. Compounds used to define group increments.

3. Results

Table 1 lists increments in the calculated electronic energy for a methylene group on going progressively from ethane to octane. The increments are highly consistent: they vary by just a few tenths of a millihartree. However, there is a perceptible alternation in the numbers; e.g., the W1BD value is −39.29347 ± 0.00001 on going from an even to an odd chain, but −39.29359 ± 0.00000 on going from an odd to an even chain. By taking (heptane − propane)/4 to define methylene, we attempt to average out this alternation. More generally, however, the high degree of constancy of the increments lends credence to the approach of adding together largely context-independent group increment energies to obtain a strain-free reference energy for a molecule.

Table 1. Calculated group increments for methylene (electronic energy plus ZPE) (hartrees).

Increment	W1BD	G4	APNO [a]	CBS-QB3	M062X [b]
ethane → propane	−39.29346	−39.27659	−39.28273	−39.22422	−39.26681
propane → butane	−39.29359	−39.27689	−39.28286	−39.22444	−39.26690
butane → pentane	−39.29346	−39.27680	−39.28285	−39.22433	−39.26666
pentane → hexane	−39.29359	−39.27696	−39.28288	−39.22447	−39.26687
hexane → heptane	−39.29348	−39.27686	−39.28289	−39.22436	−39.26663
heptane → octane		−39.27699	−39.28290	−39.22450	−39.26694

[a] CBS-APNO; [b] M062X/6-31+G(2df,p).

Table 2 lists definitions of the various group equivalents, which are generally based on the simplest example (or two, in some cases) providing the desired "type" of atom. Figure 1 shows the full set of compounds used for this purpose. There is some indeterminacy that results from the fact that one can define more reasonable atom types than corresponding examples. Following Benson, we have chosen to consider all methyl groups equivalent, as a way to address this indeterminacy. We have also included some increments that are suitable for alkynes (Ct carbons), that Benson did not originally define. Tables 3 and 4 list the values obtained for the group increments defined in Table 2 using five electronic structure methods: W1BD, G-4, CBS-QB3, CBS-APNO, and M062X/6-31G(2df,p), as enthalpies either at 0 K (Table 3) or at 298 K (Table 4). To illustrate the approach, three worked examples are provided below, and are also illustrated in Figure 2.

Table 2. Definitions for group increments.

Group	Definition
$C-(H)_3(C)$	ethane + hexane + heptane − 9 × $C-(H)_2(C)_2$
$C-(H)_2(C)_2$	(heptane − propane)/4
$C-(H)(C)_3$	isobutane − 3 × $C-(H)_3(C)$
$C-(C)_4$	neopentane − 4 × $C-(H)_3(C)$
$C_d-(H)_2$	ethene/2
$C_d-(H)(C)$	trans-2-butene/2 − $C-(H)_3(C)$
$C_d-(C)_2$	isobutene − 2 × $C-(H)_3(C)$ − $C_d-(H)_2$
$C_d-(C_d)(H)$	(1,3-butadiene − ethene)/2
$C_d-(C_d)(C)$	2-methyl-1,3-butadiene − ethene − $C_d-(C_d)(H)$ − $C-(H)_3(C)$
$C_d-(C_B)(H)$	$C_d-(C_d)(H)$
$C_d-(C_B)(C)$	α-methylstyrene − 5 × $C_B-(H)$ − $C_d-(H)_2$ − $C_B-(C_d)$ − $C-(H)_3(C)$
$C_d-(C_d)_2$	3-methylenepenta-1,4-diene − 3 × $C_d-(H)_2$ − $C_d-(C_d)(H)$
$C_B-(H)$	benzene/6
$C_B-(C)$	toluene − 5 × $C_B-(H)$ − $C-(H)_3(C)$
$C_B-(C_d)$	styrene − 5 × $C_B-(H)$ − $C_d-(H)_2$ − $C_d-(C_B)(H)$
$C_B-(C_B)$ [a]	(naphthalene − 8 × $C_B-(H)$)/2
$C-(C_d)(C)(H)_2$	1-butene − $C_d-(H)_2$ − $C_d-(H)(C)$ − $C-(H)_3(C)$
$C-(C_d)_2(H)_2$	1,4-pentadiene − 2 × $C_d-(H)_2$ − 2 × $C_d-(H)(C)$
$C-(C_d)_2(C)(H)$	3-methyl-1,4-pentadiene − 2 × $C_d-(H)_2$ − 2 × $C_d-(H)(C)$ − $C-(H)_3(C)$
$C-(C_d)(C_B)(H)_2$	allylbenzene − 5 × $C_B-(H)$ − $C_B-(C)$ − $C_d-(H)_2$ − $C_d-(H)(C)$
$C-(C_B)(C)(H)_2$	ethylbenzene − 5 × $C_B-(H)$ − $C_B-(C)$ − $C-(H)_3(C)$
$C-(C_d)(C)_2(H)$	3-methyl-1-butene − $C_d-(H)_2$ − $C_d-(H)(C)$ − 2 × $C-(H)_3(C)$
$C-(C_B)(C)_2(H)$	isopropylbenzene − 5 × $C_B-(H)$ − $C_B-(C)$ − 2 × $C-(H)_3(C)$
$C-(C_d)(C)_3$	3,3-dimethyl-1-butene − $C_d-(H)_2$ − $C_d-(H)(C)$ − 3 × $C-(H)_3(C)$
$C-(C_B)(C)_3$	tert-butylbenzene − 5 × $C_B-(H)$ − $C_B-(C)$ − 3 × $C-(H)_3(C)$
$C_t-(H)$ [a]	ethyne/2
$C_t-(C)$ [a]	(2-butyne + propyne − 3 × $C-(H)_3(C)$ − $C_t-(H)$)/2
$C-(C_t)(C)(H)_2$ [b]	(2-pentyne − 2 × $C_t-(C)$ − 2 × $C-(H)_3(C)$ + 1-butyne − propyne)/2
$C-(C_t)(C)_2(H)$ [b]	(4-methyl-2-pentyne − 2 × $C_t-(C)$ + 3-methyl-1-butyne − propyne − 4 × $C-(H)_3(C)$)/2
$C-(C_t)(C)_3$ [b]	(4,4-dimethyl-2-pentyne − 2 × $C_t-(C)$ + 3,3-dimethyl-1-butyne − propyne − 6 × $C-(H)_3(C)$)/2

[a] In fused ring compounds such as naphthalene. [b] In a departure from Benson's notation, Ct here denotes a carbon in an alkyne (triple bond).

Table 3. Calculated group increments for enthalpy at 0 K (electronic energy plus ZPE) (hartrees).

Group	W1BD	G4	CBS-APNO	CBS-QB3	M062X [a]
$C-(H)_3(C)$	−39.88450	−39.86888	−39.87385	−39.81515	−39.85354
$C-(H)_2(C)_2$	−39.29485	−39.27819	−39.28420	−39.22572	−39.26807
$C-(H)(C)_3$	−38.70779	−38.69036	−38.69736	−38.63907	−38.68527
$C-(C)_4$	−38.12171	−38.10419	−38.11260	−38.05411	−38.10416
$C_d-(H)_2$	−39.27732	−39.26094	−39.26610	−39.20832	−39.24913
$C_d-(H)(C)$	−38.69163	−38.67399	−38.68046	−38.62273	−38.66820
$C_d-(C)_2$	−38.10796	−38.08928	−38.09682	−38.03924	−38.08921
$C_d-(C_d)(H)$	−38.69403	−38.67631	−38.68284	−38.62515	−38.67072
$C_d-(C_d)(C)$	−38.10951	−38.09078	−38.09858	−38.04100	−38.09084
$C_d-(C_B)(H)$	−38.69403	−38.67631	−38.68284	−38.62515	−38.67072
$C_d-(C_B)(C)$		−38.08979	−38.09782	−38.03986	−38.08912

Table 3. Cont.

Group	W1BD	G4	CBS-APNO	CBS-QB3	M062X [a]
C_d-$(C_d)_2$	−38.10451	−38.08590	−38.09318	−38.03644	−38.08583
C_B-(H)	−38.70023	−38.68233	−38.68942	−38.63161	−38.67698
C_B-(C)	−38.11513	−38.09687	−38.10480	−38.04727	−38.09644
C_B-(C_d)	−38.11493	−38.09662	−38.10452	−38.04752	−38.09621
C_B-(C_B) [b]		−38.09858	−38.10661	−38.04901	−38.09759
C-(C_d)(C)(H)$_2$	−39.29422	−39.27752	−39.28361	−39.22506	−39.26758
C-$(C_d)_2$(H)$_2$	−39.29363	−39.27689	−39.28313	−39.22448	−39.26701
C-$(C_d)_2$(C)(H)	−38.70564	−38.68861	−38.69551	−38.63727	−38.68321
C-(C_d)(C_B)(H)$_2$		−39.27778	−39.28405	−39.22533	−39.26727
C-(C_B)(C)(H)$_2$	−39.29454	−39.27837	−39.28432	−39.22581	−39.26771
C-(C_d)(C)$_2$(H)	−38.70708	−38.68984	−38.69674	−38.63850	−38.68457
C-(C_B)(C)$_2$(H)		−38.69012	−38.69705	−38.63867	−38.68375
C-(C_d)(C)$_3$	−38.12073	−38.10332	−38.11093	−38.05327	−38.10266
C-(C_B)(C)$_3$		−38.10195	−38.10986	−38.05166	−38.09950
C_t-(H) [c]	−38.66258	−38.64518	−38.65078	−38.59372	−38.63816
C_t-(C) [c]	−38.08084	−38.06229	−38.06981	−38.01251	−38.06231
C-(C_t)(C)(H)$_2$ [c]	−39.29334	−39.27660	−39.28209	−39.22423	−39.26637
C-(C_t)(C)$_2$(H) [c]	−38.70571	−38.68838	−38.69485	−38.63723	−38.68279
C-(C_t)(C)$_3$ [c]	−39.29913	−39.28401	−39.28959	−39.23150	−39.27191

[a] M062X/6-31 + G(2df,p); [b] In fused ring compounds such as naphthalene.; [c] In a departure from Benson's notation, C_t here denotes a carbon in an alkyne (triple bond).

Table 4. Calculated group increments for enthalpy at 298 K (hartrees).

Group	W1BD	G4	CBS-APNO	CBS-QB3	M062X [a]
C-(H)$_3$(C)	−39.88235	−39.86674	−39.87172	−39.81301	−39.85142
C-(H)$_2$(C)$_2$	−39.29353	−39.27688	−39.28287	−39.22440	−39.26677
C-(H)(C)$_3$	−38.70751	−38.69006	−38.69704	−38.63879	−38.68502
C-(C)$_4$	−38.12258	−38.10499	−38.11318	−38.05497	−38.10487
C_d-(H)$_2$	−39.27532	−39.25894	−39.26411	−39.20632	−39.24714
C_d-(H)(C)	−38.69054	−38.67290	−38.67936	−38.62164	−38.66712
C_d-(C)$_2$	−38.10794	−38.08926	−38.09677	−38.03920	−38.08920
C_d-(C_d)(H)	−38.69321	−38.67549	−38.68202	−38.62433	−38.66991
C_d-(C_d)(C)	−38.10961	−38.09089	−38.09868	−38.04110	−38.09100
C_d-(C_B)(H)	−38.69321	−38.67549	−38.68202	−38.62433	−38.66991
C_d-(C_B)(C)		−38.08986	−38.09774	−38.03993	−38.08923
C_d-$(C_d)_2$	−38.10455	−38.08597	−38.09321	−38.03647	−38.08594
C_B-(H)	−38.69933	−38.68143	−38.68854	−38.63071	−38.67609
C_B-(C)	−38.11451	−38.09625	−38.10416	−38.04665	−38.09583
C_B-(C_d)	−38.11439	−38.09610	−38.10405	−38.04695	−38.09570
C_B-(C_B) [b]		−38.09824	−38.10628	−38.04867	−38.09727
C-(C_d)(C)(H)$_2$	−39.29319	−39.27648	−39.28261	−39.22405	−39.26657
C-$(C_d)_2$(H)$_2$	−39.29280	−39.27604	−39.28235	−39.22367	−39.26618
C-$(C_d)_2$(C)(H)	−38.70556	−38.68850	−38.69549	−38.63720	−38.68313
C-(C_d)(C_B)(H)$_2$		−39.27680	−39.28315	−39.22438	−39.26633
C-(C_B)(C)(H)$_2$	−39.29341	−39.27724	−39.28324	−39.22470	−39.26661
C-(C_d)(C)$_2$(H)	−38.70687	−38.68962	−38.69659	−38.63831	−38.68444
C-(C_B)(C)$_2$(H)		−38.68981	−38.69675	−38.63832	−38.68348
C-(C_d)(C)$_3$	−38.12146	−38.10402	−38.11174	−38.05401	−38.10355
C-(C_B)(C)$_3$		−38.10256	−38.11056	−38.05230	−38.10030
C_t-(H) [c]	−38.66069	−38.64325	−38.64896	−38.59183	−38.63632
C_t-(C) [c]	−38.07979	−38.06122	−38.06880	−38.01145	−38.06128
C-(C_t)(C)(H)$_2$ [c]	−39.29210	−39.27535	−39.28083	−39.22299	−39.26516
C-(C_t)(C)$_2$(H) [c]	−38.70529	−38.68795	−38.69444	−38.63682	−38.68240
C-(C_t)(C)$_3$ [c]	−39.29784	−39.28267	−39.28838	−39.23024	−39.27075

[a] M062X/6-31 + G(2df,p); [b] In fused ring compounds such as naphthalene.; [c] In a departure from Benson's notation, C_t here denotes a carbon in an alkyne (triple bond).

Example 1. *Bicyclobutane at 0 K using W1BD:*

W1BD calculation:	−155.89922
Increments:	
2 × C−(H)2(C)2	2 × −39.29485
2 × C−(H)(C)3	2 × −38.70779
Sum:	−156.00527
Difference:	0.10605 = 66.5 kcal/mol strain energy

Example 2. *[2.1.1]propellane at 298 K using G-4:*

G-4 calculation:	−233.15571
Increments:	
4 × C−(H)2(C)2	4 × −39.27688
2 × C−(C)4	2 × −38.10499
Sum:	−233.31748
Difference:	0.16178 = 101.5 kcal/mol strain energy

Example 3. *[4.4.4.4]fenestrane at 0 K using CBS-QB3:*

CBS-QB3 calculation:	−349.250562
Increments:	
4 × C−(H)2(C)2	4 × −39.22572
4 × C−(H)(C)3	4 × −38.63907
1 × C−(C)4	1 × −38.05411
Sum:	−349.51324
Difference:	0.26267 = 164.8 kcal/mol strain energy

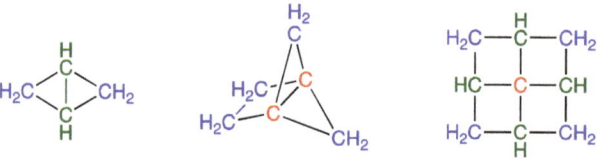

Figure 2. The three examples described in the text; blue = C−(H)2(C)2, green = C−(H)(C)3, red = C−(C)4.

Table 5 and Figure 3 show calculated strain energies for a variety of interesting hydrocarbons. Table S1 in the Supporting Information lists the group equivalents used to define the strain-free reference for each molecule. Examples have been restricted to cases in which it is reasonable to assume a single conformation is dominant, obviating the need for conformational averaging or extensive conformational searching. Some molecules that are expected to be largely strain free, such as various cyclohexane and adamantane derivatives, have purposely been included.

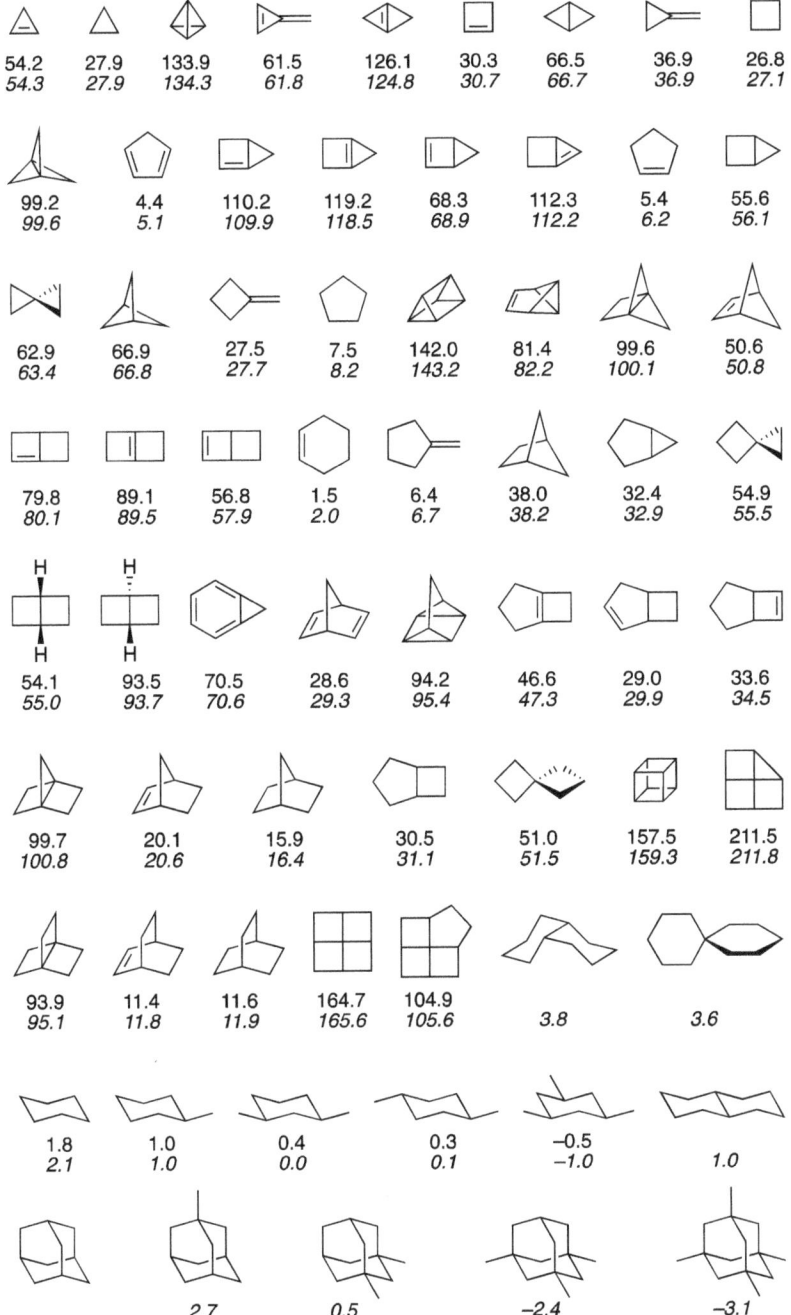

Figure 3. W1BD (normal text) and G-4 (italics) calculated strain energies of the hydrocarbons in Table 5 as enthalpies at 0 K (kcal/mol).

Table 5. Calculated strain energies of some hydrocarbons as enthalpies at 0 K (kcal/mol).

Compound	W1BD	G4	CBS-APNO	CBS-QB3	M062X [a]
cyclopropane	54.2	54.3	54.8	54.2	50.8
cyclopropane	27.9	27.9	27.6	28.1	25.4
tetrahedrane	133.9	134.3	135.9	134.4	121.8
methylenecyclopropene	61.5	61.8	61.9	61.5	55.5
bicyclo[1.1.0]but-1(3)-ene	126.1	124.8	124.4	125.4	123.4
cyclobutene	30.3	30.7	31.1	30.6	30.6
bicyclobutane	66.5	66.7	67.0	66.6	61.3
methylenecyclopropane	36.9	36.9	37.0	37.0	33.4
cyclobutane	26.8	27.1	26.3	27.0	26.7
[1.1.1]propellane	99.2	99.6	100.2	99.5	96.1
cyclopentadiene	4.4	5.1	4.8	4.7	5.9
bicyclo[2.1.0]pent-1-ene	110.2	109.9	111.0	110.3	109.7
bicyclo[2.1.0]pent-1(4)-ene	119.2	118.5		118.9	118.9
bicyclo[2.1.0]pent-2-ene	68.3	68.9	69.5	68.7	66.8
bicyclo[2.1.0]pent-4-ene	112.3	112.2	112.9	112.4	110.7
cyclopentene	5.4	6.2	6.0	5.6	6.8
bicyclo[2.1.0]pentane	55.6	56.1	56.0	55.8	53.3
spiropentane	62.9	63.4	63.3	63.2	57.5
bicyclo[1.1.1]pentane	66.9	66.8	67.1	66.0	65.5
methylenecyclobutane	27.5	27.7	27.6	27.4	27.1
cyclopentane	7.5	8.2	6.6	7.7	8.3
prismane	142.0	143.2	144.4	142.7	137.7
benzvalene	80.1	80.9	81.7	80.1	76.5
[2.1.1]propellane	99.6	100.1	100.4	99.9	98.7
bicyclo[2.1.1]hexene	50.6	50.8	51.0	49.9	52.6
bicyclo[2.2.0]hex-1-ene	79.8	80.1	80.5	79.8	80.2
Bicyclo[2.2.0]hex-1(4)-ene	89.1	89.5	90.3	88.9	90.4
Bicyclo[2.2.0]hex-2-ene	56.8	57.9	58.1	57.4	57.6
cyclohexene	1.5	2.0	1.5	1.5	2.3
methylenecyclopentane	6.4	6.7	6.1	6.3	7.1
bicyclo[2.1.1]hexane	38.0	38.2	37.7	37.3	39.3
bicyclo[3.1.0]hexane	32.4	32.9	32.1	32.4	31.1
spirohexane	54.9	55.5	55.1	55.3	52.6
cis-bicyclo[2.2.0]hexane	54.1	55.0	54.4	54.7	54.4
trans-bicyclo[2.2.0]hexane	93.5	93.7	93.5	92.9	93.8
cyclohexane	1.8	2.1	0.5	1.7	2.2
bicyclo[4.1.0]hepta-1,3,5-triene	70.5	70.6	70.7	70.1	66.6
norbornadiene	28.6	29.3	29.5	28.2	32.6
quadricycane	94.2	95.4	96.1	94.4	91.5
bicyclo[3.2.0]hept-1(5)-ene	46.6	47.3	47.2	46.3	48.7
bicyclo[3.2.0]hept-2-ene	29.0	29.9	29.8	29.2	30.8
bicyclo[3.2.0]hept-6-ene	33.6	34.5	33.7	33.7	35.2
[2.2.1]propellane	99.7	100.8	101.2	100.4	99.1
norbornene	20.1	20.6	19.1	19.5	23.2
norbornane	15.9	16.4	15.1	15.3	18.4
bicyclo[3.2.0]heptane	30.5	31.1	30.2	30.6	31.6
spiro[3.3]heptane	51.0	51.5	50.9	51.2	51.3
equatorial methylcyclohexane	1.0	1.0	−0.1	0.7	1.5
cubane	157.5	159.3	161.9	158.5	160.0
[3.4.4.4]fenestrane	211.5	211.8	213.0	211.0	207.7
[2.2.2]propellane	93.9	95.1	96.0	94.9	95.3
bicyclo[2.2.2]octene	11.4	11.8	10.4	10.9	14.1
bicyclo[2.2.2]octane	11.6	11.9	10.0	11.0	13.9
eq, eq cis-1,3-dimethylcyclohexane	0.4	0.0	−0.7	−0.1	1.0
eq, eq trans-1,4-dimethylcyclohexane	0.3	0.1	−0.6	−0.1	1.0
[4.4.4.4]fenestrane	164.7	165.6	167.1	164.8	165.9
eq,eq,eq cis-1,3,5-trimethylcyclohexane		−1.0	−1.3	−1.1	0.5
[4.4.4.5]fenestrane		105.6	106.2	104.6	106.9
adamantane		6.0	4.2	5.0	9.2
trans-decalin		1.0	−0.8	0.5	2.6
cis-decalin		3.8	2.0	3.2	5.6
1-methyladamantane		2.7	2.0	1.8	7.4
spiro[5.5]undecane		3.6	1.5	3.2	5.9
1,3-dimethyladamantane		0.5	−0.6	−0.1	5.9
1,3,5-trimethyladamantane		−2.4	−3.9	−2.8	4.3
1,3,5,7-tetramethyladamantane		−3.1	−5.3	−3.3	3.0

[a] M062X/6-31+G(2df,p).

4. Discussion

At least for the monocyclic cases, the strain energies in Table 5 and Figure 3 agree very well, typically within 1–2 kcal/mol, with previous estimates, both computational and experimental, such as those compiled by Liebman and Greenberg [2,3], Wiberg [5], Anslyn and Dougherty [29], Castaño and Notario [42], Schleyer [43], Ibrahim [44], Davis [10], Oth and Berson [45], and Doering [46], to name a few. Agreement is generally good (within 3 kcal/mol) for the more complex structures as well, although there are a few notable differences. For instance, the strain estimated here for quadricyclane (94.2 kcal/mol) agrees very closely with that reported by Doering and by Berson on an experimental basis (95–96 kcal/mol), although not at all well with that obtained by Davison on a purely computational basis (71 kcal/mol). The strain for norbornene (20.1) differs substantially from Schleyer's (27.2), but agrees well with Wiberg's (21.1). Similarly, the strain energies for spiropentane (62.9) and cubane (157.5) differ substantially from Schleyer's (65.0 and 166.0), but are fairly close to Wiberg's (63.2 and 154.7).

The spiroalkanes and cubane illustrate the principle of ring strain additivity and its limitations. The strain of spirohexane, spiro[3.3]octane, spiro[5.5]undecane, and cubane quite closely parallel the sums of the strain energies of the constituent rings: cyclopropane and cyclobutane (54.7 kcal/mol) for spirohexane (54.9 kcal/mol), twice cyclobutane (53.6 kcal/mol) for spiro[3.3]octane (51.0 kcal/mol), twice cyclohexane (4.2 kcal/mol) for spiro[5.5]undecane (3.6 kcal/mol), and six times cyclobutane (160.8 kcalmol) for cubane (157.5 kcal/mol. The strain energy of spiropentane (62.9 kcal.mol), however, significantly exceeds that of two cyclopropane rings (54.8 kcal/mol). Wiberg has noted that this happens because the central carbon is forced to adopt sp^3 hybridization, whereas in cyclopropane, the carbons are closer to sp^2 hybridization [5]. A similar phenomenon is observed for bicyclobutane, for which the computed strain energy of 66.5 kcal/mol exceeds the sum for two cyclopropane rings by 10.7 kcal/mol. The strain of [2.1.0]bicyclopentane (55.6 kcal/mol), on the other hand, closely matches the sum for cyclopropane and cyclobutane (54.7 kcal/mol) (as well as Wiberg's estimate of 54.7 kcal/mol) [5].

Prismane is a somewhat intermediate case: its strain of 142.0 kcal/mol exceeds the sum of three cyclobutanes and two cyclopropanes (136.2 kcal/mol), but only by 5.8 kcal/mol. The small-ring fenestranes, on the other hand, exhibit strain far exceeding what would be expected on the basis of ring strain additivity. That is not surprising, given the tremendous distortion of the central carbon, which is forced to be close to planar. The strain energies of [3.4.4.4], [4.4.4.4], and [4.4.4.5]fenestrane exceed the corresponding sums of the strain energies of the constituent rings by 103.2, 57.5, and 17.7 kcal/mol, respectively.

The propellanes present another interesting comparison. The strain has previously been reported to increase from 98 to 104 to 105 kcal/mol on going from [1.1.1] to [2.1.1] to [2.2.1]propellane, before dropping to 89 kcal/mol for [2.2.2]propellane [5]. This sequence seems surprising; one would expect that each replacement of a cyclopropane ring with a cyclobutane ring ought to decrease the strain by 1 kcal/mol or so, not increase it. The values computed here better match these expectations. The strain remains essentially constant, going from 99.2 to 99.6 to 99.7, along the sequence [1.1.1] to [2.1.1] to [2.2.1]propellane, before dropping to 93.9 for [2.2.2]propellane [47].

The estimate here for norbornadiene, 28.6 kcal/mol, is substantially lower than several recent estimates that are in the range of 32–35 kcal/mol, [42,43,48] although in good agreement with Doering's original experimentally-based estimate of 29.0 kcal/mol [46]. The difference results from somewhat alternative views of what the strain-free reference should be; for instance, are 1,4 interactions (such as a gauche butane interaction) to be considered part of the strain, or part of the reference against which strain is judged?

In the end, one cannot really view the differences in these strain estimates as "errors". Of course, inaccuracies in either experimental measurements or calculated energies contribute to the differences, and that can particularly be true of older calculations performed at a time when large basis sets and proper accounting of electron correlation were not feasible. However, a significant amount of the difference also originates from differences in how the strain-free reference state is defined. Philosophically, the approach taken here follows very closely that described by Schleyer in 1970 [43].

He recommended to use a wider set of parameters than just the number of CH_3, CH_2, CH, and C groups and alkene functionalities (as in the original Franklin scheme), and also to use what he termed "single conformation group increments". Using experimental enthalpies of formation obtained at normal temperatures yields values that include some contributions from conformations higher in energy than the global minimum. He argued that including these contributions resulted in a somewhat inaccurate estimate of the true strain-free energy. The same view is taken here. While, when using experimental data, it is laborious to subtract out these contributions, using computational methods makes it simple and natural not to include them in the first place. From this perspective, the strain energies presented here, and the method used to compute them, should correspond especially closely to what organic chemists intuitively mean by the concept of strain.

One could of course imagine defining an even more extensive set of group equivalents designed to take into account non-next-nearest neighbor interactions. Arguably, doing so would provide an even more precise accounting for strain, at least if these non-next-nearest neighbor interactions are not regarded as part of the strain. However, taking such an approach would greatly increase the number of group equivalents required, and likely result in only very small changes to the computed strain energies. Furthermore, it is worth noting that the reference molecules listed in Figure 1, and implicitly assumed to be strain free, were chosen so as to minimize any such non-next-nearest neighbor interactions. For instance, there are no alkane gauche interactions. The use of single, minimum-energy conformations means that the linear alkanes rigorously avoid such interactions, and the only branched alkanes (2-methylpropane and 2,2-dimethylpropane) lack a 4-carbon chain. Similarly, there are no cis alkenes. Unfortunately, there is likely some 1,3-allylic strain in 3,3-dimethyl-1-butene and in *t*-butylbenzene, as well as perhaps in 2-methy-1,3-butadiene and alpha-methylstyrene. This could lead to a slight underestimate of strain energies when the parameters that rely on these four molecules are used, insofar as these particular parameters include a small amount of inherent strain.

It is interesting that the various cyclohexane derivatives, including trans-decalin and adamantane, are not calculated to be entirely strain free. Indeed, adamantane is calculated to have a rather substantial 6 kcal/mol of strain. Schleyer explored this issue in detail in 1970, and explained the strain in all these cases as resulting from a combination of angle strain, transannular C ... C repulsion, and also an attractive interaction resulting from anti arrangements of CCCC fragments [43]. The data from the present study fit these interpretations. Roughly speaking, each cyclohexane ring provides 2 kcal/mol of strain, but each methyl that is not axial (or gauche to another methyl) reduces the strain by 1 kcal/mol. Consistent with Schleyer's explanation, each such methyl group indeed contributes two (in cyclohexane) or three (in adamantane) "anti-butane" configurations of the sort that he postulated to be stabilizing. In addition, the quaternary carbons that result from methyl substitution of the tertiary carbons in adamantane would be expected to have almost perfectly tetrahedral bond angles, thus reducing angle strain. A similar effect is likely at work in methyl-substituted cyclohexanes.

5. Conclusions

A modernized version of Wiberg's and Schleyer's computational group equivalent approach for hydrocarbon strain energies has been described, using the detailed group equivalents defined by Benson and highly accurate, modern electronic structure methods. The resulting strain energies generally agree well with previous estimates, but in some cases make more sense than earlier estimates in terms of ring strain additivity. Group equivalents are provided for just five popular and powerful methods. However, researchers desiring to use other methods can calculate corresponding equivalents using the definitions of the increments provided here, should they so desire.

Supplementary Materials: The following are available online at http://www.mdpi.com/2624-8549/2/2/347\T1\textendash360/s1, Table S1: Definitions of strain-free reference states, Table S2: Calculated enthalpies (0 K) of compounds in Figure 1, Table S3: Calculated enthalpies (298 K) of compounds In Figure 1, Table S4: Calculated enthalpies (0 K) of compounds in Figure 2, Table S5: Calculated enthalpies (298 K) of compounds in Figure 2, List S1: W1BD optimized geometries & abbreviated calculation results, List S2: G-4/SCRF optimized geometries & abbreviated calculation results.

Funding: This research received no external funding.

Conflicts of Interest: The author declares no conflict of interest.

References

1. Baeyer, A. Ueber Polyacetylenverbindungen. *Ber. Dtsch. Chem. Ges.* **1885**, *18*, 2269–2281. [CrossRef]
2. Liebman, J.F.; Greenberg, A. A Survey of Strained Organic Molecules. *Chem. Rev.* **1976**, *76*, 311–365. [CrossRef]
3. Liebman, J.F.; Greenberg, A. *Strained Organic Molecules*; Academic Press: New York, NY, USA, 1978.
4. Pitzer, K.S. Strain energies of cyclic hydrocarbons. *Science* **1945**, *101*, 672. [CrossRef] [PubMed]
5. Wiberg, K.B. The Concept of Strain in Organic Chemistry. *Angew. Chem. Int. Ed. Engl.* **1986**, *25*, 312–322. [CrossRef]
6. Dudev, T.; Lim, C. Ring Strain Energies from ab Initio Calculations. *J. Am. Chem. Soc.* **1998**, *120*, 4450–4458. [CrossRef]
7. Bach, R.D.; Dmitrenko, O. The effect of substituents on the strain energies of small ring compounds. *J. Org. Chem.* **2002**, *67*, 2588–2599. [CrossRef]
8. Bach, R.D.; Dmitrenko, O. Strain Energy of Small Ring Hydrocarbons. Influence of C-H Bond Dissociation Energies. *J. Am. Chem. Soc.* **2004**, *126*, 4444–4452. [CrossRef]
9. Bauza, A.; Quinonero, D.; Deya, P.M.; Frontera, A. Estimating ring strain energies in small carbocycles by means of the Bader's theory of atoms-in-molecules. *Chem. Phys. Lett.* **2012**, *536*, 165–169. [CrossRef]
10. Walker, J.E.; Adamson, P.A.; Davis, S.R. Comparison of calculated hydrocarbon strain energies using ab initio and composite methods. *J. Mol. Struct.* **1999**, *487*, 145–150. [CrossRef]
11. Hassenrueck, K.; Martin, H.D.; Walsh, R. Consequences of strain in $(CH)_8$ hydrocarbons. *Chem. Rev.* **1989**, *89*, 1125–1146. [CrossRef]
12. Politzer, P.; Jayasuriya, K.; Zilles, B.A. Some effects of amine substituents in strained hydrocarbons. *J. Am. Chem. Soc.* **1985**, *107*, 121–124. [CrossRef]
13. Espinosa Ferao, A. Kinetic energy density per electron as quick insight into ring strain energies. *Tetrahedron Lett.* **2016**, *57*, 5616–5619. [CrossRef]
14. George, P.; Trachtman, M.; Bock, C.W.; Brett, A.M. An alternative approach to the problem of assessing destabilization energies (strain energies) in cyclic hydrocarbons. *Tetrahedron* **1976**, *32*, 317–323. [CrossRef]
15. Howell, J.; Goddard, J.D.; Tam, W. A relative approach for determining ring strain energies of heterobicyclic alkenes. *Tetrahedron* **2009**, *65*, 4562–4568. [CrossRef]
16. Hehre, W.J.; Ditchfield, R.; Radom, L.; Pople, J.A. Molecular Orbital Theory of Electronic Structure of Organic Compounds. 5. Molecular Theory of Bond Separation. *J. Am. Chem. Soc.* **1970**, *92*, 4796–4801. [CrossRef]
17. Bachrach, S.M. The Group Equivalent Reaction: An Improved Method for Determining Ring Strain Energy. *J. Chem. Educ.* **1990**, *67*, 907–908. [CrossRef]
18. Wheeler, S.E.; Houk, K.N.; Schleyer, P.V.R.; Allen, W.D. A Hierarchy of Homodesmotic Reactions for Thermochemistry. *J. Am. Chem. Soc.* **2009**, *131*, 2547–2560. [CrossRef]
19. Benson, S.W.; Buss, J.H. Additivity Rules for the Estimation of Molecular Properties. Thermodynamic Properties. *J. Chem. Phys.* **1958**, *29*, 546–572. [CrossRef]
20. Benson, S.W. *Thermochemical Kinetics*, 2nd ed.; Wiley: New York, NY, USA, 1976.
21. Benson, S.W.; Cruickshank, F.R.; Golden, D.M.; Haugen, G.R.; O'Neal, H.E.; Rodgers, A.S.; Shaw, R.; Walsh, R. Additivity Rules for the Estimation of Thermochemical Properties. *Chem. Rev.* **1969**, *69*, 279–324. [CrossRef]
22. Eigenmann, H.K.; Golden, D.M.; Benson, S.W.J. Revised Group Additivity Parameters for Enthalpies of Formation of Oxygen-Containing Organic Compounds. *Phys. Chem.* **1973**, *77*, 1687–1691. [CrossRef]
23. Cohen, N.; Benson, S.W. Estimation of Heats of Formation of Organic Compounds by Additivity Methods. *Chem. Rev.* **1993**, *93*, 2419–2438. [CrossRef]
24. Franklin, J.L. Prediction of Heat and Free Energies of Organic Compounds. *Ind. Eng. Chem.* **1949**, *41*, 1070–1076. [CrossRef]
25. Wodrich, M.D.; Schleyer, P.V.R. New Additivity Schemes for Hydrocarbon Energies. *Org. Lett.* **2006**, *8*, 2135–2138. [CrossRef] [PubMed]
26. Gronert, S.J. An alternative interpretation of the C-H bond strengths of alkanes. *Org. Chem.* **2006**, *71*, 1209–1219. [CrossRef] [PubMed]

27. Wiberg, K.B.; Bader, R.W.; Lau, C.D.H. Theoretical Analysis of Hydrocarbon Properties. 2. Additivity of Group Properties and the Origin of Strain Energy. *J. Am. Chem. Soc.* **1987**, *109*, 1001–1012. [CrossRef]
28. Anslyn, E.V.; Dougherty, D.A. *Modern Physical Organic Chemistry*; University Science Books: Sausalito, CA, USA, 2006; pp. 79–82.
29. Wiberg, K.B.; Rablen, P.R. Increase in Strain Energy on Going from [4.4.4.5]Fenestrane to [4.4.4.4]Fenestrane: A Useful Method for Estimating the Heats of Formation of Hydrocarbons and their Derivatives from ab Initio Energies. *J. Org. Chem.* **2020**, *85*, 4981–4987. [CrossRef]
30. Wiberg, K.B. Group Equivalents for Converting Ab Initio Energies to Enthalpies of Formation. *J. Comput. Chem.* **1984**, *5*, 197–199. [CrossRef]
31. Wiberg, K.B. Structures and Energies of the Tricyclo[4.1.0.0^{1,3}]heptanes and the Tetracyclo[4.2.1.0^{2,9}0^{5,9}]nonanes. Extended Group Equivalents for Converting ab Initio Energies to Heats of Formation. *J. Org. Chem.* **1985**, *50*, 5285–5291. [CrossRef]
32. Ibrahim, M.R.; Schleyer, P.V.R. Atom Equivalents for Relating Ab Initio Energies to Enthalpies of Formation. *J. Comput. Chem.* **1985**, *6*, 157–167. [CrossRef]
33. Martin, J.M.I.; de Oliveria, G. Towards standard methods for benchmark quality ab initio thermochemistry, W1 and W2 theory. *J. Chem. Phys.* **1999**, *111*, 1843–1856. [CrossRef]
34. Curtiss, L.A.; Redfern, P.C.; Raghavachari, K. Gaussian-4 theory. *J. Chem. Phys.* **2007**, *126*, 084108. [CrossRef] [PubMed]
35. Ochterski, J.W.; Petersson, G.A.; Montgomery, J.A., Jr. A complete basis set model chemistry. V. Extensions to six or more heavy atoms. *J. Chem. Phys.* **1996**, *104*, 2598–2619. [CrossRef]
36. Montgomery, J.A., Jr.; Frisch, M.J.; Ochterski, J.W.; Petersson, G.A. A complete basis set model chemistry. VI. Use of density functional geometries and frequencies. *J. Chem. Phys.* **1999**, *110*, 2822–2827. [CrossRef]
37. Montgomery, J.A., Jr.; Frisch, M.J.; Ochterski, J.W.; Petersson, G.A. A complete basis set model chemistry. VII. Use of the minimum population localization method. *J. Chem. Phys.* **2000**, *112*, 6532–6542. [CrossRef]
38. Zhao, Y.; Truhlar, D.G. Comparative DFT study of van der Waals complexes: Rare-gas dimers, alkaline-earth dimers, zinc dimer, and zinc-rare gas dimers. *J. Phys. Chem.* **2006**, *110*, 5121–5129. [CrossRef]
39. Rablen, P.R.; McLarney, B.D.; Karlow, B.J.; Schneider, J.E. How Alkyl Halide Structure Affects E2 and S_N2 Reaction Barriers: E2 Reactions Are as Sensitive as S_N2 Reactions. *J. Org. Chem.* **2014**, *79*, 867–879. [CrossRef]
40. Frisch, M.J.; Trucks, G.W.; Schlegel, H.B.; Scuseria, G.E.; Robb, M.A.; Cheeseman, J.R.; Scalmani, G.; Barone, V.; Mennucci, B.; Peterson, G.A.; et al. *Gaussian 09*, Revision D.01; Gaussian, Inc.: Wallingford, CT, USA, 2009.
41. Frisch, M.J.; Trucks, G.W.; Schlegel, H.B.; Scuseria, G.E.; Robb, M.A.; Cheeseman, J.R.; Scalmani, G.; Barone, V.; Peterson, G.A.; Nakatsuji, H.; et al. *Gaussian 16*, Revision A.03; Gaussian, Inc.: Wallingford, CT, USA, 2016.
42. Castaño, O.; Notario, R.; Abboud, J.-L.M.; Gomperts, R.; Palmiero, R.; Frutos, L.-M. Organic Thermochemistry at High ab Initio Levels. 2. Meeting the Challenge: Standard Heats of Formation of Gaseous Norbornane, 2-Norbornene, 2,5-Norbornadiene, Cubane, and Adamantane at the G-2 Level. *J. Org. Chem.* **1999**, *64*, 9015–9018. [CrossRef]
43. Schleyer, P.V.R.; Williams, J.E.; Blanchard, K.R. The Evaluation of Strain in Hydrocarbons. The Strain in Adamantane and its Origin. *J. Am. Chem. Soc.* **1970**, *92*, 2377–2386. [CrossRef]
44. Ibrahim, M.R. Additivity of Bond Separation Energies of Hydrocarbons and Their Thermochemical Data. *J. Phys. Org. Chem.* **1990**, *3*, 126–134. [CrossRef]
45. Kabakoff, D.S.; Bünzli, J.-C.G.; Oth, J.F.M.; Hammond, W.B.; Berson, J.A. Enthalpy and Kinetics of Isomerization of Quadricyclane to Norbornadiene. Strain Energy of Quadricyclane. *J. Am. Chem. Soc.* **1975**, *97*, 1510–1512. [CrossRef]
46. Turner, R.B.; Goebel, P.; Mallon, B.J.; Doering, W.V.E.; Coburn, J.F., Jr.; Pomerantz, M. Heats of Hydrogenation. VIII. Compounds with Three- and Four-membered Rings. *J. Am. Chem. Soc.* **1968**, *90*, 4315–4322. [CrossRef]
47. Morriss, J.W.; Swarthmore College, Swarthmore, PA, USA. Personal communication, 2018. First brought this phenomenon to the author's attention.
48. Khoury, P.T.; Goddard, J.D.; Tam, W. Ring strain energies: Substituted rings, norbornanes, norbornenes and norbornadienes. *Tetrahedron* **2004**, *60*, 8103–8112. [CrossRef]

© 2020 by the author. Licensee MDPI, Basel, Switzerland. This article is an open access article distributed under the terms and conditions of the Creative Commons Attribution (CC BY) license (http://creativecommons.org/licenses/by/4.0/).

Review

Amino Acids and Peptides as Versatile Ligands in the Synthesis of Antiproliferative Gold Complexes [†]

Tina P. Andrejević [1], Biljana Đ. Glišić [1,*] and Miloš I. Djuran [2,*]

1. Faculty of Science, Department of Chemistry, University of Kragujevac, R. Domanovića 12, 34000 Kragujevac, Serbia; tina.andrejevic@pmf.kg.ac.rs
2. Serbian Academy of Sciences and Arts, Knez Mihailova 35, 11000 Belgrade, Serbia
* Correspondence: biljana.glisic@pmf.kg.ac.rs (B.Đ.G.); milos.djuran@pmf.kg.ac.rs (M.I.D.); Tel.: +381-34-336-223 (B.Đ.G.); +381-34-300-251 (M.I.D.); Fax: +381-34-335-040 (B.Đ.G. & M.I.D.)
† In Honor of Professor Bernd Giese on the Occasion of His 80th Birthday.

Received: 6 March 2020; Accepted: 24 March 2020; Published: 27 March 2020

Abstract: Gold complexes have been traditionally employed in medicine, and currently, some gold(I) complexes, such as auranofin, are clinically used in the treatment of rheumatoid arthritis. In the last decades, both gold(I) and gold(III) complexes with different types of ligands have gained considerable attention as potential antitumor agents, showing superior activity both in vitro and in vivo to some of the clinically used agents. The present review article summarizes the results achieved in the field of synthesis and evaluation of gold complexes with amino acids and peptides moieties for their cytotoxicity. The first section provides an overview of the gold(I) complexes with amino acids and peptides, which have shown antiproliferative activity, while the second part is focused on the activity of gold(III) complexes with these ligands. A systematic summary of the results achieved in the field of gold(I/III) complexes with amino acids and peptides could contribute to the future development of metal complexes with these biocompatible ligands as promising antitumor agents.

Keywords: gold complexes; amino acids; peptides; cytotoxicity

---

1. Introduction

Gold and its compounds have been used for the treatment of a wide range of diseases throughout the history of civilization [1]. The use of gold in modern medicine began with the discovery of the in vitro bacteriostatic properties of the gold(I) complex, K[Au(CN)$_2$], by the German bacteriologist Robert Koch [2]. This gold(I) complex was found to be lethal to the microorganism, *Mycobacterium tuberculosis*, which is causative agent of tuberculosis [2]. After its initial use for tuberculosis with favorable results, serious toxic side-effects were observed for K[Au(CN)$_2$] complex and the treatment was switched to the less toxic gold(I) thiolate complexes (AuSR). The mistaken belief that the *Mycobacterium tuberculosis* was also a causative agent of rheumatoid arthritis led Landé and Forestier to use gold(I) thiolate complexes for the treatment of this disease [3]. After thirty years of medicinal debate, in 1960, British Empire Rheumatism Council finally confirmed the beneficial effects of the gold(I) thiolate complexes against rheumatoid arthritis [3]. Since that time, these complexes have been widely used in the treatment of a variety of rheumatic diseases including psoriatic arthritis, juvenile arthritis, palindromic rheumatism and discoid lupus erythematosus [3]. Nowadays, chrysotherapy is an accepted part the modern medicine and refers to the use of gold-based formulations for the treatment of joint pain and inflammatory diseases [4].

Following the medicinal relevance of gold complexes for the treatment of rheumatoid arthritis, research has continued to uncover the potential of gold complexes as agents for the treatment of cancer [4–13], and various bacterial and fungal infections and tropical diseases, such as malaria, trypanosomiasis and leishmaniasis [14,15]. In some cases, gold complexes were found to be more active than the clinically used agents, e.g., cisplatin for the cancer treatment [5]. Some gold complexes showed an outstanding in vitro cytotoxicity toward cisplatin-resistant tumor cell lines, which indicates the difference in the mode of action between them and platinum-based agents [5]. Indeed, it was found that the antitumor activity of cisplatin is based on its interaction with DNA, while the antiproliferative activity of gold complexes usually involves the inhibition of enzymes, especially those containing thiol groups, such as thioredoxin reductase (TrxR) [6].

Different classes of ligands have been used for the synthesis of biologically active gold complexes, including phosphines, N-heterocyclic carbenes, thiolates, polyamines, pyridine, bipyridine, terpyridine, phenanthroline, and their derivatives, macrocyclic ligands (cyclam), porphyrins and dithiocarbamates [4–15]. Besides them, amino acids and peptides represent two important classes of ligands which are also important as building blocks of proteins and enzymes and show a wide range of biological activities [16]. As constituents of proteins, amino acids and peptides can be considered as biocompatible ligands that can deliver Au(I) ion to its biological target or, as polydentate ligands, they can stabilize the Au(III) ion, preventing its reduction to Au(I) or/and Au(0) under physiological conditions. More importantly, metal complexes with this type of ligand can be more selective toward the abnormal cells in respect to the healthy ones, due to the fact that the abnormal cells overexpress amino acids receptors and need more nutrients [17].

The aim of this review is to present the findings obtained in the field of synthesis and evaluation of gold(I) and gold(III) complexes containing amino acids and peptides moieties for their antiproliferative potential.

2. Gold(I) Complexes Containing Amino Acids and Peptides Moieties

Considering the great importance of ferrocenyl group in drug design [18], two ferrocene bioconjugates, FcCO-TrpOMe and FcCO-ProNH$_2$ (Fc = ferrocenyl, TrpOMe = methyl ester of tryptophan and ProNH$_2$ = prolinamide) were reacted with an equimolar amount of [Au(acac)(PR$_3$)] (acac = acetylacetonate, PR$_3$ = PPh$_3$, triphenylphosphine or PPh$_2$Py, 2-pyridyldiphenylphosphine) to yield gold(I) complexes, [Au(FcCO-TrpOMe-N)(PR$_3$)] (PPh$_3$ (**1**) and PPh$_2$Py (**2**)) and [Au(FcCO-ProNH$_2$-N)(PR$_3$)] (PPh$_3$ (**3**) and PPh$_2$Py (**4**)) (Figure 1) [19]. Similarly, the reaction of FcCO-MetOMe with [Au(CF$_3$SO$_3$)(PR$_3$)] (MetOMe = methyl ester of methionine) led to the formation of [Au(FcCO-MetOMe-S)(PR$_3$)]CF$_3$SO$_3$ (PPh$_3$ (**5**) and PPh$_2$Py (**6**)) complexes (Figure 1) [19]. The cytotoxicity of ferrocene bioconjugates and corresponding gold(I) complexes **1–6** was evaluated by MTT assay against two human tumor cell lines, HeLa (cervical cancer) and MCF-7 (breast cancer), and one murine cell line, N1E-115 (derived from mouse neuroblastoma C-1300) (Table 1). The evaluated gold(I) complexes **1–6** appeared to be cytotoxic against these three tumor cell lines, while the corresponding ferrocene bioconjugates used as ligands, FcCO-TrpOMe, FcCO-ProNH$_2$ and FcCO-MetOMe, did not show antiproliferative activity (IC$_{50}$ > 1000 μM, the IC$_{50}$ value is defined as concentration required to inhibit tumor cell proliferation by 50% compared to the control cells). The IC$_{50}$ values of the gold(I) complexes determined after 48 h are in range from 18 to 32 μM in HeLa cells, 15 to 52 μM in MCF-7 cells and < 10 to 54 μM in N1E-115 cells (Table 1). Among the complexes, gold(I) complex **5** with MetOMe and PPh$_3$ in its structure was shown as the most effective against the HeLa cell line; while **4**, having ProNH$_2$ and PPh$_2$Py moieties, displayed the best activity in the murine cell line, although all gold(I) complexes were less cytotoxic than the reference drug doxorubicin, with IC$_{50}$ values of approximately 1.5 μM [19]. For all compounds, the percentage of cell survival decreased with the increasing of exposure time, although the difference between 24 and 48 h was not found to be significant. Complex **4** induced the cell death through apoptosis and formation of reactive oxygen species (ROS) in tumor cells, while the gold(I) complexes did not act as DNA intercalators.

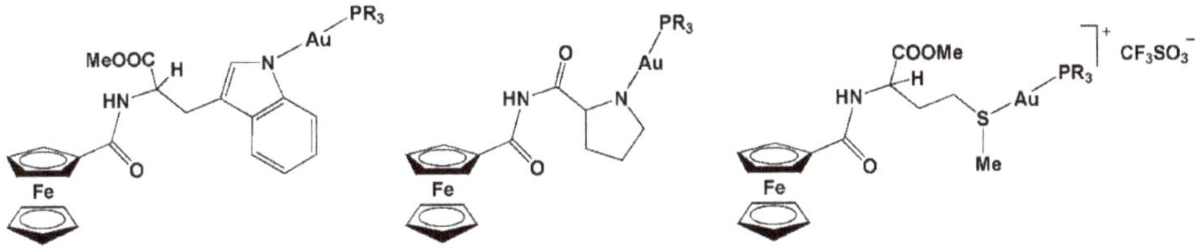

[Au(FcCO-TrpOMe-N)(PPh₃)] (1)
[Au(FcCO-TrpOMe-N)(PPh₂Py)] (2)

[Au(FcCO-ProNH₂-N)(PPh₃)] (3)
[Au(FcCO-ProNH₂-N)(PPh₂Py)] (4)

[Au(FcCO-MetOMe-S)(PPh₃)]CF₃SO₃ (5)
[Au(FcCO-MetOMe-S)(PPh₂Py)]CF₃SO₃ (6)

Figure 1. Gold(I) complexes with the ferrocene bioconjugates **1–6** showing cytotoxic activity [19].

Table 1. In vitro cytotoxic activity (IC$_{50}$, 48 h, µM) of gold(I) complexes with the ferrocene bioconjugates **1–6** [19].

Complex	Cell Line [a] HeLa	MCF-7	N1E-115
1	32 ± 1.8	15 ± 1.2	27 ± 2.2
2	22 ± 1.9	45 ± 1.7	26 ± 1.6
3	28 ± 1.6	52 ± 1.2	54 ± 2.3
4	29 ± 1.3	32 ± 2.5	<10 ± 2.2
5	18 ± 1.4	15 ± 1.8	29 ± 1.7

[a] HeLa = cervical cancer, MCF-7 = breast cancer and N1E-115 = mouse neuroblastoma C-1300.

The antiproliferative activity against different tumor cell lines was shown by gold(I) complexes obtained from the reaction of [Au(SPyCOOH)(PR₃)] complex (SPyCOOH = nicotinic acid thiolate) by the functionalization of its carboxylic group with different amino acids, ester or amide derivatives of these amino acids or with peptide moieties [20,21]. Gold(I) complexes **7–24** of the general formula [Au(SPyCOR)(PPh₃)], in which R = methyl ester of amino acid (**7–12**), amino acid (**13–18**) and amide derivative of the corresponding amino acid (**19–24**) have been structurally modified. This modification was performed by changing the type of phosphine ligand in [Au(SPyCOR)(PPh₃)] complex (PPh₂Py instead of PPh₃; complex **25**) and the nature of the coupled amino acid (R) including its structural modification or peptide functionalization (**26–32**) and increasing the number of Au(I) ions per complex unit (**33**) (Figure 2a,b) [21]. The antiproliferative activity of these gold(I) complexes was evaluated against three different human tumor cell lines, A549 (lung carcinoma), Jurkat (T-cell leukemia) and MiaPaca2 (pancreatic carcinoma), as well as against non-tumor R69 (lymphoid cell line) and 293T (embryonic kidney fibroblasts), and these cells were exposed to different concentrations of each complex for 24 h (Table 2). As can be seen from this table, the complexes **7–24** were active against the investigated tumor cell lines at low micromolar range, with Jurkat cells being the most sensitive; their IC$_{50}$ values fall in the range from 7.4 to 30.5 µM in A549, 8.2 to 27.2 µM in MiaPaca2 and 2.4 to 7.7 µM in Jurkat cells. Moreover, these complexes exhibited some selectivity for leukemia cells in respect to the non-tumor R69 cells, but this difference was not observed in the case of solid tumors. The cytotoxic activity of the ester complexes **7–12** was slightly higher than that of the corresponding precursor, [Au(SPyCOOH)(PPh₃)], in all the tested tumor cell lines, with the exception of MiaPaca2 (Table 2). The complexes containing coupled amino acids **13–18** and amide derivatives of these amino acids **19–24** were, in general, less active than the ester analogues **7–12**, although the difference in the activity is not remarkable. The proline-containing complex **12** was found to induce changes in cell and nucleus morphology, loss of the mitochondrial membrane potential, production of ROS and to inhibit the thioredoxin reductase (TrX), an enzyme which acts as a target for biologically active gold(I/III) complexes [21]. Interestingly, gold(I) species **7–24** showed much higher antiproliferative activities in vitro in the used cell lines than the cisplatin, the well-known antitumor agent used in medicine for the treatment of various cancers (Table 2) [22].

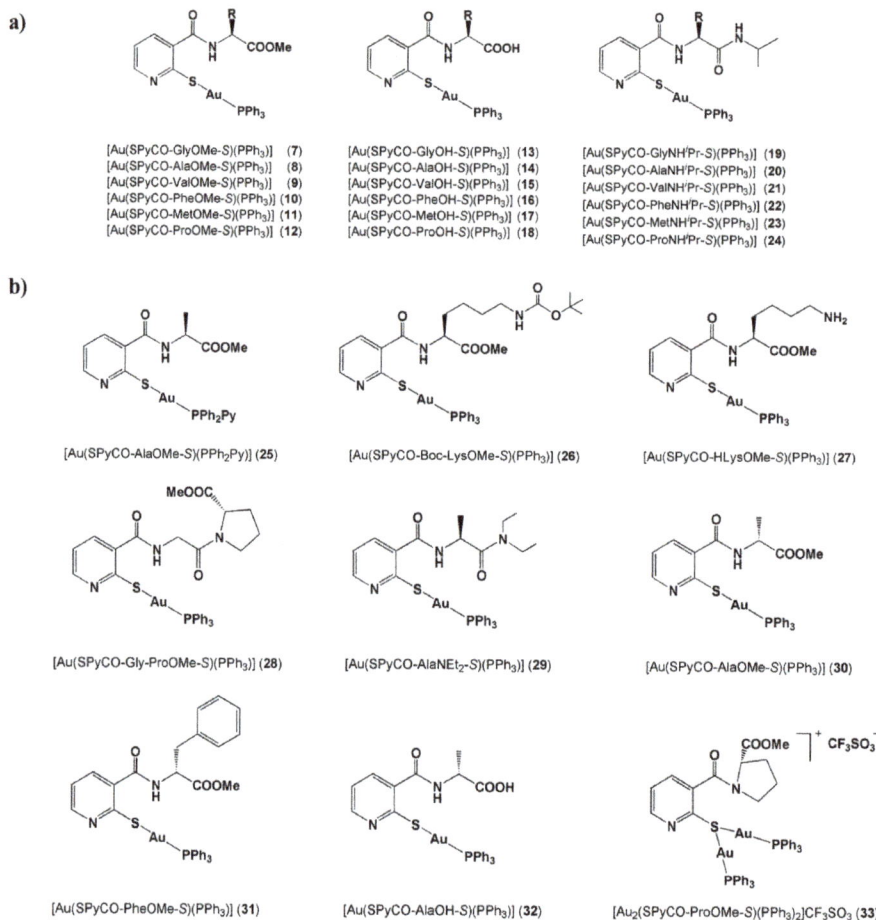

Figure 2. (a) Gold(I) complexes **7–24** of the general formula [Au(SPyCOR)(PPh3)] in which nicotinic acid thiolate is coupled with methyl ester of amino acid (**7–12**), amino acid (**13–18**) and amide derivative of the corresponding amino acid (**19–24**), and (b) complexes **25–33** obtained by different structural modifications of [Au(SPyCOR)(PPh3)] complex showing cytotoxic activity [20,21].

The IC$_{50}$ values of the structurally modified complexes **25–33** (Figure 2b) are also at low micromolar concentrations (4.1–33.5 µM in A549, 1.2–29.3 µM in MiaPaca2 and 0.9–36.5 µM in Jurkat cells), with the Jurkat cell line being, in most cases, more sensitive to these complexes than A549 and MiaPaca2 (Table 2). However, for the latter two cell lines, the complexes **25–33** are more active than cisplatin (IC$_{50}$ = 105 and 71 µM in A549 and MiaPaca2, respectively), while, in the case of Jurkat cell line, the IC$_{50}$ values for the gold(I) complexes and cisplatin are similar (IC$_{50}$ = 7.4 µM). The change of the type of phosphine ligand coordinated to the Au(I) ion (PPh$_2$Py instead of PPh$_3$; complex **25**) resulted in the same or slightly greater cytotoxicity in Jurkat and R69 cell lines, respectively, while in the remaining cell lines, lower cytotoxicity was observed in respect to the analogue complex **7**, although the differences are not significant. The coupling of lysine ester afforded complex **26** having the good antiproliferative activity, while the removal of the protective Boc (Boc is *tert*-butoxycarbonyl) group in this complex significantly decreased the cytotoxicity of the complex **27**. This was ascribed to the presence of free amino group, which acts as a strong nucleophile and can react with the other biomolecules, preventing the Au(I) ion to reach the target [21]. The gold(I) complexes having Gly-ProOMe dipeptide (**28**), tertiary amide (**29**), D-amino esters (**30** and **31**) and D-amino acids (**32**) in their structures have shown lower cytotoxicity than the corresponding analogues (Table 2). Among the investigated complexes, the best antiproliferative activity was demonstrated by the dinuclear gold(I) complex **33**, which is functionalized as ester and contains rigid proline as amino acid moiety (Figure 2b). The IC$_{50}$ values of this complex are found to be in the low micromolar range and even in the submicromolar range in the Jurkat cell line and are also lower than those for cisplatin (Table 2).

Table 2. In vitro cytotoxic activity (IC$_{50}$, 24 h, µM) of thiolate-gold(I) complexes of the general formula [Au(SPyCOR)(PPh$_3$)] in which nicotinic acid thiolate is coupled with methyl ester of amino acid (**7–12**), amino acid (**13–18**) and amide derivative of the corresponding amino acid (**19–24**) and complexes obtained by different structural modifications of [Au(SPyCOR)(PPh$_3$)] complex (**25–33**) [21].

Complex	Cell Line [a] A549	MiaPaca2	Jurkat	293T	R69
[Au(SPyCOOH)(PPh$_3$)]	15.5 ± 0.92	9.2 ± 0.28	4.6 ± 0.08	4.6 ± 0.13	16 ± 0.64
[Au(SPyCOOH)(PR$_3$)], R = methyl ester of amino acid					
7	11.5 ± 0.55	9.7 ± 0.22	3.8 ± 0.07	4.2 ± 0.08	8.6 ± 0.33
8	13.7 ± 0.71	11.0 ± 0.20	4.0 ± 0.07	2.7 ± 0.07	2.2 ± 0.08
9	10.9 ± 0.40	10.2 ± 0.25	3.3 ± 0.05	10.7 ± 0.31	19.0 ± 0.60
10	8.9 ± 0.36	12.3 ± 0.37	4.0 ± 0.08	5.5 ± 0.16	14.0 ± 0.59
11	8.2 ± 0.41	12.8 ± 0.32	4.1 ± 0.06	3.7 ± 0.07	9.6 ± 0.36
12	7.4 ± 0.34	9.4 ± 0.19	2.4 ± 0.04	10.0 ± 0.19	4.0 ± 0.13
[Au(SPyCOOH)(PR$_3$)], R = amino acid					
13	14.7 ± 0.88	8.2 ± 0.13	7.6 ± 0.11	35.2 ± 0.53	25.9 ± 1.04
14	7.7 ± 0.22	10.7 ± 0.16	3.7 ± 0.06	12.8 ± 0.27	6.1 ± 0.18
15	14.7 ± 0.20	12.3 ± 0.32	4.3 ± 0.08	11.3 ± 0.29	6.5 ± 0.25
16	15.9 ± 0.50	11.5 ± 0.29	6.7 ± 0.13	65.5 ± 1.12	33.1 ± 1.13
17	14.1 ± 0.55	14.5 ± 0.26	3.6 ± 0.06	49.3 ± 1.53	28.4 ± 1.16
18	14.3 ± 0.61	11.6 ± 0.20	7.5 ± 0.07	3.0 ± 0.08	4.0 ± 0.11
[Au(SPyCOOH)(PR$_3$)], R = amide derivative of amino acid					
19	28.3 ± 1.02	27.2 ± 0.65	3.9 ± 0.06	14.4 ± 0.42	4.9 ± 0.16
20	19.1 ± 0.67	8.1 ± 0.23	3.9 ± 0.07	8.1 ± 0.28	1.4 ± 0.04
21	14.4 ± 0.60	12.5 ± 0.32	3.8 ± 0.06	7.6 ± 0.26	5.2 ± 0.15
22	18.8 ± 0.71	14.1 ± 0.29	3.7 ± 0.06	14.6 ± 0.57	6.8 ± 0.23
23	19.4 ± 0.62	15.2 ± 0.33	5.3 ± 0.11	13.6 ± 0.40	3.0 ± 0.09
24	30.5 ± 0.82	19.2 ± 0.36	7.7 ± 0.15	5.8 ± 0.16	4.0 ± 0.16
Complexes obtained by structural modifications of [Au(SPyCOR)(PPh$_3$)]					
25	15.7 ± 0.66	17.4 ± 0.48	3.8 ± 0.04	12.0 ± 0.19	3.2 ± 0.13
26	8.3 ± 0.39	13.1 ± 0.26	3.4 ± 0.06	3.5 ± 0.11	2.8 ± 0.08
27	32.5 ± 1.24	29.3 ± 0.70	36.5 ± 0.77	>25	10.4 ± 0.35
28	18.7 ± 0.64	22.5 ± 0.67	8.6 ± 0.14	17.9 ± 0.45	15.4 ± 0.60
29	33.5 ± 1.31	>50	NT	8.3 ± 0.19	1.9 ± 0.05
30	16.5 ± 0.92	17.1 ± 0.39	4.2 ± 0.05	7.7 ± 0.24	3.0 ± 0.08
31	18.3 ± 0.75	15.1 ± 0.27	3.6 ± 0.06	>25	1.2 ± 0.02
32	>50	>50	NT	>25	10.7 ± 0.29
33	4.1 ± 0.11	1.2 ± 0.04	0.9 ± 0.07	4.5 ± 0.14	0.8 ± 0.02
Cisplatin	105 ± 0.90	71 ± 0.80	7.4 ± 0.10	14.0 ± 0.20	65.0 ± 0.92

NT—Non tested; [a] A549 = lung carcinoma, MiaPaca2 = pancreatic carcinoma, Jurkat = T-cell leukemia, R69 = lymphoid cell line and 293T = embryonic kidney fibroblasts.

A remarkable cytotoxic activity against the same human tumor cell lines (A549, Jurkat and MiaPaca2) was observed for the gold(I) complexes with cysteine-containing dipeptides **34–44** (Figure 3a and Table 3) [23]. Starting from the gold(I) complexes **34–39** of the general formula [Au(Boc-Cys-XOMe-S)(PPh$_3$)] (X = Gly, Ala, Val, Phe, Met and Pro; Boc = *tert*-butoxycarbonyl), different structural modifications of these complexes, such as changes in the phosphine ligand (PPh$_2$Py instead of PPh$_3$; **40**), introducing new amino protecting group (benzyloxycarbonyl (Z) instead of Boc; **41**), the use of non-proteinogenic rigid octahydroindole methyl ester (OicOMe; **42**) and increasing the number of Au(I) ions per complex unit (**43** and **44**), were performed in order to investigate the influence of these structural changes on the cytotoxicity of the gold(I) complexes with cysteine-containing dipeptides. As can be seen from Table 3, the IC$_{50}$ values for the complexes **34–44** are in low micromolar range, from 1.5 to 15.6 µM in A549, 0.4 to 2.2 µM in Jurkat and 0.1 to 5.4 µM in MiaPaca2 cells, being much lower than the corresponding IC$_{50}$ values for cisplatin (105, 7.4 and 71, respectively). The structural modifications leading to the formation of the complexes **40–42** (Figure 3a) resulted in almost similar activity against MiaPaca2 and A549 cells, while the introduction of an additional Au(PPh$_3$)$^+$ moiety to the complex **34** led to the formation of complex **43** (Figure 3a), which is the most potent against MiaPaca2 cell line (Table 3). On the other hand, the coordination of two Au(PPh$_3$)$^+$ fragments (complex **44**, Figure 3a) did not significantly improve the cytotoxicity of the complex [23].

Table 3. In vitro cytotoxic activity (IC$_{50}$, 24 h, µM) of gold(I) complexes **34–51** in A549, Jurkat and MiaPaca2 cell lines [23–25].

Complex	Cell Line A549	MiaPaca2	Jurkat
Complexes with cysteine-containing dipeptides			
34	1.5 ± 0.2	2.0 ± 0.2	0.9 ± 0.1
35	1.9 ± 0.1	1.9 ± 0.1	1.6 ± 0.1
36	2.3 ± 0.1	3.0 ± 0.1	2.2 ± 0.1
37	15.6 ± 0.11	5.4 ± 0.1	0.4 ± 0.1
38	4.8 ± 0.1	1.8 ± 0.1	1.7 ± 0.1
39	3.0 ± 0.1	0.7 ± 0.1	0.5 ± 0.1
40	5.0 ± 0.2	0.5 ± 0.1	0.8 ± 0.1
41	2.7 ± 0.1	1.5 ± 0.1	1.1 ± 0.1
42	2.1 ± 0.1	1.2 ± 0.1	1.5 ± 0.1
43	1.8 ± 0.1	0.1 ± 0.1	0.6 ± 0.1
44	3.5 ± 0.1	1.5 ± 0.1	0.8 ± 0.1
Complexes with 4-mercaptoproline			
45	1.8 ± 0.15	3.0 ± 0.19	0.8 ± 0.08
46	3.8 ± 0.37	6.1 ± 0.54	3.5 ± 0.32
47	>25	>25	9.3 ± 0.65
48	3.5 ± 0.29	2.3 ± 0.22	0.6 ± 0.08
49	1.9 ± 0.16	1.8 ± 0.17	0.5 ± 0.07
Complexes with N,S-heterocyclic carbenes			
50	0.4 ± 0.01	16.6 ± 0.2	6.2 ± 0.1
51	>25	24.8 ± 0.1	ca. 25
Cisplatin	105 ± 0.90	71 ± 0.80	7.4 ± 0.10

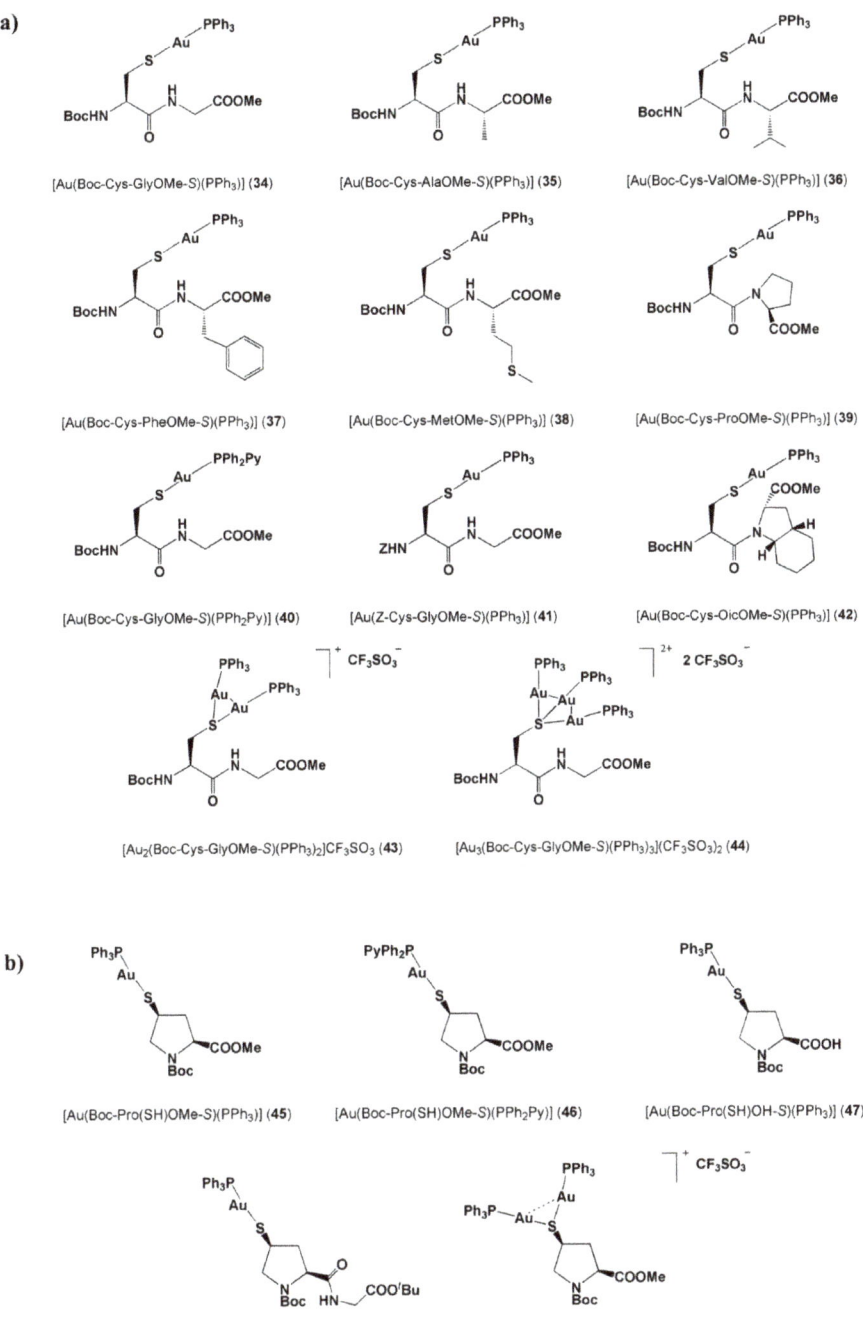

Figure 3. (a) Gold(I) complexes with cysteine-containing dipeptides **34–44** [23] and (b) with non-proteinogenic 4-mercaptoproline amino acid **45–49** [24] showing antiproliferative activity.

Similar cytotoxic activity to the gold(I) complexes with cysteine-containing dipeptides **34–44** was manifested by the gold(I) complexes bearing non-proteinogenic 4-mercaptoproline amino acid (**45–49**, Figure 3b and Table 3) [24]. This amino acid is a hybrid of proline and homocysteine, and has the properties of both amino acids, nucleophilic character and reducing properties of the thiol group of homocysteine and rigid structure of proline. N-Protected 4-mercaptoproline ester was reacted with [AuCl(PR$_3$)] to yield [Au(Boc-Pro(SH)OMe-S)(PR$_3$)] complexes (PR$_3$ = PPh$_3$ (**45**) and PPh$_2$Py (**46**)), while [Au(Boc-Pro(SH)OH-S)(PPh$_3$)] complex (**47**) was obtained after the basic hydrolysis of the amino ester moiety in **45**. The latter complex could be further transformed to the [Au(Boc-Pro(SH)-GlyO^{t}Bu-S)(PPh$_3$)] complex (**48**), while the reaction of **45** with [Au(CF$_3$SO$_3$)(PPh$_3$)] afforded the dinuclear [Au$_2$(Boc-Pro(SH)OMe-S)(PPh$_3$)$_2$]CF$_3$SO$_3$ complex (**49**) (Figure 3b). As with the case of the abovementioned complexes, the antiproliferative activity of **45–49** is mainly due to the Au(I) center. The role of the phosphine ligand is to stabilize this metal ion and to enhance the lipophilicity, allowing the crossing through the membrane, while the thiolate takes part in the substitution reactions with the biomolecules and has an influence on the complex transport or biodistribution [24]. The gold(I) complexes **45, 46, 48** and **49** showed an excellent cytotoxic activity in the investigated human tumor cell lines (A549, Jurkat and MiaPaca2), with IC$_{50}$ values being lower than 6.1 µM and, in some cases, in the nanomolar range (Table 3). Complex **45** is found to be 100, 23 and 10-fold more active than cisplatin in A549, MiaPaca2 and Jurkat cell lines, respectively, and approximately 2-fold more active than its analogue **46**, which contains PPh$_2$Py ligand instead of PPh$_3$. Similar to the abovementioned gold(I) complexes containing a thiolate ligand functionalized with several amino acids or peptide moieties [21], the formation of the complex with 4-mercaptoproline acid (**47**) decreased significantly the cytotoxic activity, probably as the consequence of higher lipophilicity of the ester group or higher reactivity of carboxylic group, preventing the complex to reach its target [24]. On the other hand, complex **48** with a dipeptide containing 4-mercaptoproline and dinuclear complex **49** have the antiproliferative activity at similar extent to the parent complex **45**.

Two gold(I) complexes **50** and **51** with N,S-heterocyclic carbenes derived from the peptides containing L-thiazolylalanine (Thz-Ala) showed good cytotoxic activity in vitro against the A549, MiaPaca2 and Jurkat cell lines (Figure 4a and Table 3) [25]. The carbene complex **50** with iodide was more efficient in all tested cell lines in respect to the thiolate-containing complex **51**, what can be the consequence of the higher lability of Au–I bond in respect to the Au–S bond and higher lipophilicity of **50**. As can be seen from Table 3, the IC$_{50}$ values of **51** in A549 cell line was higher than 25 µM and approximately 25 µM for the remaining two cells, while **50** showed an excellent cytotoxicity against the A549 cell line with the IC$_{50}$ value being in the submicromolar range.

The phenylalanine-N-heterocyclic carbene gold(I) complex **52** and its amino acid and dipeptide derivatives **53** and **54**, respectively, were evaluated for their in vitro cytotoxic potential against the human cell lines HeLa (human cervix carcinoma), HT-29 (human caucasian colon adenocarcinoma grade II) and HepG2 (human hepatocellular liver carcinoma) (Figure 4b and Table 4) [26]. These three complexes have shown moderate to good antiproliferative activity, with HeLa cells being the most sensitive. Among them, amino acid conjugate complex **53** exhibited the best activity, while the remaining two complexes showed a decrease in antitumor activity, which may be the consequence of differential uptake or different intracellular interactions [26].

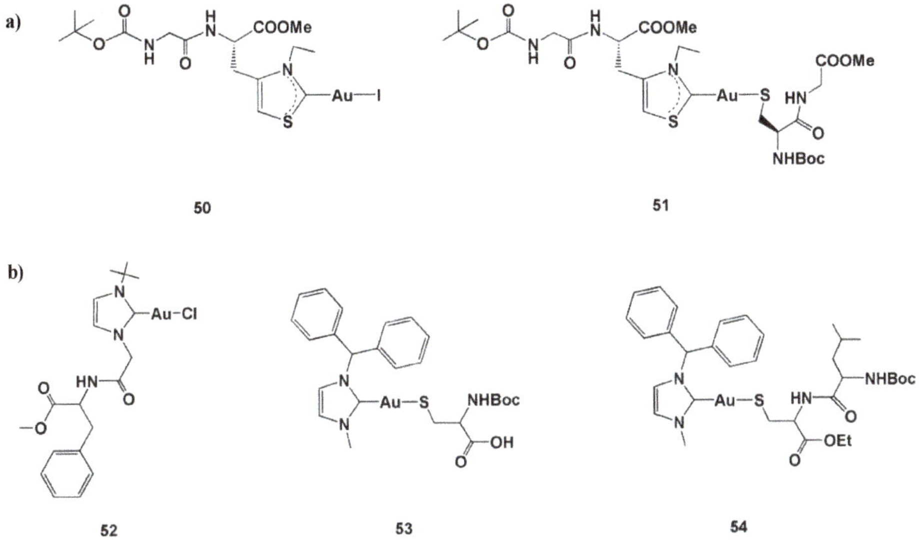

Figure 4. (a) Gold(I) complexes **50** and **51** with N,S-heterocyclic carbenes derived from the peptides containing L-thiazolylalanine [25] and (b) N-heterocyclic carbene gold(I) complexes **52–54** [26] showing antiproliferative activity.

Table 4. In vitro cytotoxic activity (IC$_{50}$, µM) of N-heterocyclic carbene gold(I) complexes **52–54** in HeLa, HT-29 and HepG2 cell lines [26].

Cell Line	HeLa		HepG2		HT-29	
Complex	CV [a]	Resazurin	CV	Resazurin	CV	Resazurin
52	45.5 ± 4.8	52.7 ± 5.0	61.4 ± 7.4	71.3 ± 5.9	63.8 ± 6.7	58.9 ± 4.3
53	3.4 ± 1.3	8.3 ± 1.4	15.2 ± 1.7	20.4 ± 0.9	10.5 ± 1.9	16.9 ± 1.7
54	17.3 ± 3	29.4 ± 1.8	28.1 ± 4.5	30.0 ± 2.6	26.8 ± 2.1	34.6 ± 1.8

[a] CV = crystal violet.

3. Gold(III) Complexes Containing Amino Acids and Peptides Moieties

The abovementioned ferrocene bioconjugate, FcCO-MetOMe, was also used for the synthesis of gold(III) species, [Au(FcCO-MetOMe-S)(C$_6$F$_5$)$_3$] (**55**; Figure 5), which was evaluated for in vitro cytotoxicity [19]. The IC$_{50}$ values determined for this complex of 87 ± 2.0, 88 ± 2.2 and 31 ± 2.4 µM in HeLa, MCF-7 and N1E-115 cell lines, respectively, are higher than the corresponding values for the gold(I) complexes with the ferrocene bioconjugates and phosphine ligands (PPh$_3$ and PPh$_2$Py) (Table 1), indicating that the presence of a phosphine ligand is important for enhancement of cytotoxic potential of gold complexes [19].

A decrease in the antiproliferative activity after oxidation of Au(I) to Au(III) was also observed for the phenylalanine-N-heterocyclic carbene gold(III) complex **56** (Figure 5) in HT-29 cell line (IC$_{50}$ = 125.8 ± 49.7 and 282.5 ± 41.8 µM determined by crystal violet and resazurin assays, respectively) [26].

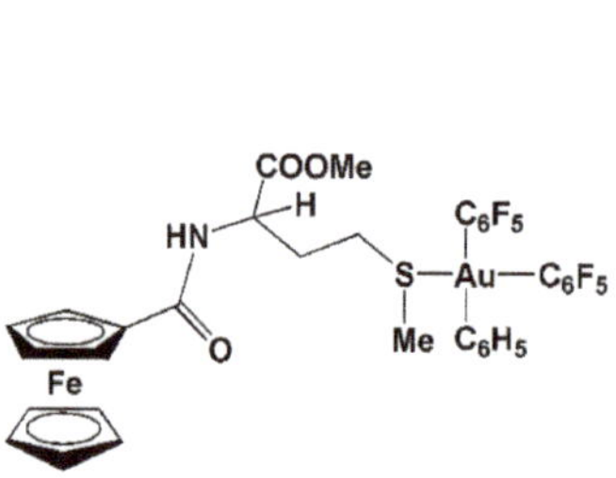

Figure 5. Structural formulas of gold(III) complexes **55** [19] and **56** [26], showing a decrease in the antiproliferative potential in comparison to the analogue gold(I) species.

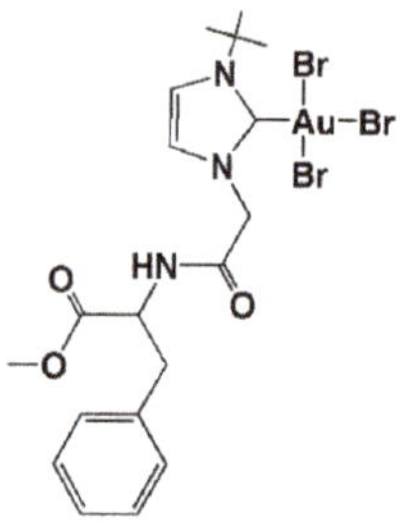

With the aim to obtain gold(III)-based peptidomimetics with anticancer properties that could target two peptide transporters, PEPT1 and PEPT2, which are upregulated in some tumor cells, Fregona et al. synthesized the complexes of the general formula [AuX$_2$(dtc-Sar-AA-O^{t}Bu)] (dtc = dithiocarbamate; AA = Gly, X = Br$^-$ (**57**)/Cl$^-$ (**58**); AA = Aib (2-aminoisobutyric acid), X = Br$^-$ (**59**)/Cl$^-$ (**60**); AA = L-Phe, X = Br$^-$ (**61**)/Cl$^-$ (**62**)) (Figure 6) [27]. The in vitro cytotoxicity of these complexes was evaluated toward the human androgen receptor-negative prostate cancer PC3 and DU145 cells, ovarian adenocarcinoma 2008 cells and the cisplatin-resistant C13 cell line, and Hodgkin's lymphoma L540 cells over 72 h, while cisplatin was used as a reference (Table 5). Among these complexes, **61** and **62** were less active, with the IC$_{50}$ values higher than cisplatin (except C13 cell line), indicating that an aromatic or highly hydrophobic fragment attached to sarcosine decreases the cytotoxicity of the gold(III) complex. On the other hand, complexes **57–60** were generally more efficient than cisplatin, with Aib-containing complex **59** being the most active towards the investigated tumor cell lines (Table 5). Importantly, **59** showed 30-fold higher activity than cisplatin in growth inhibition of cisplatin-resistant ovarian adenocarcinoma C13 cells, excluding the occurrence of cross-resistance. Both the most active Aib-containing complexes **59** and **60** exerted their cytotoxic activity within the first 24 h, while the activity of the cisplatin significantly increased with the increasing of the exposure time [27]. The fact that the two most active complexes contain 2-aminoisobutyric acid is not surprising, since this amino acid is abundant in a class of peptide antibiotics, showing anticancer and antiviral properties [28,29]. Moreover, this amino acid plays a crucial role in the biological activity of peptide antibiotics, by forcing the peptide backbone to fold into helical arrangements and providing a capability to cross and/or perturb cell membranes [30]. Apoptosis was shown to be the major mechanism of cell death in the case of prostate cancer PC3 and DU145 cells and ovarian adenocarcinoma C13 cells, for the most active complexes **59** and **60**, while, in the case of the cisplatin-sensitive 2008 cells and the Hodgkin's lymphoma L540 cell line, the majority of dead cells underwent late apoptosis/necrosis over 24 h, after exposure to the these complexes [27]. On the other hand, the remaining gold(III) complexes and cisplatin were less effective in inducing apoptosis.

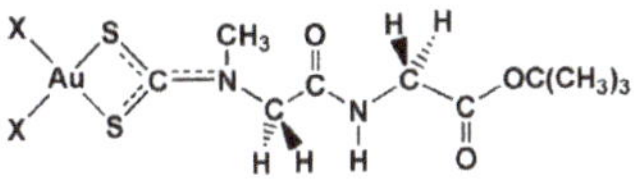

$[AuX_2(dtc\text{-}Sar\text{-}Gly\text{-}O^tBu)]$
X = Br^- (**57**)/Cl^- (**58**)

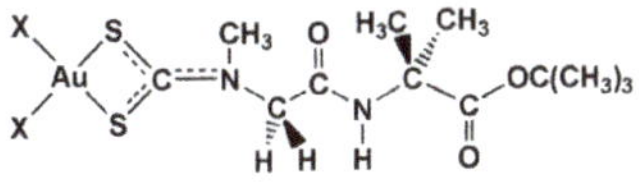

$[AuX_2(dtc\text{-}Sar\text{-}Aib\text{-}O^tBu)]$
X = Br^- (**59**)/Cl^- (**60**)

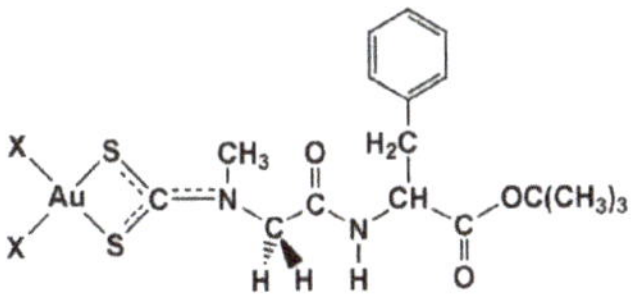

$[AuX_2(dtc\text{-}Sar\text{-}L\text{-}Phe\text{-}O^tBu)]$
X = Br^- (**61**)/Cl^- (**62**)

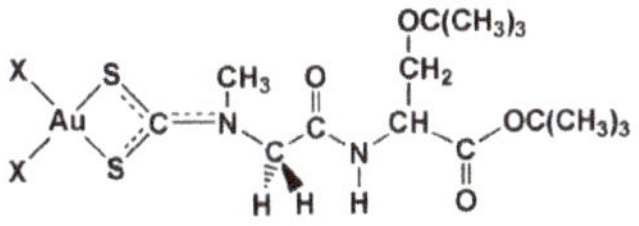

[Au(dtc-Sar-L-Ser(*t*-Bu)-O(*t*-Bu))X_2]
X = Br^- (**63**), Cl^- (**64**)

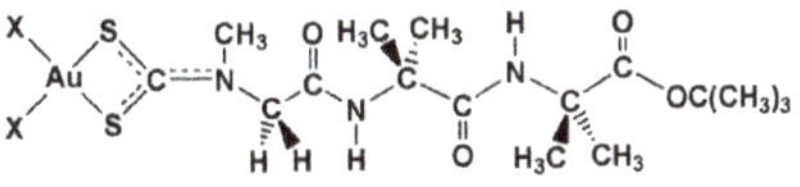

[Au(dtc-Sar-Aib_2-O(*t*-Bu))X_2]
X = Br^- (**65**), Cl^- (**66**)

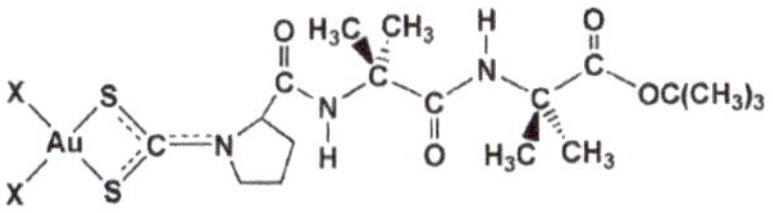

[Au(dtc-D,L-Pro-Aib_2-O(*t*-Bu))X_2]
X = Br^- (**67**), Cl^- (**68**)

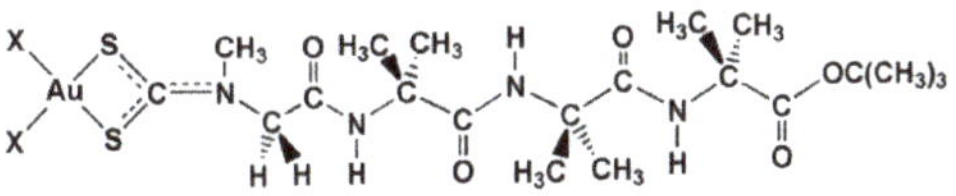

[Au(dtc-Sar-Aib_3-O(*t*-Bu))X_2]
X = Br^- (**69**), Cl^- (**70**)

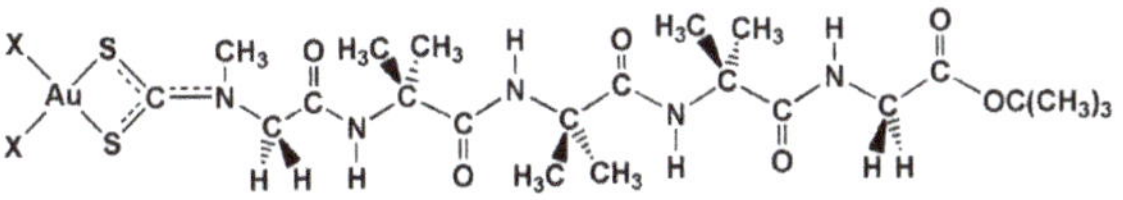

[Au(dtc-Sar-Aib_3-Gly-OEt)X_2]
X = Br^- (**71**), Cl^- (**72**)

Figure 6. Gold(III)-dithiocarbamato derivatives of oligopeptides **57–72** showing antiproliferative activity [27,31].

The same group of authors further synthesized the gold(III)-dithiocarbamato derivatives of oligopeptides, $[AuX_2(dtc\text{-}Sar\text{-}L\text{-}Ser(^tBu)\text{-}O^tBu))]$ (X = Br^- (**63**)/Cl^- (**64**)), $[AuX_2(dtc\text{-}AA\text{-}Aib_2\text{-}O^tBu)]$ (AA = Sar (sarcosine, *N*-methylglycine), X = Br^- (**65**)/Cl^- (**66**), AA = D,L-Pro, X=Br^- (**67**)/Cl^- (**68**)), $[AuX_2(dtc\text{-}Sar\text{-}Aib_3\text{-}O^tBu)]$ (X = Br^- (**69**)/Cl^- (**70**)), and $[AuX_2(dtc\text{-}Sar\text{-}Aib_3\text{-}Gly\text{-}OEt)]$ (X = Br^- (**71**)/Cl^- (**72**)) (Figure 6) and evaluated their cytotoxic activity toward four different cell lines (PC3, 2008, C13 and L540; Table 5) [31].

The IC_{50} values of the complexes determined after the exposure of L540 cells to the complexes **63–72** are in the range 1.4–5.4 μM, being similar to the corresponding value for cisplatin of 2.5 μM. The gold(III) complexes **63–68** showed antiproliferative activity comparable to or lower than cisplatin on prostate cancer and ovarian adenocarcinoma cells, while the tetra- and pentapeptide derivatives **69–72** appeared to be less effective. However, against the cisplatin-resistant C13 cell line, all these gold(III) complexes were much more active than cisplatin (Table 5). Among the complexes **63–72**, **68** containing proline and 2-aminoisobutiric acid turned out to be the most active toward all the investigated tumor cell lines, having the IC$_{50}$ values comparable to the abovementioned complex **59** [27,31].

Table 5. In vitro cytotoxic activity (IC$_{50}$, µM, 72 h) of the gold(III)-dithiocarbamato derivatives of oligopeptides **57–72** against different tumor cell lines [27,31].

Complex	Cell Line [a] PC3	DU145	2008	C13	L540
[AuX$_2$(dtc-Sar-AA-O^{t}Bu)] AA = Gly, X = Br$^-$ (**57**)/Cl$^-$ (**58**); Aib, X = Br$^-$ (**59**)/Cl$^-$ (**60**); L-Phe, X = Br$^-$ (**61**)/Cl$^-$ (**62**)					
57	1.3 ± 0.1	4.5 ± 0.9	18.0 ± 1.6	11.5 ± 1.2	2.1 ± 0.2
58	1.6 ± 0.1	2.5 ± 0.3	13.2 ± 1.1	15.9 ± 1.3	3.4 ± 0.2
59	0.8 ± 0.1	1.4 ± 0.1	4.5 ± 0.2	3.7 ± 0.3	1.5 ± 0.2
60	1.1 ± 0.1	2.2 ± 0.1	4.7 ± 0.2	5.1 ± 0.4	1.7 ± 0.3
61	16.8 ± 1.7	13.2 ± 1.2	41.2 ± 4.0	17.2 ± 1.8	16.4 ± 1.5
62	16.5 ± 1.6	15.3 ± 1.5	43.4 ± 3.8	21.5 ± 1.5	7.3 ± 0.5
[AuX$_2$(dtc-Sar-L-Ser(^{t}Bu)-O^{t}Bu))] X = Br$^-$ (**63**)/Cl$^-$ (**64**)					
63	5.2 ± 0.5	NT	17.5 ± 1.6	17.0 ± 1.6	3.4 ± 0.2
64	6.5 ± 0.7	NT	15.2 ± 1.5	16.1 ± 1.3	1.4 ± 0.1
[AuX$_2$(dtc-AA-Aib$_2$-O^{t}Bu)] AA = Sar, X = Br$^-$ (**65**)/Cl$^-$ (**66**); D,L-Pro, X = Br$^-$ (**67**)/Cl$^-$ (**68**)					
65	5.8 ± 0.6	NT	12.0 ± 1.1	15.0 ± 1.3	3.8 ± 0.4
66	5.8 ± 0.5	NT	14.2 ± 1.5	15.4 ± 1.4	4.1 ± 0.5
67	6.3 ± 0.7	NT	16.3 ± 1.4	11.5 ± 1.1	2.2 ± 0.1
68	3.0 ± 0.2	NT	8.2 ± 0.7	7.8 ± 0.5	2.2 ± 0.2
[AuX$_2$(dtc-Sar-Aib$_3$-O^{t}Bu)] X = Br$^-$ (**69**)/Cl$^-$ (**70**)					
69	16.0 ± 1.7	NT	42.8 ± 4.1	17.5 ± 1.6	3.9 ± 0.4
70	16.1 ± 0.6	NT	43.5 ± 1.2	11.5 ± 1.2	3.5 ± 0.2
[AuX$_2$(dtc-Sar-Aib$_3$-Gly-OEt)] X = Br$^-$ (**71**)/Cl$^-$ (**72**)					
71	15.0 ± 1.4	NT	29.2 ± 3.0	22.9 ± 1.9	5.4 ± 0.5
72	20.0 ± 1.9	NT	27.8 ± 2.3	35.5 ± 3.9	5.0 ± 0.8
Cisplatin	3.3 ± 0.3	4.5 ± 0.1	19.4 ± 1.2	117.2 ± 9.1	2.5 ± 0.1

NT—Non tested; [a] PC3 and DU145 cells = the human androgen receptor-negative prostate cancer cells, 2008 cells = ovarian adenocarcinoma and C13 = the parent cisplatin-resistant cell line, L540 = Hodgkin's lymphoma.

Seven gold(III) complexes with different L-histidine-containing dipeptides, [Au(Gly-L-His-N_A,N_P,N3)Cl]Cl·3H$_2$O (**73**), [Au(Gly-L-His-N_A,N_P,N3)Cl]NO$_3$·1.25H$_2$O (**74**), [Au(L-Ala-L-His-N_A,N_P,N3)Cl]NO$_3$·2.5H$_2$O (**75**), [Au(L-Ala-L-His-N_A,N_P,N3)Cl][AuCl$_4$]·H$_2$O (**76**), [Au(L-Val-L-His-N_A,N_P,N3)Cl]Cl·2H$_2$O (**77**), [Au(L-Leu-L-His-N_A,N_P,N3)Cl]Cl (**78**) and [Au(L-Leu-L-His-N_A,N_P,N3)Cl][AuCl$_4$]·H$_2$O (**79**) were evaluated for in vitro cytotoxicity against different human tumor cell lines (Figure 7 and Table 6) [32–34]. Different spectroscopic techniques confirmed that tridentate coordination of the X-L-His dipeptides (X = Gly, L-Ala, L-Val and L-Leu) through the N3 imidazole nitrogen (N3), deprotonated nitrogen of the amide bond (N_P) and to the nitrogen of the N-terminal amino group (N_A) stabilized +3 oxidation state of gold, preventing its reduction to Au(I)/Au(0) under physiological conditions. Firstly, complex **73** was tested against the tumor cell line A2780 (human ovarian carcinoma), both sensitive (A2780/S) and resistant (A2780/R) to cisplatin [32]. This complex exhibited a remarkable antiproliferative activity against A2780/S cell line (IC$_{50}$ = 5.2 ± 1.63 µM), being slightly less active than cisplatin (IC$_{50}$ = 1.6 ± 0.58 µM). Importantly, the complex retains a significant cytotoxicity on the A2780/R cell line (IC$_{50}$ = 8.5 ± 2.3 µM), having the resistance factor of only 1.6. The results of this study also showed that the Zn(II), Pd(II), Pt(II) and Co(II) complexes with the same dipeptide manifested only modest activity toward A2780/S and A2780/R cell lines, confirming that the presence of Au(III) ion was crucial for the cytotoxic effects [32].

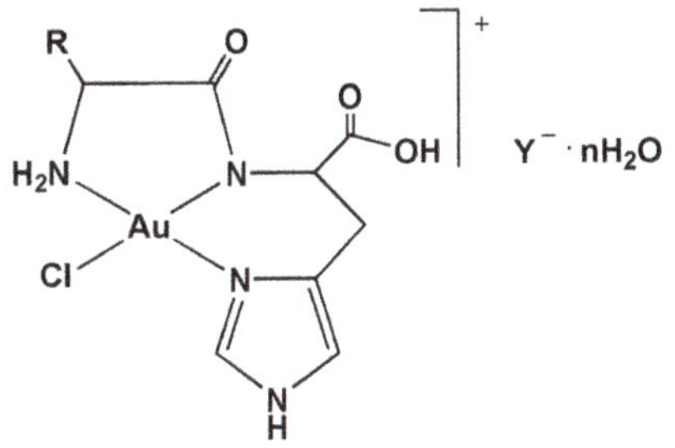

[Au(X-L-His-N_A, N_P N3)Cl]Y·nH_2O
73 – 79

Gold(III) complex	X	R	Y	n
73	Gly	H	Cl⁻	3
74	Gly	H	NO_3^-	1.25
75	L-Ala	CH_3	NO_3^-	2.5
76	L-Ala	CH_3	[$AuCl_4$]⁻	1
77	L-Val	$CH(CH_3)_2$	Cl⁻	2
78	L-Leu	$CH_2CH(CH_3)_2$	Cl⁻	0
79	L-Leu	$CH_2CH(CH_3)_2$	[$AuCl_4$]⁻	1

Figure 7. Gold(III) complexes with L-histidine-containing dipeptides **73–79**, which were evaluated for the in vitro cytotoxic potential [32–34].

Table 6. In vitro cytotoxic activity (IC_{50}, μM) of gold(III) complexes with L-histidine-containing dipeptides **73–79** against different cell lines [32–34].

Complex	Cell Line [a] MRC-5	MCF-7	HT-29	HL-60	Raji	HeLa	A549
73	>200					150	150
74	>200	19.68 ± 0.23	14.70 ± 1.36	11.93 ± 1.02	3.30 ± 0.02	>200	>200
75	>200	>100	>100	>100	>100	170	170
76	150[b]					75	30
77	>200					>200	>200
78	170					55	115
79	100					65	75
Cisplatin	0.48 ± 0.02	1.56 ± 0.26	18.60 ± 2.32	10.31 ± 2.54	2.25 ± 0.10		

[a] MRC-5 = human lung fibroblasts, MCF-7 = breast cancer, HT-29 = colon cancer, HL-60 = human promyelocytic leukemia, Raji = human Burkitt's lymphoma, HeLa = cervix cancer, and A549 = lung cancer; [b] The results are from three independent experiments, each performed in quadruplicate. SD were within 2%–5%.

In a continuation, the antiproliferative activity of **74** and **75** was evaluated against five human tumor cell lines, MCF-7 (breast cancer), HT-29 (colon cancer), HL-60 (human promyelocytic leukemia), Raji (human Burkitt's lymphoma) and one human normal cell line MRC-5 (human lung fibroblasts) [33]. These complexes are less cytotoxic than cisplatin, with the exception of **74** in the case of the HT-29 cell line (Table 6). The latter complex showed the activity against all tested human tumor cell lines, being non-toxic in the MRC-5 cell line, while the cytotoxicity of L-Ala-L-His-Au(III) complex **75** was not observed against the tested cell line.

In addition, the cytotoxicity of all seven gold(III) complexes with X-L-His dipeptides **73–79** was assessed against two human cancer, cervix (HeLa) and lung (A549), cell lines and compared to the activity against the MRC-5 cell line (Table 6) [34]. As can be seen, these complexes did not manifest great anticancer potential; however, their cytotoxicity towards normal cell line was low (IC_{50} > 100 μM). Moreover, the complexes **76** and **79** showed significant antiangiogenic activity in vivo in a zebrafish embryos model (Figure 8). Although these two complexes achieved comparable antiangiogenic effect to the clinically used auranofin and sunitinib malate at 30-fold higher concentration, the zebrafish embryos following the treatment with Au(III) complexes had no cardiovascular side effects in comparison to those upon treatment with auranofin and sunitinib malate. The binding of the gold(III) complexes to the active sites of both human and bacterial (*Escherichia coli*) thioredoxin reductases (TrxRs) was confirmed by molecular docking study, suggesting that the mechanism of biological action of these complexes can be associated with their interaction with TrxR active site [34].

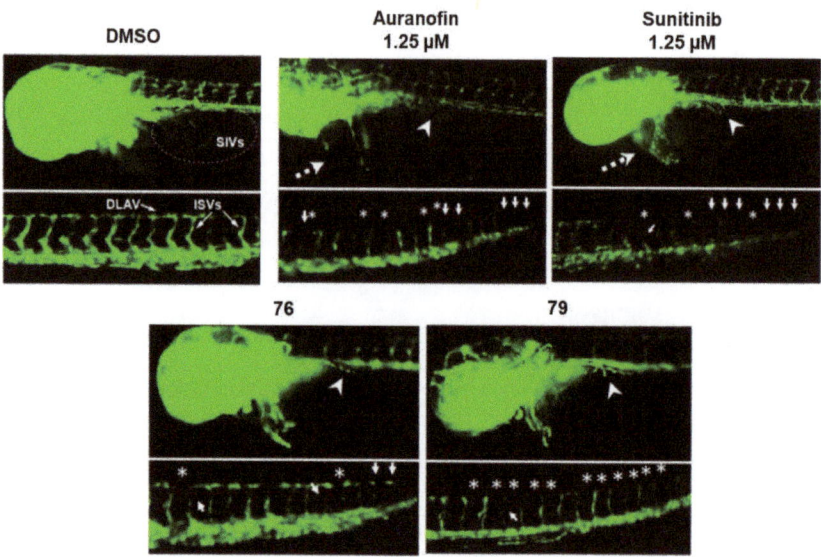

Figure 8. The effect of gold(III) complexes **76** and **79**, sunitinib malate and auranofin on subintestinal vessels (SIVs), intersegmental vessels (ISVs) and dorsal longitudinal anastomotic vessels (DLAVs) development in zebrafish embryos. Reduced SIVs (arrowhead), disrupted DLAVs (asterisk), thinner or reduced ISVs (arrow), and pericardial edema (dashed arrow) are designated. The Figure was adapted from the Reference [34].

Additionally, the antimicrobial activity of the abovementioned gold(III) complexes with L-histidine-containing dipeptides **73–79** were evaluated against the Gram-positive (*Staphylococcus aureus*, *Listeria monocytogenes*, *Enterococcus faecalis* and *Enterococcus faecium*) and Gram-negative (*Acinetobacter baumannii*) bacteria and two strains of *Candida* (*C. albicans* and *C. parapsilosis*) [34]. In most cases, the minimal inhibitory concentration (MIC) values were between 200 and 400 µM, indicating their moderate to low activity. Nevertheless, the MIC values of **74** and **79** against Gram-positive *E. faecium* and Gram-negative *A. baumannii* were found to be 80 and 100 µM, respectively. Beside these results, a study related to the antimicrobial potential of a series of gold(III) complexes differing in the ligand structure, showed that two gold(III) complexes with L-histidine-containing dipeptides **74** and **75** exhibited relatively weak effects in comparison to the other studied complexes against the investigated bacterial strains [35].

4. Conclusions

The review article summarizes the results achieved in studies on the antiproliferative activity of gold(I) and gold(III) complexes containing amino acids and peptides as biocompatible ligands that can deliver a gold ion to its target. Except the biocompatibility of amino acids and peptides, the advantage of these ligands could be the fact that the tumor cells overexpress amino acids receptors, resulting in the selectivity of the obtained gold complexes toward the tumor cells in respect to the healthy ones. From the presented data, it could be seen that a large number of gold(I) and gold(III) complexes have been evaluated in vitro against different tumor cell lines, while the in vivo studies are rather scarce. In general, gold(I) complexes have shown higher cytotoxic activity than the gold(III) species, being, in some cases, superior in respect to the well-known anticancer agent, cisplatin. Except amino acids and peptides, the gold(I) complexes usually had phosphines and *N*-heterocyclic carbenes ancillary ligands, which also contributed to their biological properties.

Gold(III) complexes with L-histidine-containing dipeptides, in which *N*-terminal amino acid is L-alanine and L-leucine, showed selectivity in terms of cancer vs. normal cell lines and achieved antiangiogenic effects comparable to the known inhibitors of angiogenesis—auranofin and sunitinib malate—without toxic-side effects, in contrast to those following auranofin and sunitinib malate treatment. These findings make them good candidates for the further development of antiangiogenic drugs.

From this review, it can be concluded that the use of amino acids and peptides as ligands for the synthesis of biologically active gold complexes has merit for the development of novel therapeutic agents for the treatment of cancer, which is a major burden of disease worldwide.

Funding: This research has been financially supported by the Ministry of Education, Science and Technological Development of the Republic of Serbia (Agreement No. 451-03-68/2020-14) and by the Serbian Academy of Sciences and Arts under strategic projects program—grant agreement No. 01-2019-F65 and project of this institution No. F128. The authors wish to thank Dr. *Jasmina Nikodinovic-Runic* (Institute of Molecular Genetics and Genetic Engineering, University of Belgrade) for her valuable comments on the manuscript.

Conflicts of Interest: The authors declare no conflicts of interest. The funders had no role in the collection, analyses or interpretation of the data, in writing of the manuscript and in the decision to publish the manuscript.

References

1. Huaizhi, Z.; Yuantao, N. China's ancient gold drugs. *Gold Bull.* **2001**, *34*, 24–29. [CrossRef]
2. Benedek, T.G. The history of gold therapy for tuberculosis. *J. Hist. Med. Allied Sci.* **2004**, *59*, 50–89. [CrossRef] [PubMed]
3. Fricker, S.P. Medical uses of gold compounds: Past, present and future. *Gold Bull.* **1996**, *29*, 53–60. [CrossRef]
4. Ott, I. On the medicinal chemistry of gold complexes as anticancer drugs. *Coord. Chem. Rev.* **2009**, *253*, 1670–1681. [CrossRef]
5. Nobili, S.; Mini, E.; Landini, I.; Gabbiani, C.; Casini, A.; Messori, L. Gold compounds as anticancer agents: Chemistry, cellular pharmacology, and preclinical studies. *Med. Res. Rev.* **2010**, *30*, 550–580. [CrossRef] [PubMed]
6. Berners-Price, S.J.; Filipovska, A. Gold compounds as therapeutic agents for human diseases. *Metallomics* **2011**, *3*, 863–873. [CrossRef]
7. Bertrand, B.; Casini, A. A golden future in medicinal inorganic chemistry: The promise of anticancer gold organometallic compounds. *Dalton Trans.* **2014**, *43*, 4209–4219. [CrossRef]
8. Nardon, C.; Boscutti, G.; Fregona, D. Beyond platinums: Gold complexes as anticancer agents. *Anticancer Res.* **2014**, *34*, 487–492.
9. Lima, J.C.; Rodríguez, L. Phosphine-Gold(I) compounds as anticancer agents: General description and mechanisms of action. *Anti-Cancer Agents Med. Chem.* **2011**, *11*, 921–928. [CrossRef]
10. Milacic, V.; Fregona, D.; Dou, Q.P. Gold complexes as prospective metal-based anticancer drugs. *Histol. Histopathol.* **2008**, *23*, 101–108.
11. Kostova, I. Gold coordination complexes as anticancer agents. *Anti-Cancer Agents Med. Chem.* **2006**, *6*, 19–32. [CrossRef] [PubMed]
12. Yeo, C.I.; Ooi, K.K.; Tiekink, E.R.T. Gold-based medicine: A paradigm shift in anti-cancer therapy? *Molecules* **2018**, *23*, 1410. [CrossRef] [PubMed]
13. Mora, M.; Gimeno, M.C.; Visbal, R. Recent advances in Gold–NHC complexes with biological properties. *Chem. Soc. Rev.* **2019**, *48*, 447–462. [CrossRef] [PubMed]
14. Glišić, B.Đ.; Djuran, M.I. Gold complexes as antimicrobial agents: An overview of different biological activities in relation to the oxidation state of the gold ion and the ligand structure. *Dalton Trans.* **2014**, *43*, 5950–5969. [CrossRef]
15. Navarro, M.; Gabbiani, C.; Messori, L.; Gambino, D. Metal-based drugs for malaria, trypanosomiasis and leishmaniasis: Recent achievements and perspectives. *Drug Discov. Today* **2010**, *15*, 1070–1078. [CrossRef] [PubMed]
16. Kastin, A.J. (Ed.) *Handbook of Biologically Active Peptides*; Elsevier: Amsterdam, The Netherland, 2006.
17. Peacock, A.F.A.; Bullen, G.A.; Gethings, L.A.; Williams, J.P.; Kriel, F.H.; Coates, J. Gold-phosphine binding to De Novo designed coiled coil peptides. *J. Inorg. Biochem.* **2012**, *117*, 298–305. [CrossRef]

18. Astruc, D. Why is ferrocene so exceptional? *Eur. J. Inorg. Chem.* **2017**, *2017*, 6–29. [CrossRef]
19. Gimeno, M.C.; Goitia, H.; Laguna, A.; Luque, M.E.; Villacampa, M.D.; Sepúlveda, C.; Meireles, M. Conjugates of ferrocene with biological compounds. Coordination to gold complexes and antitumoral properties. *J. Inorg. Biochem.* **2011**, *105*, 1373–1382. [CrossRef]
20. Gutiérrez, A.; Bernal, J.; Villacampa, M.D.; Cativiela, C.; Laguna, A.; Gimeno, M.C. Synthesis of new gold(I) thiolates containing amino acid moieties with potential biological interest. *Inorg. Chem.* **2013**, *52*, 6473–6480. [CrossRef]
21. Gutiérrez, A.; Gracia-Fleta, L.; Marzo, I.; Cativiela, C.; Laguna, A.; Gimeno, M.C. Gold(I) thiolates containing amino acid moieties. Cytotoxicity and structure–activity relationship studies. *Dalton Trans.* **2014**, *43*, 17054–17066. [CrossRef]
22. Ghosh, S. Cisplatin: The first metal based anticancer drug. *Bioorg. Chem.* **2019**, *88*, 102925. [CrossRef] [PubMed]
23. Gutiérrez, A.; Marzo, I.; Cativiela, C.; Laguna, A.; Gimeno, M.C. Highly cytotoxic bioconjugated gold(I) complexes with cysteine—containing dipeptides. *Chem. Eur. J.* **2015**, *21*, 11088–11095. [CrossRef] [PubMed]
24. Gutiérrez, A.; Cativiela, C.; Laguna, A.; Gimeno, M.C. Bioactive gold(I) complexes with 4-mercaptoproline derivatives. *Dalton Trans.* **2016**, *45*, 13483–13490. [CrossRef] [PubMed]
25. Gutiérrez, A.; Gimeno, M.C.; Marzo, I.; Metzler-Nolte, N. Synthesis, characterization, and cytotoxic activity of Au^I N,S-heterocyclic carbenes derived from peptides containing L-thiazolylalanine. *Eur. J. Inorg. Chem.* **2014**, *2014*, 2512–2519. [CrossRef]
26. Lemke, J.; Pinto, A.; Niehoff, P.; Vasylyeva, V.; Metzler-Nolte, N. Synthesis, structural characterisation and anti-proliferative activity of NHC gold amino acid and peptide conjugates. *Dalton Trans.* **2009**, 7063–7070. [CrossRef]
27. Kouodom, M.N.; Ronconi, L.; Celegato, M.; Nardon, C.; Marchiò, L.; Dou, Q.P.; Aldinucci, D.; Formaggio, F.; Fregona, D. Toward the selective delivery of chemotherapeutics into tumor cells by targeting peptide transporters: Tailored gold-based anticancer peptidomimetics. *J. Med. Chem.* **2012**, *55*, 2212–2226. [CrossRef]
28. Brown, K.L.; Hancock, R.E.W. Cationic host defense (antimicrobial) peptides. *Curr. Opin. Immunol.* **2006**, *18*, 24–30. [CrossRef]
29. Johnstone, S.A.; Gelmon, K.; Mayer, L.D.; Hancock, R.E.; Bally, M.B. In vitro characterization of the anticancer activity of membrane-active cationic peptides. I. Peptide-mediated cytotoxicity and peptide-enhanced cytotoxic activity of doxorubicin against wild-type and p-glycoprotein over-expressing tumor cell lines. *Anti-Cancer Drug Des.* **2000**, *15*, 151–160.
30. Toniolo, C.; Brückner, H. *Peptaibiotics*; Wiley-VCD: Weinheim, Germany; Wiley-VHCA: Zürich, Switzerland, 2009.
31. Kouodom, M.N.; Boscutti, G.; Celegato, M.; Crisma, M.; Sitran, S.; Aldinucci, D.; Formaggio, F.; Ronconi, L.; Fregona, D. Rational design of gold(III)-dithiocarbamato peptidomimetics for the targeted anticancer chemotherapy. *J. Inorg. Biochem.* **2012**, *117*, 248–260. [CrossRef]
32. Carotti, S.; Marcon, G.; Marussich, M.; Mazzei, T.; Messori, L.; Mini, E.; Orioli, P. Cytotoxicity and DNA binding properties of a chloro glycylhistidinate gold(III) complex (GHAu). *Chem. Biol. Interact.* **2000**, *125*, 29–38. [CrossRef]
33. Glišić, B.Đ.; Stanić, Z.D.; Rajković, S.; Kojić, V.; Bogdanović, G.; Djuran, M.I. Solution study under physiological conditions and cytotoxic activity of the gold(III) complexes with L-histidine-containing peptides. *J. Serb. Chem. Soc.* **2013**, *78*, 1911–1924. [CrossRef]
34. Warżajtis, B.; Glišić, B.Đ.; Savić, N.D.; Pavic, A.; Vojnovic, S.; Veselinović, A.; Nikodinovic-Runic, J.; Rychlewska, U.; Djuran, M.I. Mononuclear gold(III) complexes with L-histidine-containing dipeptides: Tuning the structural and biological properties by variation of the N-terminal amino acid and counter anion. *Dalton Trans.* **2017**, *46*, 2594–2608. [CrossRef] [PubMed]
35. Radulović, N.S.; Stojanović, N.M.; Glišić, B.Đ.; Randjelović, P.J.; Stojanović-Radić, Z.Z.; Mitić, K.V.; Nikolić, M.G.; Djuran, M.I. Water-soluble gold(III) complexes with N-donor ligands as potential immunomodulatory and antibiofilm agents. *Polyhedron* **2018**, *141*, 164–180. [CrossRef]

© 2020 by the authors. Licensee MDPI, Basel, Switzerland. This article is an open access article distributed under the terms and conditions of the Creative Commons Attribution (CC BY) license (http://creativecommons.org/licenses/by/4.0/).

Review

Natural and Engineered Electron Transfer of Nitrogenase

Wenyu Gu [1] and Ross D. Milton [2],*

[1] Department of Civil and Environmental Engineering, Stanford University, E250 James H. Clark Center, 318 Campus Drive, Stanford, CA 94305, USA
[2] Department of Inorganic and Analytical Chemistry, University of Geneva, Sciences II, Quai Ernest-Ansermet 30, 1211 Geneva 4, Switzerland
* Correspondence: ross.milton@unige.ch

Received: 8 April 2020; Accepted: 23 April 2020; Published: 27 April 2020

Abstract: As the only enzyme currently known to reduce dinitrogen (N_2) to ammonia (NH_3), nitrogenase is of significant interest for bio-inspired catalyst design and for new biotechnologies aiming to produce NH_3 from N_2. In order to reduce N_2, nitrogenase must also hydrolyze at least 16 equivalents of adenosine triphosphate (MgATP), representing the consumption of a significant quantity of energy available to biological systems. Here, we review natural and engineered electron transfer pathways to nitrogenase, including strategies to redirect or redistribute electron flow in vivo towards NH_3 production. Further, we also review strategies to artificially reduce nitrogenase in vitro, where MgATP hydrolysis is necessary for turnover, in addition to strategies that are capable of bypassing the requirement of MgATP hydrolysis to achieve MgATP-independent N_2 reduction.

Keywords: nitrogenase; ammonia; metalloenzyme; electron transfer; ferredoxin; flavodoxin; Fe protein; MoFe protein

---

1. Introduction to Nitrogenase

Ammonia (NH_3) is an important commodity for agricultural and chemical industries that is currently produced at over 150 million tons per year [1,2]. Currently, the majority of this NH_3 is produced from molecular hydrogen (H_2) and kinetically inert dinitrogen (N_2, bond dissociation enthalpy of +945 kJ mol^{-1}) by the Haber–Bosch process, which operates at a high temperature (~700 K) and a high pressure (~100 atm) in order to optimize NH_3 production [3]. These conditions, in combination with the production of H_2 (commonly by steam reforming of natural gas), result in the consumption of 1–2% global energy and the production of around 3% of global carbon dioxide (CO_2) emissions. Due to ever-increasing concerns and awareness of climate change, there is significant interest in the development of new catalysts for N_2 fixation. For instance, the development of a new (bio)catalytic system that operates under mild conditions could enable the decentralization of NH_3 production as a key strategy for improved environmental sustainability.

Select bacteria and archaea are able to produce an enzyme, nitrogenase, which can fix N_2 to NH_3 under mild (physiological) conditions [4]. Thus, nitrogenase is of interest to new biotechnologies and new bio-inspired N_2-fixing catalysts. Nitrogenase is a two-component metalloenzyme consisting of a reductase (iron or "Fe" protein) and a N_2-reducing protein (MoFe protein), where the name "MoFe" refers to the metals employed in its catalytic cofactor. There are two alternative nitrogenases that are dependent on V (VFe) and Fe only (FeFe), although the Mo-nitrogenase system from the soil bacterium *Azotobacter vinelandii* will serve as the model to outline nitrogenase's mechanism (Figure 1) [5].

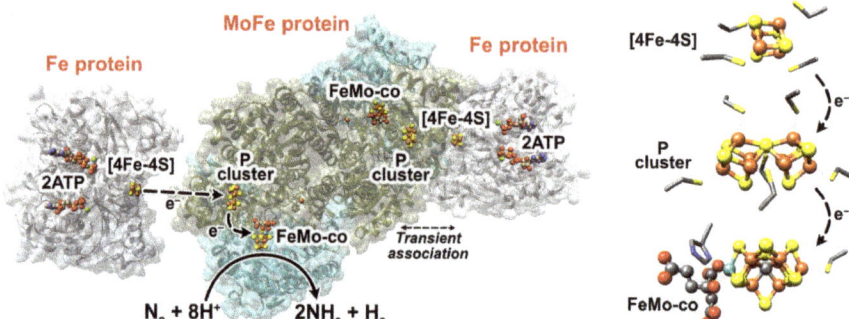

Figure 1. (**Left**) Representation of Mo-nitrogenase from *Azotobacter vinelandii*. The Fe protein is shown in gray and the MoFe protein is shown in cyan (NifD, α-subunit) and olive-green (NifK, ß-subunit). This representation was adapted from PDB:4WZA, which used non-hydrolyzable ATP analogues to form a tight Fe:MoFe protein complex. (**Right**) The FeS cofactors of Mo-dependent nitrogenase, where the coordinating residues are also shown. The homocitrate partner of the FeMo-co is shown on the left of the FeMo-co. Fe = rust, S = yellow, C = gray, N = blue, O = red, Mo = cyan, Mg = green.

The Fe protein of *A. vinelandii* is a homodimer of approximately 66 kDa in mass encoded by the *nifH* gene. Each dimer contains an adenosine triphosphate (MgATP) binding site, whereas a [4Fe-4S] iron-sulfur cluster is located between each monomer and coordinated by two cysteine (Cys) residues from each monomer [6]. The MoFe protein is a $\alpha_2\beta_2$ tetramer of approximately 240 kDa encoded by the *nifD* (α subunit) and *nifK* (ß subunit) genes. Each αß dimeric half contains a [8Fe-7S] "P" cluster that bridges both subunits (coordinated by Cys residues from each subunit) and a [7Fe-9S-C-Mo-homocitrate] FeMo-cofactor ("FeMo-co") contained within the α subunit [7,8]. The FeMo-co, the cofactor at which N_2 is reduced, is coordinated by a Cys and a histidine (His) residue. Each αß half of the MoFe protein repeatedly transiently associates with a 1e$^-$-reduced and ATP-bound Fe protein, during which a single electron is ultimately transferred from the [4Fe-4S]$^{1+}$ cluster to the FeMo-co via the P cluster [9]. The order of events that transpire during this transient association is debated, although electron transfer, 2MgATP hydrolysis (to adenosine diphosphate, 2MgADP), the release of two inorganic phosphate (P_i) equivalents, and Fe protein dissociation occur. Oxidized Fe protein is subsequently reduced by flavodoxin and/or ferredoxin in vivo or commonly dithionite (DT) in vitro, thereby restarting the "Fe protein cycle" [10,11]. The reduction potential ($E^{o\prime}$) of the [4Fe-4S]$^{2+/1+}$ couple of the Fe protein is approximately -0.30 V vs. the standard hydrogen electrode (SHE, see [12,13]), free of MgATP or MgADP [9,14]. Upon the association of MgATP, the $E^{o\prime}$ of [4Fe-4S]$^{2+/1+}$ is modulated to -0.43 V vs. SHE, which decreases further to -0.62 V when the Fe protein associates with the MoFe protein making it a more potent electron donor [14,15]. Perhaps the two most-relevant redox couples of the P cluster ($P^{N/1+}$ and $P^{1+/2+}$) both have $E^{o\prime}$'s of approximately -0.31 V vs. SHE. Further, the $E^{o\prime}$ of the as-isolated FeMo-co (M^N) and its one-electron oxidation production (M^{OX}) is approximately -0.04 V vs. SHE, while a second lower-potential redox couple (M^{RED}) is also thought to be relevant to nitrogenase's catalytic cycle, although its $E^{o\prime}$ has not been accurately determined [9].

Electron transfer from the Fe protein is currently thought to be coupled to the rate-limiting step of P_i release (25–27 s^{-1}), suggesting electron transfer takes place with a rate constant of around 13 s^{-1} [16]. Along with a study by Duval et al. in 2013, this suggests that electron transfer from the [4Fe-4S]$^{1+}$ cluster to the P cluster takes place prior to MgATP hydrolysis [17]. Finally, a deficit spending model has been proposed by which an electron is transferred from the P cluster to the FeMo-co, prior to the P cluster being reduced by the Fe protein's [4Fe-4S]$^{1+}$ cluster [18,19]. Nevertheless, the proposal that electron transfer from the Fe protein's [4Fe-4S] cluster takes place prior to the hydrolysis of MgATP suggests that artificial electron transfer to the P cluster (independent of the Fe protein) is possible. Thus significant interest has developed concerning the MgATP-independent fixation of N_2 by nitrogenase, given that the hydrolysis of each MgATP accounts for around −50 kJ mol^{-1} in vivo [20].

The reduction of N_2 to NH_3 requires 6e$^-$, although optimal N_2 fixation by nitrogenase occurs after eight transient association events of the Fe protein (and the transfer of 8e$^-$):

$$N_2 + 8e^- + 8H^+ + 16MgATP \rightarrow 2NH_3 + H_2 + 16MgADP + 16P_i \tag{1}$$

Notably, the fixation of each N_2 also results in the evolution of at least one equivalent of H_2. Lowe and Thorneley developed an early model to describe the observation of H_2 formation, which has since developed into a model that highlights the pivotal nature of a 4e$^-$-reduced FeMo-co known as the E_4 state [8,10,21–23]. By this model, the resting FeMo-co in its E_0 state accumulates individual electrons and protons during each Fe protein association event in order to reach the E_4 state at which N_2 binds and undergoes subsequent reduction. These electrons are stored as metal-hydrides (M-H) on the Fe centers. Prior to the binding of N_2, the E_{2-4} states can unproductively evolve H_2 by M-H protonation, thereby "dropping" by E_{x-2} states [21,22]. In contrast, the productive evolution of H_2 is thought to occur in order to accommodate N_2 binding and its partial reduction, by the reductive elimination (re) of H_2 from the FeMo-co E_4 state [24]. At this stage, H_2 can also undergo oxidative addition (oa) to the FeMo-co and displace N_2, which is also considered to be unfavorable [22]. Four additional Fe protein association events (transferring 4e$^-$ and hydrolyzing an additional 8MgATP) then lead to the production of $2NH_3$. This serves as a suitable model to explain the production of one equivalent of H_2 for each N_2 reduced [21,22]. Thus, H_2 can be produced by unproductive or productive pathways, with increased H_2 evolution and MgATP hydrolysis resulting from a combination of both. Questions therefore remain surrounding the reversibility or catalytic bias of the E_4 state, given that the alternative nitrogenases also appear to follow the same re mechanism while appearing to be less-efficient at N_2 fixation [5,25]. A "just-in-time" mechanism could serve as a useful model to justify the rate-limiting nature of electron transfer from the Fe protein (~13 s^{-1}) such that unproductive H_2 formation and the oa of H_2 is minimized [23]. Finally, research has also questioned the suitability of the proposed re/oa model for N_2 fixation by nitrogenase [26]. Density functional theory calculations were employed to calculate the partial charges of the FeMo-co during key turnover states, where a re mechanism was not supported and the possible involvement of hydrogen atoms (including hydrogen atom transfer, formation, and elimination steps) was highlighted.

In summary, it is clear that electron transfer within the nitrogenase complex, and by extension, to the nitrogenase complex, is of high importance to biological N_2 fixation as well as new biotechnologies and bio-inspired N_2 fixation systems. Thus, this article reviews natural and engineered electron transfer to nitrogenase, in the contexts of both in vivo and in vitro catalysis. Specifically, we discuss the nature and delivery of electrons to nitrogenase in vivo. We also review approaches to transfer electrons to nitrogenase in vitro, either through the Fe protein or independent of the Fe-protein for MgATP-decoupled catalysis by the MoFe protein.

2. In Vivo Electron Transfer to Nitrogenase

2.1. Electron Transfer from Ferredoxin and Flavodoxin

In vivo reduction of Fe protein requires low-potential electrons provided by the electron-transferring proteins flavodoxin and ferredoxin [27,28]. In vitro studies of nitrogenase typically utilize dithionite (DT, $E^{o\prime} \approx -0.66$ V vs. SHE) or Ti(III) citrate ($E^{o\prime}_{Ti(III)/(IV)} = -0.8$ V vs. SHE) as electron donors due to their ease of preparation and use [29–31]. However, recent studies have reevaluated electron transfer to the Fe protein with flavodoxin, which is crucial for a mechanistic understanding of the Fe protein cycle and of nitrogenase activity in physiological conditions [16]. It is also important to consider physiological electron donors when developing biological systems to maximize electron transfer to nitrogenase.

Flavodoxin (Fld) is a monomeric electron-transfer protein carrying a non-covalently bound flavin mononucleotide (FMN) as a redox center. It consists of a central five-stranded parallel β-sheet flanked on either side by α-helices. Based on the presence of a ~20-residue loop splitting the fifth β-strand, Flds involved in electron transfer to nitrogenase are classified as long-chain flavodoxins [32,33]. The cofactor FMN has three pertinent redox states: oxidized quinone (Ox), semiquinone (SQ), and hydroquinone (HQ), where $FMN^{Ox/SQ}$ and $FMN^{SQ/HQ}$ couples are 1e⁻ redox reactions. Although not overly abundant in aqueous solution, FMN^{SQ} is stabilized when bound to flavodoxin; further, crossed-potentials of flavins can be stabilized in proteins and can support a process named flavin-based electron bifurcation (reviewed in [34–36]). $E°$'s of Fld II from *A. vinelandii* range from −0.25 to −0.1 V (vs. SHE) for the Fld^{Ox}/Fld^{SQ} couple and from −0.5 to −0.4 V (vs. SHE) for the Fld^{SQ}/Fld^{HQ} couple [32]. The latter is one of the lowest reported potentials in the flavodoxin family [37]. Because the $E°'$ of the Fe protein is <−0.3 V (vs. SHE), the Fld^{HQ} state is expected to be the functional electron donor [27,38]. In *A. vinelandii*, Fld ($E°'_{SQ/HQ}$ = −0.46 V vs. SHE) is encoded by the gene *nifF*.

Ferredoxin (Fdx) is an FeS cluster-containing protein that was first isolated from *Clostridium pasteurianum* [39]. Fdxs coordinate FeS clusters by Cys ligands and can be divided into different groups based on the number and types of FeS clusters [40]. Fdxs that participate in electron transport to Fe protein are found to have [2Fe-2S]-type [41], [4Fe-4Fe]/[3Fe-4S]-type, or 2[4Fe-4S]-type clusters [42–44]. The [2Fe-2S]-type is normally found in cyanobacteria with the corresponding gene named *fdxH*, while the other two are found in diverse diazotrophic groups and are usually designated as *fdxN*. The possible redox states and redox potentials of Fdxs are modulated by their peptide structure (and ultimately, their coordination spheres) and occur in the ranges of −0.24 to −0.46 V, −0.05 to −0.42 V, and −0.28 to −0.68 V, for the couples [2Fe-2S]$^{2+/1+}$, [3Fe-4S]$^{1+/0}$, and [4Fe-4S]$^{2+/1+}$, respectively (vs. SHE) [45]. In comparison to Fld, Fdx is more sensitive to O_2 due to the lability of their FeS clusters. Phylogenetic analysis indicates that Fld (NifF) is enriched in diazotrophs with aerobic or facultative anaerobic life styles, and is believed to be an adaptive strategy for the diversification of Nif-nitrogenase from anaerobic to aerobic taxa during evolution [46]. Although other Fdxs are found to be involved in the assembly of nitrogenase cofactors or for protecting it against oxygen [47,48], the following discussion focuses on Fdxs involved in electron transport.

Fld forms a tight complex with Fe protein with high affinity (Figure 2). Reported dissociation constants for the pairs from *Klebsiella pneumoniae* are 13 μM (MgATP-bound Fe protein) and 49 μM (MgADP-bound Fe protein) and from *Rhodobacter capsulatus* is 0.44 μM. [49,50] It is generally believed that in the Fe protein cycle, reduction of Fe protein and exchange of MgADPs for MgATPs takes place after its dissociation from MoFe protein, implying that the oxidized Fe protein cannot be reduced when bound to MoFe protein. Further, the Fe protein is believed to interact with Fld/Fdx using the same binding interface that is used with the MoFe protein [16,33,51,52]. This hypothesis is guided by docking models that predict electrostatic interactions between the positively charged surroundings of the [4Fe-4S] cluster of Fe protein and the negatively charged surroundings of FMN of NifF from *A. vinelandii*, possibly assisted by an eight amino acid loop (residues 64–71) on NifF [16,32,52] (Figure 2). The MgADP-bound state of the Fe protein has the most complementary docking interface with Fld compared with the MgATP-bound state. Experimentally, this hypothesis is supported by results from cross-linking studies and time-resolved limited proteolysis using NifF and Fe protein from *A. vinelandii* [16,52]. Similar modeling results were obtained for Fdx, although experimental support has not yet been reported [33,41].

Early studies indicated that the specific catalytic activity of MoFe protein [27,51] and ATP/e⁻ efficiency of nitrogenase are higher when NifF is used as the reductant of Fe protein rather than DT. Second-order rate constants of Fe protein reduction are two to three orders of magnitude times higher when NifF is used [53] or included [16] as reductant compared to DT [16] or Ti (III) [53]. Both physiological reductants and chemical reductants reduce MgADP-bound Fe protein faster than its nucleotide-free form. Nevertheless, recent reexamination of the rate difference under pseudo-first order reaction conditions showed similar trends [16]. The diminished performance of DT might be partially due to the slow generation of radical anion $SO_2^{•-}$ (K_d 1.5 nM, rate constant of ~2 s⁻¹) [16].

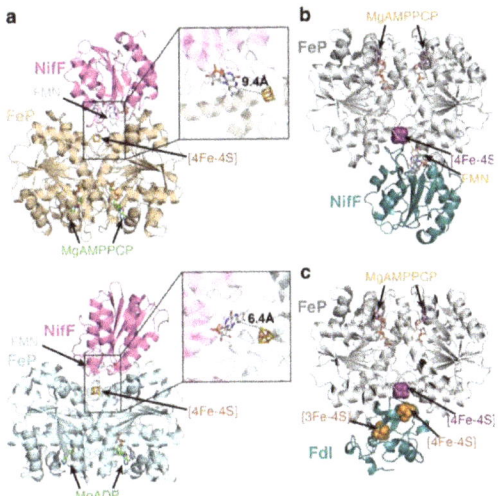

Figure 2. Docking models for reduction of the Fe protein by NifF and Fdx (FdI). (**a**) MgAMPPCP-Fe protein (top, PDB: 4WZB) and MgADP-Fe protein (bottom, PDB:1FP6) interacting with NifF (PDB:1YOB). (**b**) MgAMPPCP-Fe protein (PDB:4WZB) interacting with NifF (PDB:5K9B) and (**c**) FdI (PDB:6FDR). Hypothetical electron transfer distances between the [4Fe-4S] cluster of the Fe protein (NifH) and the FMN cofactor of NifF are shown. Reprinted (adapted) with permission from [9]. Copyright 2020 American Chemical Society. Republished with permission of American Soc for Biochemistry and Molecular Biology Copyright 2017, from [52]; permission conveyed through Copyright Clearance Center, Inc. Adapted with permission from [33]. Copyright 2017 John Wiley and Sons.

The findings above help revise kinetic models of the Fe protein cycle, as the Fe protein reduction rate has been used to estimate its dissociation from MoFe protein. When using DT, the complex dissociation rate was measured as ~6 s^{-1} [11,51] and was believed to be the limiting step of the Fe protein cycle. Yet a high rate of 759 s^{-1} was obtained when NifF was used, indicating that P$_i$ release (25–27 s^{-1}) is actually the rate-limiting step [16]. In line with this, the specific activity of nitrogenase was shown to increase by 50–170% with the presence of NifF as compared to DT alone [16,51].

Electron transfer efficiency to nitrogenase is expected to be improved when using physiological electron donors. While the 1e$^-$ reduced [4Fe-4S]$^{1+}$ of Fe protein is commonly believed to be the only physiologically relevant state [6], Watt and Reddy discovered that the [4Fe-4S]$^{1+}$ cluster can be further reduced by 1e$^-$ to a stable all-ferrous [4Fe-4S]^0 state by methyl viologen (MV) [54]. An $E^{o\prime}$ of −0.46 V vs. SHE for the 1+/0 couple was reported [54]. Since, many studies have characterized the [4Fe-4S]^0 state of Fe protein reduced by Ti(III) or Eu(II), Eu(II) complexes employed to study nitrogenase typically have $E^{o\prime}$'s ranging from −0.6 to −1.1 V vs. SHE [53,55–61]. One study reported an $E^{o\prime}$ of −0.79 V for the [4Fe-4S]$^{1+/0}$ couple [62]. Lowery et al. (2006) showed MgADP or MgATP-bound Fe protein can be reduced by FldHQ state of NifF (*A. vinelandii*) from [4Fe-4S]^1 to [4Fe-4S]^0 independent of catalysis, which supports a $E^{o\prime}$ of −0.46 V of the [4Fe-4S]^0 state and the possibility of its physiological relevance [63]. The Fe protein's [4Fe-4S]^0 state could allow two electrons to be transferred from FldHQ to Fe protein per two ATP molecules hydrolyzed, reaching a 1:1 ATP:e$^-$ ratio. This was indeed observed in a few studies using either NifF or Ti(III) [53,63,64]. Yet in most studies the ATP:e$^-$ ratio remained at 2:1, even when NifF was used as the reductant [9,16]. In contrast, DT only reduces the [4Fe-4S] cluster to the [4Fe-4S]$^{1+}$ state, and thus transfers only one e$^-$ per transient association cycle.

These findings call for further investigation into nitrogenase's natural electron donors. Identifying Fld/Fdx that directly transfers electrons to nitrogenase and to what extent could be difficult due to their redundancy in both genome and function; it is common for a bacterial or archaeal genome to encode multiple Fld/Fdxs [65]. Due to the high energetic expense of N_2 fixation, nitrogenase genes, including the ones encoding for electron transport, are often co-located and/or transcriptionally co-regulated [66,67]. Mutagenesis combined with nitrogenase activity assays in cell extracts can provide direct proof of a gene's function in electron transfer to nitrogenase. However, most diazotrophs have more than two Fld/Fdxs capable of direct electron-transfer to nitrogenase (Table 1). In rare cases such as in *K. pneumoniae*, a sole Fld NifF is the electron donor and diazotrophic growth is abolished in *nifF* deletion mutants. In comparison, *A. vinelandii* is still able to grow diazotrophically (though much is undermined) when the two electron transport components *nifF* and *fdxA* are deleted [68]. In phototrophic bacteria, Fld is commonly found to serve as an electron donor under iron depleted conditions, whereas Fdx is the main electron donor under iron replete conditions [69].

Table 1. Electron transport components required for nitrogenase in representative diazotrophs. Gene names are shown with protein names, if available.

Species	Direct Electron Donor to Fe Protein	Pathway for Reducing Electron Donor	Reference
Anabaena PCC 7120	Fdx (*fdxB, fdxH, fdxN*) Fld (*nifF* [1])	FNR (*petH*) Hydrogenase (*hupSL*) PFOR (*nifJ* [1])	[70–79]
Azotobater vinelandii	Fdx (*fdxN*) Fld (*nifF*)	FixABCX (*fixABCX*) FNR Rnf1 (*rnf*)	[68,80–82]
Clostridium pasteurianum	Fdx	PFOR (*nifJ*) Hydrogenase	[83–85]
Klebsiella pneumoniae	Fld (*nifF*)	PFOR (*nifJ*)	[86–90]
Sinorhizobium meliloti	Fdx (*fdxN*)	FixABCX (*fixABCX*)	[91–93]
Rhodobacter capsulatus	Fdx (*fdxNC*) Fld (*nifF* [1])	Rnf1 (*rnf*) Hydrogenase (*hupSL*)	[94–101]
Rhodopseudomonas palustris	Fdx (*fer1, ferN*) Fld (*fldA* [1])	FixABCX (*fixABCX*) Hydrogenase (*hupSL*)	[69,102,103]
Rhodospirillum rubrum	Fdx (*fdxN, fdxI*) Fld (*nifF* [2])	FixABCX (*fixABCX*) Hydrogenase (*hupSL*) PFOR (*nifJ* [2])	[104–109]

[1] These genes are expressed by cells growing under iron depleted conditions. [2] These genes are expressed by cells growing under anaerobic conditions in the dark.

2.2. Electron Transfer to Flavodoxin and Ferredoxin

The electrons transferred to Fld/Fdx directly come from pyruvate, NAD(P)H, or hydrogen. Five major enzyme systems capable of reducing Fld/Fdx have been identified (Figure 3, Table 1) and are discussed below.

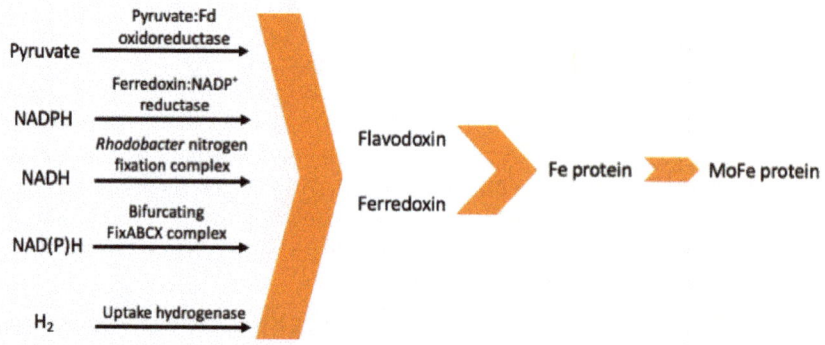

Figure 3. Schematic pathway of natural electron transfer to nitrogenase. (Fd: ferredoxin or flavodoxin).

Pyruvate-fld/fdx oxidoreductase (PFOR) catalyzes oxidization of pyruvate to acetyl-CoA and CO_2 with reduction of Fld or Fdx by using thiamine pyrophosphate and FeS clusters as cofactors. The $E^{o\prime}$ for the pyruvate cleavage is −0.5 V (vs. SHE). PFOR involved in electron transport to nitrogenase is commonly found as part of *nif* operon designated as *nifJ*. The deletion of *nifJ* in *K. pneumoniae* was found to abolish diazotrophic growth [86–88].

Fdx-NADP$^+$ Reductase (FNR) catalyzes the reversible reaction of Fdx oxidation with NADP$^+$ reduction by a flavin adenine dinucleotide (FAD) cofactor. Fdxs can possess $E^{o\prime}$'s more negative than NADPH/NADP$^+$ (−0.34 V vs. SHE), and the forward reaction (*i.e.*, Fdx oxidation) is energetically favorable. Yet, FNRs that favor the reverse reaction are found in *A. vinelandii* and cyanobacteria and are thought to support nitrogenase activity based on enzyme assay and gene coregulation [70,71,80].

Rhodobacter Nitrogen Fixation (Rnf) complex is a transmembrane ferredoxin—NAD$^+$ oxidoreductase, which catalyzes the NADH-dependent reduction of Fld/Fdx with the a depletion of electrochemical gradient [110]. The Rnf complex shares homology with Na$^+$ pumps and is distributed among diverse N_2-fixing and non-nitrogen-fixing microbes [29,110]. The disruption of genes in the *rnf* operon of *R. capsulatus* resulted in a significant decrease in nitrogenase activity and abolished its ability to grow diazotrophically [94,95].

Bifurcating FixABCX couples the endergonic reduction of Fld to the exergonic reduction of coenzyme Q ($E^{o\prime}$ = +0.01 V vs. SHE) (then to respiratory chain) at the cost of NADH by using a flavin at the bifurcating site [81,82,91,104]. This membrane protein complex was identified to have three FAD moieties, one each in FixA, FixB, and FixC, and two [4Fe-4S] clusters in FixX [81]. Disrupting the Fix system in *Rhodopseudomonas palustris*, *Rhodospirillum rubrum*, and *Sinorhizobium meliloti* completely abolishes or significantly impairs their ability to grow under N_2 fixing conditions [92,102,104]. A double mutant of Fix and RNF1 abolished diazotrophic growth in *A. vinelandii* [81].

Uptake hydrogenase (hup) is a heterodimer of the *hupA* and *hupB* (or *hupS* and *hupL*) gene products that has either a Ni-Fe or Fe-Fe active center [111]. It recaptures H_2 produced during N_2 fixation (Equation (1)) for use as electron donors cycled back to nitrogenase [96], and also seems to protect nitrogenase from oxygen, possibly by catalyzing oxyhydrogen reaction [112,113]. Overproduction of H_2 from nitrogenase was achieved by deleting the uptake hydrogenase [72,114]. The *hup* genes are co-regulated with nitrogenase through complicated and diverse regulatory pathways [83,84,97,103], which was exploited to identify H_2-overproducing variants of the Fe protein [115].

The diverse strategies to reduce Fld/Fdx by different microbes can be better understood if put into metabolic context (Table 1). Bioinformatic analysis indicates proteins that reduce Fld/Fdx with H_2 or pyruvate are enriched in anaerobes, while those with NADH/NADPH are enriched in aerobes, facultative anaerobes, and anoxygenic phototrophs [116]. Aerobic, facultatively anaerobic, and anoxygenic phototrophic diazotrophs produce NADH/NADPH, which is not considered reducing enough to provide electrons for nitrogenase; thus, FixABCX and Rnf are acquired to generate low potential electrons [46]. Specifically, oxygenic phototrophs can energize electrons to low potential using photosystem I (PSI), but the low potential electrons are not available for nitrogenase, as the latter must be spatially (forming heterocysts) or temporally (under the control of circadian clocks) separated from PSI, in which case FNR is used [70]. Interestingly, some non-heterocystous cyanobacteria show nitrogenase activity only under light conditions, implying the existence of molecular mechanisms to protect nitrogenase against oxygen evolved by photosystems [113,117,118].

3. Electron Transfer to Nitrogenase in Engineered Biological Systems

Developing and improving biological systems for N_2 fixation is greatly motivated by the need to design sustainable and economic solutions to nitrogen limitation of crop productivity. Decentralization of NH_3 production is considered one possible strategy to improve environmental sustainability. Great efforts have been made towards engineering plant-associated microbes or plants themselves to heterologously express nitrogenases [4]. From the perspective of synthetic biology, genes required for functional nitrogenase can be grouped into modules of structural genes, genes involved in biosynthesis and maturation of cofactors, and electron-transport components (ETC) [119]. ETC include Fld/Fdx and the reductase of Fld/Fdx. ETC can come from the native host to nitrogenase or be substituted by homologues from the expression host, if compatible (Table 2). In general, the former results in higher nitrogenase activity, while the latter gives the advantage of expressing fewer genes and simplifying genetic engineering [120,121]. It is important to bear in mind that there is limited crosstalk between Fld/Fdx and the Fe protein, and between Fld/Fdx and their reductases from different hosts. For example, Fdx from either *Clostridium pasteurianum* or *R. rubrum* were ineffective in coupling pyruvate oxidation to nitrogenase activity in the cell lysate of *K. pneumoniae* NifF mutant. NifF from *A. vinelandii* was only one-third as effective as NifF from *K. pneumoniae* at transferring electrons from PFOR to *Klebsiella* nitrogenases [88]. Further, in vivo activity could not be predicted by in vitro assays. For example, Fdxs from *R. capsulatus* and *S. meliloti* equally support in vitro acetylene (C_2H_2) reduction to ethylene (C_2H_4) by *R. capsulatus* nitrogenase, yet heterologous in vivo complementation by each other's Fdx was unsuccessful [122,123].

Table 2. Summary of heterologous expression of active nitrogenase in prokaryotic hosts.

Expression Host	Source of Nitrogenase Structural and Accessory Genes [1]	Genes Encoding for Electron Transport Component (ETC) to Nitrogenase	Source of ETC	Reference
E. coli JM109	*A. vinelandii* DJ [2]	*nifFJ* fldA, ydbK	*K. oxytoca* M5al *E. coli* JM109	[124]
E. coli JM109	*K. oxytoca* M5a1	*nifFJ*	*K. oxytoca* M5a1	[125]
E. coli JM109	*Paenibacillus* sp. WLY78	fldA, ydbK	*E. coli* JM109	[121]
E. coli JM109	*Paenibacillus* sp. WLY78	*nifFJ* fer or fldA, pfoAB	*K. oxytoca* M5al *Paenibacillus* sp. WLY78	[120]
E. coli MG1655	*K. oxytoca* M5al	*nifFJ*	*K. oxytoca* M5al	[126]
Pseudomonas protegens Pf-5	*Pseudomonas stutzeri* A1501	*nifFJ*	*Pseudomonas protegens* Pf-5	[127] [3]
Pseudomonas protegens Pf-5	*A. vinelandii* DJ	nifF, fixABCX, rnf1	*A. vinelandii* DJ	[128] [3]
Pseudomonas protegens Pf-5	*P. stutzeri* A1501	nifF, fdxN, rnf	*Pseudomonas protegens* Pf-5	[128]
Rhizobium sp. IRBG74	*K. oxytoca* M5al	*nifFJ*	*K. oxytoca* M5a1	[128]
Rhizobium sp. IRBG74	*R. sphaeroides* 2.4.1	rnf	*Rhizobium* sp. IRBG74	[128]
Synechocystis sp. PCC 6803	*Leptolyngbya boryana* strain dg5	fdxH petFH	*Leptolyngbya boryana* strain dg5 *Synechocystis* sp. PCC 6803	[118]
Synechocystis sp. PCC 6803	*Cyanothece* sp. ATCC 51142	fdxNHB, petFH	*Synechocystis* sp. PCC 6803	[113]

[1] MoFe-type nitrogenase (*nif*) was expressed in these studies unless otherwise noted. [2] Structural genes of Fe-only nitrogenase were expressed in combination with accessory genes of MoFe-type nitrogenase. [3] Other combinations of *nif* genes and expression hosts were explored in this study but are not listed here.

By now, there has been greater success in expressing active nitrogenases in prokaryotic hosts (Table 2). Since Dixon and Postgate conjugated N_2 fixation genes from *Klebsiella oxytoca* into *Escherichia coli* in 1972 [129], nitrogenases from various sources have been cloned and expressed in different hosts (Table 2). *E. coli* has been used as a platform to identify minimal gene requirements for active nitrogenase [121,124] and to optimize expression [125,126,130]. It was found that *fldA* (homologue to *nifF*) and *ydbK* (homologue to *nifJ*) in *E. coli* can support nitrogenase activity, although activity is significantly improved when *nifFJ* from diazotrophs are employed [121,124]. Non-diazotrophic, oxygenic cyanobacteria have also been explored as expression hosts, as they serve as a simpler model of plant chloroplasts [113,118]. The O_2 tolerance of nitrogenase was enhanced by co-expressing an uptake hydrogenase in *Synechocystis* sp. PCC 6803 [113]. Epiphytic and endophytic bacteria are desirable platforms for N_2 fixation, as they can be directly applied. Challenges here remain in the fact that different plants have specific colonizers, and that most natural diazotrophic endophytes do not express nitrogenase under the desired conditions or to a desired level [131,132]. A recent study by Ryu et al. applied different strategies to a wide range of hosts and engineered inducible promoters in combination with nitrogenase genes to tackle both of those problems [128]. Overall, heterologous nitrogenases showed significantly lower activities compared to that in their native hosts with the exception of MoFe nitrogenase from *K. oxytoca* expressed in *E. coli* [125,126,128].

Expressing nitrogenases in eukaryotic systems—either plant or yeast and green algae—as simpler models, have limited success [4]. Separate Nif components have been expressed in these systems but the formation of a fully functional nitrogenase has not been achieved [4]. Both mitochondria and chloroplasts are candidate organelles to host the expression of nitrogenase. The former provides MgATP and a low O_2 environment due to active respiration, while the latter produces abundant reduced Fdx and MgATP by photosynthesis with the byproduct of O_2 [4,133]. Several lines of evidence suggest that chloroplasts might be a more suitable choice. While both locations have Fdx-FNR-type electron transport modules, the ones in chloroplasts function as part of the photosynthesis pathway and share more similarity to those in diazotrophic bacteria. In mitochondria, the Fdx-FNR homologue modules are called adrenodoxin and NADPH-dependent adrenodoxin oxidoreductase as part of biosynthesis pathway of biotin [134–136]. Yang et al. (2017) used *E. coli* as a chassis to study the compatibility between MoFe and FeFe nitrogenases with ETC modules from plant chloroplasts, root plastids, and mitochondria. They found that Fdx-FNR from chloroplasts and root plastids can support the activities of both types of nitrogenase, while the ETC module from mitochondria could not. A hybrid module of mitochondrial FNR and the cyanobacteria Fdx could support nitrogenase activities [137]. In addition, chloroplast genomes of some plants and algae encode a nitrogenase-like enzyme called dark-operative protochlorophyllide oxidoreductase (DPOR) that participates in biosynthesis of chlorophylls and is also O_2 sensitive [138]. It was demonstrated that the Fe protein from *K. pneumoniae* nitrogenase can be expressed and functionally substitute for its homologue in DPOR in *Chlamydomonas reinhardtii*, suggesting that chloroplasts have the potential to provide a suitable environment for nitrogenase [139,140].

Alternative strategies have been explored to increase nitrogenase activities using natural diazotrophs. Several studies exploited a MoFe nitrogenase variant (α-V70A, α-H195Q) [141] to catalyze the reduction of CO_2 to methane as a way for in vivo production of biofuels in *R. palustris*. While ATP is supplied by cyclic photophosphorylation, it was found that the electron flow to nitrogenase can be enhanced by manipulating metabolic pathway or state, such as providing cells with organic alcohols, diverting electrons away from biomass synthesis by using nongrowing cells, or blocking the Calvin–Benson–Bassham cycle [142,143]. Further, the same group demonstrated that wild-type Fe-only nitrogenase in *R. palustris* reduces CO_2 simultaneously with nitrogen fixation to yield CH_4, NH_3, and H_2. Excitingly, this seems to be a universal feature for the Fe-only nitrogenases. The amount of CH_4 produced was low but sufficient to support the growth of an obligate methanotroph in co-culture with oxygen added at intervals [143]. Further studies are needed to see whether the above strategies of metabolic engineering can be applied to enhance the activity of Fe-only nitrogenase and whether similar principles can be applied to engineer other diazotrophs.

Liu et al. demonstrated the use of H_2 generated from catalytic water splitting driven by renewable energy to support diazotrophic growth of autotroph *Xanthobacter autotrophicus*. In that way the production of *X. autotrophicus* as biofertilizer or NH_3 (when glutamine synthetase was inhibited) is efficiently connected to atmospheric nitrogen. The biomass produced was applied to radishes and significantly increased the yield of radish storage roots (Figure 4) [144]. Their system achieved a nitrogen reduction turnover numbers of ~9 × 10^9 bacterial cell^{-1} and 2 × 10^6 nitrogenase^{-1} and a turnover frequency of 1.9 × 10^4 s^{-1} per bacterial cell, or ~4 s^{-1}·nitrogenase^{-1}.

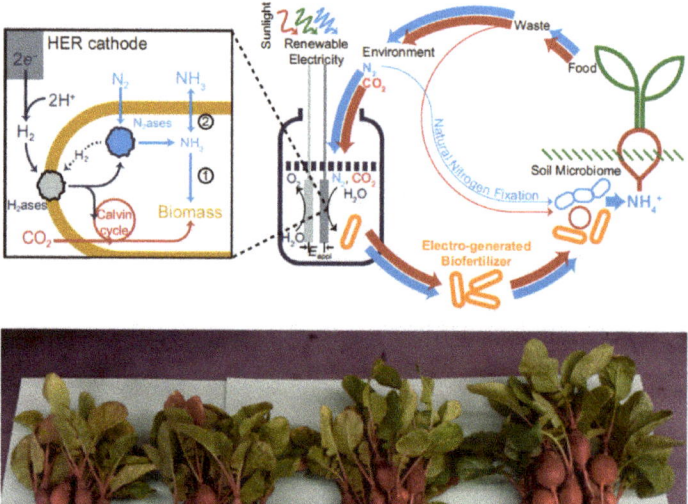

Figure 4. (**Top**) Schematic of ammonia (NH$_3$) production in a biohybrid system that produces hydrogen (H$_2$) from renewable electrical energy and sunlight. Hydrogenases within *Xanthobacter autotrophicus* subsequently oxidize the H$_2$ to ultimately supply electrons to nitrogenase for dinitrogen (N$_2$) fixation. The generated NH$_3$ is incorporated in biomass (pathway 1) or can diffuse extracellularly by inhibiting biomass formation (pathway 2). Red arrows represent carbon cycling and blue arrows represent nitrogen cycling; line widths represent the relative fluxes of these pathways. (**Bottom**) Enhanced radish growth upon biofertilization with *X. autotrophicus* grown from electricity/sunlight-sourced H$_2$. Reproduced from [144] with permission.

4. In Vitro Electron Transfer to Nitrogenase

4.1. Fe protein-Dependent Activity

For in vitro nitrogenase activity, research has sought to artificially deliver electrons to the Fe protein. Subsequently, the reduced Fe protein transfer electrons to the MoFe protein in the presence of the MgATP that is required for the Fe protein cycle. As discussed above, the most commonly employed electron donor is DT, which supports the formation of the 1e$^-$-reduced [4Fe-4S]$^{1+}$ state of the Fe protein. While DT supports nitrogenase catalysis, it can be undesirable in many ways: (i) it is considered single-use since its regeneration is complex, (ii) DT is not particularly stable in aqueous solutions, and (iii) the reduction potential of DT is dependent on its concentration and pH, which can complicate thermodynamic and kinetic interpretations of metalloenzyme properties [30].

In addition to DT, other electron mediators employed with the Fe protein include MV (and derivatives), Ti(III) citrate, Eu(II) complexes, and one of the presumed in vivo electron donors, Fld (NifF) [9]. While Ti(III) and Eu(II) complexes are useful due to their low reduction potentials, they are typically prepared as a stock of a single-use reductant. In the case of Ti(III), Seefeldt and Ensign reported in 1994 that Fe protein reduction could occur such that catalysis by the MoFe protein could be supported, where similar rates of H$^+$ and C$_2$H$_2$ reduction were observed in comparison to DT-driven assays [145]. Further, the oxidation of Ti(III) to Ti(IV) could be followed spectrophotometrically. MV and NifF have also been employed as electron donors to the Fe protein (for subsequent catalysis); however, they are typically employed alongside a relatively higher concentration of DT as the reducing agent [9,16]. As mentioned above, Yang et al. demonstrated that the reduction of nucleotide-free Fe protein by DT increased around 10-fold when NifF was included, presumably as mediator for DT reduction of the Fe protein [16]. Additionally, it was observed that the presence of either MgADP or MgATP diminished DT-driven Fe protein reduction by ~100-fold although this was alleviated by the inclusion of NifF. H_2, NADH, and KBH_4 have also been employed to transfer electrons to MV prior to catalysis by nitrogenase (in cell-free extracts of *C. pasteurianum*), although in all of the reported cases the oxidation of MV by nitrogenase cannot be followed, since the oxidized MV is regenerated in these assays [146,147].

Electrochemical Methods

Recently, MV has been utilized as the sole reductant for Fe protein-dependent nitrogenase activity assays. In 2017, Milton et al. demonstrated the use of electrochemically reduced MV (in the 1e$^-$-reduced MV$^{\bullet+}$ state) as the sole electron donor to the Fe protein [148]. First, it was demonstrated that the oxidation of MV could be followed spectrophotometrically. This was employed to rapidly determine the optimal Fe:MoFe protein ratio for MV-dependent assays (~20:1), although it was also observed that increasing MV concentrations resulted in diminished nitrogenase activity [146,148]. The nature of this has been recently ascribed to the dimerization of MV [149]. In addition to spectrophotometric activity, an electrochemical method was also employed to reduce MV in situ for nitrogenase activity assays [148]. First, this paves the way for the utilization of renewable electrical energy for "bioelectrosynthetic" N_2 reduction under ambient temperature and pressure (Figure 5).

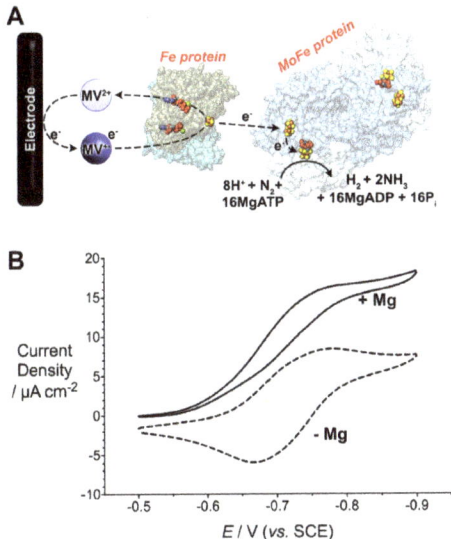

Figure 5. (**A**) Bioelectrocatalytic dinitrogen (N_2) fixation by Mo-dependent nitrogenase using methyl viologen (MV) as an electron mediator. (**B**) Cyclic voltammetric data (scan rate = 0.002 V s^{-1}) for bioelectrocatalytic N_2 fixation. The electrochemical cell contained MOPS buffer (100 mM, pH 7), a MgATP-regenerating mixture (ATP, creatine phosphate, creatine phosphokinase), 0.1 mg mL^{-1} of MoFe protein, and 20× equivalents of the Fe protein. The addition of Mg^{2+} permits MgATP hydrolysis by the Fe protein and initiates bioelectrocatalytic N_2 fixation. Reprinted (adapted) with permission from [150]. Copyright 2019 American Chemical Society.

This method was capable of electrochemically driving N_2 reduction to NH_3 where 59% of the electrons delivered to nitrogenase were directed towards N_2 fixation. This initial example reported NH_3 production at a rate of 35 nmol NH_3 min^{-1} mg^{-1} MoFe protein, corresponding to a turnover frequency (TOF) of 4.2 (single time point). In this section, TOF is calculated to be the number of moles of product formed per mole of MoFe protein (assuming, for consistency, a mass of 240 kDa for the MoFe protein) per minute. Second, the current recorded at the electrode corresponds to the rate and magnitude of electron consumption by nitrogenase (and thus, substrate reduction), which provides a method to study substrate reduction by nitrogenase in real time [151]. The subsequent optimization of this MV electrochemical approach resulted in an observed rate constant for electron flux through nitrogenase when fixing N_2 approached the expected value of 13 s^{-1} (reported as 14 s^{-1}), where rate-limiting electron transfer at 13 s^{-1} corresponds to an optimal TOF of 195 (accounting for re of H_2). Further, MV (and derivates) can also serve as an efficient electron donor to other (metallo)enzymes, such as formate dehydrogenases and hydrogenases, either for H_2/formate (HCOO$^-$) oxidation or for H$^+$/CO_2 reduction [152]. Of particular interest is the ability of MV to mediate electrons or enzymatic NADH formation [153,154]. To this end, the Minteer group has extended the use of reduced MV in bioelectrosynthetic cells to reduce NADH. In this way, the NH_3 formed by nitrogenase can be upgraded to further products of interest, such as chiral amines and chiral amino acids [155,156].

4.2. Fe Protein-Independent Activity

While nitrogenase fixes N_2 under considerably milder conditions than the Haber–Bosch process (i.e., at ambient pressure and temperature), nitrogenase can still be considered to be energy-intensive due to hydrolysis of at least 16MgATP for the fixation of each equivalent N_2. The hydrolysis of MgATP corresponds to around −50 kJ mol^{-1} in vivo; thus, there is significant interest in decoupling the Fe and MoFe proteins [20]. Not only does this present the possibility of MgATP-independent N_2 fixation, but catalytic rates could also be improved, given that the rate-limiting step of N_2 fixation is associated with the Fe protein [16]. As outlined above, evidence to supporting the idea that electron transfer between the Fe and MoFe proteins occurs prior to MgATP hydrolysis suggests that artificial reduction of the MoFe protein independent of the Fe protein (and therefore MgATP independent) could be possible. Since 2010 (when two papers submitted within two weeks of each other reported Fe protein-independent MoFe protein catalysis), chemical, photochemical, and electrochemical techniques have sought to deliver electrons to the MoFe protein for substrate reduction; prominent examples are covered below.

4.2.1. Chemical Methods for Electron Transfer

In 2010, Danyal et al. reported substrate reduction by the MoFe protein from *A. vinelandii*, for which low-potential Eu(II) was employed as the artificial reductant [157]. Eu(II) was prepared by bulk electrolytic reduction of a Eu_2O_3 ($E^{o\prime}$ = −0.36 V vs. SHE) solution, followed by the addition of a chelating polyaminocarboxylate ligand that lowered the potential of the pre-reduced Eu(II) center −0.88 or −1.14 V vs. SHE (depending on the ligand employed, reported at pH 8) [60,157]. Thus, electrochemistry was employed to prepare the reductant in batch mode, and this example can be treated as a "chemical" reduction approach. While the authors noted that N_2 fixation was not observed, the 2e$^-$-reduction of hydrazine (N_2H_4) to NH_3 was observed; further, a ß-Y98H MoFe protein mutant was required for the formation of significant quantities of NH_3 when compared to blank/control experiments. The ß-Y98 residue, located between the P cluster and FeMo-co, was previously identified as a residue that is important to electron transfer [158]. A TOF of 41 was reported for the system; further, N_2H_4 is an important substrate where its 2e$^-$-reduction to NH_3 may represent an intermediate step of N_2 fixation by nitrogenase. In 2015 Eu(II) complexes were coupled with two additional MoFe mutants, ß-F99H and α-Y64H, which also revealed enhanced N_2H_4 reduction in comparison to the WT MoFe protein [159]. The α-Y64H mutation was found to improve to TOF to 72. It was also demonstrated that these MoFe protein mutants could reduce azide (N_3^-) to NH_3, albeit it with a reduced TOF of 2 (single time point). However, only the originally reported ß-Y98H MoFe protein was found to reduce

H$^+$ to H$_2$ when coupled with a Eu(II) ethylenediaminetetraacetic acid (EDTA) complex. Of the Eu(II) complexes tested within this study, the Eu(II)–EDTA complex has the mildest reduction potential (−0.84 V vs. SHE). However, the Eu(II)–EDTA complex was selected, as minimal H$_2$ is evolved when using this Eu(II) complex in the absence of MoFe proteins. Thus, one explanation for the inability of the ß-F99H and α-Y64H proteins to reduce H$^+$ could be that the $E^{o\prime}$ of Eu(II)–EDTA complex was not reducing enough. Further, H$_2$ evolution was measured to be ~7 nmol H$_2$ min^{-1} mg^{-1} MoFe (ß-Y98H) which corresponds to a TOF of <2 (single time point). In contrast, Fe protein-derived electron transfer with a rate constant of ~13 s^{-1} loosely corresponds to the formation of ~1625 nmol H$_2$ min^{-1} mg^{-1} MoFe and a TOF of 390 (H$_2$).

In 2012, Lee et al. further demonstrated the ability of Eu(II) complexes to catalyze substrate reduction by nitrogenase [160]. However, substrate reduction was performed by variant MoFe proteins that lacked the FeMo-co and/or contained P cluster precursors (2[4Fe-4S] clusters). In *A. vinelandii*, the deletion of *nifB* leads to a MoFe protein that lacks the FeMo-co; instead, the deletion of *nifH* (Fe protein) leads to a MoFe protein that lacks the FeMo-co and has immature P-clusters (P*, 2[4Fe-4S] pairs) [6]. Remarkably, it was demonstrated that MoFe protein lacking FeMo-co could catalyze the reduction of H$^+$, C$_2$H$_2$, C$_2$H$_4$, N$_2$H$_4$, cyanide (CN$^-$), carbon monoxide (CO), and CO$_2$ to products including, H$_2$, NH$_3$, and alkanes/alkenes spanning C$_1$–C$_7$ [160]. TOFs for this system reached maxima of 0.02 (for CH$_4$ production from CN$^-$, single time point) and 0.17 (for NH$_3$ production from CN$^-$, single time point). Interestingly, the authors observed a ~2.5-fold increase in activity when using the Δ*nifH* MoFe protein vs. the Δ*nifB* MoFe protein; Jimenez-Vicente et al. recently demonstrated that the Fe protein of the V-nitrogenase system (encoded by *vnfH* and typically repressed in the presence of Mo) can substitute the NifH Fe protein in Δ*nifH* strains, which could explain the observed elevated activities [161]. In addition to the Mo-nitrogenase system, Eu(II) complexes have also been utilized to deliver electrons to the alternative V-nitrogenase system for MgATP-independent substrate reduction [162]. In 2015, Rebelein et al. demonstrated that Eu(II)-DTPA ($E^{o\prime}$ = −1.1 V vs. SHE, DTPA = diethylenetriaminepentaacetate) could function as an electron mediator to the VFe protein for CO$_2$ reduction [163]. In addition to the formation of CO and CH$_4$ formation (with a TOF for CH$_4$ production reaching 0.4×10^{-3}), C$_2$–C$_4$ hydrocarbons were also produced.

4.2.2. Photochemical Methods for Electron Transfer

In 2010, another approach for MoFe protein reduction was reported by the Tezcan group, which employed a Ru-based photosensitizer to deliver electrons to the FeMo-co through the P cluster [164]. [Ru(bpy)$_2$(phen)]$^{2+}$ possesses a long-lived photoexcited-state (*Ru^{II}), which upon quenching, results in the generation of a reducing Ru^I species that is not expected to be too dissimilar to the related [Ru(bpy)$_3$]$^+$ complex with an $E^{o\prime}$ of around −1.28 V vs. SHE [9,164,165]. The authors prepared a Cys-reactive Ru complex which was subsequently attached to a Cys mutation introduced in proximity to the P cluster (α-L158C). In this way, the Ru photosensitizer could be placed ~15 Å away from the P cluster (important for efficient electron transfer rates)—similar to the location of the Fe protein's [4Fe-4S] cluster during MoFe protein association [9]. Wild-type MoFe protein has a single solvent-exposed Cys residue; it was hypothesized that the Ru-attachment to this residue would not yield significant photo-excited electron transfer and MoFe protein catalysis. Indeed, significant substrate-reduction activity was observed following the introduction of the α-L158C mutation, where the authors first observed H$^+$ and C$_2$H$_2$ reduction at approximately 14 nmol H$_2$ min^{-1} mg^{-1} MoFe protein and 16 nmol C$_2$H$_4$ min^{-1} mg^{-1} MoFe protein (both being 2e$^-$-reductions), corresponding to TOFs of ~3.4 and ~3.8 respectively. In 2012, the authors expanded on this approach and demonstrated that photosensitized MoFe protein can also facilitate the 6e$^-$-reduction of CN$^-$ to CH$_4$ with a TOF of <0.1 [166]. However, this approach was not able to produce NH$_3$ from N$_2$ fixation at detectable quantities.

In 2016, a second photochemical approach for MoFe protein reduction was reported in which CdS nanorods were mixed with wild-type MoFe protein and N$_2$ fixation to NH$_3$ was observed [167]. In contrast to the low reduction potential afforded by the Ru-photosensitizer system, CdS nanorods offer a milder $E^{\circ\prime}$ of −0.8 V vs. SHE, which can also be accessed upon illumination with visible light. Under N_2, this system was found to produce NH_3 at 315 nmol min^{-1} mg^{-1} MoFe protein which corresponds to ~63% of NH_3 production by the Fe protein-driven, ATP-dependent activity of the MoFe protein and a TOF of 75. Recently, Harris et al. expanded on this approach by investigating electrostatic interactions between the MoFe protein and the surface of CdS nanorods modified by charged functional groups (Figure 6) [168]. In addition, the authors also prepared a flexible linker to enable the attachment of the MoFe to the CdS nanorods by a Cys residue introduced near to the P cluster (α-L158C). For the optimal configurations reported, maximal H_2 evolution rates of approximately 1250 nmol H_2 min^{-1} mg^{-1} MoFe protein were observed, which correspond to TOFs of approximately 300. In this example, N_2 fixation to NH_3 by the CdS nanorod biohybrid was not reported. In summary, precisely how the CdS biohybrid system bypasses the Fe protein and affords close-to-native TOFs is unknown. Yet, the ability to achieve N_2 fixation is promising and highly attractive to future ATP-independent NH_3 biotechnologies.

Figure 6. Harris et al. investigated the attachment of the MoFe protein to CdS nanorod surfaces: (**left**) surface-capping groups with different charges were employed, and (**right**) solvent-exposed Cys residues of the MoFe protein were used to covalently conjugate the MoFe protein to CdS nanorod surfaces. Reprinted (adapted) with permission from [168]. Copyright 2020 American Chemical Society.

4.2.3. Electrochemical Methods for Electron Transfer

Since 2016, electrochemical approaches have gained significant interest for artificial MoFe protein reduction and catalysis. Milton et al. reported the immobilization of the MoFe protein within a polymer at an electrode surface [169]. In this approach, bis(cyclopentadienyl)cobalt(II) (cobaltocene$^{1+/0}$, $E^{\circ\prime}$ = −0.96 V vs. SHE) was used as a low potential electron mediator to deliver electrode-derived electrons to the MoFe protein. In addition to H$^+$ reduction, N_3^- and nitrite (NO_2^-) reductions were also observed. As observed by Danyal et al., the use of the ß-Y98H MoFe protein mutant resulted in enhanced catalytic currents (consistent with enhanced catalysis): Faradaic efficiencies (FEs) of 35% and 101% and TOFs of 12 and 40 were reported for N_3^- and NO_2^- reduction (single time points). This method provided a method by which electron transfer to (and thus, catalysis of) the MoFe protein could be observed in real time. Hu et al. also adopted this approach to demonstrate that the MoFe and FeFe proteins could reduce CO_2 to $HCOO^-$ with respective FEs of 9% and 32%. Khadka et al. also demonstrated that this approach could be employed to investigate the rate-limiting step of the MoFe when catalyzing H$^+$ reduction [170]. In this study, the cobaltocene concentration was titrated such that electron transfer to the MoFe protein was not limiting, and bioelectrocatalytic H$^+$ reduction to H_2 was evaluated as a function of catalytic current. Following the addition of increasing D_2O fractions to the electrochemical setup, a decrease in the catalytic current was observed, and this was attributed to a kinetic isotope effect. This was further exemplified by the use of a range of MoFe protein mutants and FeMo-co-transporting/storage proteins. Supporting density functional theory (DFT) calculations led to the conclusion that the protonation of a M-H at the FeMo-co is the rate-limiting step of H_2 evolution (within the MoFe protein component). In 2018, Cai et al. also employed this approach to investigate catalysis of the VFe protein for H^+ and CO_2 reduction [171]. In addition to unsubstituted cobaltocene, mono-cobaltocene ([Co(Cp)(CpCOOH)], $E^{o\prime}$ = −0.79 V vs. SHE) and di-carboxy cobaltocene ([Co(CpCOOH)$_2$], $E^{o\prime}$ = −0.65 V vs. SHE) derivatives were also investigated as electron mediators for the VFe protein. After the passage of 4 C of charge within an electrochemical cell that contained the VFe and protein and dicarboxy-cobaltocene, ~850 µmol H_2 was detected µmol^{-1} VFe protein (corrected to control experiments). Upon the addition of bicarbonate (HCO_3^-) as the CO_2 source, the passage of 4 C of charge led to the formation of ~20 nmol C_2H_4 µmol^{-1} VFe and ~30 nmol propene (C_3H_6) µmol^{-1} VFe protein (corrected to control experiments).

An additional approach to electrochemical MoFe protein reduction was reported in 2016, wherein Paengnakorn et al. reported the reduction H^+ by the MoFe protein also entrapped within a polymeric matrix [61]. A mixture of three Eu complexes was employed to mediate electron transfer to the MoFe protein, with E^o's spanning −0.63 to −1.09 V vs. SHE (at pH 8). Excitingly, this approach was coupled with an attenuated total reflectance infrared (ATR-IR) spectroelectrochemical cell such that substrate binding at the FeMo-co could be interrogated as a function of potential. Interestingly, the binding of CO to the FeMo-co (which does not typically affect H^+ reduction) was found to commence at potentials of ~<−0.7 V vs. SHE. H_2 formation was also confirmed by gas chromatography resulting in a TOF of 14 for the ß-F99H mutant MoFe protein; this mutant and the ß-Y98H mutant (TOF = 13) MoFe proteins showed elevated electrocatalytic currents over the wild-type MoFe protein (TOF = 7) (single time points).

In contrast to the use of electron mediators to deliver electrons to the MoFe protein, Hickey et al. reported on the ability to directly deliver electrons to the MoFe protein when immobilized at electrode surfaces [172,173]. This approach represents an exciting new approach by which the MoFe protein (and the VFe and FeFe proteins) can be explored in greater detail, as this permits the direct observation of the redox state and reduction potentials of enzymatic cofactors. This approach permitted NH_3 production with TOFs of around 0.1–0.6, alongside a TOF of 2.6 for H_2 formation (under Ar, single time point) [172]. Interestingly, this article also identified that the background production of H_2 by the electrode (at low potentials) is inhibitory to the fixation of N_2 by nitrogenase, which should be avoided in future electrochemical investigation of nitrogenase.

5. Conclusions

As discussed, nitrogenase is the only enzyme known to reduce N_2 to NH_3. Thus, many streams of research are underway to understand how nitrogenase fixes N_2 and to engineer (semi)biological systems for NH_3 production. The understanding of nitrogenase's N_2 fixation mechanism could lead to the design of new bio-inspired catalysts with improved efficiencies, or lead to biotechnologies that exploit nitrogenase's reactivity to produce NH_3 (such as directing renewable electricity towards MgATP-independent N_2 fixation by the MoFe protein). The ability of engineered plants or symbiotic systems to produce their own NH_3 also represents a significant milestone. As presented here, many pathways (both natural and engineered) exist, or are being developed, to deliver electrons to nitrogenase for N_2 fixation. Nevertheless, ambiguity surrounds the possibility of the Fe protein undergoing the transfer of 2e$^-$ during each transient association with the MoFe protein, which has the potential to halve nitrogenase's MgATP hydrolysis requirement. Further, a range of natural and unnatural electron donors are being explored for electron delivery. In the case of natural electron donors, it is still necessary to improve the efficiencies of electron transfer systems in non-diazotrophic hosts to support higher nitrogenase activity. In artificial in vitro systems, it is unclear as to why certain systems support H^+ reduction by nitrogenase (representing necessary M-H formation at the FeMo-co) but N_2 is not observed. For example, how are photochemical methods (such as CdS nanorods) able to support N_2 fixation, while other photochemical methods (i.e., Ru-based photosensitizers) are not? Similarly, it remains unclear as to how mediated electron delivery to the MoFe protein (or VFe/FeFe proteins) cannot currently support N_2 fixation (although H^+, CO_2, N_3^-, and NO_2^- reduction has been achieved), whereas direct electron transfer appears to facilitate NH_3 production. Thus, further research into nitrogenase is required in a multitude of areas; hence, multi-disciplinary research efforts surely present the best opportunity (and are clearly necessary) to advance nitrogenase research and understanding.

One method of improving environmental sustainability is the decentralization of key global-scale processes. To that end, nitrogenase-based technologies present opportunities for on-demand NH_3 production. However, much development is still required to overcome some of the limitations surrounding the utilization of nitrogenase. For example, MgATP hydrolysis presents a significant energetic requirement, and the sensitivity of nitrogenase to O_2 also inhibits deployment. Further, nitrogenase assembly and maturation is complex and limits in vitro utilization. Nevertheless, perhaps the most promising approach for nitrogenase deployment in new biotechnologies is presented by Liu et al., where photoelectrically generated H_2 feeds a N_2-fixing microbe for NH_3 production and/or biomass accumulation [144].

Author Contributions: W.G. and R.D.M. conceived and wrote the manuscript. All authors have read and agreed to the published version of the manuscript.

Funding: R.D.M. acknowledges funding from the University of Geneva.

Conflicts of Interest: The authors declare no conflict of interest.

References and Note

1. Foster, S.L.; Bakovic, S.I.P.; Duda, R.D.; Maheshwari, S.; Milton, R.D.; Minteer, S.D.; Janik, M.J.; Renner, J.N.; Greenlee, L.F. Catalysts for nitrogen reduction to ammonia. *Nat. Catal.* **2018**, *1*, 490–500. [CrossRef]
2. Montoya, J.H.; Tsai, C.; Vojvodic, A.; Nørskov, J.K. The Challenge of Electrochemical Ammonia Synthesis: A New Perspective on the Role of Nitrogen Scaling Relations. *ChemSusChem* **2015**, *8*, 2180–2186. [CrossRef] [PubMed]
3. Chen, J.G.; Crooks, R.M.; Seefeldt, L.C.; Bren, K.L.; Bullock, R.M.; Darensbourg, M.Y.; Holland, P.L.; Hoffman, B.; Janik, M.J.; Jones, A.K.; et al. Beyond fossil fuel–driven nitrogen transformations. *Science* **2018**, *360*, eaar6611. [CrossRef]
4. Burén, S.; Rubio, L.M. State of the art in eukaryotic nitrogenase engineering. *FEMS Microbiol. Lett.* **2017**, *365*. [CrossRef]
5. Jasniewski, A.J.; Lee, C.C.; Ribbe, M.W.; Hu, Y. Reactivity, Mechanism, and Assembly of the Alternative Nitrogenases. *Chem. Rev.* **2020**. accepted. [CrossRef]
6. Jasniewski, A.; Sickerman, N.; Hu, Y.; Ribbe, M. The Fe Protein: An Unsung Hero of Nitrogenase. *Inorganics* **2018**, *6*, 25. [CrossRef]
7. Lancaster, K.M.; Roemelt, M.; Ettenhuber, P.; Hu, Y.; Ribbe, M.W.; Neese, F.; Bergmann, U.; DeBeer, S. X-ray emission spectroscopy evidences a central carbon in the nitrogenase iron-molybdenum cofactor. *Science* **2011**, *334*, 974–977. [CrossRef]
8. Burgess, B.K.; Lowe, D.J. Mechanism of Molybdenum Nitrogenase. *Chem. Rev.* **1996**, *96*, 2983–3012. [CrossRef]
9. Rutledge, H.L.; Tezcan, F.A. Electron Transfer in Nitrogenase. *Chem. Rev.* **2020**. accepted. [CrossRef]
10. Thorneley, R.; Lowe, D. Kinetics and mechanism of the nitrogenase enzyme system. In *Molybdenum Enzymes*; Spiro, T., Ed.; Wiley-Interscience: New York, NY, USA, 1985; pp. 221–284.
11. Thorneley, R.N.; Lowe, D.J. Nitrogenase of Klebsiella pneumoniae. Kinetics of the dissociation of oxidized iron protein from molybdenum-iron protein: Identification of the rate-limiting step for substrate reduction. *Biochem. J.* **1983**, *215*, 393–403. [CrossRef]
12. Reduction potentials reported in the literature have been converted to the standard hydrogen electrode (SHE) for this review, by the following adjustments: $E_{SHE} = E_{NHE} = E_{Ag/AgCl(satd.)} + 0.197$ V $= E_{SCE} + 0.242$ V. SCE = saturated calomel electrode [13].
13. Bard, A.J.; Faulkner, L.R. *Electrochemical Methods: Fundamentals and Applications*; Wiley: New York, NY, USA, 2001; Chapter 1: Introduction and overview of electrode processes.
14. Lough, S.; Burns, A.; Watt, G.D. Redox reactions of the nitrogenase complex from azotobacter vinelandii. *Biochemistry* **1983**, *22*, 4062–4066. [CrossRef]
15. Lanzilotta, W.N.; Seefeldt, L.C. Changes in the midpoint potentials of the nitrogenase metal centers as a result of iron protein-molybdenum-iron protein complex formation. *Biochemistry* **1997**, *36*, 12976–12983. [CrossRef] [PubMed]

16. Yang, Z.Y.; Ledbetter, R.; Shaw, S.; Pence, N.; Tokmina-Lukaszewska, M.; Eilers, B.; Guo, Q.; Pokhrel, N.; Cash, V.L.; Dean, D.R.; et al. Evidence That the Pi Release Event Is the Rate-Limiting Step in the Nitrogenase Catalytic Cycle. *Biochemistry* **2016**, *55*, 3625–3635. [CrossRef] [PubMed]
17. Duval, S.; Danyal, K.; Shaw, S.; Lytle, A.K.; Dean, D.R.; Hoffman, B.M.; Antony, E.; Seefeldt, L.C. Electron transfer precedes ATP hydrolysis during nitrogenase catalysis. *Proc. Natl. Acad. Sci. USA* **2013**, *110*, 16414–16419. [CrossRef] [PubMed]
18. Danyal, K.; Dean, D.R.; Hoffman, B.M.; Seefeldt, L.C. Electron Transfer within Nitrogenase: Evidence for a Deficit-Spending Mechanism. *Biochemistry* **2011**, *50*, 9255–9263. [CrossRef]
19. Seefeldt, L.C.; Hoffman, B.M.; Peters, J.W.; Raugei, S.; Beratan, D.N.; Antony, E.; Dean, D.R. Energy Transduction in Nitrogenase. *Acc. Chem. Res.* **2018**, *51*, 2179–2186. [CrossRef]
20. Berg, J.M.; Tymoczko, J.L.; Gatto, G.J.; Stryer, L. *Biochemistry*, 8th ed.; W.H. Freeman and Company: New York, NY, USA, 2015.
21. Rohde, M.; Sippel, D.; Trncik, C.; Andrade, S.L.A.; Einsle, O. The Critical E4 State of Nitrogenase Catalysis. *Biochemistry* **2018**, *57*, 5497–5504. [CrossRef]
22. Hoffman, B.M.; Dean, D.R.; Seefeldt, L.C. Climbing nitrogenase: Toward a mechanism of enzymatic nitrogen fixation. *Acc. Chem. Res.* **2009**, *42*, 609–619. [CrossRef]
23. Buscagan, T.M.; Rees, D.C. Rethinking the Nitrogenase Mechanism: Activating the Active Site. *Joule* **2019**, *3*, 2662–2678. [CrossRef]
24. Lukoyanov, D.; Barney, B.M.; Dean, D.R.; Seefeldt, L.C.; Hoffman, B.M. Connecting nitrogenase intermediates with the kinetic scheme for N 2 reduction by a relaxation protocol and identification of the N 2 binding state. *Proc. Natl. Acad. Sci. USA* **2007**, *104*, 1451–1455. [CrossRef]
25. Harris, D.F.; Lukoyanov, D.A.; Kallas, H.; Trncik, C.; Yang, Z.Y.; Compton, P.; Kelleher, N.; Einsle, O.; Dean, D.R.; Hoffman, B.M.; et al. Mo-, V-, and Fe-Nitrogenases Use a Universal Eight-Electron Reductive-Elimination Mechanism to Achieve N2 Reduction. *Biochemistry* **2019**, *58*, 3293–3301. [CrossRef] [PubMed]
26. Dance, I. Misconception of reductive elimination of H2, in the context of the mechanism of nitrogenase. *Dalton Trans.* **2015**, *44*, 9027–9037. [CrossRef] [PubMed]
27. Yates, M.G. Electron transport to nitrogenase in Azotobacter chroococcum: Azotobacter flavodoxin hydroquinone as an electron donor. *FEBS Lett.* **1972**, *27*, 63–67. [CrossRef]
28. Mortenson, L.E. Nitrogen Fixation: Role of Ferredoxin in Anaerobic Metabolism. *Annu. Rev. Microbiol.* **1963**, *17*, 115–138. [CrossRef]
29. Saeki, K. Electron Transport to Nitrogenase: Diverse Routes for a Common Destination. In *Genetics and Regulation of Nitrogen Fixation in Free-Living Bacteria*; Klipp, W., Masepohl, B., Gallon, J.R., Newton, W.E., Eds.; Springer Science & Business Media: Berlin, Germany, 2006; pp. 257–290. [CrossRef]
30. Mayhew, S.G. The Redox Potential of Dithionite and SO−2 from Equilibrium Reactions with Flavodoxins, Methyl Viologen and Hydrogen plus Hydrogenase. *Eur. J. Biochem.* **1978**, *85*, 535–547. [CrossRef]
31. Uppal, R.; Incarvito, C.D.; Lakshmi, K.V.; Valentine, A.M. Aqueous spectroscopy and redox properties of carboxylate-bound titanium. *Inorg. Chem.* **2006**, *45*, 1795–1804. [CrossRef]
32. Alagaratnam, S.; van Pouderoyen, G.; Pijning, T.; Dijkstra, B.W.; Cavazzini, D.; Rossi, G.L.; Van Dongen, W.M.A.M.; van Mierlo, C.P.M.; van Berkel, W.J.H.; Canters, G.W. A crystallographic study of Cys69Ala flavodoxin II from Azotobacter vinelandii: Structural determinants of redox potential. *Protein Sci.* **2005**, *14*, 2284–2295. [CrossRef]
33. Segal, H.M.; Spatzal, T.; Hill, M.G.; Udit, A.K.; Rees, D.C. Electrochemical and structural characterization of Azotobacter vinelandii flavodoxin II. *Protein Sci.* **2017**, *26*, 1984–1993. [CrossRef]
34. Buckel, W.; Thauer, R.K. Flavin-Based Electron Bifurcation, Ferredoxin, Flavodoxin, and Anaerobic Respiration With Protons (Ech) or NAD+ (Rnf) as Electron Acceptors: A Historical Review. *Front. Microbiol.* **2018**, *9*, 401. [CrossRef]
35. Nitschke, W.; Russell, M.J. Redox bifurcations: Mechanisms and importance to life now, and at its origin. *BioEssays* **2012**, *34*, 106–109. [CrossRef]
36. Buckel, W.; Thauer, R.K. Flavin-Based Electron Bifurcation, A New Mechanism of Biological Energy Coupling. *Chem. Rev.* **2018**, *118*, 3862–3886. [CrossRef]
37. Steensma, E.; Heering, H.A.; Hagen, W.R.; Van Mierlo, C.P.M. Redox properties of wild-type, Cys69Ala, and Cys69Ser Azotobacter vinelandii flavodoxin II as measured by cyclic voltammetry and EPR spectroscopy. *Eur. J. Biochem.* **1996**, *235*, 167–172. [CrossRef]

38. Duyvis, M.G.; Wassink, H.; Haaker, H. Nitrogenase of Azotobacter vinelandii: Kinetic analysis of the Fe protein redox cycle. *Biochemistry* **1998**, *37*, 17345–17354. [CrossRef]
39. Mortenson, L.E.; Valentine, R.C.; Carnahan, J.E. An electron transport factor from Clostridium pasteurianum. *Biochem. Biophys. Res. Commun.* **1962**, *7*, 448–452. [CrossRef]
40. Matsubara, H.; Saeki, K. Structural and functional diversity of ferredoxins and related proteins. *Adv. Inorg. Chem.* **1992**, *38*, 223–280. [CrossRef]
41. Chatelet, C.; Meyer, J. Mapping the interaction of the [2Fe-2S] Clostridium pasteurianum ferredoxin with the nitrogenase MoFe protein. *Biochim. Biophys. Acta-Protein Struct. Mol. Enzymol.* **2001**, *1549*, 32–36. [CrossRef]
42. Stout, C.D. Structure of the iron-sulphur clusters in Azotobacter ferredoxin at 4.0 A resolution. *Nature* **1979**, *279*, 83–84. [CrossRef]
43. Stout, C.D. Refinement of the 7 Fe ferredoxin from Azotobacter vinelandii at 1.9 Å resolution. *J. Mol. Biol.* **1989**, *205*, 545–555. [CrossRef]
44. Stout, C.D. Crystal structures of oxidized and reduced Azotobacter vinelandii ferredoxin at pH 8 and 6. *J. Biol. Chem.* **1993**, *268*, 25920–25927. [CrossRef]
45. Liu, J.; Chakraborty, S.; Hosseinzadeh, P.; Yu, Y.; Tian, S.; Petrik, I.; Bhagi, A.; Lu, Y. Metalloproteins containing cytochrome, iron-sulfur, or copper redox centers. *Chem. Rev.* **2014**, *114*, 4366–4369. [CrossRef]
46. Boyd, E.S.; Garcia Costas, A.M.; Hamilton, T.L.; Mus, F.; Peters, J.W. Evolution of molybdenum nitrogenase during the transition from anaerobic to aerobic metabolism. *J. Bacteriol.* **2015**, *197*, 1690–1699. [CrossRef]
47. Hu, Y.; Ribbe, M.W. Biosynthesis of the Metalloclusters of Nitrogenases. *Annu. Rev. Biochem.* **2016**, *85*, 455–483. [CrossRef]
48. Shethna, Y.I. Non-heme iron (iron-sulfur) proteins of Azotobacter vinelandii. *Biochim. Biophys. Acta Bioenerg.* **1970**, *205*, 58–62. [CrossRef]
49. Thorneley, R.N.F.; Deistung, J. Electron-transfer studies involving flavodoxin and a natural redox partner, the iron protein of nitrogenase. Conformational constraints on protein-protein interactions and the kinetics of electron transfer within the protein complex. *Biochem. J.* **1988**, *253*, 587–595. [CrossRef]
50. Hallenbeck, P.C.; Gennaro, G. Stopped-flow kinetic studies of low potential electron carriers of the photosynthetic bacterium, Rhodobacter capsulatus: Ferredoxin I and NifF. *Biochim. Biophys. Acta-Bioenerg.* **1998**, *1365*, 435–442. [CrossRef]
51. Hageman, R.V.; Burris, R.H.; Hageman, R.V. Kinetic Studies on Electron Transfer and Interaction between Nitrogenase Components from Azotobacter vinelandii†. *Biochemistry* **1978**, *17*, 4117–4124. [CrossRef]
52. Pence, N.; Tokmina-Lukaszewska, M.; Yang, Z.Y.; Ledbetter, R.N.; Seefeldt, L.C.; Bothner, B.; Peters, J.W. Unraveling the interactions of the physiological reductant flavodoxin with the different conformations of the Fe protein in the nitrogenase cycle. *J. Biol. Chem.* **2017**, *292*, 15661–15669. [CrossRef]
53. Erickson, J.A.; Nyborg, A.C.; Johnson, J.L.; Truscott, S.M.; Gunn, A.; Nordmeyer, F.R.; Watt, G.D. Enhanced efficiency of ATP hydrolysis during nitrogenase catalysis utilizing reductants that form the all-ferrous redox state of the Fe protein. *Biochemistry* **1999**, *38*, 14279–14285. [CrossRef]
54. Watt, G.D.; Reddy, K.R.N. Formation of an all ferrous Fe4S4 cluster in the iron protein component of Azotobacter vinelandii nitrogenase. *J. Inorg. Biochem.* **1994**, *53*, 281–294. [CrossRef]
55. Nyborg, A.C.; Johnson, J.L.; Gunn, A.; Watt, G.D. Evidence for a two-electron transfer using the all-ferrous Fe protein during nitrogenase catalysis. *J. Biol. Chem.* **2000**, *275*, 39307–39312. [CrossRef]
56. Nyborg, A.C.; Erickson, J.A.; Johnson, J.L.; Gunn, A.; Truscott, S.M.; Watt, G.D. Reactions of Azotobacter vinelandii nitrogenase using Ti(III) as reductant. *J. Inorg. Biochem.* **2000**, *78*, 371–381. [CrossRef]
57. Angove, H.C.; Yoo, S.J.; Burgess, B.K.; Münck, E. Mössbauer and EPR evidence for an all-ferrous Fe4S4 cluster with S = 4 in the Fe protein of nitrogenase. *J. Am. Chem. Soc.* **1997**, *119*, 8730–8731. [CrossRef]
58. Angove, H.C.; Yoo, S.J.; Münck, E.; Burgess, B.K. An all-ferrous state of the Fe protein of nitrogenase. Interaction with nucleotides and electron transfer to the MoFe protein. *J. Biol. Chem.* **1998**, *273*, 26330–26337. [CrossRef]
59. Strop, P.; Takahara, P.M.; Chiu, H.J.; Angove, H.C.; Burgess, B.K.; Rees, D.C. Crystal structure of the all-ferrous [4Fe-4S]0 form of the nitrogenase iron protein from Azotobacter vinelandii. *Biochemistry* **2001**, *40*, 651–656. [CrossRef]
60. Vincent, K.A.; Tilley, G.J.; Quammie, N.C.; Streeter, I.; Burgess, B.K.; Cheesman, M.R.; Armstrong, F.A. Instantaneous, stoichiometric generation of powerfully reducing states of protein active sites using Eu(ii) and polyaminocarboxylate ligands. *Chem. Commun.* **2003**, 2590–2591. [CrossRef]

61. Paengnakorn, P.; Ash, P.A.; Shaw, S.; Danyal, K.; Chen, T.; Dean, D.R.; Seefeldt, L.C.; Vincent, K.A. Infrared spectroscopy of the nitrogenase MoFe protein under electrochemical control: Potential-triggered CO binding. *Chem. Sci.* **2017**. [CrossRef]
62. Guo, M.; Sulc, F.; Ribbe, M.W.; Farmer, P.J.; Burgess, B.K. Direct assessment of the reduction potential of the [4Fe-4S]1+/0 couple of the Fe protein from Azotobacter vinelandii. *J. Am. Chem. Soc.* **2002**, *124*, 12100–12101. [CrossRef]
63. Lowery, T.J.; Wilson, P.E.; Zhang, B.; Bunker, J.; Harrison, R.G.; Nyborg, A.C.; Thiriot, D.; Watt, G.D. Flavodoxin hydroquinone reduces Azotobacter vinelandii Fe protein to the all-ferrous redox state with a S = 0 spin state. *Proc. Natl. Acad. Sci. USA* **2006**, *103*, 17131–17136. [CrossRef]
64. Hardy, R.W.F.; Knight, E. Reductant-dependent adenosine triphosphatase of nitrogen-fixing extracts of azotobacter vinelandii. *BBA-Enzymol. Biol. Oxid.* **1966**, *122*, 520–531. [CrossRef]
65. Reyntjens, B.; Jollie, D.R.; Stephens, P.J.; Gao-Sheridan, H.S.; Burgess, B.K. Purification and characterization of a fixABCX-linked 2[4Fe-4S] ferredoxin from Azotobacter vinelandii. *J. Biol. Inorg. Chem.* **1997**, *2*, 595–602. [CrossRef]
66. Hamilton, T.L.; Ludwig, M.; Dixon, R.; Boyd, E.S.; Dos Santos, P.C.; Setubal, J.C.; Bryant, D.A.; Dean, D.R.; Peters, J.W. Transcriptional profiling of nitrogen fixation in Azotobacter vinelandii. *J. Bacteriol.* **2011**, *193*, 4477–4486. [CrossRef]
67. Jacobson, M.R.; Brigle, K.E.; Bennett, L.T.; Setterquist, R.A.; Wilson, M.S.; Cash, V.L.; Beynon, J.; Newton, W.E.; Dean, D.R. Physical and genetic map of the major nif gene cluster from Azotobacter vinelandii. *J. Bacteriol.* **1989**, *171*, 1017–1027. [CrossRef]
68. Martin, A.E.; Burgess, B.K.; Iismaa, S.E.; Smartt, C.T.; Jacobson, M.R.; Dean, D.R. Construction and characterization of an Azotobacter vinelandii strain with mutations in the genes encoding flavodoxin and ferredoxin I. *J. Bacteriol.* **1989**, *171*, 3162–3167. [CrossRef]
69. Fixen, K.R.; Pal Chowdhury, N.; Martinez-Perez, M.; Poudel, S.; Boyd, E.S.; Harwood, C.S. The path of electron transfer to nitrogenase in a phototrophic alpha-proteobacterium. *Environ. Microbiol.* **2018**, *20*, 2500–2508. [CrossRef]
70. Bothe, H.; Schmitz, O.; Yates, M.G.; Newton, W.E. Nitrogen Fixation and Hydrogen Metabolism in Cyanobacteria. *Microbiol. Mol. Biol. Rev.* **2010**, *74*, 529–551. [CrossRef]
71. Razquin, P.; Schmitz, S.; Peleato, M.L.; Fillat, M.F.; Gómez-Moreno, C.; Böhme, H. Differential activities of heterocyst ferredoxin, vegetative cell ferredoxin, and flavodoxin as electron carriers in nitrogen fixation and photosynthesis in Anabaena sp. *Photosynth. Res.* **1995**, *43*, 35–40. [CrossRef]
72. Masukawa, H.; Mochimaru, M.; Sakurai, H. Disruption of the uptake hydrogenase gene, but not of the bidirectional hydrogenase gene, leads to enhanced photobiological hydrogen production by the nitrogen-fixing cyanobacterium Anabaena sp. PCC 7120. *Appl. Microbiol. Biotechnol.* **2002**, *58*, 618–624. [CrossRef]
73. Masepohl, B.; Schölisch, K.; Görlitz, K.; Kutzki, C.; Böhme, H. The heterocyst-specific fdxH gene product of the cyanobacterium Anabaena sp. PCC 7120 is important but not essential for nitrogen fixation. *Mol. Gen. Genet.* **1997**, *253*, 770–776. [CrossRef]
74. Jones, K.M.; Buikema, W.J.; Haselkorn, R. Heterocyst-specific expression of patB, a gene required for nitrogen fixation in Anabaena sp. strain PCC 7120. *J. Bacteriol.* **2003**, *185*, 2306–2314. [CrossRef]
75. Böhme, H.; Haselkorn, R. Molecular cloning and nucleotide sequence analysis of the gene coding for heterocyst ferredoxin from the cyanobacterium Anabaena sp. strain PCC 7120. *MGG Mol. Gen. Genet.* **1988**, *214*, 278–285. [CrossRef]
76. Böhme, H.; Schrautemeier, B. Comparative characterization of ferredoxins from heterocysts and vegetative cells of Anabaena variabilis. *BBA-Bioenerg.* **1987**, *891*, 1–7. [CrossRef]
77. Lázaro, M.C.; Fillat, M.F.; Gómez-Moreno, C.; Peleato, M.L. A Possible Role of Flavodoxin in Nitrogen Fixation in Heterocysts from Anabaena. In *Nitrogen Fixation*; Springer: Dordrecht, The Netherlands, 1991; pp. 431–436. [CrossRef]
78. Bauer, C.C.; Scappino, L.; Haselkorn, R. Growth of the cyanobacterium Anabaena on molecular nitrogen: NifJ is required when iron is limited. *Proc. Natl. Acad. Sci. USA* **1993**, *90*, 8812–8816. [CrossRef]
79. Schrautemeier, B.; Böhme, H. A distinct ferredoxin for nitrogen fixation isolated from heterocysts of the cyanobacterium Anabaena variabilis. *FEBS Lett.* **1985**, *184*, 304–308. [CrossRef]
80. Isas, J.M.; Yannone, I.S.M.; Burgess, B.K. Azotobacter vinelandii NADPH:ferredoxin reductase cloning, sequencing, and overexpression. *J. Biol. Chem.* **1995**, *270*, 21258–21263. [CrossRef]

81. Ledbetter, R.N.; Garcia Costas, A.M.; Lubner, C.E.; Mulder, D.W.; Tokmina-Lukaszewska, M.; Artz, J.H.; Patterson, A.; Magnuson, T.S.; Jay, Z.J.; Duan, H.D.; et al. The Electron Bifurcating FixABCX Protein Complex from Azotobacter vinelandii: Generation of Low-Potential Reducing Equivalents for Nitrogenase Catalysis. *Biochemistry* **2017**, *56*, 4177–4190. [CrossRef]
82. Klugkist, J.; Haaker, H.; Veeger, C. Studies on the mechanism of electron transport to nitrogenase in Azotobacter vinelandii. *Eur. J. Biochem.* **1986**, *155*, 41–46. [CrossRef]
83. Therien, J.B.; Artz, J.H.; Poudel, S.; Hamilton, T.L.; Liu, Z.; Noone, S.M.; Adams, M.W.W.; King, P.W.; Bryant, D.A.; Boyd, E.S.; et al. The physiological functions and structural determinants of catalytic bias in the [FeFe]-hydrogenases CpI and CpII of Clostridium pasteurianum strain W5. *Front. Microbiol.* **2017**, *8*. [CrossRef]
84. Chen, J.S.; Blanchard, D.K. Isolation and properties of a unidirectional H2-oxidizing hydrogenase from the strictly anaerobic N2-fixing bacterium Clostridium pasteurianum W5. *Biochem. Biophys. Res. Commun.* **1978**, *84*, 1144–1150. [CrossRef]
85. Wahl, R.C.; Orme-Johnson, W.H. Clostridial pyruvate oxidoreductase and the pyruvate-oxidizing enzyme specific to nitrogen fixation in Klebsiella pneumoniae are similar enzymes. *J. Biol. Chem.* **1987**, *262*, 10489–10496.
86. Hill, S.; Kavanagh, E.P. Roles of nifF and nifJ gene products in electron transport to nitrogenase in Klebsiella pneumoniae. *J. Bacteriol.* **1980**, *141*, 470–475. [CrossRef]
87. Nieva-Gomez, D.; Roberts, G.P.; Klevivkis, S.; Brill, W. Electron transport to nitrogenase in Klebsiella pneumoniae. *Proc. Natl. Acad. Sci. USA* **1980**, *77*, 2555–2558. [CrossRef]
88. Shah, V.K.; Stacey, G.; Brill, W.J. Electron transport to nitrogenase. Purification and characterization of pyruvate:flavodoxin oxidoreductase. The nifJ gene product. *J. Biol. Chem.* **1983**, *258*, 12064–12068. [PubMed]
89. St John, R.T.; Johnston, H.M.; Seidman, C.; Garfinkel, D.; Gordon, J.K.; Shah, V.K.; Brill, W.J. Biochemistry and genetics of Klebsiella pneumoniae mutant strains unable to fix N2. *J. Bacteriol.* **1975**, *121*, 759–765. [CrossRef] [PubMed]
90. Roberts, G.P.; MacNeil, T.; MacNeil, D.; Brill, W.J. Regulation and characterization of protein products coded by the nif (nitrogen fixation) genes of Klebsiella pneumoniae. *J. Bacteriol.* **1978**, *136*, 267–279. [CrossRef] [PubMed]
91. Earl, C.D.; Ronson, C.W.; Ausubel, F.M. Genetic and structural analysis of the Rhizobium meliloti fixA, fixB, fixC, and fixX genes. *J. Bacteriol.* **1987**, *169*, 1127–1136. [CrossRef] [PubMed]
92. Ruvkun, G.B.; Sundaresan, V.; Ausubel, F.M. Directed transposon Tn5 mutagenesis and complementation analysis of rhizobium meliloti symbiotic nitrogen fixation genes. *Cell* **1982**, *29*, 551–559. [CrossRef]
93. Klipp, W.; Reiländer, H.; Schlüter, A.; Krey, R.; Pühler, A. The Rhizobium meliloti fdxN gene encoding a ferredoxin-like protein is necessary for nitrogen fixation and is cotranscribed with nifA and nifB. *MGG Mol. Gen. Genet.* **1989**, *216*, 293–302. [CrossRef]
94. Schmehl, M.; Jahn, A.; Meyer zu Vilsendorf, A.; Hennecke, S.; Masepohl, B.; Schuppler, M.; Marxer, M.; Oelze, J.; Klipp, W. Identification of a new class of nitrogen fixation genes in Rhodobacter capsalatus: A putative membrane complex involved in electron transport to nitrogenase. *MGG Mol. Gen. Genet.* **1993**, *241*, 602–615. [CrossRef]
95. Jeong, H.S.; Jouanneau, Y. Enhanced nitrogenase activity in strains of Rhodobacter capsulatus that overexpress the rnf genes. *J. Bacteriol.* **2000**, *182*, 1208–1214. [CrossRef]
96. Elsen, S.; Dischert, W.; Colbeau, A.; Bauer, C.E. Expression of uptake hydrogenase and molybdenum nitrogenase in Rhodobacter capsulatus is coregulated by the RegB-RegA two-component regulatory system. *J. Bacteriol.* **2000**, *182*, 2831–2837. [CrossRef]
97. Vignais, P.M.; Elsen, S.; Colbeau, A. Transcriptional regulation of the uptake [NiFe]hydrogenase genes in Rhodobacter capsulatus. In Proceedings of the Biochemical Society Transactions; University of Reading, Reading, UK, 24–29 August 2004; pp. 28–32.
98. Jouanneau, Y.; Meyer, C.; Naud, I.; Klipp, W. Characterization of an fdxN mutant of Rhodobacter capsulatus indicates that ferredoxin I serves as electron donor to nitrogenase. *BBA-Bioenerg.* **1995**, *1232*, 33–42. [CrossRef]
99. Yakunin, A.F.; Hallenbeck, P.C. Purification and characterization of pyruvate oxidoreductase from the photosynthetic bacterium Rhodobacter capsulatus. *Biochim. Biophys. Acta-Bioenerg.* **1998**, *1409*, 39–49. [CrossRef]
100. Saeki, K.; Suetsugu, Y.; Tokuda, K.I.; Miyatake, Y.; Young, D.A.; Marrs, B.L.; Matsubara, H. Genetic analysis of functional differences among distinct ferredoxins in Rhodobacter capsulatus. *J. Biol. Chem.* **1991**, *266*, 12889–12895. [PubMed]

101. Gennaro, G.; Hübner, P.; Sandmeier, U.; Yakunin, A.F.; Hallenbeck, P.C. Cloning, characterization, and regulation of nifF from Rhodobacter capsulatus. *J. Bacteriol.* **1996**, *178*, 3949–3952. [CrossRef]
102. Huang, J.J.; Heiniger, E.K.; McKinlay, J.B.; Harwood, C.S. Production of hydrogen gas from light and the inorganic electron donor thiosulfate by Rhodopseudomonas palustris. *Appl. Environ. Microbiol.* **2010**, *76*, 7717–7722. [CrossRef] [PubMed]
103. Rey, F.E.; Oda, Y.; Harwood, C.S. Regulation of uptake hydrogenase and effects of hydrogen utilization on gene expression in Rhodopseudomonas palustris. *J. Bacteriol.* **2006**, *188*, 6143–6152. [CrossRef]
104. Edgren, T.; Nordlund, S. The fixABCX Genes in Rhodospirillum rubrum Encode a Putative Membrane Complex Participating in Electron Transfer to Nitrogenase. *J. Bacteriol.* **2004**, *186*, 2052–2060. [CrossRef]
105. Yoch, D.C.; Arnon, D.I.; Sweeney, W.V. Characterization of two soluble ferredoxins as distinct from bound iron sulfur proteins in the photosynthetic bacterium Rhodospirillum rubrum. *J. Biol. Chem.* **1975**, *250*, 8330–8336.
106. Yoch, D.C.; Arnon, D.I. Comparison of two ferredoxins from Rhodospirillum rubrum as electron carriers for the native nitrogenase. *J. Bacteriol.* **1975**, *121*, 743–745. [CrossRef]
107. Edgren, T.; Nordlund, S. Electron transport to nitrogenase in Rhodospirillum rubrum: Identification of a new fdxN gene encoding the primary electron donor to nitrogenase. *FEMS Microbiol. Lett.* **2005**, *245*, 345–351. [CrossRef]
108. Kern, M.; Klipp, W.; Klemme, J.H. Increased nitrogenase-dependent H2 photoproduction by hup mutants of Rhodospirillum rubrum. *Appl. Environ. Microbiol.* **1994**, *60*, 1768–1774. [CrossRef]
109. Edgren, T.; Nordlund, S. Two pathways of electron transport to nitrogenase in Rhodospirillum rubrum: The major pathway is dependent on the fix gene products. *FEMS Microbiol. Lett.* **2006**, *260*, 30–35. [CrossRef]
110. Biegel, E.; Schmidt, S.; González, J.M.; Müller, V. Biochemistry, evolution and physiological function of the Rnf complex, a novel ion-motive electron transport complex in prokaryotes. *Cell. Mol. Life Sci.* **2011**, *68*, 613–634. [CrossRef]
111. Vignais, P.M.; Colbeau, A. Molecular biology of microbial hydrogenases. *Curr. Issues Mol. Biol.* **2004**, *6*, 159–188.
112. Zhang, X.; Sherman, D.M.; Shermana, L.A. The uptake hydrogenase in the unicellular diazotrophic cyanobacterium cyanothece sp. strain PCC 7822 protects nitrogenase from oxygen toxicity. *J. Bacteriol.* **2014**, *196*, 840–849. [CrossRef]
113. Liu, D.; Liberton, M.; Yu, J.; Pakrasi, H.B.; Bhattacharyya-Pakrasi, M. Engineering nitrogen fixation activity in an oxygenic phototroph. *MBio* **2018**, *9*, e01029-18. [CrossRef]
114. Lindberg, P.; Schütz, K.; Happe, T.; Lindblad, P. A hydrogen-producing, hydrogenase-free mutant strain of Nostoc punctiforme ATCC 29133. *Int. J. Hydrogen Energy* **2002**, *27*, 1291–1296. [CrossRef]
115. Barahona, E.; Jiménez-Vicente, E.; Rubio, L.M. Hydrogen overproducing nitrogenases obtained by random mutagenesis and high-throughput screening. *Sci. Rep.* **2016**, *6*, 38291. [CrossRef]
116. Poudel, S.; Colman, D.R.; Fixen, K.R.; Ledbetter, R.N.; Zheng, Y.; Pence, N.; Seefeldt, L.C.; Peters, J.W.; Harwood, C.S.; Boyd, E.S. Electron transfer to nitrogenase in different genomic and metabolic backgrounds. *J. Bacteriol.* **2018**, *200*, e00757-17. [CrossRef]
117. Bergman, B.; Gallon, J.R.; Rai, A.N.; Stal, L.J. N2 fixation by non-heterocystous cyanobacteria. *FEMS Microbiol. Rev.* **1997**, *19*, 139–185. [CrossRef]
118. Tsujimoto, R.; Kotani, H.; Yokomizo, K.; Yamakawa, H.; Nonaka, A.; Fujita, Y. Functional expression of an oxygen-labile nitrogenase in an oxygenic photosynthetic organism. *Sci. Rep.* **2018**, *8*, 7380. [CrossRef]
119. Yang, J.; Xie, X.; Yang, M.; Dixon, R.; Wang, Y.P. Modular electron-transport chains from eukaryotic organelles function to support nitrogenase activity. *Proc. Natl. Acad. Sci. USA* **2017**, *114*, E2460–E2465. [CrossRef] [PubMed]
120. Li, X.X.; Liu, Q.; Liu, X.M.; Shi, H.W.; Chen, S.F. Using synthetic biology to increase nitrogenase activity. *Microb. Cell Factories* **2016**, *15*, 43. [CrossRef]
121. Wang, L.; Zhang, L.; Liu, Z.; Zhao, D.; Liu, X.; Zhang, B.; Xie, J.; Hong, Y.; Li, P.; Chen, S.; et al. A Minimal Nitrogen Fixation Gene Cluster from Paenibacillus sp. WLY78 Enables Expression of Active Nitrogenase in Escherichia coli. *PLoS Genet.* **2013**, *9*, e1003865. [CrossRef]
122. Masepohl, B.; Kutsche, M.; Riedel, K.U.; Schmehl, M.; Klipp, W.; Pühler, A. Functional analysis of the cysteine motifs in the ferredoxin-like protein FdxN of Rhizobium meliloti involved in symbiotic nitrogen fixation. *MGG Mol. Gen. Genet.* **1992**, *233*, 33–41. [CrossRef]
123. Riedel, K.-U.; Jouanneau, Y.; Masepohl, B.; Pühler, A.; Klipp, W. A Rhizobium Meliloti Ferredoxin (FdxN) Purified from Escherichia Coli Donates Electrons to Rhodobacter Capsulatus Nitrogenase. *Eur. J. Biochem.* **1995**, *231*, 742–746. [CrossRef]

124. Yang, J.; Xie, X.; Wang, X.; Dixon, R.; Wang, Y.P. Reconstruction and minimal gene requirements for the alternative iron-only nitrogenase in Escherichia coli. *Proc. Natl. Acad. Sci. USA* **2014**, *111*, E3718–E3725. [CrossRef]
125. Yang, J.; Xie, X.; Xiang, N.; Tian, Z.-X.; Dixon, R.; Wang, Y.-P. Polyprotein strategy for stoichiometric assembly of nitrogen fixation components for synthetic biology. *Proc. Natl. Acad. Sci. USA* **2018**, *115*, E8509–E8517. [CrossRef]
126. Smanski, M.J.; Bhatia, S.; Zhao, D.; Park, Y.J.; Woodruff, L.B.A.; Giannoukos, G.; Ciulla, D.; Busby, M.; Calderon, J.; Nicol, R.; et al. Functional optimization of gene clusters by combinatorial design and assembly. *Nat. Biotechnol.* **2014**, *32*, 1241–1249. [CrossRef]
127. Setten, L.; Soto, G.; Mozzicafreddo, M.; Fox, A.R.; Lisi, C.; Cuccioloni, M.; Angeletti, M.; Pagano, E.; Díaz-Paleo, A.; Ayub, N.D. Engineering Pseudomonas protegens Pf-5 for Nitrogen Fixation and its Application to Improve Plant Growth under Nitrogen-Deficient Conditions. *PLoS ONE* **2013**, *8*, e63666. [CrossRef]
128. Ryu, M.H.; Zhang, J.; Toth, T.; Khokhani, D.; Geddes, B.A.; Mus, F.; Garcia-Costas, A.; Peters, J.W.; Poole, P.S.; Ané, J.M.; et al. Control of nitrogen fixation in bacteria that associate with cereals. *Nat. Microbiol.* **2020**, *5*, 314–330. [CrossRef]
129. Dixon, R.A.; Postgate, J.R. Genetic transfer of nitrogen fixation from klebsiella pneumoniae to escherichia coli. *Nature* **1972**, *237*, 102–103. [CrossRef]
130. Wang, X.; Yang, J.G.; Chen, L.; Wang, J.L.; Cheng, Q.; Dixon, R.; Wang, Y.P. Using Synthetic Biology to Distinguish and Overcome Regulatory and Functional Barriers Related to Nitrogen Fixation. *PLoS ONE* **2013**, *8*, e68677. [CrossRef]
131. James, E.K. Nitrogen fixation in endophytic and associative symbiosis. *Field Crops Res.* **2000**, *65*, 197–209. [CrossRef]
132. Arsene, F.; Katupitiya, S.; Kennedy, I.R.; Elmerich, C. Use of lacZ fusions to study the expression of nif genes of Azospirillum brasilense in association with plants. *Mol. Plant-Microbe Interact.* **1994**, *7*, 748–757. [CrossRef]
133. Beatty, P.H.; Good, A.G. Future prospects for cereals that fix nitrogen. *Science* **2011**, *333*, 416–417. [CrossRef]
134. Fukuyama, K. Structure and function of plant-type ferredoxins. *Photosynth. Res.* **2004**, *81*, 289–301. [CrossRef]
135. Arakaki, A.K.; Ceccarelli, E.A.; Carrillo, N. Plant-type ferredoxin-NADP + reductases: A basal structural framework and a multiplicity of functions. *FASEB J.* **1997**, *11*, 133–140. [CrossRef]
136. Estabrook, R.W.; Suzuki, K.; Ian Mason, J.; Baron, J.; Taylor, W.E.; Simpson, E.R.; Purvis, J.; McCarthy, J. Adrenodoxin: An Iron-Sulfur Protein of Adrenal Cortex Mitochondria. In *Biological Properties*; Academic Press: Cambridge, MA, USA, 1973; pp. 193–223. [CrossRef]
137. Hanke, G.; Mulo, P. Plant type ferredoxins and ferredoxin-dependent metabolism. *Plant Cell Environ.* **2013**, *36*, 1071–1084. [CrossRef]
138. Reinbothe, C.; El Bakkouri, M.; Buhr, F.; Muraki, N.; Nomata, J.; Kurisu, G.; Fujita, Y.; Reinbothe, S. Chlorophyll biosynthesis: Spotlight on protochlorophyllide reduction. *Trends Plant Sci.* **2010**, *15*, 614–624. [CrossRef]
139. Cheng, Q.; Day, A.; Dowson-Day, M.; Shen, G.F.; Dixon, R. The Klebsiella pneumoniae nitrogenase Fe protein gene (nifH) functionally substitutes for the chlL gene in Chlamydomonas reinhardtii. *Biochem. Biophys. Res. Commun.* **2005**, *329*, 966–975. [CrossRef]
140. Ivleva, N.B.; Groat, J.; Staub, J.M.; Stephens, M. Expression of active subunit of nitrogenase via integration into plant organelle genome. *PLoS ONE* **2016**, *11*, e0160951. [CrossRef]
141. Yang, Z.Y.Z.-Y.; Moure, V.R.R.; Dean, D.R.R.; Seefeldt, L.C.C. Carbon dioxide reduction to methane and coupling with acetylene to form propylene catalyzed by remodeled nitrogenase. *Proc. Natl. Acad. Sci. USA* **2012**, *109*, 19644–19648. [CrossRef]
142. Fixena, K.R.; Zhenga, Y.; Harrisb, D.F.; Shawb, S.; Yangb, Z.Y.; Deanc, D.R.; Seefeldtb, L.C.; Harwooda, C.S. Light-driven carbon dioxide reduction to methane by nitrogenase in a photosynthetic bacterium. *Proc. Natl. Acad. Sci. USA* **2016**, *113*, 10163–10167. [CrossRef]
143. Zheng, Y.; Harris, D.F.; Yu, Z.; Fu, Y.; Poudel, S.; Ledbetter, R.N.; Fixen, K.R.; Yang, Z.Y.; Boyd, E.S.; Lidstrom, M.E.; et al. A pathway for biological methane production using bacterial iron-only nitrogenase. *Nat. Microbiol.* **2018**, *3*, 281–286. [CrossRef]
144. Liu, C.; Sakimoto, K.K.; Colón, B.C.; Silver, P.A.; Nocera, D.G. Ambient nitrogen reduction cycle using a hybrid inorganic-biological system. *Proc. Natl. Acad. Sci. USA* **2017**, *114*, 6450–6455. [CrossRef]
145. Seefeldt, L.C.; Ensign, S.A. A continuous, spectrophotometric activity assay for nitrogenase using the reductant Titanium(III) citrate. *Anal. Biochem.* **1994**, *221*, 379–386. [CrossRef]

146. D'Eustachio, A.J.; Hardy, R.W.F. Reductants and electron transport in nitrogen fixation. *Biochem. Biophys. Res. Commun.* **1964**, *15*, 319–323. [CrossRef]
147. Ware, D.A. Nitrogenase of klebsiella pneumoniae: Interaction with viologen dyes as measure by acetylene reduction. *Biochem. J.* **1972**, *130*, 301–302. [CrossRef]
148. Milton, R.D.; Cai, R.; Abdellaoui, S.; Leech, D.; De Lacey, A.L.; Pita, M.; Minteer, S.D. Bioelectrochemical Haber–Bosch Process: An Ammonia-Producing H2/N2 Fuel Cell. *Angew. Chem.-Int. Ed.* **2017**, *56*, 2680–2683. [CrossRef]
149. Badalyan, A.; Yang, Z.Y.; Hu, B.; Luo, J.; Hu, M.; Liu, T.L.; Seefeldt, L.C. An Efficient Viologen-Based Electron Donor to Nitrogenase. *Biochemistry* **2019**, *58*, 4590–4595. [CrossRef]
150. Milton, R.D.; Minteer, S.D. Nitrogenase Bioelectrochemistry for Synthesis Applications. *Acc. Chem. Res.* **2019**, *52*, 3351–3360. [CrossRef]
151. Badalyan, A.; Yang, Z.Y.; Seefeldt, L.C. A Voltammetric Study of Nitrogenase Catalysis Using Electron Transfer Mediators. *ACS Catal.* **2019**, *9*, 1366–1372. [CrossRef]
152. Cadoux, C.M.; Milton, R.D. Recent enzymatic electrochemistry for reductive reactions. *ChemElectroChem* **2020**. accepted. [CrossRef]
153. Quah, T.; Milton, R.D.; Abdellaoui, S.; Minteer, S.D. Bioelectrocatalytic NAD+/NADH inter-conversion: Transformation of an enzymatic fuel cell into an enzymatic redox flow battery. *Chem. Commun.* **2017**, *53*, 8411–8414. [CrossRef]
154. Alkotaini, B.; Abdellaoui, S.; Hasan, K.; Grattieri, M.; Quah, T.; Cai, R.; Yuan, M.; Minteer, S.D. Sustainable Bioelectrosynthesis of the Bioplastic Polyhydroxybutyrate: Overcoming Substrate Requirement for NADH Regeneration. *ACS Sustain. Chem. Eng.* **2018**, *6*, 4909–4915. [CrossRef]
155. Chen, H.; Cai, R.; Patel, J.; Dong, F.; Chen, H.; Minteer, S.D. Upgraded Bioelectrocatalytic N2 Fixation: From N2 to Chiral Amine Intermediates. *J. Am. Chem. Soc.* **2019**, *141*, 4963–4971. [CrossRef]
156. Chen, H.; Prater, M.B.; Cai, R.; Dong, F.; Chen, H.; Minteer, S.D. Bioelectrocatalytic Conversion from N2 to Chiral Amino Acids in a H2/α-keto Acid Enzymatic Fuel Cell. *J. Am. Chem. Soc.* **2020**. accepted. [CrossRef]
157. Danyal, K.; Inglet, B.S.; Vincent, K.A.; Barney, B.M.; Hoffman, B.M.; Armstrong, F.A.; Dean, D.R.; Seefeldt, L.C. Uncoupling nitrogenase: Catalytic reduction of hydrazine to ammonia by a MoFe protein in the absence of Fe protein-ATP. *J. Am. Chem. Soc.* **2010**, *132*, 13197–13199. [CrossRef]
158. Peters, J.W.; Fisher, K.; Newton, W.E.; Dean, D.R. Involvement of the P Cluster in Intramolecular Electron Transfer within the Nitrogenase MoFe Protein. *J. Biol. Chem.* **1995**, *270*, 27007–27013. [CrossRef]
159. Danyal, K.; Rasmussen, A.J.; Keable, S.M.; Inglet, B.S.; Shaw, S.; Zadvornyy, O.A.; Duval, S.; Dean, D.R.; Raugei, S.; Peters, J.W.; et al. Fe Protein-Independent Substrate Reduction by Nitrogenase MoFe Protein Variants. *Biochemistry* **2015**, *54*, 2456–2462. [CrossRef]
160. Lee, C.C.; Hu, Y.; Ribbe, M.W. ATP-independent substrate reduction by nitrogenase P-cluster variant. *Proc. Natl. Acad. Sci. USA* **2012**, *109*, 6922–6926. [CrossRef]
161. Jimenez-Vicente, E.; Yang, Z.Y.; Keith Ray, W.; Echavarri-Erasun, C.; Cash, V.L.; Rubio, L.M.; Seefeldt, L.C.; Dean, D.R. Sequential and differential interaction of assembly factors during nitrogenase MoFe protein maturation. *J. Biol. Chem.* **2018**, *293*, 9812–9823. [CrossRef]
162. Sickerman, N.S.; Hu, Y.; Ribbe, M.W. Activation of CO2 by Vanadium Nitrogenase. *Chem.-Asian J.* **2017**, *12*, 1985–1996. [CrossRef]
163. Rebelein, J.G.; Hu, Y.; Ribbe, M.W. Widening the Product Profile of Carbon Dioxide Reduction by Vanadium Nitrogenase. *ChemBioChem* **2015**, *16*, 1993–1996. [CrossRef]
164. Roth, L.E.; Nguyen, J.C.; Tezcan, F.A. ATP- and Iron−Protein-Independent Activation of Nitrogenase Catalysis by Light. *J. Am. Chem. Soc.* **2010**, *132*, 13672–13674. [CrossRef]
165. Gray, H.B.; Maverick, A.W. Solar chemistry of metal complexes. *Science* **1981**, *214*, 1201–1205. [CrossRef]
166. Roth, L.E.; Tezcan, F.A. ATP-Uncoupled, Six-Electron Photoreduction of Hydrogen Cyanide to Methane by the Molybdenum–Iron Protein. *J. Am. Chem. Soc.* **2012**, *134*, 8416–8419. [CrossRef]
167. Brown, K.A.; Harris, D.F.; Wilker, M.B.; Rasmussen, A.; Khadka, N.; Hamby, H.; Keable, S.; Dukovic, G.; Peters, J.W.; Seefeldt, L.C.; et al. Light-driven dinitrogen reduction catalyzed by a CdS:nitrogenase MoFe protein biohybrid. *Science* **2016**, *352*, 448–450. [CrossRef]
168. Harris, A.W.; Harguindey, A.; Patalano, R.E.; Roy, S.; Yehezkeli, O.; Goodwin, A.P.; Cha, J.N. Investigating Protein-Nanocrystal Interactions for Photodriven Activity. *ACS Appl. Bio Mater.* **2020**, *3*, 1026–1035. [CrossRef]

169. Milton, R.D.; Abdellaoui, S.; Khadka, N.; Dean, D.R.; Leech, D.; Seefeldt, L.C.; Minteer, S.D. Nitrogenase bioelectrocatalysis: Heterogeneous ammonia and hydrogen production by MoFe protein. *Energy Environ. Sci.* **2016**, *9*, 2550–2554. [CrossRef]
170. Khadka, N.; Milton, R.D.; Shaw, S.; Lukoyanov, D.; Dean, D.R.; Minteer, S.D.; Raugei, S.; Hoffman, B.M.; Seefeldt, L.C. Mechanism of Nitrogenase H2 Formation by Metal-Hydride Protonation Probed by Mediated Electrocatalysis and H/D Isotope Effects. *J. Am. Chem. Soc.* **2017**, *139*, 13518–13524. [CrossRef]
171. Cai, R.; Milton, R.D.; Abdellaoui, S.; Park, T.; Patel, J.; Alkotaini, B.; Minteer, S.D. Electroenzymatic C-C bond formation from CO2. *J. Am. Chem. Soc.* **2018**, *140*, 5041–5044. [CrossRef] [PubMed]
172. Hickey, D.P.; Lim, K.; Cai, R.; Patterson, A.R.; Yuan, M.; Sahin, S.; Abdellaoui, S.; Minteer, S.D. Pyrene hydrogel for promoting direct bioelectrochemistry: ATP-independent electroenzymatic reduction of N2. *Chem. Sci.* **2018**, *9*, 5172–5177. [CrossRef] [PubMed]
173. Hickey, D.P.; Cai, R.; Yang, Z.Y.; Grunau, K.; Einsle, O.; Seefeldt, L.C.; Minteer, S.D. Establishing a Thermodynamic Landscape for the Active Site of Mo-Dependent Nitrogenase. *J. Am. Chem. Soc.* **2019**, *141*, 17150–17157. [CrossRef] [PubMed]

© 2020 by the authors. Licensee MDPI, Basel, Switzerland. This article is an open access article distributed under the terms and conditions of the Creative Commons Attribution (CC BY) license (http://creativecommons.org/licenses/by/4.0/).

Article

Appropriate Buffers for Studying the Bioinorganic Chemistry of Silver(I) †

Lucille Babel *, Soledad Bonnet-Gómez and Katharina M. Fromm *

Chemistry Department, University of Fribourg, Chemin du Musée 9, 1700 Fribourg, Switzerland; sole82@bluewin.ch
* Correspondence: lucille.simond@unifr.ch (L.B.); katharina.fromm@unifr.ch (K.M.F.);
   Tel.: +41-2630-087-32 (K.M.F.)
† Dedicated to the radical chemist Prof. Bernd Giese on behalf of his 80th birthday.

Received: 21 February 2020; Accepted: 18 March 2020; Published: 22 March 2020

Abstract: Silver(I) is being largely studied for its antimicrobial properties. In parallel to that growing interest, some researchers are investigating the effect of this ion on eukaryotes and the mechanism of silver resistance of certain bacteria. For these studies, and more generally in biology, it is necessary to work in buffer systems that are most suitable, i.e., that interact least with silver cations. Selected buffers such as 4-(2-hydroxyethyl)-1-piperazineethane sulfonic acid (HEPES) were therefore investigated for their use in the presence of silver nitrate. Potentiometric titrations allowed to determine stability constants for the formation of (Ag(Buffer)) complexes. The obtained values were adapted to extract the apparent binding constants at physiological pH. The percentage of metal ions bound to the buffer was calculated at this pH for given concentrations of buffer and silver to realize at which extent silver was interacting with the buffer. We found that in the micromolar range, HEPES buffer is sufficiently coordinating to silver to have a non-negligible effect on the thermodynamic parameters determined for an analyte. Morpholinic buffers were more suitable as they turned out to be weaker complexing agents. We thus recommend the use of MOPS for studies of physiological pH.

Keywords: silver; buffer; association constant; HEPES

1. Introduction

A well-known list of buffers was published between 1966 and 1980, called Good's buffers, for their use in biological systems [1]. This list contains essentially sterically hindered amines that aim to replace common buffers used in biology such as imidazole, sodium phosphate and sodium citrate. Indeed, these previously employed buffers are inadequate for certain experiments because of their reactivity towards small molecules (ATP), metal ions, or because of their toxicity for the cells [2–8]. For example, a phosphate buffer leads to precipitates with many cations and is known to inhibit or enhance certain reactions of a cellular system [2,3]. Imidazole is a very good complexing ligand for many metal cations and, due to its similar structure, could replace histidine residues in metal binding proteins [9–11]. Good's buffers on the contrary were believed to be largely inactive towards the cell metabolism and thus should not interact with any biological molecule and/or metal ions. Nevertheless, since this list was established, many studies have proved that most of these sterically hindered tertiary amine-based buffers are able to coordinate slightly some metal ions [12,13]. Therefore, binding constants determined for other ligands could be affected by the presence of these buffers, which are usually in large excess compared to the ligand to ensure a stable pH, hence it is a necessity to know these values. A correction can then be applied to the thermodynamic model to take into consideration the effect of the buffer. To limit the effect of this correction, careful consideration of the metal ions in solution and the concentration of the buffer is necessary prior to use. For example, complexation of copper(II) by buffers was thoroughly studied over recent years and it was shown that

Good's buffers coordinate the metal ion with variable but non-negligible affinities of $3 \leq \log K_{Cu,L} \leq 5$ [14–16]. However, most of the studies found in the literature concern divalent metal cations and little is known on monovalent ones [17,18]. Moreover, publications on the morpholinic and piperazinic family of buffers are sometimes concluding to contradictory results [12].

Our group is interested in the use of silver as an antimicrobial agent. Silver is used in in vitro studies to investigate e.g., the silver resistance mechanism of some bacteria or in studies investigating toxicity and/or antimicrobial properties of silver agents, yet appropriate buffers for this kind of studies are lacking in the literature. We have recently been studying peptide models inspired by the protein SilE, a protein of the silver efflux pump in Gram negative bacteria, which is able to bind a large amount of silver(I) [19,20]. In this case, phosphate buffer could not be used because of the immediate formation of the poorly soluble silver phosphate salt.

HEPES contains N-donors and is not innocent with respect to silver(I) as shown by a crystal structure of a HEPES-silver(I) complex [21]. Two nitrogen atoms from the piperazine moieties of HEPES molecules as well as two oxygen atoms from the alcohol and sulfonate functions coordinate the silver ion in a distorted tetrahedral geometry. However, the binding affinity was not quantified.

Herein, we determined the affinity of HEPES for silver ions in order to quantize the buffer effect. In comparison, we also studied the effect of other buffers that were expected to possess the least interaction with silver ions (Scheme 1) to find out which one would be ideal for studies with silver(I) in biological media.

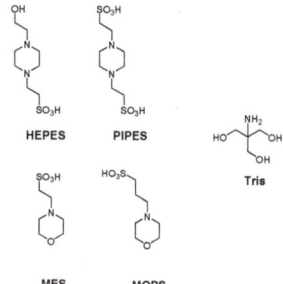

Scheme 1. Structures of buffers investigated for their affinities with silver ions.

2. Materials and Methods

Silver nitrate AgNO$_3$ was purchased from Carlo Erba reagents (RPE, Analytical 99+%). 4-(2-hydroxyethyl)-1-piperazineethane sulfonic acid (HEPES), 3-(N-morpholino)propanesulfonic acid (MOPS), tris(hydroxymethyl)aminomethane (Tris) (Roche), sodium nitrate NaNO$_3$ and potassium hydrogen phthalate (KHP) (Merck) were purchased from Sigma-Aldrich. Piperazine-1,4-bis(2-ethanesulfonic acid) (PIPES) and 2-(N-morpholino)ethanesulfonic acid (MES) were purchased from Roth. Nitric acid was purchased from Fluka and NaOH pellets from Acros. HNO$_3$ 0.1 M stock solution in 0.1 M NaNO$_3$ was standardized towards KHP (0.4 g) where the equivalence point is followed with the help of phenolphthalein indicator. NaOH 0.2 M stock solutions in 0.1 M NaNO$_3$ were standardized with stock solution of HNO$_3$ 0.1 M and used within two weeks to avoid carbonate formation. Buffers and silver nitrate were dissolved at a concentration of 0.05 M in 0.1 M NaNO$_3$. PIPES was insoluble in water, and NaOH had to be added up to a 0.069 M concentration (1.4 eq.).

Buffers were titrated manually in presence of 0.1 M NaNO$_3$ at 296 K over the pH range of 2–11 (HNO$_3$ was added to obtain the starting pH of 2) with NaOH 0.2 M as titrant. Changes in pH were monitored with a glass electrode (Primatrode with NTC Methrom, combined glass-Ag/AgCl electrode), calibrated daily with standard buffers at pH 4 and 7. Titrations were conducted in triplicates for each buffer at three different concentrations between 2 and 12 mM with a sample volume of 50 mL. Silver nitrate was added at three different ratios from 0.2 to 1.0 equivalents compared to the buffer. The titration data were analyzed using the SUPERQUAD software according to equilibriums defined in Appendix A. Mass spectrometry was performed on an ESI-MS Bruker Esquire HCT in H_2O/MeOH solution (0.8:0.2) on the positive and negative mode with each buffer adjusted at pH 7 and 0.5 equivalent of silver nitrate.

3. Results

Acid dissociation constants were first determined without silver (Table 1, Figures 1 and S1) [22–25]. In the presence of silver nitrate, titrations were stopped at pH 8.0 because silver hydroxide and silver oxide are known to precipitate above this pH. The titration curve for HEPES with silver was found to have a lower plateau compared to HEPES alone, likely due to the coordination of HEPES to silver ions (Figure 1).

Table 1. Acid dissociation constants pK_{an} (n= number of protons dissociated, see Figure S6 to visualize equilibrium considered) and complexation constants $\beta_{1,m}^{Ag,B}$ (m = number of buffer molecules bound by one silver ion for the formation of the complex [Ag(B)$_m$], see Figure S12 for proposed structures) obtained for the different buffers with potentiometric titrations and comparison with literature (L = HEPES, PIPES, MOPS, MES, Tris).

Buffer	pK_{an} (23 °C) [a]	pK_{an} Literature (25 °C)	$\log(\beta_{1,m}^{Ag,B})$ (23 °C) [b]
HEPES	7.46(1), 3.07(2)	7.45(1) [23], 3.0(1) [24]	2.36(2)
PIPES	6.65(1), 2.54(2), 1.3(4)	6.71(1) [23]	1.95(3)
MOPS	7.03(1)	7.09(1) [23]	1.1(1)
MES	6.00(2)	6.07(1) [23]	1.69(8)
Tris	8.24(1)	8.08(1) [25]	3.1(2), 6.5 (1)

[a] Acid dissociation constants fitted on titration points between pH 2.0 and 11.5. [b] Stability constants fitted on titration points between pH 2.0 and 8.0.

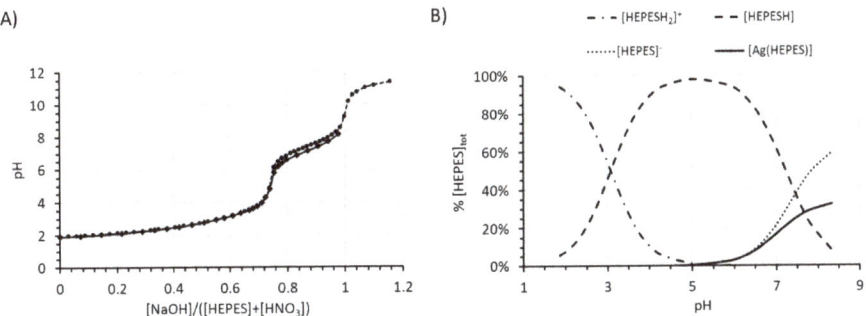

Figure 1. (A) Titration curves obtained for HEPES (7 mM) without (circles, dashed line) and in presence of silver nitrate (diamonds, plain line, 0.7 eq., 5 mM). (B) Speciation diagram according to pH for the species involving HEPES buffer.

We supposed the formation of a complex with one silver ion per HEPES ligand, based on the crystal structure obtained by Bilinovich et al. in 2011 resolved as a 1D coordination polymer with alternating HEPES and silver ions (Scheme 2) [21]. As solid-state structures do not always reflect the speciation in solution, three different ratios of silver to HEPES were tested. The titration curve fitted well (a 1) to the formation of a 1:1 complex and gave a stability constant of $\log(K_{1,1}^{Ag,HEPES})$ = 2.36(2) (Table 1). A stability constant for the formation of a hypothetical complex [Ag(HEPESH)]$^+$ with protonated HEPES in acid medium could be excluded as the fitting immediately results in negative values when considering this equilibrium. Thus, the protonated complex [Ag(HEPESH)]$^+$ is unlikely to form in solution.

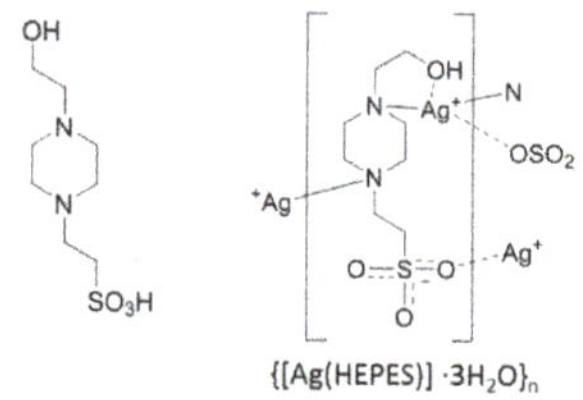

Scheme 2. Structure of HEPES in its neutral form and schematic representation of the crystal structure obtained in presence of silver(I).

Due to the non-negligible amount of silver bound to HEPES buffer, we investigated other buffers in the same way as well: another piperazine type buffer PIPES, and two morpholine type buffers MES and MOPS as well as Tris buffer (Figure S2–S5). Indeed, morpholinic and piperazinic families were selected to be the most innocent buffers because they contain bulky tertiary amines and a low number of other weakly coordinating groups (alcohols, sulfonates). Indeed, at physiological pH, these two families were considered to be suitable buffers due their weak complexation ability with other metals [12]. Tris buffer, which is widely used in biology, was expected to yield higher binding constants with silver ions due to the weak steric hindrance of the amine.

Stoichiometry of the complexes was proposed according to mass spectra and by testing various models for the determination of binding constants. Nevertheless, m/z signals in the positive and the negative modes for silver complexes were not observed for HEPES, PIPES, MES and MOPS buffers either because these are polymeric species or because the major species is neutral (Figures S7–S10). For the Tris complex, a 2:1 species was observed with two ligands around one metallic center $[Ag(Tris)_2]^+$ (Figure S11). For the determination of silver binding constants, larger errors were obtained when considering $[Ag(MES)_2]^-$ or $[Ag_2(PIPES)]$ (Table S1) and negative values were found for $[Ag(MOPS)_2]^-$, so we decided to give only one stability constant for the formation of the $[Ag(L)]$ complexes, with L = PIPES, MOPS, MES (Table 1 and Figure S12).

4. Discussion

Acid dissociation constants for the buffers alone were in good agreement with data from the literature (Table 1), confirming the validity of our measurements [22–25]. For the titration experiments with silver ions, the stability constant obtained for HEPES $\log(K_{1,1}^{Ag,HEPES}) = 2.36(2)$ was lower than the value obtained for the 1:1 complex of HEPES with copper(II) $\log(K_{1,1}^{Cu,HEPES}) = 3.22(2)$ [15]. This trend is expected as copper(II) is usually presenting greater affinities with nitrogen ligands due to its higher charge density [26,27].

Given the relatively high value for a buffer considered to be innocent of $\log(K_{1,1}^{Ag,HEPES}) = 2.36(2)$ for the silver-HEPES complex, we simulated how binding constants of an analyte binding silver would be affected by the presence of HEPES buffer (Table 2, Appendix B). The decrease on stability constants that would be measured without taking the silver-HEPES complex into account depend on the concentration of the analyte and the relative stoichiometry with the buffer. However, these effects are still quite weak on the logarithmic scale of the stability constants, except when working at high concentrations, i.e., using Nuclear Magnetic Resonance (NMR) spectroscopy to obtain the stability constants.

Table 2. Apparent binding constants $\log(K_{app,1,1}^{Ag,L})$ corrected for the effect of buffer for various real values of binding constants $\log(K_{1,1}^{Ag,L})$ (L = peptide or analyte investigated for its complexation to silver, B = HEPES buffer at pH 7.4, 40 equivalents) and at different concentrations. Percentage of decrease is indicated in parenthesis.

	$\log(K_{app,1,1}^{Ag,L})$			
$\log(K_{1,1}^{Ag,L})$	6.4	4.0	3.0	2.0
[L] = 10 µM, [B] = 0.4 mM	6.38 (−0.3%)	3.98 (−0.4%)	2.98 (−0.6%)	1.98 (−0.9%)
[L] = 100 µM, [B] = 4 mM	6.25 (−2.3%)	3.85 (−3.7%)	2.85 (−4.9%)	1.85 (−7.3%)
[L] = 500 µM, [B] = 20 mM	5.92 (−7.5%)	3.52 (−11.9%)	2.52 (−15.9%)	1.52 (−23.8%)

In our previous study of SilE in presence of HEPES, the fact that HEPES binds to silver ions can be neglected. Binding constants of SilE model peptides with silver ions were indeed determined in presence of HEPES, but using a competitor with known binding affinity for silver ions. The competitor was then similarly affected by the buffer as the peptide ligand. The stability constant of the competitor was itself calculated in competition with imidazole (whose contribution to the thermodynamic equilibrium was taken into consideration).

Looking for evidence for the stoichiometry of the complexes formed with silver, published crystal structures were examined (Scheme 3) [28–30] but no structures were found for MOPS or Tris [31]. Triethanolamine buffer (TEOA) yields a [Ag(TEOA)$_2$]$^+$ complex, and this, together with the linear [Ag(NH$_3$)$_2$]$^+$ complex, suggests the possible formation of a complex [Ag(Tris)$_2$]$^+$ [32]. Interestingly, for the crystal structures of the silver-PIPES and silver-MES complexes, the silver(I) ions always has at least a coordination number of four (Table S2). The silver ion is typically maintained by two quite strong coordination bonds, preferentially with nitrogen atoms, and by two weaker secondary bonding interactions with oxygen atoms of sulfonate and alcohol groups. The silver-MES complex includes a benzimidazole ligand (Bz) together with the complexation of MES buffer. According to these structures, one could expect a 2:1 silver to buffer ratio for PIPES and a 1:2 ratio for MOPS, MES and Tris. The stoichiometry was confirmed by mass spectrometry for Tris buffer where the complex [Ag(Tris)$_2$]$^+$ was clearly identified as the main species in solution (Figure S11). Indeed, only the model with a 1:1 silver/buffer complex was working while fitting potentiometric data. Possible second binding constants are likely too weak to be precisely determined (Figure S12).

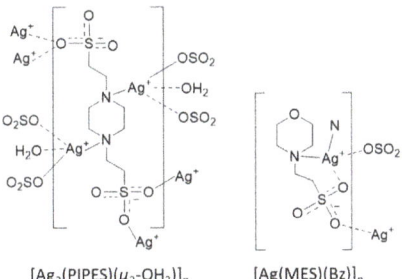

Scheme 3. Crystal structures obtained for PIPES [28] and MES [29] buffers in presence of silver(I).

Unsurprisingly, the primary amine Tris is the strongest silver binder in this study and the stability constant obtained is comparable to other amine ligands such as ethanolamine [33–35]. This value is in line with other studies at different ionic strengths and temperatures that have been quantifying the interaction between Tris buffer and silver(I) [36,37], validating our approach. Morpholine type buffers were less coordinating than piperazine type buffers, as expected by previous results on unsubstituted morpholine [33] and piperazine [38] molecules. MOPS turned out to be clearly the least coordinating buffer of the buffer series studied here (Figure S13). Compared to other metal ions, a lower first stability constant was obtained for silver (I) compared to the ones for copper(II), and similar to nickel(II) or cobalt(II) as found for amines in the literature [26,27,37,39].

To fully benefit from these results and apply them to the standard conditions of a titration (i.e., at constant pH, maintained with a buffer), stability constants were corrected to take into account the partial protonation of the buffer ligand (Figure 2A, Appendix C). The apparent binding constants are slightly decreased compared to the original values, especially when working at high concentrations. Please note that accurate determination of stability constants lower than $\log(K_{app,1,1}^{Ag,L}) = 3$ will ultimately necessitate the use of higher concentrations for the analyte and so for the buffer in order to see the association process. At these concentrations, and according to the third line of Table 2, buffer complexation cannot be neglected. Only high stability constants ($\log(K_{app,1,1}^{Ag,L}) \geq 4$) can thus be determined when using buffers. Another way to see the effect of the buffer on metal ion interactions is to calculate the amount of silver(I) ions bound to the buffer (Figure 2B, Appendix C). According to this percentage, a high proportion of silver ions -more than 90% for Tris buffer- would be complexed by the buffer. Fortunately, when measuring high stability constants at low concentrations for an analyte, the fact that silver ions are not free but bound to the buffer does not affect much the formation of the silver-analyte complex.

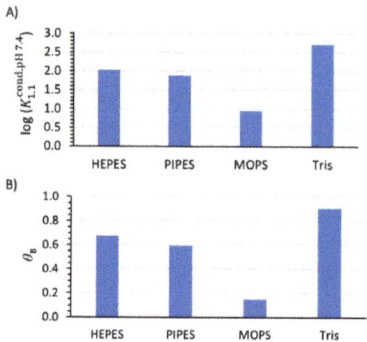

Figure 2. (**A**) Conditional (or apparent) stability constant for the complexation of silver(I) to the buffer at physiological pH 7.4 (for molecules comprised in their buffer range). (**B**) Fraction of silver bound to the buffer (total concentration of buffer 20 mM and silver 1 mM) at pH 7.4.

One could also decide to work at a lower pH (so MES could be considered, but not Tris, Figure S14 and S15B) or to work at different concentrations of buffers (Figure S15A). In the buffer range of the molecules studied here, whatever the conditions, MOPS was always the most suitable buffer. MES, HEPES and PIPES had similar coordination strength regarding silver(I) ions. They can reasonably be used if taking into consideration partial complexation to the buffer for accurate determination of stability constants of ligand/silver complexes.

In conclusion, between pH 6.5 and 7.9, MOPS would be recommended for the studies necessitating the use of silver(I) as it was the less coordinating buffer.

Supplementary Materials: The following are available online at http://www.mdpi.com/2624-8549/2/1/193\T1\textendash202/s1. Figure S1: Titration curves obtained for HEPES without and in presence of silver nitrate in solution, Figure S2: Titration curves obtained for PIPES without and in presence of silver nitrate in solution, Figure S3: Titration curves obtained for MOPS without and in presence of silver nitrate in solution, Figure S4: Titration curves obtained for MES without and in presence of silver nitrate in solution, Figure S5: Titration curves obtained for Tris without and in presence of silver nitrate in solution, Figure S6: Acid dissociation equilibriums considered in the present study for the different buffers, Figure S7: Mass spectra in positive and negative mode for HEPES buffer with silver nitrate, Figure S8: Mass spectra in positive and negative mode for PIPES buffer with silver nitrate, Figure S9: Mass spectra in positive and negative mode for MOPS buffer with silver nitrate, Figure S10: Mass spectra in positive and negative mode for MES buffer with silver nitrate, Figure S11: Mass spectra in positive mode for Tris buffer with silver nitrate, Figure S12: Proposed structures of complexes formed with silver. This stoichiometry was retained for determination of stability constants, Figure S13: Logarithm of stability constants for the first complexation of silver(I) on ligands B (B= buffer studied in this paper), Figure S14: Logarithm of conditional (or apparent) stability constants for the first complexation of silver(I) on buffers at a fixed pH value pH = 6.7, Figure S15: Fraction of silver bound to the buffer, Table S1: Stability constants obtained when considering other equilibrium than the one for the formation of [Ag(L)] (complex [Ag$_2$(PIPES)] or [Ag(MES)$_2$]$^-$, Table S2: Bond distances (Ag-donor atom), average bond valences ($v_{Ag,N1X2}$ and $v_{Ag,O3-5}$) and total atom valence (V_{Ag}) in the molecular structures of [Ag$_x$(Buffer)$_m$].

Author Contributions: Conceptualization, L.B.; methodology, L.B.; validation, L.B.; and K.M.F.; formal analysis, L.B.; investigation, L.B. and S.B.-G.; resources, L.B.; data curation, L.B.; writing—original draft preparation, L.B.; writing—review and editing, L.B. and K.M.F.; visualization, L.B.; supervision, K.M.F.; funding acquisition, K.M.F. All authors have read and agreed to the published version of the manuscript.

Funding: This research was funded by Swiss National Science Foundation and BNF Universität Bern program.

Acknowledgments: I would like to acknowledge Jihane Hankache for mass spectra and Aurélien Crochet for search of crystallographic structures.

Conflicts of Interest: The authors declare no conflict of interest. The funders had no role in the design of the study; in the collection, analyses, or interpretation of data; in the writing of the manuscript, or in the decision to publish the results.

Appendix A. Thermodynamic Equilibrium Used to Fit Potentiometric Titrations

$$\text{BufferH}_n \rightleftarrows \text{BufferH}_{n-1} + \text{H}^+ \quad K_{an} = \frac{[\text{BH}_{n-1}][\text{H}^+]}{[\text{BH}_n]} \tag{A1}$$

For acid dissociation constant K_{an}, there are one to three constants depending on the sum of amine group (one) and the number of sulfonates groups present in the buffer molecule.

A stability constant was then fitted with the fully deprotonated buffer according to Equation (A2).

$$\text{Buffer} \rightleftarrows [\text{Ag}(\text{Buffer})_m] + \text{H}^+ \quad \beta_{1,m}^{Ag,B} = \frac{[AgB_m]}{[B]^m[Ag]} \tag{A2}$$

For all buffers, $m = 1$ except in the case of Tris buffer where there are two constants for $m = 1$ and $m = 2$. For conversion between cumulative constants and stepwise constants (as usually found in literature):

$$\beta_{1,1}^{Ag,B} = K_{1,1}^{Ag,B} \text{ and } K_{1,2}^{Ag,B} = \beta_{1,2}^{Ag,B} / \beta_{1,1}^{Ag,B} \tag{A3}$$

The presence of a complex [Ag(BufferH)] was tested for the fitting of titration curves for all buffers but could not lead to any reliable results (constants were systematically negative). Thus, we consider that this complex was unlikely to be formed in solution.

Appendix B. Calculation of Apparent Binding Constants of a Ligand Binding Silver

$$\text{Ag}^+ + \text{L} \rightleftarrows [\text{AgL}] \quad K_{1,1}^{Ag,L} = \frac{[AgL]}{[L][Ag]} \tag{A4}$$

We define an apparent binding constant which will be the one obtained if not considering the buffer-silver complexation:

$$K_{app,1,1}^{Ag,L} = \frac{[AgL]}{[L]([Ag]_{tot} - [AgL])} = \frac{[AgL]}{([L]_{tot} - [AgL])([Ag]_{tot} - [AgL])} \tag{A5}$$

The mass balance equation for the total concentration of silver is expressed in Equation (A6):

$$[Ag]_{tot} = [Ag] + [AgL] + [Ag(\text{HEPES})] \tag{A6}$$

Concentration of silver complexes can be expressed according to the binding constants:

$$[Ag(HEPES)] = \frac{K_{1,1}^{Ag,B}}{1 + \sum_{i=1}^{n} \beta_{an}(10^{-pH})^n} \cdot \frac{[Ag][HEPES]_{tot}}{1 + \frac{K_{1,1}^{Ag,B}}{1 + \sum_{i=1}^{n} \beta_{an}(10^{-pH})^n} \cdot [Ag]} \tag{A7}$$

$$[AgL] = K_{1,1}^{Ag,L} \cdot \frac{[Ag][L]_{tot}}{1 + K_{1,1}^{Ag,L} \cdot [Ag]} \tag{A8}$$

Introducing Equations (A7) and (A8) in Equation (A6), we obtain an expression of total silver concentration as a function of silver free concentration [Ag].

$$[Ag]_{tot} = [Ag] + \frac{K_{1,1}^{Ag,B}}{1 + \sum_{i=1}^{n} \beta_{an}(10^{-pH})^n} \cdot \frac{[Ag][HEPES]_{tot}}{1 + \frac{K_{1,1}^{Ag,B}}{1 + \sum_{i=1}^{n} \beta_{an}(10^{-pH})^n} \cdot [Ag]} + K_{1,1}^{Ag,L} \cdot \frac{[Ag][L]_{tot}}{1 + K_{1,1}^{Ag,L} \cdot [Ag]} \tag{A9}$$

This concentration is optimized to minimize the difference between the actual concentration $[Ag]_{tot}$ and the one calculated by Equation (A9). Once the concentration of free silver [Ag] at hand, the apparent binding constant can be calculated from Equation (A8) and reintroducing in Equation (A5).

Appendix C. Calculation of Conditional Stability Constants at a Certain pH and Calculation of Percentage of Metal Bound to the Buffer θ_B

Conditional stability constants are defined as the apparent binding constants of the complex between silver(I) and the buffer B at a certain pH value. Thus, we considered that a certain part of the buffer is not coordinating silver(I) as it is protonated but it is still considered in the equilibrium as shown in Equation (A10):

$$K_{1,1}^{cond,pH\,cst} = \frac{[AgB]}{[Ag]([B] + \sum_{i=1}^{n}[BH_n])} = \frac{K_{1,1}^{Ag,B}}{1 + \sum_{i=1}^{n} \beta_{an}(10^{-pH})^n} \tag{A10}$$

For the calculation of the concentration of species and the percentage of metal bound to the buffer, we first established the mass balance equations:

$$[B]_{tot} = [B] + \sum_{i=1}^{m} m[AgB_m] + \sum_{i=1}^{n}[H_n B] \tag{A11}$$

$$[B]_{tot} = [B] + \sum_{i=1}^{m} m[AgB_m] + \sum_{i=1}^{n}[H_n B] \tag{A12}$$

Then we rearrange Equation (A6) and silver total concentration to express the concentration of free silver:

$$[Ag] = \frac{[Ag]_{tot}}{1 + \sum_{i=1}^{m} \beta_{1,m}^{Ag,B}[B]^m} \tag{A13}$$

Free concentration of silver was then reintroduced in Equation (A11):

$$[B]_{tot} = [B]\left(1 + \sum_{i=1}^{m} \frac{m \cdot \beta_{1,m}^{Ag,B}[Ag]_{tot}[B]^{m-1}}{1 + \sum_{i=1}^{m} \beta_{1,m}^{Ag,B}[B]^m} + \sum_{i=1}^{n} \beta_{an}[H]^n\right) \tag{A14}$$

We assumed a certain value for the concentration of the free ligand [B] to obtain a value of [B]$_{tot, calc.}$ with Equation (A14) at a certain pH value. Difference between the calculated total ligand concentration and the one set in the experiment was minimized by tuning the value of free ligand [B].

Once the parameter of free ligand/buffer [B] has been optimized, one could calculate the concentration of complexed species [AgB$_m$] with concentration of free metal being determined with Equation (A12):

$$[AgB_m] = \beta_{1,m}^{Ag,B}[Ag][B]^m \tag{A15}$$

The percentage of metal bound to the buffer is then calculated according to Equation (A16):

$$\theta_B = \frac{\sum_{i=1}^{m}[AgB_m]}{[Ag]_{tot}} = \frac{\sum_{i=1}^{m}\beta_{1,m}^{Ag,B}[Ag][B]^m}{[Ag]_{tot}} \tag{A16}$$

References

1. Good, N.E.; Winget, G.D.; Winter, W.; Connolly, T.N.; Izawa, S.; Singh, R.M.M. Hydrogen Ion Buffers for Biological Research*. *Biochemistry* **1966**, *5*, 467–477. [CrossRef]
2. Ferguson, W.J.; Braunschweiger, K.I.; Braunschweiger, W.R.; Smith, J.R.; McCormick, J.J.; Wasmann, C.C.; Jarvis, N.P.; Bell, D.H.; Good, N.E. Hydrogen ion buffers for biological research. *Anal. Biochem.* **1980**, *104*, 300–310. [CrossRef]
3. Blanchard, J.S. Buffers for enzymes. In *Methods in Enzymology*; Academic Press: New York, NY, USA, 1984; Volume 104, pp. 404–414. [CrossRef]
4. Grady, J.K.; Chasteen, N.D.; Harris, D.C. Radicals from "Good's" buffers. *Anal. Biochem.* **1988**, *173*, 111–115. [CrossRef]
5. Renganathan, M.; Bose, S.J.P.R. Inhibition of photosystem II activity by Cu++ ion. Choice of buffer and reagent is critical. *Photosynth. Res.* **1990**, *23*, 95–99. [CrossRef] [PubMed]
6. Nagira, K.; Hayashida, M.; Shiga, M.; Sasamoto, K.; Kina, K.; Osada, K.; Sugahara, T.; Murakami, H. Effects of organic pH buffers on a cell growth and an antibody production of human-human hybridoma HB4C5 cells in a serum-free culture. *Cytotechnology* **1995**, *17*, 117–125. [CrossRef] [PubMed]
7. Russell, D.W.; Sambrook, J. *Molecular Cloning: A Laboratory Manual*, 3rd ed; Cold Spring Harbor Laboratory Press: New York, NY, USA, 2001.
8. Koerner, M.M.; Palacio, L.A.; Wright, J.W.; Schweitzer, K.S.; Ray, B.D.; Petrache, H.I. Electrodynamics of lipid membrane interactions in the presence of zwitterionic buffers. *Biophys. J.* **2011**, *101*, 362–369. [CrossRef]
9. McRee, D.E.; Jensen, G.M.; Fitzgerald, M.M.; Siegel, H.A.; Goodin, D.B. Construction of a bisaquo heme enzyme and binding by exogenous ligands. *Proc. Natl. Acad. Sci. USA* **1994**, *91*, 12847–12851. [CrossRef]
10. Barrick, D. Replacement of the Proximal Ligand of Sperm Whale Myoglobin with Free Imidazole in the Mutant His-93.fwdarw.Gly. *Biochemistry* **1994**, *33*, 6546–6554. [CrossRef]
11. Hirst, J.; Wilcox, S.K.; Ai, J.; Moënne-Loccoz, P.; Loehr, T.M.; Goodin, D.B. Replacement of the Axial Histidine Ligand with Imidazole in Cytochrome c Peroxidase. 2. Effects on Heme Coordination and Function. *Biochemistry* **2001**, *40*, 1274–1283. [CrossRef]
12. Ferreira, C.M.H.; Pinto, I.S.S.; Soares, E.V.; Soares, H.M.V.M. (Un)suitability of the use of pH buffers in biological, biochemical and environmental studies and their interaction with metal ions—A review. *RSC Adv.* **2015**, *5*, 30989–31003. [CrossRef]
13. Vasconcelos, M.T.S.D.; Azenha, M.A.G.O.; Lage, O.M. Electrochemical Evidence of Surfactant Activity of the Hepes pH Buffer Which May Have Implications on Trace Metal Availability to Culturesin Vitro. *Anal. Biochem.* **1996**, *241*, 248–253. [CrossRef]
14. Mash, H.E.; Chin, Y.-P.; Sigg, L.; Hari, R.; Xue, H. Complexation of Copper by Zwitterionic Aminosulfonic (Good) Buffers. *Anal. Chem.* **2003**, *75*, 671–677. [CrossRef]
15. Sokołowska, M.; Bal, W. Cu(II) complexation by "non-coordinating" N-2-hydroxyethylpiperazine-N'-2-ethanesulfonic acid (HEPES buffer). *J. Inorg. Biochem.* **2005**, *99*, 1653–1660. [CrossRef]
16. Zawisza, I.; Rózga, M.; Poznański, J.; Bal, W. Cu(II) complex formation by ACES buffer. *J. Inorg. Biochem.* **2013**, *129*, 58–61. [CrossRef]

17. Fischer, B.E.; Häring, U.K.; Tribolet, R.; Sigel, H. Metal Ion/Buffer Interactions. *Eur. J. Biochem.* **1979**, *94*, 523–530. [CrossRef] [PubMed]
18. Scheller, K.H.; Abel, T.H.J.; Polanyi, P.E.; Wenk, P.K.; Fischer, B.E.; Sigel, H. Metal Ion/Buffer Interactions. *Eur. J. Biochem.* **1980**, *107*, 455–466. [CrossRef] [PubMed]
19. Chabert, V.; Hologne, M.; Sénèque, O.; Crochet, A.; Walker, O.; Fromm, K.M. Model peptide studies of Ag+ binding sites from the silver resistance protein SilE. *Chem. Commun.* **2017**, *53*, 6105–6108. [CrossRef] [PubMed]
20. Chabert, V.; Hologne, M.; Sénèque, O.; Walker, O.; Fromm, K.M. Alpha-helical folding of SilE models upon Ag(His)(Met) motif formation. *Chem. Commun.* **2018**, *54*, 10419–10422. [CrossRef] [PubMed]
21. Bilinovich, S.M.; Panzner, M.J.; Youngs, W.J.; Leeper, T.C. Poly[[{[mu]3-2-[4-(2-hydroxyethyl)piperazin-1-yl]ethanesulfonato}silver(I)] trihydrate]. *Acta Crystallogr. E* **2011**, *67*, m1178–m1179. [CrossRef] [PubMed]
22. Goldberg, R.N.; Kishore, N.; Lennen, R.M. Thermodynamic Quantities for the Ionization Reactions of Buffers. *J. Phys. Chem. Ref. Data* **2002**, *31*, 231–370. [CrossRef]
23. Fukada, H.; Takahashi, K. Enthalpy and heat capacity changes for the proton dissociation of various buffer components in 0.1 M potassium chloride. *Proteins* **1998**, *33*, 159–166. [CrossRef]
24. Kitamura, Y.; Itoh, T. Reaction volume of protonic ionization for buffering agents. Prediction of pressure dependence of pH and pOH. *J. Solut. Chem.* **1987**, *16*, 715–725. [CrossRef]
25. Bates, R.G.; Pinching, G.D. Dissociation Constants of Weak Bases from Electromotive-Force Measurements of Solutions of Partially Hydrolyzed Salts. *J. Res. Natl. Bur. Stand.* **1949**, *43*, 519–526. [CrossRef]
26. Stability Constants Available on Joint Expert Speciation System. Available online: http://jess.murdoch.edu.au/rawrxnbiny.shtml (accessed on 22 March 2020).
27. Smith, R.M.; Martell, A.E. *Critical Stability Constants*; Springer: Boston, MA, USA, 1989. [CrossRef]
28. Daofeng, S.; Rong, C.; Yucang, L.; Maochun, H. Self-Assembly of A Novel Sulphonate Silver(I) Complex. *Chem. Lett.* **2002**, *31*, 198–199. [CrossRef]
29. Jiang, H.; Ma, J.-F.; Zhang, W.-L.; Liu, Y.-Y.; Yang, J.; Ping, G.-J.; Su, Z.-M. Metal–Organic Frameworks Containing Flexible Bis(benzimidazole) Ligands. *Eur. J. Inorg. Chem.* **2008**, *2008*, 745–755. [CrossRef]
30. Sun, D.; Cao, R.; Bi, W.; Li, X.; Wang, Y.; Hong, M. Self-Assembly of Novel Silver Polymers Based on Flexible Sulfonate Ligands. *Eur. J. Inorg. Chem.* **2004**, *2004*, 2144–2150. [CrossRef]
31. Our own attempts to crystallize complexes of Tris, MES or MOPS with silver were unsuccessful mostly due to the reduction of silver.
32. Kumar, R.; Obrai, S.; Kaur, A.; Hundal, M.S.; Meehnian, H.; Jana, A.K. Synthesis, crystal structure investigation, DFT analyses and antimicrobial studies of silver(i) complexes with N,N,N',N''-tetrakis(2-hydroxyethyl/propyl) ethylenediamine and tris(2-hydroxyethyl)amine. *New J. Chem.* **2014**, *38*, 1186–1198. [CrossRef]
33. Bruehlman, R.J.; Verhoek, F.H. The Basic Strengths of Amines as Measured by the Stabilities of Their Complexes with Silver Ions. *J. Am. Chem. Soc.* **1948**, *70*, 1401–1404. [CrossRef]
34. Datta, S.P.; Grzybowski, A.K. 222. The stability constants of the silver complexes of some aliphatic amines and amino-acids. *J. Chem. Soc.* **1959**, 1091–1095. [CrossRef]
35. Poucke, L.C.V.; Eeckhaut, Z. Stability Constants of Ag(I) Complexes of Some Amino-Alcohols. *Bull. Soc. Chim. Belg.* **1972**, *81*, 363–366. [CrossRef]
36. Canepari, S.; Carunchio, V.; Castellano, P.; Messina, A. Protonation and silver(I) complex-formation equilibria of some amino-alcohols. *Talanta* **1997**, *44*, 2059–2067. [CrossRef]
37. Perrin, L.D. *Stability Constants of Metal-ion Complexes. Part B. Organic Ligands. IUPAC Chem. Data Set No. 22*; Pergamon Press: Oxford, UK, 1979.
38. Houngbossa, K.; Enea, G.B.E.O. Étude thermodynamique de la complexation de l'argent par la pipérazine et quelques-uns de ses dérivés en milieu aqueux. *Thermochim. Acta* **1973**, *6*, 215–222. [CrossRef]
39. Wyrzykowski, D.; Pilarski, B.; Jacewicz, D.; Chmurzyński, L. Investigation of metal–buffer interactions using isothermal titration calorimetry. *J. Therm. Anal. Calorim.* **2013**, *111*, 1829–1836. [CrossRef]

© 2020 by the authors. Licensee MDPI, Basel, Switzerland. This article is an open access article distributed under the terms and conditions of the Creative Commons Attribution (CC BY) license (http://creativecommons.org/licenses/by/4.0/).

Review

Giant Polymer Compartments for Confined Reactions

Elena C. dos Santos, Alessandro Angelini, Dimitri Hürlimann, Wolfgang Meier * and Cornelia G. Palivan *

Department of Chemistry, University of Basel, Mattenstrasse 24a, BPR 1096, 4002 Basel, Switzerland; e.dossantos@unibas.ch (E.C.d.S.); alessandro.angelini@unibas.ch (A.A.); dimitri.huerlimann@unibas.ch (D.H.)
* Correspondence: wolfgang.meier@unibas.ch (W.M.); cornelia.palivan@unibas.ch (C.G.P.); Tel.: +41-61-207-38-02 (W.M.); +41-61-207-38-39 (C.G.P.)

Received: 25 April 2020; Accepted: 8 May 2020; Published: 12 May 2020

Abstract: In nature, various specific reactions only occur in spatially controlled environments. Cell compartment and subcompartments act as the support required to preserve the bio-specificity and functionality of the biological content, by affording absolute segregation. Inspired by this natural perfect behavior, bottom-up approaches are on focus to develop artificial cell-like structures, crucial for understanding relevant bioprocesses and interactions or to produce tailored solutions in the field of therapeutics and diagnostics. In this review, we discuss the benefits of constructing polymer-based single and multicompartments (capsules and giant unilamellar vesicles (GUVs)), equipped with biomolecules as to mimic cells. In this respect, we outline key examples of how such structures have been designed from scratch, namely, starting from the application-oriented selection and synthesis of the amphiphilic block copolymer. We then present the state-of-the-art techniques for assembling the supramolecular structure while permitting the encapsulation of active compounds and the incorporation of peptides/membrane proteins, essential to support in situ reactions, e.g., to replicate intracellular signaling cascades. Finally, we briefly discuss important features that these compartments offer and how they could be applied to engineer the next generation of microreactors, therapeutic solutions, and cell models.

Keywords: artificial cells; biomimicry; polymer GUVs; polymer capsules; single compartments; multicompartments

---

1. Introduction

Compartmentalization produces a remarkably efficient organization of membranes and biomolecules that is essential to cope with the complexity of metabolic reactions in cells, and whose stability and functions are vital [1]. Inspired by natural biocompartments, significant efforts have been made to produce compartments that mimic cells and organelles, either in terms of their membrane properties or of the reactivity of encapsulated biomolecules [2]. Micrometer-sized vesicles, namely giant unilamellar vesicles, GUVs for short, are preferably used in this context, since their size and architecture can mimic cells, such as to extract information regarding reactions in a bio-relevant confined space. In addition, they allow for real time visualization of the membrane structure (providing information regarding membrane fluidity and integrity), and of biochemical reactions and enzymatic crowding effects that occur within a controlled and simplified environment, yet still preserving defined characteristics of cells. Lipid based compartments are straightforward systems for mimicking a cell/organelle membrane, nevertheless, their mechanical instability and the presence of membrane defects are limiting factors. One elegant way of introducing robustness to compartments, and at the same time of expanding new membrane properties, is the use of compartments made of copolymers. With the progress in polymer chemistry, numerous amphiphilic block copolymers have been synthesized with a variety of compositions, block ratios and functions [3]. Due to a greater chemical versatility compared to lipids, block copolymers increase the opportunities to achieve desired self-assembled morphologies made of membranes with tailored properties and excellent biocompatibility. The architecture of such compartments—whether they are micro- or nano-sized—offers three different regions: the inner cavity for encapsulating hydrophilic molecules, the membrane for insertion of hydrophobic molecules, and the external membrane surface, for attachment of specific molecules [4,5] and eventually, for immobilization onto external functional surfaces [6,7]. From a topological point of view, nano-sized compartments, such as small layer-by-layer (LbL) capsules [8], polymersomes [9] or liposomes [10], can be designed as to mimic organelles, the cellular subcompartments. They have been used as nano-scale catalytic compartments, serving as to produce various desired compounds, as artificial peroxisomes [11], acting in tandem to support cascade reactions or as subcompartments inside GUVs to allow development of multicompartment systems [12]. When nanocompartments are encapsulated inside polymer giants in combination with active compounds, they are able to communicate among them to allow reactions, similarly to intracellular organization [13]. Permeability of membranes (either in a single or in multicompartments) favors molecular transport (enzyme substrates and products) and can be achieved in various ways, resulting from the chemical nature of the copolymer [14], by insertion of peptides [15,16] and membrane proteins [17,18]. A schematic of the most common compositions that polymer single and multicompartments can attain is presented in Figure 1.

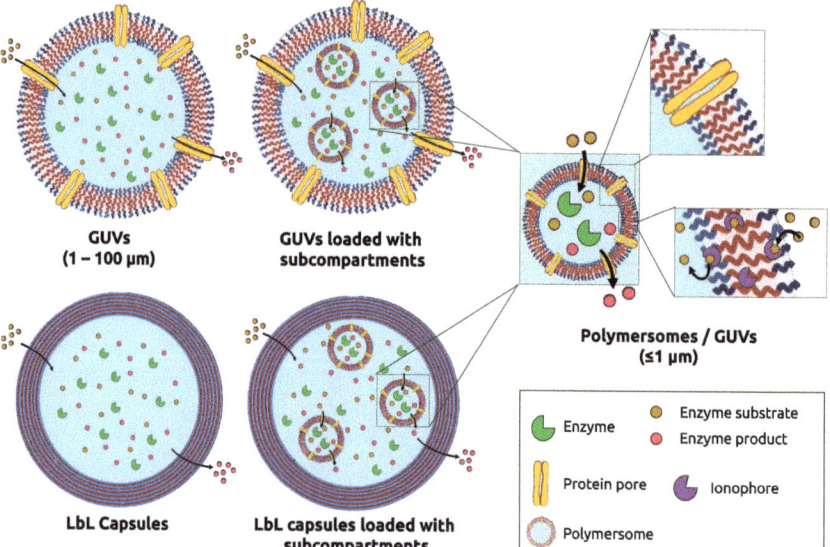

Figure 1. Schematic representation of the different types of polymer compartments (polymer giant unilamellar vesicles (GUVs) and layer-by-layer (LbL) capsules), showing their diversity in terms of size, arrangements and the different types of biomolecules, including their possible locations within the assemblies.

This review presents micrometer-sized compartments either as single or as multicompartment reaction space, as powerful tools for biomimicry, lowering the degree of complexity to enable studies on targeted processes. We first introduce the synthesis of amphiphilic block copolymers through the various known polymerization techniques as building blocks of such compartments and indicate the conditions and properties required to support in situ reactions. Different methods for the preparation of these vesicular structures and their combination with active compounds (e.g., enzymes and peptides/membrane proteins) are presented together with the crucial points ensuring an efficient compartmentalization for desired applications. Permeabilization methods will not be described in this review, as they were already discussed extensively elsewhere [19,20]. We rather explore the biomimetic approach of these compartments equipped with peptides/membrane proteins to render them permeable for molecular flow and containing active compounds/subcompartments. Reactions inside confined spaces at microscale allow studying and better understanding of natural mechanisms of such reactions and their role inside them to support applications in various domains, as sensing, therapeutics and catalysis.

2. Polymers as Building Blocks of Micrometer-Sized Compartments

The progress in polymer chemistry gave access to a variety of polymers with tailored properties and excellent biocompatibility, thus serving to select specific components, where the precise role of each leads to a well-controlled system. Two or more chemically different polymeric domains, covalently bound together are defined as block copolymers. More specifically, amphiphilic block copolymers are composed of both hydrophilic and hydrophobic blocks, often named as diblocks (AB), triblocks (ABA or ABC) or multiblocks (ABCBA, ABCD, etc.). According to the required properties, amphiphilic block copolymers are built/designed by combining specific types of hydrophilic and hydrophobic blocks.

2.1. Amphiphilic Block Copolymers as Building Blocks for Generation of GUVs

In order to prepare amphiphilic block copolymers, controlled polymerization techniques are commonly used: Atom transfer radical polymerization (ATRP) [21], reversible addition fragmentation chain transfer (RAFT) [22,23], ionic polymerization and combinations thereof [3,24]. Typically, sequential chain extension can be used, in which a first block is polymerized using the aforementioned techniques, forming the so-called macro-initiator. Immediate addition of a second monomer leads to chain-extension, yielding a diblock copolymer. This approach allows the adjustment of each block length by terminating the corresponding chain extension according to the desired degree of polymerization. Tri- or multiblock copolymers can be obtained analogously either by sequential chain extension (asymmetric ABA, ABC, ABCD, etc.) or by a bifunctional initiator (symmetric ABA, ABCBA, etc.). Monomer conversion as well as the living character of the polymer chain are fundamental parameters to be considered among each chain extension. In ionic polymerizations, high monomer conversions can easily be reached by maintaining narrow polydispersity [25]. For example, poly(ethylene oxide)-*block*-polybutadiene (PEO-*b*-PBD) has been successfully synthesized by anionic polymerization in a two steps sequential monomer addition [26–28]. Prior to the addition of the second monomer, modifications are required as for poly(ethylene oxide)-*block*-poly(ethyl ethylene) (PEO-*b*-PEE), in which a PBD precursor is first hydrogenated to yield the PEE macroinitiator. Subsequently, ethylene oxide is polymerized to obtain the diblock copolymer [29]. In another study, poly(acrylic acid)-*block*-polybutadiene (PAA-*b*-PBD) was prepared by sequentional addition of butadiene and *tert*-butylacrylate, followed by hydrolysis to its acid form [28]. The combination of different polymerization techniques is another possibility. Namely, the preparation of poly(dimethyl sulfoxide) (PDMS) by anionic polymerization was followed by activation and cationic ring-opening polymerization of 2-methyl-2-oxazoline (MOXA) monomers to obtain poly(dimethyl sulfoxide)-*block*-poly(2-methyl-2-oxazoline) (PDMS-*b*-PMOXA) diblock copolymers (Scheme 1) [30].

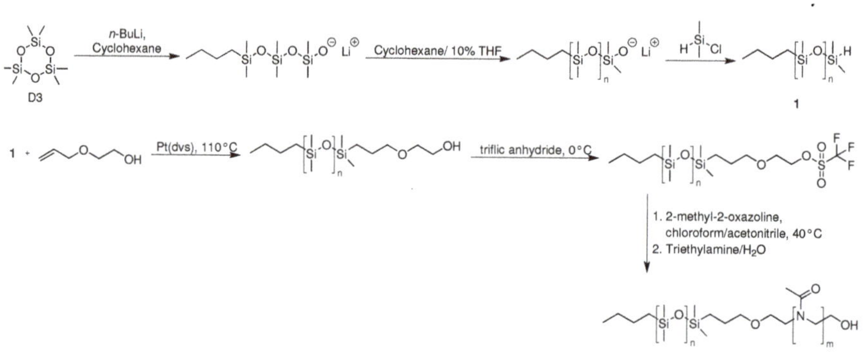

Scheme 1. Synthesis route of PDMS-*b*-PMOXA combining both types of ionic polymerizations. Reprinted with permission from [30]. Copyright © 2014 American Chemical Society.

Ionic polymerization techniques are limited due to their high sensitivity to impurities. Hence, the solvent choice is important and the preparation of each reactant has to be handled very carefully to reach the desired purity grade. Controlled radical polymerization techniques (CRP), such as ATRP or RAFT, have recently been developed and have provided interesting and more versatile alternatives for the production of block copolymers. Both techniques require an initiator and the polymerization is governed by an equilibrium between an active species and a dormant one. The latter is constantly re-initiated in order to form the active species responsible for propagation through the addition of monomers. With these techniques, it is usually recommended not to exceed monomer conversions of 90%, above which the probability of termination is higher, risking to form dead chains unable to continue chain-extension [22]. As an example, poly(ethylene oxide)-*block*-poly(4″-acryloyloxybutyl 2,5-bis(4′-butyloxybenzoyloxy)benzoate) (PEO-*b*-PA444) was obtained from PEO modified to a macroinitiator for ATRP on which PA444 has been polymerized [31]. By using RAFT, a diblock copolymer poly(pentafluorophenyl acrylate)-*block*-poly(*n*-butyl acrylate) (PFPA-*b*-P*n*BA) was firstly prepared using the appropriate chain transfer agent (CTA), followed by modification to yield the amphiphilic glycopolymer PNβGluEAM-*b*-PBA [32]. More recently, polymerization induced self-assembly (PISA) allowed the preparation of poly(ethylene oxide)-*block*-poly(2-hydroxypropyl methacrylate) (PEO-*b*-PHPMA). In a suitable solvent, a solution of monomer feeds the growing chain on the PEO macroinitiator, producing an amphiphile that gradually self-assembles into structures, while polymerization is ongoing and leading to turbidity in the medium (Figure 2) [33].

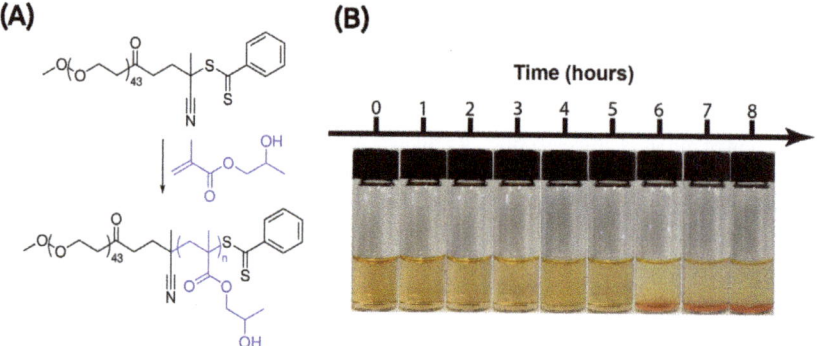

Figure 2. (**A**) RAFT polymerization of 2-hydroxypropyl methacrylate controlled by a PEO macroinitiator (**B**) Reaction mixture throughout the PISA polymerization process. Adapted with permission from [33]. Copyright © 2017 Springer Nature.

Although, the possibility of combining synthetic approaches broadens the library of accessible polymers, chemists still need to work hard on the quantitative attachment of the re-initation site for the next polymerization, which is highly recommended to prevent purification difficulties. To circumvent this problem, two or more homopolymers can be connected together using coupling reactions such as Diels-Alder, copper-catalyzed azide-alkyne cycloaddition (CuAAC) or thiol-ene click chemistry, thus offering an increased number of possibilities. To illustrate, PAA-b-PBD has been prepared by combining poly(*tert*-butyl acrylate) (P*t*BA) and PBD homopolymers, both synthesized beforehand. A hydrolysis step leads to the final diblock [34]. Poly(dimethylsiloxane)-*block*-poly(ethylene oxide) (PDMS-b-PEO) diblock copolymers were synthesized using ring-opening polymerization of hexamethylcyclotrisiloxane to obtain PDMS-N_3 and further coupling with PEO-Alkyne chains via click chemistry [35]. However, some reactive conditions can require high temperatures or metal catalysts, which might not be suitable for biomedical applications [36,37]. Moreover, complete end-group functionalization and equimolar ratios of both homopolymers are required, preventing the challenging removal of unreacted homopolymers. Increasing the number of blocks introduces more challenges, especially in re-initiation, purification and finding suitable solvent for all the blocks.

The self-assembly of amphiphilic block copolymers in solution leads to the formation of many different assemblies including spherical, cylindrical, gyroidal and lamellar structures [38]. These assemblies are directly influenced by intrinsic molecular parameters of the amphiphilic block copolymers and the conditions in which the self-assembly process takes place (concentration of the copolymer, presence of solvents, temperature, etc). In this respect, the hydrophilic to the total mass ratio (f) calculated as the ratio of the molar mass of the hydrophilic block to the total molar mass of the copolymer is an important parameter, which governs the resulting supramolecular assembly. Vesicular structures are typically obtained for f values ranging from 0.20 to 0.40. Another molecular parameter influencing the self-assembly into different assemblies is the packing parameter ($p = v/a_0 l_c$; v = volume of the hydrophobic part, a_0 = contact area of the head group, l_c = length of the hydrophobic part) that describes the degree of curvature from the membrane. For low packing parameter values ($0 < p < 0.5$), the curvature gradually decreases from high to medium, resulting in the formation of spherical or cylindrical micelles, respectively. For higher values ($0.5 < p < 1$), the curvature of the membrane is considerably low, which is more favorable for vesicular structures. The dispersity, D, of the copolymer is affecting the size distribution of the formed vesicles: a narrow dispersity typically leads to uniform-sized polymersomes, whilst on the opposite, a more polydisperse population of vesicles is obtained [39–41].

2.2. Polymers as Building Blocks for Generation of Polymer Capsules

There are a few works that produced polymer capsules via methods originating from the LbL deposition, e.g., single-step polymer adsorption, surface polymerization and ultrasonic assembly [42]. However the vast majority have employed purely the LbL assembly technique [43], where different polymer segments are alternately deposited and adsorbed. These layers are typically formed by homopolymers. The wide range of polymers provides capsules with a variety of walls, as a result of adjusting important parameters, such as composition, permeability and surface functionality of the capsules [44]. Nevertheless, such polymers must have functional groups capable of providing electrostatic interactions or hydrogen bonds. For electrostatic interactions, polyelectrolytes having anionic or cationic groups in their side chains are used, poly(styrene sulfonate) (PSS) or poly(allylamine hydrochloride) (PAH), respectively [45,46]. In the case of polymers forming hydrogen bonds, the side chains are composed of functional groups called "hydrogen-bond receptors", which have at least one lone pair (carbonyl, ether, hydroxyl, amino, imino, and nitrile groups), like polyvinylpyrrolidone (PVP), or "hydrogen-bond donors" represented by the presence of a hydrogen atom covalently bound to a more electronegative atom (hydroxyl, amino, and imino groups), like poly(methacrylic acid) (PMAA). These interactions are fundamental for the formation and maintenance of the layers during the LbL deposition.

3. Technologies for Engineering Polymer Single and Multicompartments in Combination with Biomolecules

Important features of supramolecular assemblies, designed as functional single or multicompartments to accommodate active compounds, are highly dependent on the preparation methods. Aiming at obtaining the desired structures with optimized characteristics as, size and size distribution, membrane composition and specific functionalities, biomolecular content inside cavities, etc., appropriate procedures need to be selected [47].

3.1. Polymer GUVs

A wide selection of methods to generate polymer vesicular structures are available; ranging from the fairly established bulk techniques, as electroformation and film rehydration, to more automated and high-throughput ones as microfluidcs, currently still underused in the domain of cell mimicry.

3.1.1. Bulk Techniques

Electroformation

Electroformation, the most common method to obtain GUVs, involves the swelling of the amphiphilic polymer film in the presence of an electric field. The dry copolymer film is deposited on conductive indium tin oxide (ITO) coated glass slides and is subjected to an alternating sin-wave electric current while it is rehydrated in aqueous solution. The former contains the desired biomolecules to be encapsulated or incorporated inside the GUVs core or membrane, respectively. The electric field induces a periodic electroosmotic movement of the water in between the individual bilayer lamellae in the film, causing the vesicle detachment from the substrate surface [47] as represented in Figure 3B. This method was successfully employed in many different occasions [14,16,48,49]. In particular, Itel et al. [14] formed giants consisting of diblocks (PMOXA-b-PDMS) or triblocks (PMOXA-b-PDMS-b-PMOXA), yielding membranes with thicknesses ranging from 5–30 nm, which can represent 2–10 times that of the phospholipids. Albeit this feature can contribute to a hydrophobic mismatch between membrane thickness and the size of the proteins of more than 5 times, (PMOXA-b-PDMS) offered enough flexibility and fluidity to facilitate the membrane protein insertion [50]. To enable reactions, Lomora et al. [51] produced GUVs of different poly(2-methyloxazoline)-b-poly(dimethylsiloxane)-b-poly(2-methyloxazoline) (PMOXA$_x$-PDMS$_y$-PMOXA$_x$) triblock copolymers. These were equipped with a peptide (Gramicidine, gA) for inducing a selective monovalent ion permeability. Another example was the formation of GUVs with PEO-12 dimethicone, in which permeability was induced by the addition of calcimycin, an ionophore that enabled the transport of Ca^{2+} selectively, serving for the in situ mineralization of calcium carbonate [52]. Nevertheless, this method is not recommended for charged amphiphilic copolymers due to electrostatic interactions, which might affect the self-assembly process [40].

Film-Rehydration

A more suitable technique to circumvent the problem of electrostatic interactions is the direct rehydration of a thin polymer film to form the GUVs. For example, this method succeeded in forming GUVs made of a mixture of PMOXA$_5$-b-PDMS$_{58}$-b-PMOXA$_5$ and the negatively charged PDMS$_{65}$-b-heparin copolymers as a mimic for heparan sulfate, known to be exposed on the plasma membrane of most cell types [12]. In the film rehydration method, the block copolymers are first dissolved in an appropriate organic solvent, followed by evaporation either with a stream of nitrogen or by applying vacuum in a rotary evaporator. Rehydration takes place by pouring aqueous solution to the dried film, resulting in the detachment of the GUVs from the substrate surface, (Figure 3A). In general, the desired hydrophilic biomolecular content is encapsulated into the GUVs cavity by mixing it to the rehydration buffer solution. Whereas, as shown by Belluati et al. [15], the hydrophobic ion channels can be inserted in different steps of the hydration processes, e.g.: (i) blended and co-dried with the copolymer film, (ii) added to the rehydration buffer or (iii) added to the pre-formed GUVs suspension (*ex post*). Moreover, aiming at obtaining functional compartments and to follow a reaction in situ, Garni et al. [18] simultaneously encapsulated a model enzyme horseradish peroxidase (HRP) inside the polymer GUVs and inserted a channel porin, Outer membrane protein F double mutant (OmpF-M), by adding these biomolecules to the rehydration buffer during the formation process. Self-assembly process of GUVs by film-rehydration and electroformation does not produce GUVs with homogeneous sizes, instead a mixture of GUVs in the size range of 1 to 40 µm is formed. In case smaller sizes and narrower size distribution are required, the polymer giants suspension can be subjected to additional processes [53], as freeze-thawing, sonication or extrusion through a polycarbonate membrane [16,51]. Dialysis and size exclusion chromatography are alternative steps to obtaining a relatively monodisperse population.

Solvent Switch/Exchange

With the solvent switch method, the supramolecular assembly is induced by adding water drop-wise into a dissolved and molecularly dispersed polymer organic solution, thus, gradually exchanging the organic solvent with water. The turbid solution that is formed is immediately quenched by being poured slowly into an excess of water under continuous stirring. Finally, the organic solvent is removed from the solution via dialysis [54]; an important step especially when envisaging biomimetic applications [55]. However, due to the possible denaturation and degradation caused by traces of organic solvents, such a method may be incompatible with sensitive molecules, limiting their use in biomedical applications [40]. As it has been demonstrated by Daubian et al. [56], depending on the chemical nature of the amphiphile, the solvent switch method may perform faster than the film-rehydration, especially when many metastable phases of the block copolymer can be formed, leading mostly to less aggregates. [57]. With this technique, GUVs are assembled via nucleation and growth of unimers [58,59]. Due to the solvent exchange, the great number of unimers formed deplete the unimers in solution reaching rapidly phase equilibrium, and thus are prevented to grow to larger sizes. GUVs produced hence are the smallest ($\approx$1 µm), and can be tuned to form polymersomes on the nanoscale [57].

3.1.2. Microfluidics

Double Emulsion Method

Microfluidic techniques allow for the production of defined polymer stabilized water-oil-water (w/o/w) double emulsions, which are used as templates for generating GUVs. Double emulsion formation proceeds when the inner aqueous phase, containing the biological solution is enveloped by the organic phase, typically consisting of the diblock copolymer dissolved in a volatile and water-immiscible solvent, which breaks up into double emulsions, due to shear caused by the external aqueous phase (Figure 3B) [60–62]. These GUVs have narrow size distributions, with mean sizes ranging from 10–100 µm, which are highly dependent on the microdevice channel sizes and junctions (where generally droplets are formed) [63,64]. To form GUVs from double emulsions, the amphiphile chains are brought together by evaporating the volatile solvent, forming the bilayer membrane. While the complete removal of the organic phase might not be trivial and implies a limitation, this method allows for efficient encapsulation of large amounts of water soluble biomolecules [64]. Thus, its employment is vastly recommended when high-efficiency encapsulation is required, e.g., for loading enzymes and pore-forming proteins within GUVs for mimicking cells. Despite essential contribution on the development of such compartments has been made, there exists only one example where biological machinery (i.e., an aqueous mixture containing *E. coli* ribosomal extract and MreB DNA plasmid) was encapsulated into semi-permeable poly(ethylene oxide)-*block*-poly(lactic acid) (PEO-*b*-PLA) GUVs for carrying out protein expression [65].

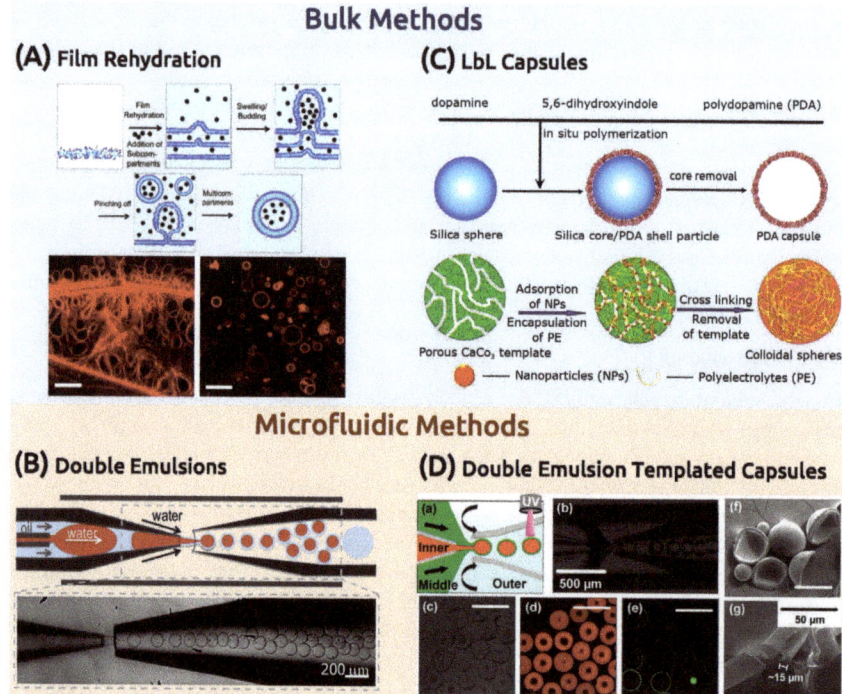

Figure 3. Engineering strategies for constructing polymeric single and multi-compartments, capsules and polymer-based giant unilamellar vesicles (GUVs). (**A**) Mechanism of polymer GUV detachment from the substrate surface by the film-rehydration method. Adapted with permission from Thamboo et al. [12]. Copyright © 2019 Wiley-VCH. (**B**) Double emulsion droplets formed in a microfluidic capillary device, which serve as templates for producing GUVs. Adapted with permission from do Nascimento et al. [64]. Copyright © 2016 American Chemical Society. (**C**) Polymer microcapsules produced via layer-by-layer (LbL) deposition onto hard colloidal sacrificial templates. Mechanism using Silica particles, adapted with permission from Yan et al. [66]. Copyright © 2012 Wiley-VCH. Mechanism using $CaCO_3$ particles, adapted with permission from Postma et al. [67]. Copyright © 2009 American Chemical Society. (**D**) Polymer microcapsules produced via double emulsion technique, followed by UV polymerization. Adapted with permission from Xie et al. [68]. Copyright © 2017 American Chemical Society.

3.2. Polymer Capsules

Differently from GUVs, polymer capsules require the employment of either a soft or a hard template. They have operated as delivery vehicles, since they also allow for the selective diffusion of reagents/reaction products; yet their use as microreactors for mimicking cells has been limited.

Layer-by-layer Microcapsules

Fabrication of polymer microcapsules involves multiple synthetic steps and compositional complexity for the particular application. The LbL technique requires the use of a colloidal particle as a sacrificial template, which plays a pivotal role, since it determines the capsule size and shape, and most importantly the biomolecular encapsulation method. Soft sacrificial templates have been employed, including the commercial ones: poly(methyl methacrylate) (PMMA) and polystyrene (PS), however they do not allow for the pre-loading of the active components, hampering the microcapsules application for therapeutics, due to low reproducibility of the diffusion process involved in the post-loading method [69]. Instead, when employing hard sacrificial templates, e.g., calcium carbonate [66] or silica [67], encapsulation of enzymes and sensitive dyes is reached via their concurrent precipitation with the template, ensuring a high loading efficiency [70,71]. Decomposition of these templates is then induced for the creation of the inner cavity loaded with the specific biomolecule (Figure 3C). With respect to the outer shell, two polymers interacting by electrostatic forces or hydrogen bonding are deposited alternately on the template before it is dissolved to obtain the hollow sphere. Typically, PVP and PMAA which interact via hydrogen bonding at pH values below the pKa of PMAA are used for this technique. Using PMAA, the stability of these capsules can be extended to physiologically relevant pH by crosslinking. The resulting pure PMAA hydrogel capsules are biodegradable, nontoxic, semipermeable and thus well suited for biomedical applications. More recent studies replace the labour intense LbL assembly of PMAA/PVP capsules by polydopamine shells that are deposited on the template in a single step [72].

Double Emulsion Templated Microcapsules

Opposite to the conventional fabrication methods, where multiple laborious synthetic steps must be satisfied, microfluidics offers an alternative technique for a rapid, with low polydispersity and highly reproducible production of polymer microcapsules. To this aim, double emulsion droplets, formed following the same procedure aforementioned, serve as non-sacrificial templates [73]. For generating polymer microcapsules, flowing droplets are subjected to UV irradiation and thus continuously and rapidly polymerize (Figure 3D). Here, the oil phase contains a photocurable polymer and a photoinitiatior dissolved in a water miscible organic solvent [69]. For biomedical applications, poly(ethylene glycol) diacrylate (PEGDA) microcapsules of around 15 µm were produced and allowed for the diffusion of molecules as large as heparin labeled with Fluorescein isothiocyanate (FITC) ($M_W \approx 10$ kD) [68]. These results demonstrate the biosensing ability and the promising versatility of microfluidics for the preparation of microreactors.

3.3. Building Multicompartments

Multicompartments are considered as an advance towards functional models for eukaryotic cells and their cellular organelles, which are able to perform multiple, chemically incompatible, enzymatic reactions simultaneously by separating them in subcompartments. Multicompartment vesicles were pioneered when the so called vesosomes (liposomes encapsulated inside larger liposomes) were first developed [74]. This process was promptly transferred to synthetic polymeric assemblies, such as polymeric vesicles or LbL capsules, resulting in all conceivable combinations. Multicompartments consist mainly of bigger outer compartments that can be loaded with different kinds of subcompartments, as subsequently detailed.

3.3.1. Loading Polymeric GUVS with Subcompartments

The encapsulation of subcompartments as, small polymersomes, micelles or liposomes, but also nanoparticles, is usually attained during the polymer GUV self-assembly. Each subcompartment can be previously equipped with biomolecules and/or the biomolecules can be encapsulated together with the mixture of empty subcompartments. For example, GUVs of polystyrene-b-poly(L-isocyanoalanine (2-thiophen-3-yl-ethyl) amide) (PS-b-PIAT) were prepared by the solvent switch method, using as aqueous phase, a mixture of the cyanine-5 conjugated immunoglobulin G proteins (Cy5-IgG) and a suspension of smaller polymersomes made of PMOXA-b-PDMS-b-PMOXA, previously generated by film rehydration and equipped with green fluorescent protein (GFP) [75]. Co-localized red and green fluorescence emission measurements were used to compute that only 45% of the supramolecular assemblies resulted in multicompartments. Marguet et al. [76] also demonstrated the generation of polymer multicompartments based on the emulsion-centrifugation method. The inner polymersomes were formed by nanoprecipitation of poly(trimethylene carbonate)-b-poly(L-glutamic acid) (PTMC-b-PGA), and subsequently loaded in GUVs made of polybutadiene-b-poly(ethylene oxide) (PBD-b-PEO) by emulsion–centrifugation. By using such technique, yet for formation of giant liposomes, the loading efficiency reached up to 98% [49]. Regardless of the method used, the obtained structures will always consist of a combination of single and multicompartments. Double emulsion microfluidics has also been used to form multicompartments made of PEO-b-PLA diblock-copolymers for both the inner and the outer membranes [77]. Despite promised control over the number of the inner polymersomes by solely adjusting the flow rates, no loading efficiency was reported.

3.3.2. Layer-by-Layer Multicompartments

LbL multicompartments are constructed with either one smaller LbL capsule as single subcompartment (shell-in-shell structure) [78] or thousands of subcompartments that are deposited onto the template during the preparation of the micron-sized outer capsule. In this regard, the subcompartments may comprise small LbL capsules, polymersomes [9] or liposomes [10], with the former being used for the majority of the LbL multicompartments. The LbL deposition offers the control over the spatial positioning of the subcompartments. Depending on the polymers used for the precursor or separation layer, they either stay attached to the inner walls of the LbL capsule or become "free-floating" after template removal [79]. If only one hemisphere of the template is exposed to the subunit deposition, Janus type multicompartments can also be prepared by the LbL approach [80]. As for single compartments, it is possible to encapsulate the biological content inside the subunits or the lumen of the main compartment, in addition, it can be also found within or outside of the membranes. Replacement of the liposomal subcompartments with polymersomes offers the possibility to address challenges, as prolonged stability of the subcompartments to sustain activity of the encapsulated enzyme. However, examples for polymersome subunits in LbL capsules remain scarce [9].

4. Vesicular Compartments for In Situ Reactions

Biomimicry offers strategies for the creation of vesicular compartments with incorporated peptides/membrane proteins and encapsulated active compounds providing an approach for various applications. Polymeric compartments with encapsulated cargo have been employed in imaging, sensing, therapeutics, as artificial cells, etc. So far, such compartments were almost solely assembled by film rehydration, electroformation, and LbL.

4.1. Reactions inside Single Compartments

4.1.1. GUVs

Reconstruction of biological structures and processes can be achieved with a bottom-up approach using GUVs. Encapsulation of enzymes inside the cavities of GUVs is an emerging way to fabricate artificial environments that mimic the complexity of cells by introducing similar functionalities. The resulting GUVs serve as platforms to visualize biological processes in real time, contributing to our understanding of human cells, which in turn promotes new developments of biomedical applications [81,82]. Since the permeability of polymeric GUVs is essential for in situ reactions, one biomimicry approach is to equip them with peptides/membrane proteins to allow molecular transport through the membrane. Up to now, there are only few examples of polymeric GUVs with incorporated membrane proteins/peptides and they are primarily based on PMOXA-b-PDMS-b-PMOXA triblock copolymer membranes. For example, the permeability of GUVs, to selectively transport Ca^{2+} ions, was attained by inserting several ionophores: calcimycin [52], Lasalocid A, and N,N-dicyclohexyl-N',N"-dioctadecyl-3-oxapentane-1,5-diamide [83]. Furthermore, Gramicidine (gA) allowed Na^+ and K^+ ions to specifically pass the membranes of GUVs [16]. The hydrophobic mismatch of pore length and membrane represented a barrier to membranes thicker than 12.1 nm, whereas thinner membranes facilitated successful gA insertion. The bee venom melittin was inserted into various PMOXA-b-PDMS-b-PMOXA membranes. The insertion process and the resulting functionality of the peptide have been related to the membrane curvature [15]. Besides, the membrane protein OmpF was successfully reconstituted in membranes of GUVs allowing an enzymatic reaction inside the cavity, which was monitored in real time with a confocal microscope (Figure 4) [18].

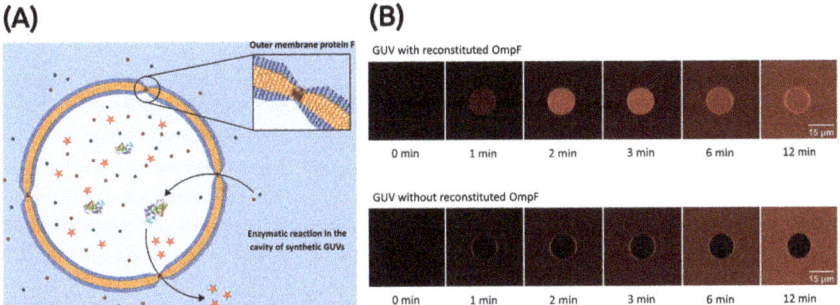

Figure 4. Reaction inside single polymer GUVs. (**A**) Schematic representation of a polymeric GUV equipped with the membrane protein OmpF. Substrates and products diffuse through the membrane, thus enabling an enzymatic reaction. (**B**) Fluorescence micrographs of a single GUV recorded at several time points after addition of the substrates showing the difference of GUVs with and without reconstituted OmpF. Adapted with permission from Garni et al. [18]. Copyright © 2018 American Chemical Society.

4.1.2. Layer-by-Layer Microcapsules

LbL capsules with an encapsulated enzyme offer various possibilities in sensing and imaging [84]. The preparation of (PSS/PAH)$_4$/PSS shell structures, the co-encapsulation of urease and the pH sensitive fluorophore enabled the quantification of urea on a single capsule level [70]. Continuing with this approach Kazakova et al. [71] managed to encapsulate lactate oxidase, peroxidase, or glucose oxidase, with respective sensitive dyes to detect lactate, oxygen, and glucose levels [71]. In another (PSS/PAH)$_4$ system the detection of oxaloacetic acid with NADH as cofactor was possible. Thus, the efficacy of an enzyme fluorophore coupled system was demonstrated (Figure 5) [85]. Magnetic polydopamine capsules enhanced the activity after reusing and the long-term stability of the encapsulated Candida Rugosa Lipase compared to the free enzyme [86]. Reuse is a key factor for potential application in industry. Moreover, further attempts are required to validate the performance of these systems in vitro and in vivo. Another application of LbL capsules is therapeutics: e.g., microcapsules with encapsulated L-Asparaginase in poly-L-arginine and dextran sulfate layers were tested in vitro on leukemic cell lines resulting in a decreased proliferation [87]. LbL enables convenient encapsulation of enzymes in one single particle. However, their semi-permeable membrane allowing unspecific transport of small molecules is rather a deficiency, that needs to be overcome for future applications.

4.2. Reactions inside Multicompartmentalized Structures

4.2.1. GUV Multicompartments

In biological cells, evolution has developed the system of subcompartmentalization (cellular organelles) within individual cells in order to allow specific reactions to take place in a spatially defined manner. This is an efficient solution, as many reactions (e.g., protein lysis, electron transport) require very specific conditions (e.g., low pH, proton gradient) to occur. Careful application of biomimicry principles allows integral cell mimics as combining nano- and microstructures with biomolecules. In this respect, biomolecule equipped polymersomes have previously been shown to be functional as artificial organelles both in vitro and in vivo where they supported the natural cellular metabolism, and have even been shown to function "on demand" in a life-like manner [11,88]. Artificial cell mimics have been designed by constructing synthetic multicompartmentalized systems. The most significant examples were found when using a larger polymer-based GUV loaded either with smaller nano-sized liposomes or smaller nano-sized polymersomes [89]. Synthetic GUVs based on (PBD$_{46}$-b-PEO$_{30}$) loaded with hydrophilic dyes, liposomes (DPPC), and polymersomes (PBD$_{23}$-b-PEO$_{14}$) allowed fast, selectively triggered release due to a light-induced increase in osmotic pressure, resulting in rupture of the GUVs [90]. Examples of polymeric GUVs acting as artificial cell-mimics are still scarce, but more complex multicompartmentalized GUVs exist. Namely, where enzymes and membrane protein equipped polymersomes coexisted in the GUVs inner cavity [12]. By applying the principle of multicompartmentalization, an artificial cell mimic is created with subcompartmentalized polymersomes acting as artificial organelles. This allows cascade reactions to occur successively due to the segregation of enzymes in different subcompartments. The proximity among subcompartments provided by larger GUVs, facilitates the diffusion of reagents and reaction products, while confining the enzymes to their individual subcompartments. PBD-b-PEO polymer GUVs can mimic structural and functional eukaryotic cells by encapsulating enzyme-filled intrinsically semi-permeable PS-b-PIAT polymer nanoreactors together with free enzymes and substrates to fulfill a three-enzyme cascade reaction inside the multicompartmentalized structures [13]. Although this study represents an important step towards artificial cells, it only reports the fluorescent product of the reaction, without providing detailed information about localization of the enzymes. In addition, such examples lack the complexity of cells because they are mainly developed with only few functional elements and by using buffer medium. For the creation of an artificial cell mimic by multicompartmentalization, there are requirements still not fulfilled. The selective permeabilization of every membrane of the involved compartments, which allows for a higher control of the diffusion of substrates and products across the membranes and a more complex medium mimicking the cytoplasm represent advancements not yet provided. Systems addressing this question are multicompartmentalized GUVs with stimuli-responsive and non-responsive subcompartments (Figure 6). With an external signaling molecule passively diffusing through the GUV's membrane, inducing the disassembly of the stimuli-responsive nanoparticle and the release of the entrapped cargo (peptides or enzyme substrates). These molecules allowed a selective ion flux through the GUV's membrane or an enzymatic reaction inside the GUV [12].

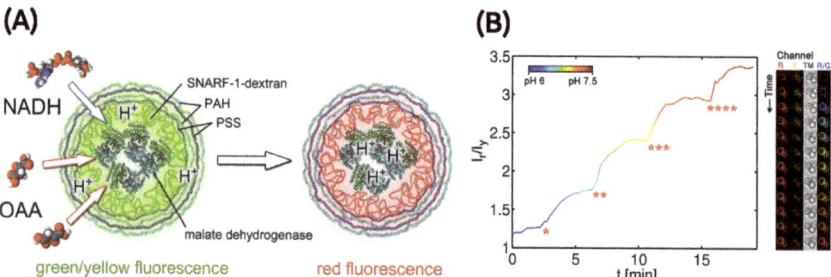

Figure 5. Reaction inside single polymer microcapsules. (**A**) Schematic illustration of an oxaloacetic acid (OAA) or nicotinamide adenine dinucleotide (NADH) sensing microcapsule. The encapsulated pH-sensitive fluorescent dye (seminaphtharhodafluor (SNARF-1)-dextran) responds to a decrease in local proton concentration caused by the enzymatic reaction. (**B**) Reaction kinetics demonstrating NADH as the limiting factor. The first dose of substrate is added at (*) and then added gradually, after the plateau was reached, from (**)-(****). The corresponding micrographs on the right hand side represent the reaction of one capsule (red (R), yellow (Y), transmission (TM), and the false-colored ratio I_r/I_y (R/G)). Adapted from Harimech et al. [85] under the terms of CC BY 3.0.

More and more experimental successes in combination with vesicle engineering techniques are leading into a new era of complexity in artificial cells. These have attracted increasing attention as substituents for living cells. Polymer GUVs and polymersomes offer an ideal platform to engineer cell mimics, allowing reactions to take place in compartmentalized spaces. Meanwhile, they remain stable for longer periods compared to their lipid-based counterparts. With the beforehand mentioned functionalization with membrane proteins and peptides, a life-like functionalization of membranes is approved.

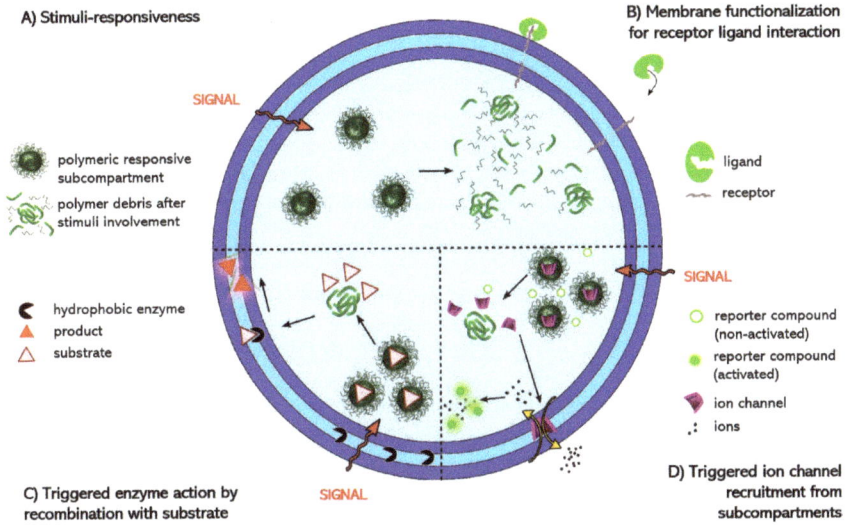

Figure 6. Multicompartmentalized GUV with reduction sensitive and ion channel recruiting modular subcompartments (**A–D**). Dithiothreitol was used as a triggering signal (red arrow). Reprinted With permission from Thamboo et al. [12]. Copyright © 2019 Wiley-VCH.

4.2.2. Layer-by-Layer Polymer Capsules as Multicompartments

Multicompartmentalization allows for separation in preparation and modularity in formulation to increase also functionality in theranostics [91,92]. LbL assembled capsules prepared of polyvinylpyrrolidone (PVP) and thiolated poly(methacrylic acid) (PMAA) containing smaller crosslinked capsules showed possibilities in catalysis and drug delivery [8]. PVP and tannic acid (TA) LbL capsules filled with POEGMA$_{26}$-b-PDPA$_{50}$ polymersomes loaded with pDNA could release the cargo in response to pH changes [9]. Microfluidics have been improving the development of such attractive microreactor systems with increased complexity and modularity. Encapsulation of glucose oxidase (GOx) conjugated on quantum dots or on gold nanorods (NR) into PEO-based microreactors acting as glucose biosensor, while amine functionalized NRs were employed as heparin sensors. GOx oxidized glucose to gluconic acid and H_2O_2 leading to fluorescence quenching by the quantum dots providing a sensitivity in the range of glucose sensors for diabetes diagnostics [68]. However, most capsules only work in defined pH-ranges and do not resist enormous local pH-changes, which would be necessary for in vivo applications [84].

5. Conclusions and Outlook

Biological systems as cells are highly compartmentalized across several length scales. Their precise features, as biomolecule compartmentalization through attachment to membranes and cytoskeleton scaffolds, lateral organization on membrane rafts, as well as compartmentalization in membrane-bound or protein-based organelles, are current subject of study. Understanding their underlying mechanisms opens exciting avenues in many application fields, notably in material science, biotechnology and medicine. As we have seen, efforts at achieving this are accelerating and synthetic approaches to mimic increasingly intricate biological structures are being developed.

This review demonstrates the relevance of polymer-based systems with special focus on polymer GUVs and capsules, for addressing the challenge of eukaryotic cell biomimicry. We only reviewed examples of supramolecular structures, whose membrane is equipped with peptides and membrane proteins since they genuinely represent bioinspired catalytic compartments where the membranes are mimicking biomembranes.

The progress in polymer chemistry gave access to a variety of polymers with tailored properties as biodegradability and non-toxicity that leads to enhanced structural properties and with the significant fluidity, necessary to cope with the insertion of peptides and membrane proteins. Due to their large variety of chemical composition and functionalization, a greater versatility for further improvement of their properties concerning the desired application can be achieved, regarding stability, loading efficiency, intervesicular interaction, and selective permeability for specific substrates. Besides the polymer, the selection of an appropriate preparation method for engineering the supramolecular structures in single and multicompartments to conform biochemical reactions, is detrimental. Although electroformation and film rehydration are the most commonly used techniques, both of them do not produce homogeneous sizes of GUVs. Alternatively, microfluidics became a serious candidate, showing great potential to generate polymer GUVs with narrow size distribution and controlled biomolecular content at high-throughput.

GUVs loaded with enzymes inside their cavities and equipped with peptides/membrane proteins acting as "gates" for the diffusion of molecules through the synthetic membrane, constitute complex reaction spaces. Careful application of biomimicry principles allows integral cell mimics as combining nano- and microstructures with biomolecules. The inner compartments (the organelle mimics) that are loaded in the outer one (the cell membrane mimic) do not necessarily need to encapsulate the same content, where the various contents can even be incompatible or act synergistically. As a result, combinatory drug delivery becomes possible in one single vector. Such systems serve multifold purposes. Besides the aforementioned design of an artificial cell, allowing systematic studies of biological phenomena in simplified environments, they are explored as (compartmentalized) microreactors, where segregation of the catalytic steps in separated compartments allows distinct chemical environment (e.g., different pH, redox states, presence of cofactors, etc.) to couple enzymatic reaction steps that would otherwise inhibit one another. Because of the GUV cell-size, real time imaging of the fluorescence activity of model enzymes can be monitored and used for enhancing diagnosis capabilities. Lastly, polymer capsules have been aimed for use in therapeutic applications, by encapsulating hydrophilic molecules in the aqueous core for enzyme therapy and controlled drug release. In addition, such structures can improve the therapeutic index of drugs by influencing drug absorption and metabolism, and can extent drug half-life as well as reduce toxicity. Although an effort at outlining a roadmap for the field had been made, there will surely be many new developments that will take this research area to unforeseen directions. From a material perspective, the advanced control in polymer synthesis and self-assembly that is available nowadays would certainly bring a real breakthrough in this cell biomimicry field. Hence, it is expected a growing interest in such biomimetic approaches to soon offer many new opportunities in drug delivery, cell-sized reactors, biosensors and imaging for therapeutics, which will offer better communication and interaction with living systems.

Author Contributions: E.C.d.S., A.A. and D.H. carried out literature research and wrote the manuscript. W.M. and and C.G.P. provided additional guidance and assisted in finalizing the manuscript. All authors reviewed the final version of the manuscript and approved it for publication.

Funding: The authors acknowledge financial support from the Swiss National Science Foundation, NCCR-MSE, and University of Basel.

Conflicts of Interest: The authors declare no conflict of interest.

List of Abbreviation

ATRP	Atom transfer radical polymerization
CRP	Controlled radical polymerization
CTA	Chain transfer agent
CuAAC	Copper-catalyzed azide-alkyne cycloaddition
Cy5-IgG	Cyanine-5 conjugated immunoglobulin G proteins
DNA	Deoxyribonucleic acid
DPPC	Dipalmitoylphosphatidylcholine
E. coli	Escherichia coli
FITC	Fluorescein isothiocyanate
gA	Gramicidine
GFP	Green fluorescent protein
GOx	Glucose oxidase
GUV	Giant unilamellar vesicle
HRP	Horseradish peroxidase
ITO	Indium tin oxide
LbL	Layer-by-layer
MOXA	2-methyl-2-oxazoline
NAD	Nicotinamide adenine dinucleotide
NADH	Nicotinamide adenine dinucleotide (reduced)
n-BuLi	n-butyllithium
NR	Nanorod
OAA	Oxaloacetic acid
OmpF	Outer membrane protein F
OmpF-M	Outer membrane protein double mutant
PA444	Poly(4"-acryloyloxybutyl 2,5-bis(4'-butyloxybenzoyloxy)benzoate)
PAA	Poly(acrylic acid)
PAH	Poly(allylamine hydrochloride)
PBD	Polybutadiene
PDA	Polydopamine
PDEAEMA	Poly(2-(diethylamino)ethyl methacrylate)
PDMS	Poly(dimethyl sulfoxide)
PDPA	Poly(2-(diisopropylamino)-ethyl methacrylate)
PEE	Poly(ethyl ethylene)
PEG/PEO	Poly(ethylene glycol)/poly(ethylene oxide)
PEGDA	Poly(ethylene glycol) diacrylate
PFPA	Poly(pentafluorophenyl acrylate)
PGA	Poly(L-glutamic acid)
PHPMA	Poly(2-hydroxypropyl methacrylate)
PIAT	Poly(L-isocyanoalanine(2-thiophen-3-yl-ethyl)amide)
PISA	Polymerization induced self-assembly

PLA	Poly(lactic acid)
PMA	Poly(methyl acrylate)
PMAA	Poly(methacrylic acid)
PMOXA	Poly(2-methyl-2-oxazoline)
PnBA	Poly(n-butyl acrylate)
PNIPAM	Poly(N-isopropylacrylamide)
POEGMA	Poly(oligo(ethylene glycol) methacrylate)
PPS	Poly(propylene sulfide)
PS	Polystyrene
PSBA	Poly(styrene boronic acid)
PSS	Poly(styrene sulfonate)
PtBA	Poly(tert-butyl acrylate)
PTMC	Poly(trimethylene carbonate)
PVP	Polyvinylpyrrolidone
RAFT	Reversible addition fragmentation chain transfer
SNARF-1	Seminaphtharhodafluor
TA	Tannic acid
THF	Tetrahydrofuran

References

1. Chen, A.H.; Silver, P.A. Designing biological compartmentalization. *Trends Cell Biol.* **2012**, *22*, 662–670. [CrossRef]
2. Palivan, C.G.; Fischer-Onaca, O.; Delcea, M.; Itel, F.; Meier, W. Protein–polymer nanoreactors for medical applications. *Chem. Soc. Rev.* **2012**, *41*, 2800–2823. [CrossRef]
3. Feng, H.; Lu, X.; Wang, W.; Kang, N.G.; Mays, J.W. Block Copolymers: Synthesis, Self-Assembly, and Applications. *Polymers* **2017**, *9*, 494. [CrossRef]
4. Antonietti, M.; Förster, S. Vesicles and Liposomes: A Self-Assembly Principle Beyond Lipids. *Adv. Mater.* **2003**, *15*, 1323–1333. [CrossRef]
5. Messager, L.; Gaitzsch, J.; Chierico, L.; Battaglia, G. Novel aspects of encapsulation and delivery using polymersomes. *Curr. Opin. Pharmacol.* **2014**, *18*, 104–111. [CrossRef] [PubMed]
6. Rigo, S.; Gunkel-Grabole, G.; Meier, W.; Palivan, C.G. Surfaces with Dual Functionality through Specific Coimmobilization of Self-Assembled Polymeric Nanostructures. *Langmuir* **2019**, *35*, 4557–4565, doi:10.1021/acs.langmuir.8b02812. [CrossRef] [PubMed]
7. Wu, D.; Rigo, S.; Di Leone, S.; Belluati, A.; Constable, E.C.; Housecroft, C.E.; Palivan, C.G. Brushing the surface: Cascade reactions between immobilized nanoreactors. *Nanoscale* **2020**, *12*, 1551–1562. [CrossRef] [PubMed]
8. Kulygin, O.; Price, A.D.; Chong, S.F.; Staedler, B.; Zelikin, A.N.; Caruso, F. Subcompartmentalized Polymer Hydrogel Capsules withSelectively Degradable Carriers and Subunits. *Small* **2010**, *6*, 1558–1564. [CrossRef] [PubMed]
9. Lomas, H.; Johnston, A.P.R.; Such, G.K.; Zhu, Z.; Liang, K.; Koeverden, M.P.V.; Alongkornchotikul, S.; Caruso, F. Polymersome-Loaded Capsules for Controlled Release of DNA. *Small* **2011**, *7*, 2109–2119. [CrossRef]
10. Stadler, B.; Chandrawati, R.; Price, A.D.; Chong, S.F.; Breheney, K.; Postma, A.; Connal, L.A.; Zelikin, A.N.; Caruso, F. A Microreactor with Thousands of Subcompartments: Enzyme-Loaded Liposomes within Polymer Capsules. *Angew. Chem.* **2009**, *48*, 4359–4362. [CrossRef]
11. Tanner, P.; Balasubramanian, V.; Palivan, C.G. Aiding nature's organelles: Artificial peroxisomes play their role. *Nano Lett.* **2013**, *13*, 2875–2883. [CrossRef]
12. Thamboo, S.; Najer, A.; Belluati, A.; von Planta, C.; Wu, D.; Craciun, I.; Meier, W.; Palivan, C.G. Mimicking Cellular Signaling Pathways within Synthetic Multicompartment Vesicles with Triggered Enzyme Activity and Induced Ion Channel Recruitment. *Adv. Funct. Mater.* **2019**, *29*, 1904267. [CrossRef]
13. Peters, R.J.R.W.; Marguet, M.; Marais, S.; Fraaije, M.W.; van Hest, J.C.M.; Lecommandoux, S. Cascade Reactions in Multicompartmentalized Polymersomes. *Angew. Chem. Int. Ed.* **2014**, *53*, 146–150. [CrossRef] [PubMed]

14. Itel, F.; Chami, M.; Najer, A.; Lorcher, S.; Wu, D.L.; Dinu, I.A.; Meier, W. Molecular Organization and Dynamics in Polymersome Membranes: A Lateral Diffusion Study. *Macromolecules* **2014**, *47*, 7588–7596. [CrossRef]
15. Belluati, A.; Mikhalevich, V.; Yorulmaz Avsar, S.; Daubian, D.; Craciun, I.; Chami, M.; Meier, W.P.; Palivan, C.G. How Do the Properties of Amphiphilic Polymer Membranes Influence the Functional Insertion of Peptide Pores? *Biomacromolecules* **2020**, *21*, 701–715. [CrossRef] [PubMed]
16. Lomora, M.; Garni, M.; Itel, F.; Tanner, P.; Spulber, M.; Palivan, C.G. Polymersomes with engineered ion selective permeability as stimuli-responsive nanocompartments with preserved architecture. *Biomaterials* **2015**, *53*, 406–414. [CrossRef]
17. Baumann, P.; Spulber, M.; Fischer, O.; Car, A.; Meier, W. Investigation of Horseradish Peroxidase Kinetics in an "Organelle-Like" Environment. *Small* **2017**, *13*, 1603943. [CrossRef]
18. Garni, M.; Einfalt, T.; Goers, R.; Palivan, C.G.; Meier, W. Live Follow-Up of Enzymatic Reactions Inside the Cavities of Synthetic Giant Unilamellar Vesicles Equipped with Membrane Proteins Mimicking Cell Architecture. *ACS Synth. Biol.* **2018**, *7*, 2116–2125. [CrossRef]
19. Garni, M.; Thamboo, S.; Schoenenberger, C.A.; Palivan, C.G. Biopores/membrane proteins in synthetic polymer membranes. *Biochim. Biophys. Acta (BBA) Biomembr.* **2017**, *1859*, 619–638. [CrossRef]
20. Yorulmaz Avsar, S.; Kyropoulou, M.; Di Leone, S.; Schoenenberger, C.A.; Meier, W.P.; Palivan, C.G. Biomolecules Turn Self-Assembling Amphiphilic Block Co-polymer Platforms Into Biomimetic Interfaces. *Front. Chem.* **2019**, *6*, 645. [CrossRef]
21. Ribelli, T.G.; Lorandi, F.; Fantin, M.; Matyjaszewski, K. Atom Transfer Radical Polymerization: Billion Times More Active Catalysts and New Initiation Systems. *Macromol. Rapid Commun.* **2019**, *40*, 1800616. [CrossRef] [PubMed]
22. Perrier, S. 50th Anniversary Perspective: RAFT Polymerization—A User Guide. *Macromolecules* **2017**, *50*, 7433–7447. [CrossRef]
23. Keddie, D.J. A guide to the synthesis of block copolymers using reversible-addition fragmentation chain transfer (RAFT) polymerization. *Chem. Soc. Rev.* **2014**, *43*, 496–505. [CrossRef] [PubMed]
24. Hadjichristidis, N.; Pitsikalis, M.; Iatrou, H. Synthesis of Block Copolymers. In *Block Copolymers I*; Abetz, V., Ed.; Springer: Berlin/Heidelberg, Germany, 2005; pp. 1–124.
25. Herzberger, J.; Niederer, K.; Pohlit, H.; Seiwert, J.; Worm, M.; Wurm, F.R.; Frey, H. Polymerization of Ethylene Oxide, Propylene Oxide, and Other Alkylene Oxides: Synthesis, Novel Polymer Architectures, and Bioconjugation. *Chem. Rev.* **2016**, *116*, 2170–2243. [CrossRef] [PubMed]
26. Förster, S.; Krämer, E. Synthesis of PB-PEO and PI-PEO Block Copolymers with Alkyllithium Initiators and the Phosphazene Base t-BuP4. *Macromolecules* **1999**, *32*, 2783–2785. [CrossRef]
27. Nuss, H.; Chevallard, C.; Guenoun, P.; Malloggi, F. Microfluidic trap-and-release system for lab-on-a-chip-based studies on giant vesicles. *Lab Chip* **2012**, *12*, 5257–5261. [CrossRef] [PubMed]
28. Christian, D.A.; Tian, A.; Ellenbroek, W.G.; Levental, I.; Rajagopal, K.; Janmey, P.A.; Liu, A.J.; Baumgart, T.; Discher, D.E. Spotted vesicles, striped micelles and Janus assemblies induced by ligand binding. *Nat. Mater.* **2009**, *8*, 843–849. [CrossRef]
29. Hillmyer, M.A.; Bates, F.S. Synthesis and Characterization of Model Polyalkane-Poly(ethylene oxide) Block Copolymers. *Macromolecules* **1996**, *29*, 6994–7002. [CrossRef]
30. Wu, D.; Spulber, M.; Itel, F.; Chami, M.; Pfohl, T.; Palivan, C.G.; Meier, W. Effect of Molecular Parameters on the Architecture and Membrane Properties of 3D Assemblies of Amphiphilic Copolymers. *Macromolecules* **2014**, *47*, 5060–5069. [CrossRef]
31. Mabrouk, E.; Cuvelier, D.; Pontani, L.L.; Xu, B.; Lévy, D.; Keller, P.; Brochard-Wyart, F.; Nassoy, P.; Li, M.H. Formation and material properties of giant liquid crystal polymersomes. *Soft Matter* **2009**, *5*, 1870–1878. [CrossRef]
32. Kubilis, A.; Abdulkarim, A.; Eissa, A.M.; Cameron, N.R. Giant Polymersome Protocells Dock with Virus Particle Mimics via Multivalent Glycan-Lectin Interactions. *Sci. Rep.* **2016**, *6*, 32414. [CrossRef] [PubMed]
33. Albertsen, A.N.; Szymański, J.K.; Pérez-Mercader, J. Emergent Properties of Giant Vesicles Formed by a Polymerization-Induced Self-Assembly (PISA) Reaction. *Sci. Rep.* **2017**, *7*, 41534. [CrossRef] [PubMed]
34. Meeuwissen, S.A.; Bruekers, S.M.C.; Chen, Y.; Pochan, D.J.; van Hest, J.C.M. Spontaneous shape changes in polymersomes via polymer/polymer segregation. *Polym. Chem.* **2014**, *5*, 489–501. [CrossRef]

35. Fauquignon, M.; Ibarboure, E.; Carlotti, S.; Brûlet, A.; Schmutz, M.; Le Meins, J.F. Large and Giant Unilamellar Vesicle(s) Obtained by Self-Assembly of Poly(dimethylsiloxane)-b-poly(ethylene oxide) Diblock Copolymers, Membrane Properties and Preliminary Investigation of Their Ability to Form Hybrid Polymer/Lipid Vesicles. *Polymers* **2019**, *11*, 2013. [CrossRef]
36. Agrahari, V.; Agrahari, V. Advances and applications of block-copolymer-based nanoformulations. *Drug Discov. Today* **2018**, *23*, 1139–1151. [CrossRef]
37. Qi, Y.; Li, B.; Wang, Y.; Huang, Y. Synthesis and sequence-controlled self-assembly of amphiphilic triblock copolymers based on functional poly(ethylene glycol). *Polym. Chem.* **2017**, *8*, 6964–6971. [CrossRef]
38. Discher, D.E.; Ahmed, F. POLYMERSOMES. *Annu. Rev. Biomed. Eng.* **2006**, *8*, 323–341. [CrossRef]
39. Blanazs, A.; Armes, S.P.; Ryan, A.J. Self-Assembled Block Copolymer Aggregates: From Micelles to Vesicles and their Biological Applications. *Macromol. Rapid Commun.* **2009**, *30*, 267–277. [CrossRef]
40. Garni, M.; Wehr, R.; Avsar, S.Y.; John, C.; Palivan, C.; Meier, W. Polymer membranes as templates for bio-applications ranging from artificial cells to active surfaces. *Eur. Polym. J.* **2019**, *112*, 346–364. [CrossRef]
41. Mai, Y.; Eisenberg, A. Self-assembly of block copolymers. *Chem. Soc. Rev.* **2012**, *41*, 5969–5985. [CrossRef]
42. Cui, J.; Van Koeverden, M.P.; Müllner, M.; Kempe, K.; Caruso, F. Emerging methods for the fabrication of polymer capsules. *Adv. Colloid Interface Sci.* **2014**, *207*, 14–31. [CrossRef] [PubMed]
43. Zhang, X.; Xu, Y.; Zhang, X.; Wu, H.; Shen, J.; Chen, R.; Xiong, Y.; Li, J.; Guo, S. Progress on the layer-by-layer assembly of multilayered polymer composites: Strategy, structural control and applications. *Prog. Polym. Sci.* **2019**, *89*, 76–107. [CrossRef]
44. Becker, A.L.; Johnston, A.P.R.; Caruso, F. Layer-By-Layer-Assembled Capsules and Films for Therapeutic Delivery. *Small* **2010**, *6*. [CrossRef] [PubMed]
45. Sukhorukov, G.B.; Antipov, A.A.; Voigt, A.; Donath, E.; Möhwald, H. pH-Controlled Macromolecule Encapsulation in and Release from Polyelectrolyte Multilayer Nanocapsules. *Macromol. Rapid Commun.* **2001**, *22*, 44–46. [CrossRef]
46. Georgieva, R.; Moya, S.; Hin, M.; Mitlöhner, R.; Donath, E.; Kiesewetter, H.; Möhwald, H.; Bäumler, H. Permeation of Macromolecules into Polyelectrolyte Microcapsules. *Biomacromolecules* **2002**, *3*, 517–524. [CrossRef] [PubMed]
47. Walde, P.; Cosentino, K.; Engel, H.; Stano, P. Giant vesicles: Preparations and applications. *Chembiochem* **2010**, *11*, 848–65. [CrossRef]
48. Lim, S.; de Hoog, H.P.; Parikh, A.; Nallani, M.; Liedberg, B. Hybrid, Nanoscale Phospholipid/Block Copolymer Vesicles. *Polymers* **2013**, *5*, 1102–1114. [CrossRef]
49. Pautot, S.; Frisken, B.J.; Weitz, D.A. Production of Unilamellar Vesicles Using an Inverted Emulsion. *Langmuir* **2003**, *19*, 2870–2879. [CrossRef]
50. Itel, F.; Najer, A.; Palivan, C.G.; Meier, W. Dynamics of Membrane Proteins within Synthetic Polymer Membranes with Large Hydrophobic Mismatch. *Nano Lett.* **2015**, *15*, 3871–3878. [CrossRef]
51. Lomora, M.; Itel, F.; Dinu, I.A.; Palivan, C.G. Selective ion-permeable membranes by insertion of biopores into polymersomes. *Phys. Chem. Chem. Phys.* **2015**, *17*, 15538–15546. [CrossRef]
52. Picker, A.; Nuss, H.; Guenoun, P.; Chevallard, C. Polymer vesicles as microreactors for bioinspired calcium carbonate precipitation. *Langmuir* **2011**, *27*, 3213–3218. [CrossRef] [PubMed]
53. Kita-Tokarczyk, K.; Grumelard, J.; Haefele, T.; Meier, W. Block copolymer vesicles—Using concepts from polymer chemistry to mimic biomembranes. *Polymer* **2005**, *46*, 3540–3563. [CrossRef]
54. Fetsch, C.; Gaitzsch, J.; Messager, L.; Battaglia, G.; Luxenhofer, R. Self-Assembly of Amphiphilic Block Copolypeptoids – Micelles, Worms and Polymersomes. *Sci. Rep.* **2016**, *6*, 33491. [CrossRef] [PubMed]
55. Yildiz, M.E.; Prud'homme, R.K.; Robb, I.; Adamson, D.H. Formation and characterization of polymersomes made by a solvent injection method. *Polym. Adv. Technol.* **2007**, *18*, 427–432. [CrossRef]
56. Daubian, D.; Gaitzsch, J.; Meier, W. Synthesis and complex self-assembly of amphiphilic block copolymers with a branched hydrophobic poly(2-oxazoline) into multicompartment micelles, pseudo-vesicles and yolk/shell nanoparticles. *Polym. Chem.* **2020**, *11*, 1237–1248. [CrossRef]
57. Dionzou, M.; Morère, A.; Roux, C.; Lonetti, B.; Marty, J.D.; Mingotaud, C.; Joseph, P.; Goudounèche, D.; Payré, B.; Léonetti, M.; Mingotaud, A.F. Comparison of methods for the fabrication and the characterization of polymer self-assemblies: What are the important parameters? *Soft Matter* **2016**, *12*, 2166–2176. [CrossRef]
58. Pearson, R.T.; Warren, N.J.; Lewis, A.L.; Armes, S.P.; Battaglia, G. Effect of pH and Temperature on PMPC–PDPA Copolymer Self-Assembly. *Macromolecules* **2013**, *46*, 1400–1407. [CrossRef]

59. Fernyhough, C.; Ryana, A.J.; Battaglia, G. pH controlled assembly of a polybutadiene–poly(methacrylic acid) copolymer in water: Packing considerations and kinetic limitations. *Soft Matter* **2009**, *5*, 1674–1682. [CrossRef]
60. Lorenceau, E.; Utada, A.S.; Link, D.R.; Cristobal, G.; Joanicot, M.; Weitz, D.A. Generation of Polymerosomes from Double-Emulsions. *Langmuir* **2005**, *21*, 9183–9186. [CrossRef]
61. Shum, H.C.; Lee, D.; Yoon, I.; Kodger, T.; Weitz, D.A. Double Emulsion Templated Monodisperse Phospholipid Vesicles. *Langmuir* **2008**, *24*, 7651–7653. [CrossRef]
62. Shum, H.C.; Kim, J.W.; Weitz, D.A. Microfluidic Fabrication of Monodisperse Biocompatible and Biodegradable Polymersomes with Controlled Permeability. *J. Am. Chem. Soc.* **2008**, *130*, 9543–9549. [CrossRef] [PubMed]
63. Deshpande, S.; Caspi, Y.; Meijering, A.E.C.; Dekker, C. Octanol-assisted liposome assembly on chip. *Nat. Commun.* **2016**, *7*, 1–9. [CrossRef] [PubMed]
64. Do Nascimento, D.F.; Arriaga, L.R.; Eggersdorfer, M.; Ziblat, R.; Marques, M.d.F.V.; Reynaud, F.; Koehler, S.A.; Weitz, D.A. Microfluidic Fabrication of Pluronic Vesicles with Controlled Permeability. *Langmuir* **2016**, *32*, 5350–5355. [CrossRef] [PubMed]
65. Martino, C.; Kim, S.; Horsfall, L.; Abbaspourrad, A.; Rosser, S.J.; and. David A. Weitz, J.C. Protein Expression, Aggregation, and Triggered Release from Polymersomes as Artificial Cell-like Structures. *Angew. Chem.* **2012**, *51*, 6416–6420. [CrossRef] [PubMed]
66. Yan, X.; Li, J.; Möhwald, H. Templating Assembly of Multifunctional Hybrid Colloidal Spheres. *Adv. Mater.* **2012**, *24*, 2663–2667. [CrossRef]
67. Postma, A.; Yan, Y.; Wang, Y.; Zelikin, A.N.; Tjipto, E.; Caruso, F. Self-Polymerization of Dopamine as a Versatile and Robust Technique to Prepare Polymer Capsules. *Chem. Mater.* **2009**, *21*, 3042–3044. [CrossRef]
68. Xie, X.; Zhang, W.; Abbaspourrad, A.; Ahn, J.; Bader, A.; Bose, S.; Vegas, A.; Lin, J.; Tao, J.; Hang, T.; et al. Microfluidic Fabrication of Colloidal Nanomaterials-Encapsulated Microcapsules for Biomolecular Sensing. *Nano Lett.* **2017**, *17*, 2015–2020. [CrossRef]
69. Larrañaga, A.; Lomora, M.; Sarasua, J.; Palivan, C.; Pandit, A. Polymer capsules as micro-/nanoreactors for therapeutic applications: Current strategies to control membrane permeability. *Prog. Mater. Sci.* **2017**, *90*, 325–357. [CrossRef]
70. Kazakova, L.I.; Shabarchina, L.I.; Sukhorukov, G.B. Co-encapsulation of enzyme and sensitive dye as a tool for fabrication of microcapsule based sensor for urea measuring. *Phys. Chem. Chem. Phys.* **2011**, *13*, 11110–11117. [CrossRef]
71. Kazakova, L.I.; Shabarchina, L.I.; Anastasova, S.; Pavlov, A.M.; Vadgama, P.; Skirtach, A.G.; Sukhorukov, G.B. Chemosensors and biosensors based on polyelectrolyte microcapsules containing fluorescent dyes and enzymes. *Anal. Bioanal. Chem.* **2013**, *405*, 1559–1568. [CrossRef]
72. Hosta-Rigau, L.; York-Duran, M.J.; Zhang, Y.; Goldie, K.N.; Städler, B. Confined Multiple Enzymatic (Cascade) Reactions within Poly(dopamine)-based Capsosomes. *ACS Appl. Mater. Interfaces* **2014**, *6*, 12771–12779. [CrossRef]
73. Chen, H.; Man, J.; Li, Z.; Li, J. Microfluidic Generation of High-Viscosity Droplets by Surface-Controlled Breakup of Segment Flow. *ACS Appl. Mater. Interfaces* **2017**, *9*, 21059–21064. [CrossRef] [PubMed]
74. Kisak, E.T.; Coldren, B.; Zasadzinski, J.A. Nanocompartments Enclosing Vesicles, Colloids, and Macromolecules via Interdigitated Lipid Bilayers. *Langmuir* **2002**, *18*, 284–288. [CrossRef]
75. Fu, Z.; Ochsner, M.A.; de Hoog, H.P.M.; Tomczak, N.; Nallani, M. Multicompartmentalized polymersomes for selective encapsulation of biomacromolecules. *Chem. Commun.* **2011**, *47*, 2862–2864. [CrossRef] [PubMed]
76. Marguet, M.; Edembe, L.; Lecommandoux, S. Polymersomes in Polymersomes: Multiple Loading and Permeability Control. *Angew. Chem. Int. Ed.* **2012**, *51*, 1173–1176. [CrossRef] [PubMed]
77. Kim, S.H.; Shum, H.C.; Kim, J.W.; Cho, J.C.; Weitz, D.A. Multiple Polymersomes for Programmed Release of Multiple Components. *J. Am. Chem. Soc.* **2011**, *133*, 15165–15171. [CrossRef] [PubMed]
78. Kreft, O.; Prevot, M.; Möhwald, H.; Sukhorukov, G.B. Shell-in-Shell Microcapsules: A Novel Tool for Integrated, Spatially Confined Enzymatic Reactions. *Angew. Chem.* **2007**, *46*, 5605–5608. [CrossRef]
79. Hosta-Rigau, L.; Chung, S.F.; Postma, A.; Chandrawati, R.; Caruso, B.S.F. Capsosomes with "Free-Floating" Liposomal Subcompartments. *Angew. Chem.* **2011**, *23*, 4082–4087. [CrossRef]
80. Schattling, P.; Dreier, C.; Städler, B. Janus subcompartmentalized microreactors. *Soft Matter* **2015**, *11*, 5327–5335. [CrossRef]

81. Küchler, A.; Yoshimoto, M.; Luginbühl, S.; Mavelli, F.; Walde, P. Enzymatic reactions in confined environments. *Nat. Nanotechnol.* **2016**, *11*, 409. [CrossRef]
82. Jeong, S.; Nguyen, H.T.; Kim, C.H.; Ly, M.N.; Shin, K. Toward Artificial Cells: Novel Advances in Energy Conversion and Cellular Motility. *Adv. Funct. Mater.* **2020**, *30*, 1907182. [CrossRef]
83. Sauer, M.; Haefele, T.; Graff, A.; Nardin, C.; Meier, W. Ion-carrier controlled precipitation of calcium phosphate in giant ABA triblock copolymer vesicles. *Chem. Commun.* **2001**, 2452–2453. [CrossRef] [PubMed]
84. Zhao, S.; Caruso, F.; Dähne, L.; Decher, G.; De Geest, B.G.; Fan, J.; Feliu, N.; Gogotsi, Y.; Hammond, P.T.; Hersam, M.C.; et al. The Future of Layer-by-layer Assembly: A Tribute to ACS Nano Associate Editor Helmuth Möhwald. *ACS Nano* **2019**, *13*, 6151–6169. [CrossRef] [PubMed]
85. Harimech, P.K.; Hartmann, R.; Rejman, J.; del Pino, P.; Rivera-Gil, P.; Parak, W.J. Encapsulated enzymes with integrated fluorescence-control of enzymatic activity. *J. Mater. Chem. B* **2015**, *3*, 2801–2807. [CrossRef] [PubMed]
86. Hou, C.; Wang, Y.; Zhu, H.; Zhou, L. Formulation of robust organic–inorganic hybrid magnetic microcapsules through hard-template mediated method for efficient enzyme immobilization. *J. Mater. Chem. B* **2015**, *3*, 2883–2891. [CrossRef] [PubMed]
87. Karamitros, C.S.; Yashchenok, A.M.; Möhwald, H.; Skirtach, A.G.; Konrad, M. Preserving Catalytic Activity and Enhancing Biochemical Stability of the Therapeutic Enzyme Asparaginase by Biocompatible Multilayered Polyelectrolyte Microcapsules. *Biomacromolecules* **2013**, *14*, 4398–4406. [CrossRef]
88. Einfalt, T.; Witzigmann, D.; Edlinger, C.; Sieber, S.; Goers, R.; Najer, A.; Spulber, M.; Onaca-Fischer, O.; Huwyler, J.; Palivan, C.G. Biomimetic artificial organelles with in vitro and in vivo activity triggered by reduction in microenvironment. *Nat. Commun.* **2018**, *9*, 1127. [CrossRef]
89. Marguet, M.; Bonduelle, C.; Lecommandoux, S. Multicompartmentalized polymeric systems: Towards biomimetic cellular structure and function. *Chem. Soc. Rev.* **2013**, *42*, 512–529. [CrossRef]
90. Peyret, A.; Ibarboure, E.; Tron, A.; Beauté, L.; Rust, R.; Sandre, O.; McClenaghan, N.D.; Lecommandoux, S. Polymersome Popping by Light-Induced Osmotic Shock under Temporal, Spatial, and Spectral Control. *Angew. Chem. Int. Ed.* **2017**, *56*, 1566–1570. [CrossRef]
91. Delcea, M.; Yashchenok, A.; Videnova, K.; Kreft, O.; Möhwald, H.; Skirtach, A.G. Multicompartmental micro-and nanocapsules: Hierarchy and applications in biosciences. *Macromol. Biosci.* **2010**, *10*, 465–474. [CrossRef]
92. Xiong, R.; Soenen, S.J.; Braeckmans, K.; Skirtach, A.G. Towards theranostic multicompartment microcapsules: In-situ diagnostics and laser-induced treatment. *Theranostics* **2013**, *3*, 141. [CrossRef] [PubMed]

© 2020 by the authors. Licensee MDPI, Basel, Switzerland. This article is an open access article distributed under the terms and conditions of the Creative Commons Attribution (CC BY) license (http://creativecommons.org/licenses/by/4.0/).

Article

Combining the Sensitivity of LAMP and Simplicity of Primer Extension via a DNA-Modified Nucleotide

Moritz Welter and Andreas Marx *

Department of Chemistry & Konstanz Research School Chemical Biology, University of Konstanz, 78457 Konstanz, Germany; moritz.welter@uni-konstanz.de
* Correspondence: andreas.marx@uni-konstanz.de; Tel.: +49-7531-88-5139

Received: 16 April 2020; Accepted: 11 May 2020; Published: 14 May 2020

Abstract: LAMP is an approach for isothermal nucleic acids diagnostics with increasing importance but suffers from the need of tedious systems design and optimization for every new target. Here, we describe an approach for its simplification based on a single nucleoside-5′-O-triphosphate (dNTP) that is covalently modified with a DNA strand. We found that the DNA-modified dNTP is a substrate for DNA polymerases in versatile primer extension reactions despite its size and that the incorporated DNA indeed serves as a target for selective LAMP analysis.

Keywords: modified nucleotide; DNA polymerase; LAMP; primer extension

1. Introduction

Despite the widespread use of PCR-based amplification, the drawback of this technology, however, is its need for temperature cycling. Many attempts have been made to develop isothermal amplification methods that do not require heating of the double-stranded nucleic acid for the separation of templates [1]. These methods include strand-displacement amplification (SDA) [2], rolling circle amplification (RCA) [3], and helicase-dependent amplification (HDA) [4]. Another important method for nucleic acids diagnostics is loop mediated isothermal amplification (LAMP) [5,6]. LAMP relies on auto-cycling strand displacement DNA synthesis that is performed by a DNA polymerase with high strand displacement activity and a set of two specially designed inner and two outer primers (for details of the method, see Supplementary Materials Figure S1 in the ESI).

The reaction can be monitored by e.g., the addition of dyes used for nucleic acid staining [7] or turbidity analysis of precipitating magnesium phosphate [8]. Furthermore, the use of low-buffered reaction mixtures allows amplification monitoring with pH sensitive indicator dyes by the naked eye, as during the DNA polymerase reaction a proton is released for each nucleotide incorporation [9]. LAMP assays have also been adapted for many applications e.g., to cover genotyping and RNA detection [10–13]. However, in order to work as intended, the primers required for the amplification have to be carefully designed to meet particular requirements in regards to their melting temperature, spacing, and concentrations [5]. Thus, the design of suitable LAMP primers has to be optimized for every target that can be tedious, even when done with specific design software. While setting up the LAMP reaction, further problems are described such as the amplification of non-template controls, which drastically impedes the reliability of LAMP assays [14,15].

2. Materials and Methods

General

All reagents and solvents were obtained from Sigma-Aldrich (Darmstadt, Germany) and used without further purification. All synthetic reactions were performed under an inert atmosphere. Flash chromatography was performed using Merck silica gel G60 (230–400 mesh, Darmstadt, Germany)

and Merck precoated plates (silica gel 60 F254) were used for TLC. Anion-exchange chromatography was performed on an Äkta Purifier (GE Healthcare, Chicago, IL, USA) with a DEAE Sephadex™ A-25 (GE Healthcare Bio-Sciences, (GE Healthcare, Chicago, IL, USA) column using a linear gradient (0.1 M–1.0 M) of triethylammonium bicarbonate buffer (TEAB, pH 7.5). Reversed phase high pressure liquid chromatography (RP-HPLC, Shimadzu, Kyoto, Japan) for the purification of compounds was performed using a Shimadzu system having LC8a pumps and a Dynamax UV-1 detector (RP-HPLC, Shimadzu, Kyoto, Japan). A VP 250/16 NUCLEODUR C18 HTec, 5 µm (Macherey-Nagel, D) column and a gradient of acetonitrile in 50 mM TEAA buffer were used. All compounds purified by RP-HPLC were obtained as their triethylammonium salts after repeated freeze-drying. NMR spectra were recorded on Bruker Avance III 400 (^{1}H: 400 MHz, ^{13}C: 101 MHz, ^{31}P: 162 MHz, Billerica, MA; USA) spectrometer. The solvent signals were used as references and the chemical shifts converted to the TMS scale and are given in ppm (δ). HR-ESI-MS spectra were recorded on a Bruker Daltronics microTOF II. KlenTaq DNA polymerase was expressed and purified as described before. [16] T4 polynucleotide kinase (PNK) was purchased from New England BioLabs (Ipswich, MA; USA). [γ-^{32}P] ATP was purchased from Hartmann Analytics (Braunschweig, Deutschland) and natural dNTPs from Thermo Scientific (Waltham, MA, USA).

Synthesis of nucleoside triphosphate 2. To a solution of 18.9 µmol (10.3 mg) 5-(aminopentynyl)-2′-deoxyuridinetriphosphate tetrabutylammonium salt (1) [16] in 1 mL DMF, 94.5 µmol (5 eq., 9.5 mg) of NEt$_3$ were added. In parallel, 37.8 µmol (2 eq, 11.2 mg) of 16- azidohexadecanoic acid, 94.5 µmol (5 eq, 9.5 mg) NEt$_3$ and 37.8 µmol HATU (2 eq., 14.4 mg) were dissolved in 1 mL DMF and stirred for 30 min. Both mixtures were then combined and stirred at room temperature for additional 12 h. The solvent was removed under reduced pressure and the residual oil was subjected to C18-RP-HPLC (95% 50 mM triethylammonium acetate (TEAA) buffer to 100% MeCN). 2 was obtained in 58% yield as determined by Nanodrop ND1000 spectrometer with ε (290 nm) = 13,300 M^{-1}·cm^{-1}. The compound was diluted in MilliQ water and kept as a 10 mM stock solution at −20 °C.

Analysis of 2: ^{1}H NMR (400 MHz, Methanol-d4) δ 8.01 (s, 1H, H-C(6)), 6.26 (t, ^3J = 6.8 Hz, 1H, H-C(1′)), 4.63–4.57 (m, 1H, H-C(3′)), 4.34–4.26 (m, 1H, H-C(5′a)), 4.23–4.17 (m, 1H, H-C(5′b)), 4.11–4.06 (m, 1H, H-C(4′)), 3.29 (t, ^3J = 6.9, 2H, H-C(L16)-), 3.22 (q, ^3J = 7.1, 17H, −CH2CH2CH2NH-, Et3N), 2.46 (t, ^3J = 6.9 Hz, 2H, H-C(L2)), 2.32–2.25 (m, 2H, H-C(2′)), 2.1 (t, ^3J = 7.5, 2H, −CH2CH2CH2NH-), 2.20 (t, ^3J = 7.6 Hz, 2H, H-C(L2)), 1.79 (p, ^3J = 6.8 Hz, 2H, −CH2CH2CH2NH-), 1.60 (p, ^3J = 6.8 Hz, 4H, H-C(L3+15)), 1.45–1.27 (m, 51H, H-C(4-14), Et3N). 31P NMR (243 MHz, Methanol-d4): δ = −10.44 (d, ^2J = 20.5 Hz), −11.33 (d, ^2J = 21.3 Hz), −23.72 (t, ^2J = 21.3 Hz). HR-ESI-MS (m/z): [M − H]$^-$ = calcd: 827.2552; found: 827.2562.

Preparation of dT^{15LAMP}TP. The split LAMP target sequence was ligated using T4 DNA ligase and a splint oligonucleotide. 1 nmol of LAMP_TARGET_A and LAMP_TARGET_B (10 µM, Biomers.net) were mixed with 2 nmol (20 µM) of the splint oligonucleotide in a total volume of 98 µL of 1× T4 ligase buffer provided by the manufacturer (NEB). The mixture was heated to 95 °C for 2 min and slowly cooled down to 25 °C. Subsequently, 2 µL of T4 Ligase (800 U) were added and the reaction was incubated at 16 °C overnight. The mixture was then diluted to 200 µL with MilliQ water and subjected to 95 °C for 5 min. Ion-exchange HPLC was performed at 85 °C column temperature using 100 µL of the solution on an analytical HPLC system with a semi-preparative Thermo Scientific™ Dionex™ DNAPac™ PA100 column and a gradient from IEX-HPLC buffer A (25 mM Tris-HCl, pH 8) to IEX-HPLC buffer B (25 mM Tris-HCl, 0.5 M sodium perchlorate, pH 8). Peaks demonstrating an absorbance at λ = 260 nm were collected and pooled in Amicon 4 centrifugal filters. After repeated washing with MilliQ water, the ligated LAMP target was transferred to a 1.5 mL reaction tube and absorbance was measured by NanoDrop ND-1000 spectrometry at 260 nm with ε (403 nm) = 1,752,400 M^{-1} cm^{-1}. To conjugate the 5′-DBCO labeled LAMP target with compound **2**, the above generated oligonucleotide was incubated with 10 eq of the nucleotide in 1× PBS (pH 7.4) overnight. IEX-HPLC and Amicon purification were repeated to yield **dT^{15LAMP}TP**.

Primer extension (PEx) in solution with dT^{15LAMP}TP. To 1× polymerase buffer (50 mM Tris-HCl, 16 mM ammonium sulfate, 2.5 mM magnesium chloride, 0.1% Tween 20, pH 9.2), 150 nM 5'-^{32}P-labeled primer and 200 nM template were added. The mixture was annealed at 95 °C for 5 min. Subsequently, DNA polymerase was added (100 nM KlenTaq DNA polymerase) and the reaction was started by addition of the dNTP (1 µM final concentration). Time points were collected by quenching 2 µL of the reaction mixture with 10 µL stopping solution (80% v/v formamide, 20 mM EDTA, 0.025% w/v bromophenol blue, 0.025% w/v xylene cyanol). Denaturing polyacrylamide gels (9%) were prepared by polymerization of a solution of urea (8.3 M) and bisacrylamide/acrylamide (9%) in TBE buffer using ammonium peroxodisulfate (APS, 0.08%) and N,N,N',N'-tetramethylethylene-diamine (TEMED, 0.04%). Immediately after addition of APS and TEMED, the solution was filled in a sequencing gel chamber (Bio-Rad) and left for polymerization for at least 45 min. After addition of TBE buffer (1×) to the electrophoresis unit, gels were pre-warmed by electrophoresis at 100 W for 30 min and samples were added and separated during electrophoresis (100 W) for approximately 1.5 h. The gel was transferred to Whatman filter paper, dried at 80 °C in vacuo using a gel dryer, and exposed to an imager screen. Readout was performed with a molecular imager FX.

LAMP assay in solution. To avoid contaminations, all LAMP reactions were pipetted with Biosphere filter tips. The LAMP target used for the positive controls was pipetted with a second pipette set. Initial LAMP reactions were performed according to the conditions reported by Tanner and co-workers [9] with 8 U Bst 2.0 WarmStart® DNA Polymerase (NEB) DNA polymerase, 1× SYBR I, 0.2/0.4/1.6 µM LAMP primers (outer/loop/inner), 10 nM 5'-DBCO LAMP target (positive control), 350 µM/dNTP, 65 °C in 1× isothermal amplification buffer (NEB) with 8 mM MgSO4. LAMP reactions in optimized conditions were carried out using 200 µM dNTPs, 0.2/0.4/1.6 µM primers (outer/loop/inner), 1× SYBR I, 4 U of Isotherm2G DNA polymerase (myPOLS Biotec, Konstanz, Germany) and 0.1 nM 5'-DBCO LAMP target (positive control) in a total of 10 µL of 1× Isotherm2G buffer at 55 °C for the indicated amount of time. Fluorescence of SYBR I was measured in 1 min intervals minute in a Bio-Rad CFX384 Touch™ Real-Time PCR Detection System. Following the amplification, melting point measurement was carried out with a gradient from 55 °C to 95 °C in 0.5 °C steps.

Primer extension (PEx) and LAMP assay using immobilized primers. Pierce Streptavidin coated 8-well strips (ThermoFisher, Waltham, MA, USA) were washed twice with 200 µL of 1× plate washing buffer. Subsequently, 1 µL of 500 µM 5'-biotin BRAF primer in 100 µL 1× PBS buffer were added to each well. After 15 min of incubation at room temperature, the primer solution was removed and 200 µL of 1 mM (D)-+-biotin in 1× PBS were added. After 5 min of incubation, the liquid was removed and the wells were washed once with 200 µL of PBS buffer and twice with 200 µL of 1× KTq reaction buffer. Following this, 50 µL of PEx reaction mixture (100 nM KlenTaq DNA polymerase, 200 nM BRAF template, 200 nM dT^{15LAMP}TP in a total of 50 µL 1× reaction buffer (50 mM Tris-HCl, 16 mM ammonium sulfate, 2.5 mM magnesium chloride, 0.1% Tween 20, pH 9.2) were applied, the wells were sealed with PCR foil seal and incubated at 55 °C (measured with a digital thermometer) in a shallow water bath on a thermal block for 30 min. The supernatant was removed and the wells were washed first twice with 100 µL 1× reaction buffer, then three times with 1× PBS buffer and finally rinsed for 2 min under a water tap with MilliQ water. All liquid was removed and the wells were finally washed with 200 µL of 1× Isotherm2G DNA polymerase buffer. 50 µL of LAMP reaction mixture (200 µM dNTPs, 0.2/0.4/1.6 µM primers (outer/loop/inner), 1× SYBR I, 4 U Isotherm2G DNA polymerase) were employed in each well. For the positive control, 2 nM 5'-DBCO LAMP targets were added. The wells were incubated at 55 °C for the primer extension. Amplification was stopped by rapidly cooling the wells to 0 °C in an ice bath. Samples were instantly collected and run on a 2.5% agarose gel. The gel was read out using GelRed staining under UV light on a Chemidoc™ XRS system (Bio-Rad, Hercules, CA, USA).

3. Results

Here, we describe an approach towards the simplification of LAMP reactions based on a nucleoside-5′-O-triphosphate (dNTP) that is modified with a DNA strand serving as a LAMP target (Figure 1).

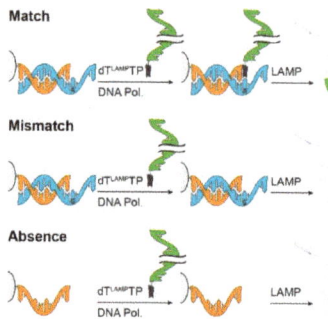

Figure 1. Depiction of the approach explored in this study. A primer is immobilized on a solid support via biotin-streptavidin interaction. After annealing of the template sequence, DNA polymerase and the LAMP template-modified dNTP (dT^{LAMP}TP) are added. After incubation, the unbound conjugate is removed by repeated washing and the LAMP reaction mixture is added afterwards. Only in cases where matched primer template duplexes are present (top), the LAMP reaction is expected to be positive.

The LAMP template-modified dNTP (dT^{LAMP}TP) may be a substrate for DNA polymerases in sequence selective primer extension reactions on solid support. Single nucleotide incorporation is highly sequence selective [17,18] and allows discrimination of single nucleotide variations by the mere difference of incorporation efficiencies of a matched versus mismatched nucleotide. In turn, the immobilized LAMP template (green, Figure 1) can be targeted by strand displacement-proficient DNA polymerases in LAMP reactions. This approach holds promise offering the advantage that—in principle—with one single LAMP sequence an infinite number of targets can be analyzed without tedious redesign of the LAMP sequence, since the immobilized primer strand (orange, Figure 1) is responsible for the sequence selective capture of the target (blue, Figure 1).

The approach depicted above is based on the covalent connection of the LAMP template to a nucleotide. A LAMP template, however, has to harbour six distinct binding sites in fixed spacing. Typical sequences hence consist of around 200 or more nucleotides (nt), which vastly exceed the oligonucleotide-modified nucleotides that have been reported for successful incorporation and which were modified with up to 40 nucleotides [19]. We chose a LAMP template (sequence see Table 1) with a 245 nt sequence taken from the genome of the Lambda phage [9,20]. The sequence was shortened by 61 nucleotides in the middle area and split up into two halves with lengths of 91 nt and 93 nt. For conjugation to the nucleotide, the oligonucleotide representing the 5′-end of the target sequence was equipped with a 5′-dibenzocyclooctyne (DBCO) modification (Figure 2). The 3′-half of the sequence was phosphorylated on its 5′-end to allow splint ligation with T4 DNA ligase, which was carried out at 16 °C in presence of two equivalents of a 30nt splint.

Table 1. Employed DNA sequences.

LAMP Backward Inner Primer	5'-d (GAG AGA ATT TGT ACC ACC TCC CAC CGG GCA CAT AGC AGT CCT AGG GAC AGT)
LAMP backward loop primer	5'-d (ACC ATC TAT GAC TGT ACG CC)
LAMP backward outer primer	5'-d (GGA CGT TTG TAA TGT CCG CTC C)
LAMP forward inner primer	5'-d (CAG CCA GCC GCA GCA CGT TCG CTC ATA GGA GAT ATG GTA GAG CCG C)
LAMP forward loop primer	5'-d (CTG CAT ACG ACG TGT CT)
LAMP forward outer primer	5'-d (GGC TTG GCT CTG CTA ACA CGT T)
LAMP_Splint	5'-d (GGC TGG CTG TCC AGT GAG AGA ATT TGT ACC)
LAMP_Target_A	5'-DBCO-d (GGA CGT TTG TAA TGT CCG CTC CGG CAC ATA GCA GTC CTA GGG ACA GTG GCG TAC AGT CAT AGAT GGT CGG TGG GAG GTG GTA CAA ATT CTC TC)
LAMP_Target_B	5'-P-d (ACT GGA CAG CCA GCC GCA GCA CGT TCC TGC ATA CGA CGT GTC TGC GGC TCT ACC ATA TCT CCT ATG AGC AAC GTG TTA GCA GAG CCA AGC C)
5'-biotin BRAF primer	5'-biotin-d (TTT TTT TTT TTT TTT TTT TGA CCC ACT CCA TCG AGA TTT C)
BRAF template (DNA)	5'-d (TGC CTG GTG TTT GGG AGA AAT CTC GAT GGA GTG GGT C)

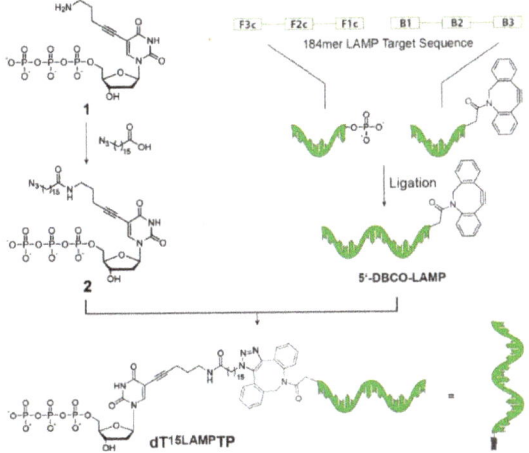

Figure 2. Synthesis of LAMP template-conjugated dT^{15LAMP}TP. Left: reaction conditions: 1, DMF, HATU, Et3N, rt, 12 h; right: the two modified halves of the LAMP template are ligated by splint ligation with T4 DNA ligase yielding a 5'-DBCO-modified 184mer. The click reaction between the azide-functionalized nucleotide and the LAMP template is carried out in PBS buffer at room temperature 12 h and the product is purified by ion-exchange HPLC.

To react with the 5'-DBCO modified oligonucleotide that harbors the LAMP sequence, an azide-functionalized nucleotide was prepared starting from the known dTTP analog 1 (Figure 2) [16]. We chose a linker length that has been demonstrated before to be suitable for appending large "cargo" to dNTPs without greatly compromising DNA polymerase activity [19]. Employment of 16-azidohexadecanoic acid [21], HATU, and Et3N in DMF yields compound 2. Conjugation of 2 and the LAMP template by strain-promoted 1,3-dipol cycloaddition (SPAAC) [22] was achieved in PBS buffer and the product dT^{15LAMP}TP was purified by ion exchange HPLC and centrifugal filtration.

Next, primer extension experiments were conducted with dT^{15LAMP}TP in comparison with natural dTTP and the dTTP derivative 2 with KlenTaq DNA polymerase using a template containing the B type raf kinase (BRAF) T1796A point mutation, which is strongly associated with carcinogenesis [23].

After incubation, samples of the primer extension (PEx) reaction were quenched and analyzed by denaturing polyacrylamide gel electrophoresis (PAGE). For dTTP, the expected shift for single nucleotide incorporation (Figure 3) was observed.

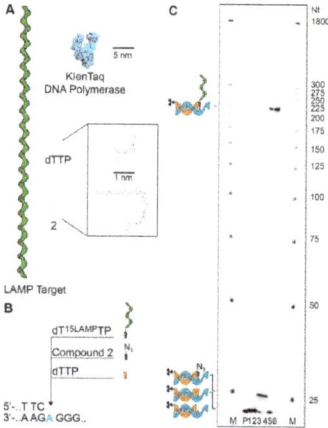

Figure 3. Primer extension experiment employing the dT^{15LAMP}TP. (**A**) The LAMP target sequence, KlenTaq DNA polymerase, natural dTTP and compound 2 drawn to scale. (**B**) Partial sequence of the incorporation site in the BRAF sequence context and the three different nucleotides used in this experiment. (**C**) PEx experiment with KlenTaq DNA polymerase at 55 °C with 1 μM dT^{15LAMP}TP. P: Primer, 1: dTTP, 1 min, 2: dTTP, 30 min; 3: Compound **2**, 1 min, 4: Compound 2, 30 min; 5: dT^{15LAMP}TP, 1 min, 6: dT^{15LAMP}TP, 30 min; M: Marker.

Processing of compound **2** and incorporation of the respective modified nucleotide into the nascent DNA strand led to a pronounced shift of the product by PAGE analysis due to the long alkyl chain-modification impeding migration through the gel matrix. Finally, the usage of dT^{15LAMP}TP led to a very pronounced shift of the product in PAGE analysis, similar to that observed when protein-conjugated nucleotides and shorter oligonucleotide-conjugated nucleotides were used [19,21,24–26]. The band corresponding to the LAMP sequence-conjugated nucleotide runs at approx. 225 nt, which is consistent with the combined size of the 21-mer primer, the 184 nt LAMP sequence, and the connecting alkyl linker. Therefore, not only was the conjugation between the LAMP sequence and the nucleotide confirmed, but it was also shown that this DNA polymerase is able to incorporate nucleotides equipped with ssDNA, being considerably longer than the sequence context used for incorporation.

With the LAMP target-modified nucleotide dT^{15LAMP}TP in hand, the LAMP reaction itself was optimized. The assay was conducted as reported in the original publication with Bst 2.0 DNA polymerase, 350 μM for each dNTP, and 0.2/0.4/1.6 μM primers (outer/loop/inner, respectively) at 65 °C and monitored by real-time SYBR green I fluorescence detection [9]. Using these conditions, the positive control containing 10 nM of the ligated LAMP target was amplified at a cycle quantification value (Cq) of 15.4, but all non-template controls (NTC) showed a similar amplification ranging from Cq 26.4 to 46.6 (Figure 4A). To ensure that this behavior was not caused by a contamination with LAMP target, the experiment was repeated several times with freshly prepared reagents. However, amplification within non-template controls was persistent in all runs. Similar issues were reported in other studies with false positive amplification in LAMP assays [14,15].

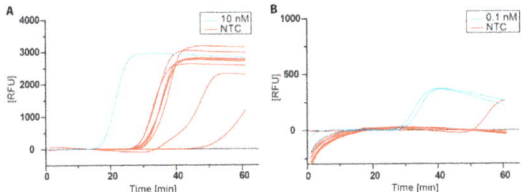

Figure 4. Real-time monitoring of the LAMP with SYBR green I using dT$^{15\text{LAMP}}$TP. (**A**) LAMP Assay using the conditions reported by Tanner et al. [9] with Bst 2.0 DNA polymerase at 65 °C, 0.2/0.4/ 1.6 µM outer/loop/inner primers, 8 mM MgSO4 and 350 µM dNTP each. (**B**) Optimized conditions with Isotherm2G DNA Polymerase at 55 °C, 0.2/0.4/1.6 µM outer/loop/inner primers, 2 mM MgSO4 and 200 µM dNTP each.

To overcome these issues, a screening for appropriate LAMP conditions was carried out including different DNA polymerases, incubation temperatures, primer ratios and concentrations of dNTPs, Mg^{2+}, SYBR green I and betaine. In the end, the assay conditions were changed to Isotherm2G DNA polymerase, 55 °C reaction temperature, 200 µM dNTP each with SYBR green I and remaining primer concentrations at their original levels (0.2/0.4/1.6 µM outer/loop/inner primer). Primers were denatured prior to the addition to the master mix in order to minimize the effect of primer dimers or the presence of any self-primed secondary structure. With the optimized conditions, 0.1 nM of the ligated LAMP target were detected at Cq 28 with no to minimal false positive reactions, which was amplified at sufficiently delayed time points (Cq 51, Figure 4B).

Having optimized the LAMP conditions, we next investigated immobilized primers in a primer extension of the LAMP target-modified nucleotide and subsequent LAMP reaction. Therefore, we used streptavidin coated 8-well plates as the solid phase to immobilize biotin-modified primer strands. The wells were first incubated with 5′ biotinylated primer in PBS and subsequently blocked with biotin. After several washing steps, 50 µL of primer extension reaction mixture containing 100 nM KlenTaq DNA polymerase, 1 µM dT$^{15\text{LAMP}}$PTP, and 200 nM BRAF template was added (Figure 5, lane 1).

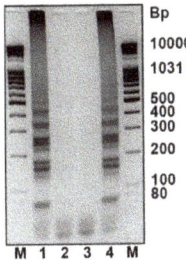

Figure 5. LAMP detection of DNA targets. A PEx reaction employing dT$^{15\text{LAMP}}$TP and KlenTaq DNA polymerase was performed on an immobilized primer in the presence or absence of the matched PEx template. Subsequently, the wells were washed and a LAMP reaction mixture was added. Samples of each well were taken and analysed on a 2.5% agarose gel. Picture colours were inverted to improve contrast. 1: 200 nM (10 pmol) matched BRAF template, 2: no BRAF template, 3: 200 nM BRAF template but natural dTTP instead of dT$^{15\text{LAMP}}$TP, 4: no PEx reactions but LAMP reaction spiked with LAMP target sequence as a positive control, M: marker.

Controls were set up without a BRAF template (lane 2) or with dTTP instead of the LAMP target modified nucleotide dT$^{15\text{LAMP}}$TP (lane 3). Following 30 min of incubation at 55 °C, the reaction mixture was removed. After the plates were washed intensively with 1× KlenTaq reaction buffer, PBS buffer, MilliQ water and 1× Isotherm2G reaction buffer, the LAMP reaction mixture was applied. Furthermore, an additional sample (Figure 5, lane 4) was treated equally as the other samples, but incubated in KlenTaq reaction buffer instead of a primer extension reaction mixture. Here, the LAMP reaction was spiked with 2 nM of LAMP target to serve as a positive control. The LAMP reaction was incubated at 55 °C for 20 min and stopped by rapid cooling of the plates in ice water. Samples were taken and directly subjected to analysis by agarose gel electrophoresis. Upon agarose gel electrography analysis, a ladder-like pattern of bands was observed for the positive reaction (Figure 5, lane 1). The pattern is consistent with the positive control in well 4, which proves the specific amplification starting from the LAMP target. No amplification was observed for both negative controls analysed in lanes 2 and 3. Hence, a LAMP reaction can be utilized to detect the presence of a PEx template using the LAMP target-conjugated dTTP derivative dT$^{15\text{LAMP}}$TP.

4. Conclusions

In summary, a shortened LAMP target sequence derived from the genome of the Lambda phage was generated and conjugated to an azide-functionalized dTTP derivative via click chemistry to yield dT$^{15\text{LAMP}}$TP. Primer extension experiments with the LAMP conjugate revealed that KlenTaq DNA polymerase is able to incorporate the modified nucleotide into a primer strand in spite of the length of the attached oligonucleotide. Next, dT$^{15\text{LAMP}}$TP was employed in primer extension reactions in solid phase in which the modification, if covalently connected to the primer, served as a reporter for the presence of the template in the primer extension reaction. Indeed, we found that amplification starting from the immobilized LAMP template was only observed if the template for the preceding primer extension was present in the reaction due to processing of dT$^{15\text{LAMP}}$TP. Thus, the results demonstrate proof-of-concept that the robustness and simple setup of primer extension-based assays with the rapid and sensitive amplification of LAMP is feasible. The herein depicted results might advance and simplify LAMP-based applications.

Supplementary Materials: The following are available online at http://www.mdpi.com/2624-8549/2/2/490\T1\textendash498/s1, Figure S1: Nucleic acid amplification by LAMP. The two inner primers (green in Step I and orange in Step II) comprise of a site complementary to a sequence in the target oligonucleotide and a 5′ overhang that is complementary to a site within the elongated primer (F1c, B1c). After the inner primer is elongated, the outer primer (black) binds upstream of the inner primer at the target sequence and its elongation by the strand displacement DNA polymerase releases the prolonged inner primer (Step I). The procedure is then repeated at the other side of the released, elongated inner primer (Step II), generating a new sequence that is similar to the target sequence, but instead of the outer primer binding site, it is now on both sides equipped with a sequence complementary to an area inside the oligonucleotide (orange, Step III). Annealing of these complementary sequences will lead to a dumb bell-like structure in which first, the self-primed 3′ end is elongated by the DNA polymerase to open the dumb bell-end on the other side and second, the annealing and elongation of new inner primer releases the stem-loop generated in the first step (Step IV). Thus, a new self-primed 3′-end is formed with which the cycle of self-primed elongation and release by an inner primer is continued. In the end, a mixture of stem-loop like DNA concatemers and cauliflower-like structures with various repeat counts are obtained (Step V).

Author Contributions: Conceptualization, M.W. and A.M.; methodology, M.W. and A.M.; validation, M.W. and A.M.; formal analysis, M.W.; investigation, M.W.; resources, A.M.; writing—original draft preparation, M.W. and A.M.; writing—review and editing, M.W. and A.M.; visualization, M.W. and A.M.; supervision, A.M.; project administration, A.M.; funding acquisition, A.M. All authors have read and agreed to the published version of the manuscript.

Funding: This research was funded by DFG Deutsche Forschungsgemeinschaft, grant number MA 2288/16-2.

Acknowledgments: We thank Samra Ludmann for assisting in the preparation of the manuscript.

Conflicts of Interest: The authors declare no conflict of interest.

References

1. Yan, L.; Zhou, J.; Zheng, Y.; Gamson, A.S.; Roembke, B.T.; Nakayama, S.; Sintim, H.O. Isothermal amplified detection of DNA and RNA. *Mol. Biosyst.* **2014**, *10*, 970–1003. [CrossRef] [PubMed]
2. Walker, G.T.; Fraiser, M.S.; Schram, J.L.; Little, M.C.; Nadeau, J.G.; Malinowski, D.P. Strand displacement amplification—An isothermal, in vitro DNA amplification technique. *Nucleic Acids Res.* **1992**, *20*, 1691–1696. [CrossRef] [PubMed]
3. Fire, A.; Xu, S.Q. Rolling replication of short DNA circles. *Proc. Natl. Acad. Sci. USA* **1995**, *92*, 4641–4645. [CrossRef]
4. Vincent, M.; Xu, Y.; Kong, H. Helicase dependent isothermal DNA amplification. *EMBO* **2004**, *5*, 795–800. [CrossRef]
5. Notomi, T.; Okayama, H.; Masubuchi, H.; Yonekawa, T.; Watanabe, K.; Amino, N.; Hase, T. Loop-mediated isothermal amplification of DNA. *Nucleic Acids Res.* **2000**, *28*, E63. [CrossRef]
6. Notomi, T.; Mori, Y.; Tomita, N.; Kanda, H. Loop-mediated isothermal amplification (LAMP): Principle, features, and future prospects. *J. Microbiol.* **2015**, *53*, 1–5. [CrossRef]
7. Quyen, T.L.; Ngo, T.A.; Bang, D.D.; Madsen, M.; Wolff, A. Classification of multiple DNA dyes based on inhibition effects on real-time loop-mediated isothermal amplification (LAMP): Prospect for point of care setting. *Front. Microbiol.* **2019**, *10*, 2234. [CrossRef]
8. Mori, Y.; Nagamine, K.; Tomita, N.; Notomi, T. Detection of loop-mediated isothermal amplification reaction by turbidity derived from magnesium pyrophosphate formation. *Biochem. Biophys. Res. Commun.* **2001**, *289*, 150–154. [CrossRef]
9. Tanner, N.A.; Zhang, Y.H.; Evans, T.C. Visual detection of isothermal nucleic acid amplification using pH-sensitive dyes. *Biotechniques* **2015**, *58*, 59–68. [CrossRef]
10. Ikeda, S.; Takabe, K.; Inagaki, M.; Funakoshi, N.; Suzuki, K. Detection of gene point mutation in paraffin sections using in situ loop-mediated isothermal amplification. *Pathol. Int.* **2007**, *57*, 594–599. [CrossRef]
11. Iwasaki, M.; Yonekawa, T.; Otsuka, K.; Suzuki, W.; Nagamine, K.; Hase, T.; Tatsumi, K.I.; Horigome, T.; Notomi, T.; Kanda, H. Validation of the loop-mediated isothermal amplification method for single nucleotide polymorphism genotyping with whole blood. *Genome Lett.* **2003**, *2*, 119–126. [CrossRef]
12. Chen, H.; Zhang, J.; Ma, L.; Ma, Y.-P.; Ding, Y.-Z.; Liu, X.; Chen, L.; Zhang, Y.; Liu, Y.-S.; Ma, L.-Q. Rapid pre-clinical detection of classical swine fever by reverse transcription loop-mediated isothermal amplification. *Mol. Cell. Probes* **2009**, *23*, 71–74. [CrossRef]
13. Pham, H.M.; Nakajima, C.; Ohashi, K.; Onuma, M. Loop-mediated isothermal amplification for rapid detection of Newcastle Disease Virus. *J. Clin. Microbiol.* **2005**, *43*, 1646–1650. [CrossRef]
14. Suleman, E.; Mtshali, M.S.; Lane, E. Investigation of false positives associated with loop-mediated isothermal amplification assays for detection of Toxoplasma gondii in archived tissue samples of captive felids. *J. Vet. Diagn. Investig.* **2016**, *28*, 536–542. [CrossRef]
15. Nagai, K.; Horita, N.; Yamamoto, M.; Tsukahara, T.; Nagakura, H.; Tashiro, K.; Shibata, Y.; Watanabe, H.; Nakashima, K.; Ushio, R.; et al. Diagnostic test accuracy of loop-mediated isothermal amplification assay for Mycobacterium tuberculosis: Systematic review and meta-analysis. *Sci. Rep.* **2016**, *6*, 39090. [CrossRef]
16. Bergen, K.; Steck, A.L.; Strütt, S.; Baccaro, A.; Welte, W.; Diederichs, K.; Marx, A. Structures of KlenTaq DNA polymerase caught while incorporating C5-modified pyrimidine and C7-modified 7-deazapurine nucleoside triphosphates. *J. Am. Chem. Soc.* **2012**, *134*, 11840–11843. [CrossRef]
17. Ganai, R.A.; Johansson, E. DNA Replication-A Matter of Fidelity. *Mol. Cell.* **2016**, *62*, 745–755. [CrossRef]
18. Ishino, S.; Ishino, Y. DNA polymerases as useful reagents for biotechnology—The history of developmental research in the field. *Front. Microbiol.* **2014**, *5*, 465. [CrossRef]
19. Baccaro, A.; Steck, A.L.; Marx, A. Barcoded nucleotides. *Angew. Chem. Int. Ed.* **2012**, *51*, 254–257. [CrossRef]
20. Nagamine, K.; Hase, T.; Notomi, T. Accelerated reaction by loop-mediated isothermal amplification using loop primers. *Mol. Cell. Probes* **2002**, *16*, 223–229. [CrossRef]
21. Balintova, J.; Welter, M.; Marx, A. Antibody–nucleotide conjugate as a substrate for DNA polymerases. *Chem. Sci.* **2018**, *9*, 7122–7125. [CrossRef] [PubMed]
22. Agard, N.J.; Prescher, J.A.; Bertozzi, C.R. A strain-promoted [3 + 2] azide–alkyne cycloaddition for covalent modification of biomolecules in living systems. *J. Am. Chem. Soc.* **2004**, *126*, 15046–15047. [CrossRef] [PubMed]

23. Davies, H.; Bignell, G.R.; Cox, C.; Stephens, P.; Edkins, S.; Clegg, S.; Teague, J.; Woffendin, H.; Garnett, M.J.; Bottomley, W.; et al. Mutations of the BRAF gene in human cancer. *Nature* **2002**, *417*, 949–954. [CrossRef] [PubMed]
24. Welter, M.; Verga, D.; Marx, A. Sequence-specific incorporation of enzyme-nucleotide chimera by DNA polymerases. *Angew. Chem. Int. Ed.* **2016**, *55*, 10131–10135. [CrossRef]
25. Verga, D.; Welter, M.; Steck, A.L.; Marx, A. DNA polymerase-catalyzed incorporation of nucleotides modified with a G-quadruplex-derived DNAzyme. *Chem. Commun.* **2015**, *51*, 7379–7381. [CrossRef]
26. Verga, D.; Welter, M.; Marx, A. Sequence selective naked-eye detection of DNA harnessing extension of oligonucleotide-modified nucleotides. *Bioorg. Med. Chem. Lett.* **2016**, *26*, 841–844. [CrossRef]

© 2020 by the authors. Licensee MDPI, Basel, Switzerland. This article is an open access article distributed under the terms and conditions of the Creative Commons Attribution (CC BY) license (http://creativecommons.org/licenses/by/4.0/).

Article

On the Importance of the Thiazole Nitrogen in Epothilones: Semisynthesis and Microtubule-Binding Affinity of Deaza-Epothilone C

Adriana Edenharter [1], Lucie Ryckewaert [1], Daniela Cintulová [1], Juan Estévez-Gallego [2], José Fernando Díaz [2] and Karl-Heinz Altmann [1],*

[1] ETH Zürich, Department of Chemistry and Applied Biosciences, Institute of Pharmaceutical Sciences, CH-8093 Zürich, Switzerland; adriana.edenharter@gmx.com (A.E.); lucie.ryckewaert@yahoo.com (L.R.); daniela.cintulova@tuwien.ac.at (D.C.)
[2] Centro de Investigaciones Biológicas Margarita Salas, Consejo Superior de Investigaciones Científicas, 28040 Madrid, Spain; jeg@cib.csic.es (J.E.-G.); fer@cib.csic.es (J.F.D.)
* Correspondence: karl-heinz.altmann@pharma.ethz.ch

Received: 18 April 2020; Accepted: 21 May 2020; Published: 23 May 2020

Abstract: Deaza-epothilone C, which incorporates a thiophene moiety in place of the thiazole heterocycle in the natural epothilone side chain, has been prepared by semisynthesis from epothilone A, in order to assess the contribution of the thiazole nitrogen to microtubule binding. The synthesis was based on the esterification of a known epothilone A-derived carboxylic acid fragment and a fully synthetic alcohol building block incorporating the modified side chain segment and subsequent ring-closure by ring-closing olefin metathesis. The latter proceeded with unfavorable selectivity and in low yield. Distinct differences in chemical behavior were unveiled between the thiophene-derived advanced intermediates and what has been reported for the corresponding thiazole-based congeners. Compared to natural epothilone C, the free energy of binding of deaza-epothilone C to microtubules was reduced by ca. 1 kcal/mol or less, thus indicating a distinct but non-decisive role of the thiazole nitrogen in the interaction of epothilones with the tubulin/microtubule system. In contrast to natural epothilone C, deaza-epothilone C was devoid of antiproliferative activity in vitro up to a concentration of 10 µM, presumably due to an insufficient stability in the cell culture medium.

Keywords: binding affinity; epothilones; deaza-epothilone; microtubules; structure-activity relationship; thiophene

1. Introduction

Epothilones A and B (**Epo A** and **B**) (Figure 1) are 16-membered, polyketide-derived macrolides that were discovered in 1987 by Reichenbach et al. in the context of a screening for new antifungal agents from the soil-dwelling myxobacterium *Sorangium cellulosum* Soce 90 [1,2]. A number of closely related, but less prevalent polyketides, like epothilones C (**Epo C**) and D (**Epo D**) were later identified in larger scale fermentations of *S. cellulosum* [3].

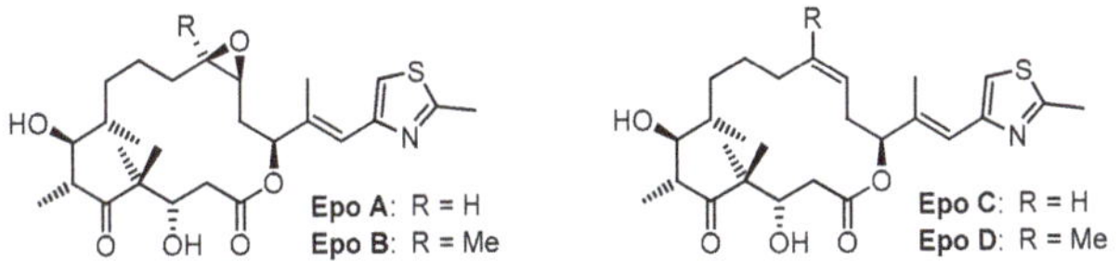

Figure 1. Molecular structures of natural epothilones A–D (**Epo A–D**).

Epo A and B were subsequently found to be highly cytotoxic in the 60-cell line panel of the National Cancer Institute, although the mechanistic underpinnings of this effect remained unknown at the time [4]. Interest in epothilones then surged in 1995, when Bollag et al. demonstrated that **Epo A** and **B** were new microtubule-stabilizing agents and, thus, inhibited cell proliferation by a taxol-like mechanism [5]. At the time of discovery of their mode of action, **Epo A** and **B** were the only non-taxane compound class known to stabilize microtubules, but in contrast to taxol they retained almost full activity (i.e., IC_{50} values in the nM range) against multidrug-resistant cancer cells expressing the P-glycoprotein efflux pump or harboring tubulin mutations [6,7].

Numerous total syntheses of natural epothilones have been reported in the literature, and these efforts have been reviewed extensively [8–13]. Based on the chemistry developed in the context of the total synthesis work, hundreds of synthetic analogues of epothilones have been prepared for structure-activity relationship (SAR) studies and with the objective to deliver compounds with an improved overall pharmacological profile [14,15]. In addition, semisynthetic approaches have been pursued to explore the epothilone-like structural space, in particular by Höfle and co-workers. (This work is summarized in ref. [16]). The most important semisynthetic epothilone derivative is the **Epo B** lactam ixabepilone, which is approved by the FDA (under the tradename Ixempra®) for the treatment of metastatic or locally advanced breast cancer [17]. At least eight additional epothilone-type agents have been advanced to clinical trials in oncology, including the natural product **Epo A** [14,15], and **Epo D** has also been investigated in Phase I clinical trials for Alzheimer's disease [18]. The development of most of these compounds has been discontinued (including the development of **Epo D** for Alzheimer's disease), but an analog termed utidelone (or UTD1) is currently being studied in Phase III clinical trials against breast cancer (in combination with capecitabine) [19].

While numerous side chain-modified epothilone analogs have been investigated as part of comprehensive SAR studies [15,16], a specific question that has been addressed only indirectly relates to the importance of the N-atom in the thiazole ring for microtubule binding and cellular potency. Thus, Nicolaou and co-workers have shown that among the three possible **pyridyl-Epo B** variants (Figure 2), the isomer with the N-atom in the position *ortho* to the vinyl linker between the pyridine ring and the macrolactone core (i.e., ***o*-Pyr-Epo B**) is the most potent with regard to both the promotion of tubulin polymerization and the inhibition of cancer cell growth in vitro [20]. At the same time, the *meta* and *para* isomers (***m*-Pyr-Epo B** and ***p*-Pyr-Epo B**, respectively) still retained a significant tubulin-polymerizing capacity, which reflects the ability of a compound to stabilize microtubules.

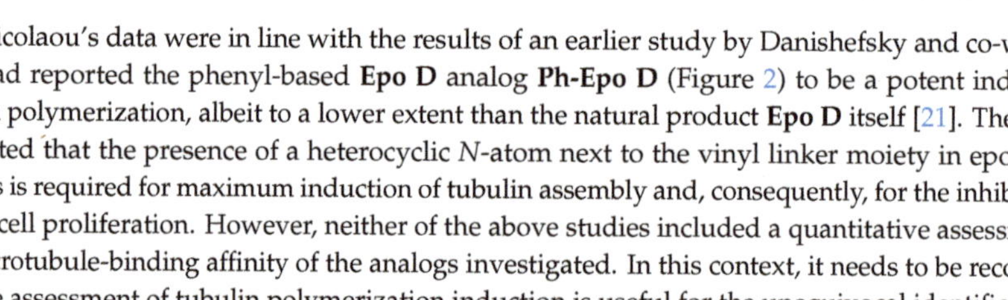

Figure 2. Molecular structures of the three possible **pyridyl-Epo B** variants and of **phenyl-Epo D**.

Nicolaou's data were in line with the results of an earlier study by Danishefsky and co-workers who had reported the phenyl-based **Epo D** analog **Ph-Epo D** (Figure 2) to be a potent inducer of tubulin polymerization, albeit to a lower extent than the natural product **Epo D** itself [21]. These data suggested that the presence of a heterocyclic N-atom next to the vinyl linker moiety in epothilone analogs is required for maximum induction of tubulin assembly and, consequently, for the inhibition of cancer cell proliferation. However, neither of the above studies included a quantitative assessment of the microtubule-binding affinity of the analogs investigated. In this context, it needs to be recognized that the assessment of tubulin polymerization induction is useful for the unequivocal identification of compounds with *poor* tubulin assembly properties, but is less suited for the high-resolution quantitative differentiation between potent assembly inducers. In our own work, we have demonstrated that the microtubule-binding affinities of quinoline-based epothilone analogs (Figure 3) [22] that incorporate the side chain N-atom in the "natural" position (i.e., *m*-**Quin-Epo B** and *m*-**Quin-Epo D**) are ca. one order of magnitude higher than those of the corresponding regioisomers *p*-**Quin-Epo B** and *p*-**Quin-Epo D**, respectively (K_b's of 92×10^7 and 88×10^7 for *m*-**Quin-Epo B** and *m*-**Quin-Epo D**, respectively, vs. 6.9×10^7 and 6.1×10^7 for *p*-**Quin-Epo B** and *p*-**Quin-Epo D**). Quite intriguingly, and for reasons not understood, the difference in the microtubule-binding affinity between *m*-**Quin-Epo D** and *p*-**Quin-Epo D** translates into a corresponding difference in the cellular activity, while the epoxides *m*-**Quin-Epo B** and *p*-**Quin-Epo B** are virtually equipotent (and highly active).

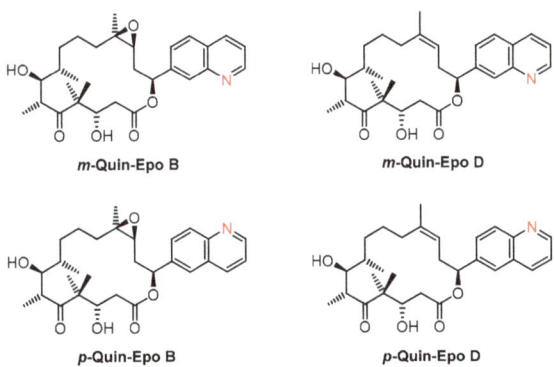

Figure 3. Molecular structures of quinoline-derived epothilone analogs.

While our data for quinoline-based epothilone analogs seem to re-enforce the conclusions derived from the earlier studies by Nicolaou and Danishefsky on pyridyl- and phenyl-based epothilone analogs, respectively, they need to be interpreted with some care, in light of the significant overall modification of the side chain vs. natural epothilones. Finally, the recent X-ray crystal structure of a tubulin-**Epo A** complex [23] invokes a hydrogen bond between the thiazole nitrogen and the side chain hydroxy group of βThr297. Overall, the available experimental data suggest that the presence of a properly positioned heterocyclic N-atom in side chain-modified epothilone analogs is required, in order to maximize interactions with the tubulin/microtubule system. At the same time, the magnitude of the effect associated with the simple removal of this nitrogen from the *natural* side chain, quite surprisingly, has never been assessed. We have thus been interested for some time in the synthesis of a thiophene-containing analog of a natural epothilone and the determination of its microtubule-binding affinity in comparison with the natural parent structure. In this paper, we describe the results of these efforts.

2. Materials and Methods

Detailed protocols for the synthesis of new compounds and the associated analytical data can be found in the Supplementary Materials.

3. Results and Discussion

3.1. Synthesis of *Deaza-Epo C* (5)

The initial synthetic target of our work was **deaza-Epo A** (1), which we planned to access from ketone **2** by means of the Wittig or Horner–Woodsworth–Emmons (HWE) reaction (Scheme 1). Ketone **2** can be accessed from **Epo A** by TMS-protection and ozonolysis [24]; and its conversion into TMS-protected **3** via the transformation of the ketone moiety into the corresponding vinyl boronic acid, followed by iodination and Stille coupling, had been described earlier by Höfle and co-workers [25].

However, no experimental protocols for this reaction sequence are provided in ref. [25], and no yields are reported for the iodination and Stille coupling steps. On the other hand, we had been successful ourselves in the elaboration of ketone **2** into epothilone A analogs bearing modified thiazole, pyrimidine or pyridine moieties in place of the natural thiazole heterocycle by means of HWE chemistry (Hauenstein & Altmann, unpublished experiments; see also refs. [26,27]). Unfortunately, the attempted HWE coupling of **2** with phosphonate **4** did not yield any of the desired olefin (Scheme 1).

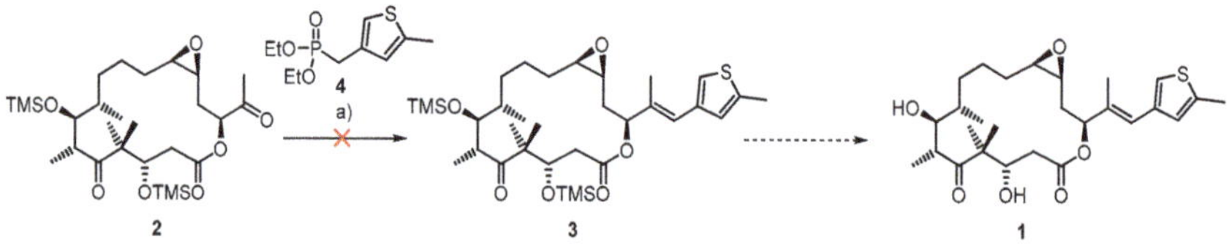

Scheme 1. Reagents and conditions: a) nBuLi, −78 °C to RT, 0%.

In light of a previous report by Höfle and co-workers, who had been unable of convert **2** into **phenyl-Epo A** or to reconstruct **Epo A** from **2** by a variety of olefination methods [24], we did not further pursue the direct elaboration of **2** into **1**. Instead, we turned our attention to a different aspect of the chemistry of **Epo A** that had been unveiled by Höfle's work on semisynthetic epothilone derivatives and that involves the targeted removal of the C13–C15 segment of the macrocycle to generate acid **8** [28] (Scheme 2). The latter was then to be esterified with alcohol **7**, and the resulting diene would be cyclized by ring-closing olefin metathesis (RCM) in analogy to previous work on the total synthesis of **Epo A** [29–31]. Deprotection would then yield **deaza-Epo C (5)**.

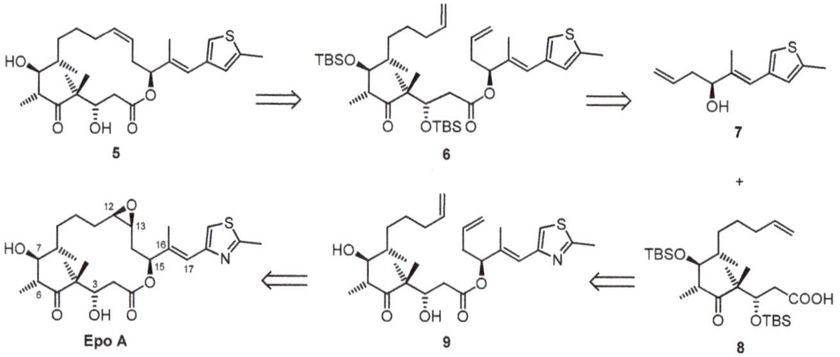

Scheme 2. Retrosynthesis of deaza-Epo C (5).

When elaborating this strategy, we were cognizant of the fact that the conversion of the **Epo C** analog **5** into **1** might be impaired by the competing epoxidation of the C16–C17 double bond, which is more nucleophilic than in **Epo C**, and/or oxidation at sulfur [32]. At the same time, we also felt that our question about the effects of the removal of the thiazole nitrogen could be addressed by comparing the microtubule-binding affinity of **5** with that of **Epo C**, as the latter had also been reported to be a potent inducer of tubulin polymerization [33] (although no microtubule binding data for the compound exist in the literature).

In the forward direction, the synthesis of acid **8** commenced with the deoxygenation of **Epo A** with 3-methylbenzo[d]thiazole-2(3H)-selenone (**10**), prepared according to Calo and co-workers by refluxing methylbenzothiazolium iodide and elemental selenium in pyridine [34] (Scheme 3).

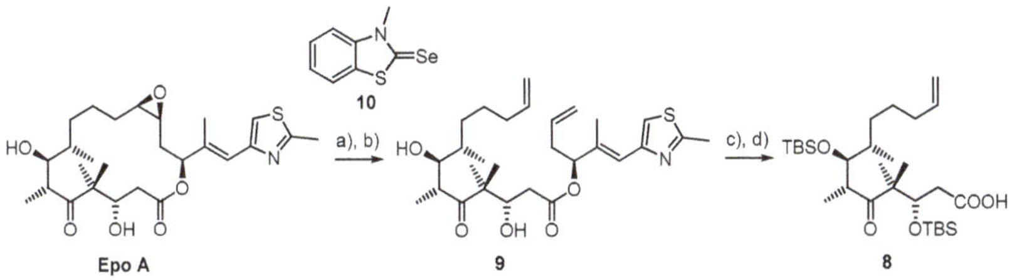

Scheme 3. Reagents and conditions: a) **10**, TFA, DCM, 6 h, RT, 43%; b) Hoveyda–Grubbs-II cat., DCM, 20 h, RT, 61%; c) i) 2,6-lutidine, DCM, 10 min, RT, ii) TBSOTf, 5 min, 110 °C, microwave; d) LiOH, H$_2$O, iPrOH, 5 min, 150 °C, microwave, 61% over two steps.

The deoxygenation reaction gave **Epo C** in highly variable yields, due to incomplete conversion and the formation of an unknown side product that was difficult to remove. It was found eventually, however, that the subsequent ring-opening reaction proceeded equally well without the prior removal of this impurity, thus obviating the need for the tedious purification of the intermediate **Epo C**. The ring-opening of the macrocycle with ethylene in the presence of the Grubbs–Hoveyda II catalyst furnished diene **9** (as described [28]), which was converted into acid **8** by sequential treatment with TBSOTf and LiOH at elevated temperature under microwave conditions in 61% overall yield.

The synthesis of alcohol **7** started from 3-thienylmethanol (**11**), which was TBS-protected; deprotonation of TBS-ether **12** with nBuLi (1.1 equiv) and reaction with a two-fold excess of MeI then provided TBS-ether **13** in 62% yield (Scheme 4). The deprotection of **13** with TBAF, followed by oxidation of the ensuing free alcohol under Swern conditions, afforded the desired aldehyde **15** in 71% overall yield. The Wittig reaction of **15** with **16** then delivered homologated aldehyde **17** (71% yield).

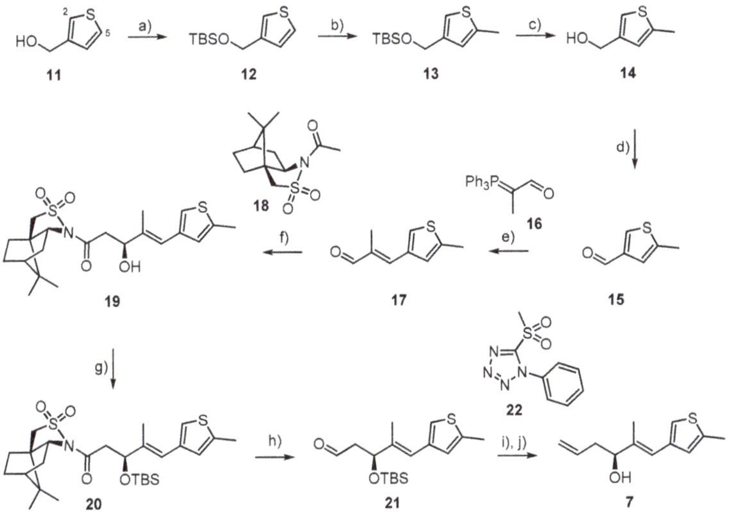

Scheme 4. Reagents and conditions: a) TBSCl, imidazole, DMF, 5 h, 45 °C, quant.; b) nBuLi, MeI, THF, 18 h, −35 °C to RT, 62%; c) TBAF, THF, 3 h, RT; d) oxalyl chloride, DMSO, DCM, NEt$_3$, 2 h, −78 °C to RT, 71% (2 steps); e) **16**, benzene, 23 h, 103 °C, 71%; f) **18**, dibutylboron triflate, DIPEA, DCM, 4 h, −78 °C to 0 °C, 51%; g) TBSCl, imidazole, DMF, 4.5 h, 45 °C, 79%; h) DIBAL-H, DCM, 2 h, −78 °C, 80%; i) **22**, NaHMDS, THF, 2 h, −78 °C, 88%; j) TBAF, THF, 5 h, 0 °C to RT, 94%.

The reaction of **17** with the dibutylboron enolate of acetylsultam **18**, obtained by the successive treatment of the latter with dibutylboron triflate and DIPEA [22,35], proceeded with only moderate selectivity, to afford a ca. 2:1 mixture of aldol products, in favor of the desired isomer **19** [36]. Preparing the dibutylboron triflate in situ from triethylborane and trifluoromethanesulfonic acid did not lead to an improved *dr*. Although tedious, isomer separation was possible by column chromatography, and **19** was finally obtained as a single isomer in 51% yield; the latter was then protected as its TBS-ether **20**. The reductive cleavage of the auxiliary with DIBAL-H, followed by Julia–Kocienski olefination of the ensuing aldehyde **21** with sulfone **22** and subsequent TBS-deprotection, gave homoallylic alcohol **7** in ca. 8.5% overall yield for the 10-step sequence from 3-thienylmethanol (**11**). The attempted Wittig olefination of aldehyde **21** with methyltriphenylphosphonium bromide in combination with various bases had been found previously to induce the elimination of TBSOH (Cintulová & Altmann, unpublished).

The esterification of alcohol **7** with acid **8** under Yamaguchi conditions [37] at RT furnished diene **6** in moderate but acceptable yields of 44%–65% (Scheme 5).

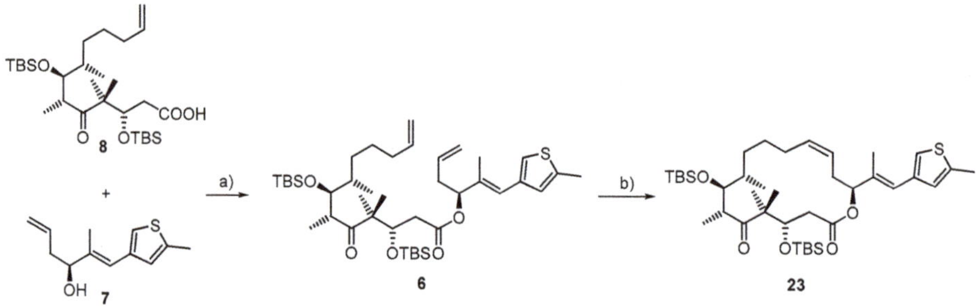

Scheme 5. a) Reagents and conditions: NEt$_3$, 2,4,6-trichlorobenzoyl chloride (TCBC), DMAP, 3 h, RT, 44%–65%; b) Grubbs II cat., toluene, 40 °C, 17%.

A series of screening experiments was then performed on an analytical scale to identify the best conditions for the crucial ring-closure reaction, including a range of solvents (DCM, benzene, toluene, THF) and different metathesis catalysts (Grubbs I and II, Hoveyda–Grubbs II, Grubbs III) [38]. Not unexpectedly, the reaction under all conditions investigated suffered from low selectivity; low selectivity has also been reported for the RCM of bis-TBS-protected **Epo C** with the Grubbs I catalyst [29–31]). However, in contrast to the latter, which delivered the E/Z isomeric mixture of macrocycles in high yield, diene **6** appeared to be of limited stability under the reaction conditions and/or to be consumed by alternative reaction pathways. Overall, the screening experiments, unfortunately, did not provide consistent guidance on how to maximize the yield of the desired macrocycle **23**. Ultimately, the reaction was conducted with 0.1 equiv. of Grubbs II catalyst in toluene at 40 °C and quenched before complete conversion of the starting material. On a 20 mg scale, these conditions provided 3.5 mg of slightly impure **23** (17%) together with 6.9 mg (33%) of the corresponding 12,13-E isomer. When the reaction was carried out for 3 h at reflux, the E isomer of **23** was obtained in 51% isolated yield.

The difficulties with the RCM of diene **6** were then further aggravated by the fact that no conditions could be identified that would have allowed for the clean deprotection of **23**, including conditions that have been employed successfully in the deprotection of bis-TBS-protected **Epo C** [29–31]. These findings, which point to a distinct instability of **23** (or **5**) compared to (protected) **Epo C**, called for a change to a protecting group more easily removable than TBS. An obvious candidate that would meet this requirement was the TMS group and we felt that the corresponding acid **8a** (Figure 4) could be readily available from **9** in analogy to the preparation of **8**, by simply substituting TMSOTf for TBSOTf.

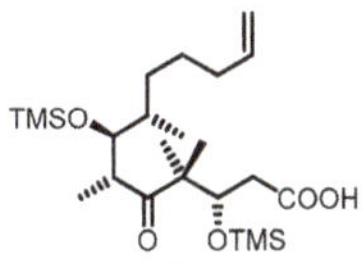

Figure 4. Molecular structures of acid **8a**.

Unfortunately, the treatment of **9** with TMSOTf under a variety of conditions did not lead to complete conversion to the bis-TMS ether **8a**, even if a large excess of TMSOTf was employed (up to 75-fold); in addition, the retro aldol cleavage of the C3–C4 bond was frequently observed as a side reaction [39]. The purification of acid **8a** by silica gel chromatography was possible to some extent, but was complicated by the limited stability of the compound under the chromatographic conditions. In contrast to **8a**, mono-TMS-ether **26** (Scheme 6) could be obtained in 53% yield after column chromatography, when TMSOTf and 2,6-lutidine were premixed in DCM before the addition of **9** to the reaction mixture (Scheme 6). While the use of **26** presented its own problems in the subsequent esterification step (*vide infra*), sufficient quantities of this material could be produced to complete the synthesis of the target structure.

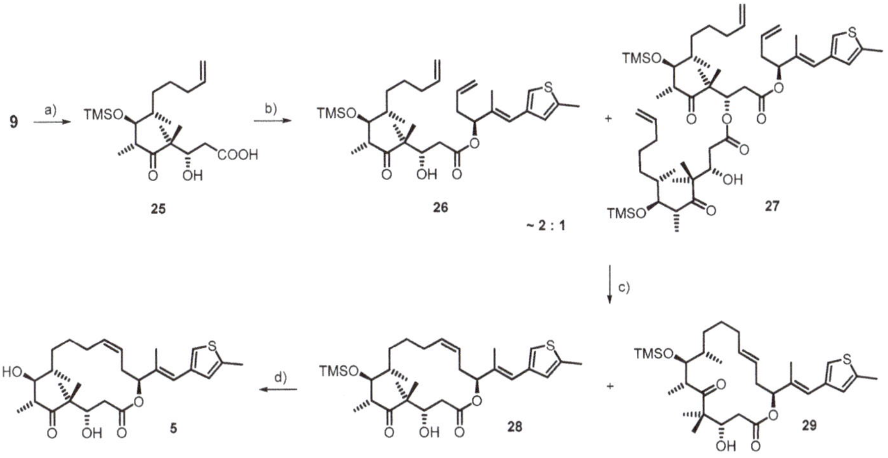

Scheme 6. Reagents and conditions: a) TMSOTf, 2,6-lutidine, DCM, 1.5 h, 0 °C to RT, 53%; b) **7**, TCBC, NEt$_3$, THF, DMAP, 2 h, 0 °C to RT, 39% for the 2:1 mixture (not separated) of **26** (26%) and **27** (13%); c) Grubbs I cat., toluene, RT, 18 h, 13% (**28**) and 19% (**29**); d) PPTS, EtOH, 0 °C, 90 min, 67%.

Thus, the reaction of **26** with **7** under Yamaguchi conditions gave the desired ester **26** and the dimeric structure **27** [40] in a ca. 2:1 ratio and with an overall yield of 39%. The separation of **26** and **27** was possible, if tedious, but was better performed after the RCM step. Intriguingly, two attempts at the Yamaguchi esterification of doubly TMS-protected acid **8a** only led to slow decomposition, with alcohol **7** being recovered in almost quantitative yields.

RCM with 60 mg of the 2:1 mixture of **26** and **27** with Grubbs I catalyst gave 4.5 mg (13%) of Z-product **28** and 6.6 mg (19%) of E-isomer **29** after two preparative RP-HPLC runs. Furthermore, 6.9 mg (19%) of diene **26** were recovered; no other products were characterized. The configurational assignment of the C12–C13 double bond in **28** and **29** was based on *J* coupling constants obtained via NMR decoupling experiments. For Z-product **28**, a coupling constant of 10.9 Hz was determined, while E-isomer **29** showed a coupling constant of 15.3 Hz (see the Supporting Material). With the protected macrocycle in hand, removal of the TMS moiety was then attempted with citric acid (MeOH, RT, 12 h), as this had proven successful for other TMS-protected epothilone analogs [27]; however, the mass of **5** was not detected in the reaction mixture, and RP-HPLC indicated the formation of multiple products. In contrast, the final deprotection of **28** with PPTS in EtOH gratifyingly afforded deaza-**Epo C** (**5**) in 67% yield after HPLC purification (Scheme 6).

3.2. Biochemical and Cellular Assessment

With deaza-**Epo C** (**5**) in hand, we assessed the binding of the compound to cross-linked microtubules at different temperatures in comparison with natural **Epo C**. As can be seen from the data presented in Table 1, for temperatures up to 35 °C, the binding constant of deaza-**Epo C** (**5**) for microtubules is ca. 4-fold to 6-fold lower than that of **Epo C**, corresponding to a free energy difference ΔΔG of ca. 0.7 kcal/mol to 1 kcal/mol. Thus, the loss in binding free energy incurred by the replacement of the thiazole nitrogen in **Epo C** by a CH group is rather modest and appears to be comparable with the loss in binding energy observed upon removal of the epoxide oxygen from **Epo A** (to form **Epo C**) (Table 1) or the **Epo B**→**Epo D** transformation [41]. The difference appears somewhat less pronounced than for the quinoline-based **Epo D** analogs *o*-**Quin-Epo D** and *p*-**Quin Epo D** (Figure 3), where the difference in binding constants is >10-fold; however, the differences are small and should not be overinterpreted. At the same time, the data for the quinoline-based epothilones depicted in Figure 3 suggest that the difference in binding free energy observed here between **Epo C** and deaza-**Epo C** (**5**) can most likely be extrapolated to epothilones A, B and D and their corresponding deaza analogs. For temperatures above 35 °C, a marked drop in the microtubule-binding affinity of deaza-**Epo C** (**5**) was observed (K_b of 5×10^5 at 42 °C); the effect is significantly more pronounced than for **Epo C** (K_b of 88×10^5 at 42 °C). The cause for this behavior is unclear at this point, but may be related to the limited chemical stability of the compound **5** at higher temperatures (*vide infra*).

Table 1. Binding constants of deaza-**Epo C** (**5**), **Epo C** and **Epo A** for stabilized microtubules. [1]

Compound	K_b [10^7 M^{-1}]		
	26 °C	30 °C	35 °C
5	0.39 ± 0.03	0.33 ± 0.02	0.37 ± 0.05
Epo C	1.46 ± 0.33	1.19 ± 0.15	1.93 ± 0.27
Epo A [2]	7.48 ± 1.00	5.81 ± 1.08	3.63 ± 0.51

[1] Association constant K_b with glutaraldehyde-stabilized microtubules at different temperatures, determined as described in ref. [42]. Numbers are average values from three independent experiments ± standard deviation.
[2] From ref. [41].

Epothilone analogs with similar K_b's as deaza-**Epo C** (**5**) have been reported to exhibit sub-µM antiproliferative activity [26], and it was, therefore, surprising that **5** showed no growth inhibition of the human cancer cell lines MCF7 (breast) or A549 (lung) up to a concentration of 10 µM. In contrast, and as expected from the literature, **Epo C** inhibited the growth of MCF7 and A549 cells with IC$_{50}$ values of 9 nM and 103 nM, respectively [33]. These findings triggered experiments on the stability of **5** in cell culture medium, which revealed that the compound was degraded with a half-life of less than 2 h; this is significantly lower than the half-life of **Epo C** under identical conditions (see the Supporting Material). Due to limitations in the sensitivity of the analytical system, the experiments had to be carried out at a 100 µM concentration, which is 10-fold higher than the highest concentration tested in the growth inhibition experiments. While our stability data, thus, are largely qualitative in nature, they do indicate that the limited stability of deaza-**Epo C** (**5**) in cell culture medium may contribute to the lack of cellular potency.

4. Conclusions

Deaza-epothilone C (**5**), which incorporates a thiophene heterocycle in place of the thiazole moiety in natural epothilones, has been prepared by semisynthesis from epothilone A. The synthesis of building blocks **8** (from epothilone A) and **7** (from 3-thienylmethanol) was accomplished with reasonable efficiency; the synthesis of the alternatively protected acids **8a** and **23** proved to be more difficult. Likewise, the elaboration of these building blocks into deaza-epothilone C was hampered by significant difficulties, which (partly) reflect distinct differences in the chemical behavior between the thiophene-containing intermediates and their thiazole-derived congeners. As a consequence, deaza-epothilone C (**5**) was obtained from **7** in only 2.3% overall yield (three steps). Nevertheless, sufficient material could be procured to assess the binding of **5** to microtubules and its cellular activity. While the replacement of the thiazole nitrogen by a simple CH group causes a drop in the the free energy of binding of **5** relative to epothilone C, in agreement with structural data for the tubulin/**Epo A** complex and also previous (binding) data for quinoline-based epothilone analogs, the magnitude of the decrease (between 0.7 and 1 kcal/mol) is rather modest. Thus, the data indicate that, in principle, thiophene-derived analogs of epothilone A or B (or macrocycle-modified variants thereof) should be high-affinity microtubule binders. However, high-affinity microtubule-binding of such analogs may not translate into potent cellular activity, as indicated by the reduced stability (compared to epothilone C) and lack of cellular activity observed for deaza-epothilone C (**5**). At the same time, stability may be enhanced by appropriate modification of the thiophene ring (e.g., by the replacement of the methyl group by a quasi-isosteric electron-withdrawing chlorine atom). These questions will have to be clarified in future experiments.

Supplementary Materials: The following are available online at http://www.mdpi.com/2624-8549/2/2/499\T1\textendash509/s1, synthesis protocols and analytical data for all new compounds, Figure S1: Decoupling experiments with **28**, Figure S2: Decoupling experiments with **29**, Figure S3: Stability of **deaza-Epo C** (**5**) in Dulbecco's Modified Eagle Medium (DMEM) over a 2 h time period, Figure S4: Stability of **deaza-Epo C** (**5**) in DMEM over a 24 h time period, Figure S5: Stability of **Epo C** in DMEM over a 2 h time period, Figure S6: Stability of **Epo C** in DMEM over a 24 h time period.

Author Contributions: Conceptualization, K.-H.A.; methodology, K.-H.A., A.E., J.F.D.; formal analysis, A.E., K.-H.A., J.F.D.; investigation, A.E., L.R., D.C., J.E.-G.; data curation, A.E., J.E.-G.; writing, K.-H.A.; supervision, K.-H.A.; funding acquisition, J.F.D. All authors have read and agreed to the published version of the manuscript.

Funding: This work was supported by institutional funding from ETH Zürich (A.E., K.-H.A., L.R., D.C.) and by Ministerio de Economia y Competitividad grants BFU2016-75319-R to J.F.D. (both AEI/FEDER, UE) and H2020-MSCA-ITN-2019 860070 TUBINTRAIN to J.F.D. and a donation from Club deportivo Escuela Hungaresa de Pontevedra. No funds were received to cover the costs for publication in open access.

Acknowledgments: The authors acknowledge networking contributions by the COST Action CM1407 "Challenging organic syntheses inspired by nature - from natural products chemistry to drug discovery". We are indebted to Bernhard Pfeiffer (ETHZ) for NMR support, and to Xiangyang Zhang, Louis Bertschi, Rolf Häfliger and Oswald Greter (ETHZ) for HRMS spectra. Kurt Hauenstein is acknowledged for excellent technical assistance.

Conflicts of Interest: The authors declare no conflict of interest.

References and Notes

1. Gerth, K.; Bedorf, N.; Höfle, G.; Irschik, H.; Reichenbach, H. Epothilons A and B: Antifungal and Cytotoxic Compounds from Sorangium cellulosum (Myxobacteria). Production, Physico-chemical and Biological Properties. *J. Antibiot.* **1996**, *49*, 560–563. [CrossRef]
2. Höfle, G.; Reichenbach, H. Epothilone, a myxobacterial metabolite with promising antitumor activity. In *Anticancer Agents from Natural Products*; Cragg, G.M., Kingston, D.G.I., Newman, D.J., Eds.; Taylor & Francis: Boca Raton, FL, USA, 2005; pp. 413–450. [CrossRef]
3. Hardt, I.H.; Steinmetz, H.; Gerth, K.; Sasse, F.; Reichenbach, H.; Höfle, G. New Natural Epothilones from Sorangium cellulosum, Strains So ce90/B2 and So ce90/D13: Isolation, Structure Elucidation, and SAR Studies. *J. Nat. Prod.* **2001**, *64*, 847–856. [CrossRef] [PubMed]
4. Höfle, G.H.; Bedorf, N.; Steinmetz, H.; Schomburg, D.; Gerth, K.; Reichenbach, H. Antibiotics from gliding bacteria. 77. Epothilone A and B-novel 16-membered macrolides with cytotoxic activity: Isolation, crystal structure, and conformation in solution. *Angew. Chem. Int. Ed.* **1996**, *35*, 1567–1569. [CrossRef]

5. Bollag, D.M.; McQueney, P.A.; Zhu, J.; Hensens, O.; Koupal, L.; Liesch, J.; Goetz, M.; Lazarides, E.; Woods, C.M. Epothilones, a new class of microtubule-stabilizing agents with a Taxol-like mechanism of action. *Cancer Res.* **1995**, *55*, 2325–2333.
6. Kowalski, R.J.; Giannakakou, P.; Hamel, E. Activities of the microtubule-stabilizing agents epothilones A and B with purified tubulin and in cells resistant to paclitaxel (Taxol). *J. Biol. Chem.* **1997**, *272*, 2534–2541. [CrossRef] [PubMed]
7. Altmann, K.-H.; Wartmann, M.; O'Reilly, T. Epothilones and related structures-a new class of microtubule inhibitors with potent in vivo antitumor activity. *Biochim. Biophys. Acta* **2000**, *1470*, M79–M91. [CrossRef]
8. Nicolaou, K.C.; Roschangar, F.; Vourloumis, D. Chemical Biology of Epothilones. *Angew. Chem. Int. Ed.* **1998**, *37*, 2014–2045. [CrossRef]
9. Harris, C.R.; Danishefsky, S.J. Complex target-oriented synthesis in the drug discovery process: A case history in the dEpoB series. *J. Org. Chem.* **1999**, *64*, 8434–8456. [CrossRef]
10. Nicolaou, K.C.; Ritzén, A.; Namoto, K. Recent developments in the chemistry, biology and medicine of the epothilones. *Chem. Commun.* **2001**, 1523–1535. [CrossRef]
11. Altmann, K.-H. The merger of natural product synthesis and medicinal chemistry: On the chemistry and chemical biology of epothilones. *Org. Biomol. Chem.* **2004**, *2*, 2137–2151. [CrossRef]
12. Watkins, E.B.; Chittiboyina, A.G.; Avery, M.A. Recent Developments in the Syntheses of the Epothilones and Related Analogues. *Eur. J. Org. Chem.* **2006**, 4071–4084. [CrossRef]
13. Nicolaou, K.C. The Chemistry-Biology-Medicine Continuum and the Drug Discovery and Development Process in Academia. *Chem. Biol.* **2014**, *21*, 1039–1045. [CrossRef] [PubMed]
14. Altmann, K.-H.; Pfeiffer, B.; Arseniyadis, S.; Pratt, B.A.; Nicolaou, K.C. The chemistry and biology of epothilones-The wheel keeps turning. *ChemMedChem* **2007**, *2*, 397–423. [CrossRef] [PubMed]
15. Altmann, K.H.; Kinghorn, A.D.; Höfle, G.; Müller, R.; Prantz, K. *For a Book on Epothilones cf.: The Epothilones: An Outstanding Family of Anti-Tumor Agents (Progress Chem. Org. Nat. Prod. 90)*; Kinghorn, A.D., Falk, H., Kobayashi, J., Eds.; Springer Wien: New York, NY, USA, 2009; ISBN 978-3-211-78207-1. [CrossRef]
16. Altmann, K.-H.; Gaugaz, F.Z.; Schiess, R. Diversity through semisynthesis: The chemistry and biological activity of semisynthetic epothilone derivatives. *Mol. Divers.* **2011**, *15*, 383–399. [CrossRef] [PubMed]
17. Lechleider, R.J.; Kaminskas, E.; Jiang, X.; Aziz, R.; Bullock, J.; Kasliwal, R.; Harapanhalli, R.; Pope, S.; Sridhara, R.; Leighton, J.; et al. Ixabepilone in Combination with Capecitabine and as Monotherapy for Treatment of Advanced Breast Cancer Refractory to Previous Chemotherapies. *Clin. Cancer Res.* **2008**, *14*, 4378–4384. [CrossRef]
18. Barten, D.M.; Fanara, P.; Andorfer, C.; Hoque, N.; Wong, P.Y.A.; Husted, K.H.; Cadelina, G.W.; DeCarr, L.B.; Yang, L.; Liu, V.; et al. Hyperdynamic microtubules, cognitive deficits, and pathology are improved in tau transgenic mice with low doses of the microtubule-stabilizing agent BMS-241027. *J. Neurosci.* **2012**, *32*, 7137–7145. [CrossRef]
19. Zhang, P.; Sun, T.; Zhang, Q.; Yuan, Z.; Jiang, Z.; Wang, X.J.; Cui, S.; Teng, Y.; Hu, X.-C.; Yang, J.; et al. Utidelone plus capecitabine versus capecitabine alone for heavily pretreated metastatic breast cancer refractory to anthracyclines and taxanes: A multicentre, open-label, superiority, phase 3, randomised controlled trial. *Lancet Oncol.* **2017**, *18*, 371–383. [CrossRef]
20. Nicolaou, K.C.; Scarpelli, R.; Bollbuck, B.; Werschkun, B.; Pereira, M.M.A.; Wartmann, M.; Altmann, K.-H.; Zaharevitz, D.; Gussio, R.; Giannakakou, P. Chemical synthesis and biological properties of pyridine epothilones. *Chem. Biol.* **2000**, *7*, 593–599. [CrossRef]
21. Su, D.-S.; Meng, D.; Bertinato, P.; Balog, A.; Sorensen, E.J.; Danishefsky, S.J.; Zheng, Y.-H.; Chou, T.C.; He, L.; Horwitz, S.B. Total synthesis of (-)-epothilone B: An extension of the Suzuki coupling method and insights into structure-activity relationships of the epothilones. *Angew. Chem. Int. Ed.* **1997**, *37*, 757–759. [CrossRef]
22. Dietrich, S.; Lindauer, R.; Stierlin, C.; Gertsch, J.; Matesanz, R.; Notararigo, S.; Díaz, J.F.; Altmann, K.-H. Epothilone Analogs with Benzimidazole and Quinoline Side Chains: Chemical Synthesis, Antiproliferative Activity, and Interactions with Tubulin. *Chem. Eur. J.* **2009**, *15*, 10144–10157. [CrossRef]
23. Prota, A.E.; Bargsten, K.; Zurwerra, D.; Field, J.J.; Díaz, J.F.; Altmann, K.-H.; Steinmetz, M.O. Molecular Mechanism of Action of Microtubule-Stabilizing Anticancer Agents. *Science* **2013**, *339*, 587–590. [CrossRef]
24. Sefkow, M.; Kiffe, M.; Schummer, D.; Höfle, G. Oxidative and reductive transformations of epothilone A. *Bioorg. Med. Chem. Lett.* **1998**, *8*, 3025–3030. [CrossRef]

25. Höfle, G.; Glaser, N.; Leibold, T.; Karama, U.; Sasse, F.; Steinmetz, H. Semisynthesis and degradation of the tubulin inhibitors epothilone and tubulysin. *Pure Appl. Chem.* **2003**, *75*, 167–178. [CrossRef]
26. Erdélyi, M.; Navarro-Vázquez, A.; Pfeiffer, B.; Kuzniewski, C.N.; Felser, A.; Widmer, T.; Gertsch, J.; Pera, B.; Díaz, J.F.; Altmann, K.-H.; et al. The binding mode of side chain- and C3-modified epothilones to tubulin. *ChemMedChem* **2010**, *5*, 911–920. [CrossRef]
27. Schiess, R.; Gertsch, J.; Schweizer, W.B.; Altmann, K.-H. Stereoselective Synthesis of 12,13-Cyclopropyl-Epothilone B and Side-Chain-Modified Variants. *Org. Lett.* **2011**, *13*, 1436–1439. [CrossRef]
28. Karama, U.; Höfle, G. Synthesis of epothilone 16,17-alkyne analogs by replacement of the C13-C15(O)-ring segment of natural epothilone C. *Eur. J. Org. Chem.* **2003**, 1042–1049. [CrossRef]
29. Schinzer, D.; Limberg, A.; Bauer, A.; Böhm, O.M.; Cordes, M. Total synthesis of (-)-epothilone A. *Angew. Chem. Int. Ed. Engl.* **1997**, *36*, 523–524. [CrossRef]
30. Nicolaou, K.C.; He, Y.; Vourloumis, D.; Vallberg, H.; Zang, Z. Total synthesis of epothilone A: The olefin metathesis approach. *Angew. Chem. Int. Ed. Engl.* **1997**, *36*, 166–168. [CrossRef]
31. Balog, A.; Meng, D.; Kamenecka, T.; Bertinato, P.; Su, D.S.; Sorensen, E.J.; Danishefsky, S.J. Total synthesis of (-)-epothilone A. *Angew. Chem. Int. Ed. Engl.* **1996**, *35*, 2801–2803. [CrossRef]
32. Nicolaou, K.C.; He, Y.; Vourloumis, D.; Vallberg, H.; Roschangar, F.; Sarabia, F.; Ninkovic, S.; Yang, Z.; Trujillo, J.I. The Olefin Metathesis Approach to Epothilone A and Its Analogs. *J. Am. Chem. Soc.* **1997**, *119*, 7960–7973. [CrossRef]
33. Altmann, K.-H.; Flörsheimer, A.; Bold, G.; Caravatti, G.; End, N.; Wartmann, M. Natural product-based drug discovery-Epothilones as lead structures for the development of new anticancer agents. *Chimia* **2004**, *58*, 686–690. [CrossRef]
34. Calo, V.; Lopez, L.; Mincuzzi, A.; Pesce, G. 3-Methyl-2-selenoxobenzothiazole, a New Reagent for the Stereospecific Deoxygenation of Epoxides and the Desulfurization of Episulfides into Olefins. *Synthesis* **1976**, 200–201. [CrossRef]
35. Bond, S.; Perlmutter, P. N-Acetylbornane-10,2-sultam: A Useful, Enantiomerically Pure Acetate Synthon for Asymmetric Aldol Reactions. *J. Org. Chem.* **1997**, *62*, 6397–6400. [CrossRef]
36. The (S)-configuration of the major isomer is inferred from the known stereochemical outcome of acetate aldol reactions with N-acetylbornane-10,2-sultam 18 related (see refs. [22,35]).
37. Inanaga, J.; Hirata, K.; Saeki, H.; Katsuki, T.; Yamaguchi, M. A Rapid Esterification by Means of Mixed Anhydride and Its Application to Large-ring Lactonization. *Bull. Chem. Soc. Jpn.* **1979**, *52*, 1989–1993. [CrossRef]
38. Lecourt, C.; Dhambri, S.; Allievi, L.; Sanogo, Y.; Zeghbib, N.; Ben Othman, R.; Lannou, M.-I.; Sorin, G.; Ardisson, J. Natural products and ring-closing metathesis: Synthesis of sterically congested olefins. *Nat. Prod. Rep.* **2018**, *35*, 105–124. [CrossRef]
39. Niggemann, J.; Michaelis, K.; Frank, R.; Zander, N.; Höfle, G. Natural product-derived building blocks for combinatorial synthesis. Part 1. Fragmentation of natural products from myxobacteria. *J. Chem. Soc. Perkin Trans. 1* **2002**, 2490–2503. [CrossRef]
40. The structural assignment of **27** is tentatively based on ^{1}H-NMR analysis and MS.
41. Buey, R.M.; Díaz, J.F.; Andreu, J.M.; O'Brate, A.; Giannakakou, P.; Nicolaou, K.C.; Sasmal, P.K.; Ritzén, A.; Namoto, K. Interaction of Epothilone Analogs with the Paclitaxel Binding Site: Relationship between Binding Affinity, Microtubule Stabilization, and Cytotoxicity. *Chem. Biol.* **2004**, *11*, 225–236. [CrossRef]
42. Matesanz, R.; Barasoain, I.; Yang, C.; Wang, L.; Li, X.; De Ines, C.; Coderch, C.; Gago, F.; Jiménez-Barbero, J.; Andreu, J.M.; et al. Optimization of taxane binding to microtubules. Binding affinity dissection and incremental construction of a high-affinity analogue of paclitaxel. *Chem. Biol.* **2008**, *15*, 573–585. [CrossRef]

© 2020 by the authors. Licensee MDPI, Basel, Switzerland. This article is an open access article distributed under the terms and conditions of the Creative Commons Attribution (CC BY) license (http://creativecommons.org/licenses/by/4.0/).

MDPI
St. Alban-Anlage 66
4052 Basel
Switzerland
Tel. +41 61 683 77 34
Fax +41 61 302 89 18
www.mdpi.com

Chemistry Editorial Office
E-mail: chemistry@mdpi.com
www.mdpi.com/journal/chemistry

www.ingramcontent.com/pod-product-compliance
Lightning Source LLC
LaVergne TN
LVHW070141100526
838202LV00015B/1870